Origins of Man

John Buettner-Janusch DUKE UNIVERSITY

JOHN WILEY & SONS, INC.

New York · London · Sydney

Origins of Man

PHYSICAL ANTHROPOLOGY

Library of Congress Catalog Card Number: 66–14128
Printed in the United States of America

To my students, past and future,
without whom this book
would never have been written.

Preface

THE BOOK BEFORE YOU is about physical anthropology—the origins and evolution of man. Physical anthropology is one of the established subdivisions of anthropology—the study of man. It has become a habit, in some circles, to begin the account of man's origins with the primordial planetary soup (20 per cent organic material), continue with a sequential account of the fossil record and the confused chapter on hominid fossils, and end with a few words about the unpleasant prospects our planet faces now that its dominant mammalian life form is able to release vast sources of energy that it is unable to control and use intelligently. This book has molecules in it; there are fossils; there are living primates; and there are some mildly pessimistic remarks about the probable future course of man's evolution. Its theme is the evolution of man as a sequence of events in the evolution of the Primates. I examine man as a member of the order Primates, a zoological object. I show how the evolution of man is studied with the perspectives of anatomy, paleontology, genetics, cytogenetics, ethology, archeology, and biochemistry. My point of view is that man is a natural object and that physical anthropology offers reasonable, testable, even provocative explanations for his relationship to and differentiation from his primate relatives.

I have divided this book into two major parts. Part 1 contains a discussion of evolutionary theory and systematics and descriptions of our fossil relatives and the various major groups of living primates. I have summarized many of the views we hold about the course of evolution of the entire order Primates. Part 2 is devoted principally to one of the major species of the order Primates, *Homo sapiens*—his culture, population and biochemical genetics—and to the effect of evolutionary processes on human populations. I have attempted to present many of the data and theories, as many as possible in a book of this size, that are part of the physical-anthropological approach to man. I have adhered to the point of view of modern genetics, insofar as possible or relevant.

Man is a natural object, as much a part of nature as mosses, amoebae, trees, whelks, viruses, bacteria, reptiles, or monkeys. Unless anthropologists understand man as a part of nature they tend to operate within the intellectual realm of topical journalism or literary studies. It is my hope that I have presented the achievements of physical anthropology as part of anthropologists' attempts to understand man from the naturalistic point of view.

The book is based upon a one-year course called Human Evolution that I taught for six years. This course was designed for undergraduate and beginning graduate students. The material in Part 1 was taught during the first semester, Part 2 during the second. Students were told to use source material. The suggested readings and sources at the end of each chapter serve at least three purposes. They provide works the reader may use to expand his understanding of the topics treated in the various chapters. They introduce the reader to some of the vast literature of anthropology, and they provide a list of the sources from which data included in the chapter were taken. The reader will find that points of view other than mine, some of which I do not consider valid or important, or with which I disagree, are presented in these additional readings.

There are many facts in this book—facts relevant to the story of human evolution, of man as part of the general evolutionary process. But I must warn the reader that facts are not hard, clear, and self-evident. They are not hard little objects that yield obvious conclusions and theories when a sufficient number are gathered, counted, sorted, and listed. Facts are elusive, and they are subject to much interpretation. Indeed, facts themselves are often the products of interpretations. Interpretations of facts, fitting together of relevant facts, and relationships of facts to general propositions are important in physical anthropology today. A list of a number of bits of fossilized bones, with measurements, locales, and dates, really is not very meaningful unless some attempt is made to organize and present explanations. I have attempted to provide such explanations and generalizations. Facts do not speak for themselves—ever. No doubt in some cases I have gone far beyond the limits some of my colleagues would like to see imposed upon the scientific imagination. In some cases I am sure that I have not been imaginative enough to suit them. In other cases I know that I have not agreed with the views of some of my eminent colleagues. I trust the reader will find the generalizations, the theories, of sufficient interest to stimulate him to further exploration of physical anthropology and to help him understand the many descriptions and discussions of human evolution, fossil man, and living primates, so often in the popular scientific literature of today.

Anthropology as a whole is basically a biological science, although physical anthropology is often considered the biological part. Anthropology has its own intellectual tradition which grew out of biology. It has absorbed ideas, methods, and personnel from sociology, the humanities, and literary journalism as well. Anthropologists have developed an eclectic and vigorous science. At the same time it must be admitted that anthropology, at least in some localities, has become a subject precious and obscure, taught and practiced by dilettantes. The influence of the humanities has not been wholly distracting, but we must be aware that in our development into an eclectic and vigorous science we have had to endure some harmful side effects along with the useful lessons from the humanities.

The core of any sound program in anthropology is the study of the origins and evolution of man and the examination of man as a part of nature. Yet there is no question that social and cultural anthropology must draw closer and closer to physical anthropology. Physical anthropology need not become a kind of social biology, nor need it become human biology, while we adhere to the essential biological foundations of anthropology. I have tried to suggest throughout this book that we must view primate evolution as occurring by means of natural selection. Although it is difficult to do so, we should investigate the selective events that led to man's development of culture. Culture, anthropology's compulsive concern, is certainly a biological event, although independent of biology to an astonishing extent. Why did culture appear in only a single lineage, that of the genus *Homo?* Answers commonly given range from the ludicrous to the evasive, and the question remains incompletely answered in this book. Another incompletely answered question in physical anthropology concerns the extent to which the sociocultural rules under which humans must live and by which they are organized affect their biology. Human populations are organized along sociocultural lines. Some of the most significant contributions that the various parts of anthropology can make to each other and to science in general are to be found in studies of the genetic structure of these human populations. I have discussed some aspects of this vast problem in Part 2.

One eminent colleague suggested that, because three-noun titles are in vogue for books in biological and social sciences (he has written one so titled), I should call this book *Monkeys, Men, and Mysticism.* His reasoning was, perhaps, more apt than most physical anthropologists would care to admit. He told me that a book on physical anthropology must be about monkeys, fossil and living, since they are the biological root from which man evolved. It must be about man, since anthropology is about man. And, in light of the kinds of theories about human evolution current today and for the last hundred years, it must certainly have some mysticism. Despite the valuable association between anthropologists and humanists, one of the most serious defects in the relationship has been the inability of most humanists to understand, much less adopt, the naturalistic view of man. (A number of anthropologists find it difficult also.) Even today humanists assert the uniqueness of man, the special quality of his "mind." They insist upon the inability to account for symbolic behavior, intellect, and culture by naturalistic processes, many of them simple.

I chose *Origins of Man* as the title of this book, for I think it symbolizes the proposition that physical anthropology is organized by evolutionary theory and that its subject matters are the evolutionary history of the fossil primates and the comparative biology of the living primates. The dominant points of view, in both paleontological and neontological studies, should be those of population biology, ecology, and population genetics. I hope that this book will show that there is a great deal known about primate and human primate evolution, that little of it need be mystically treated, and, when we put together some of the things we know, that a reasonably sensible, coherent, provocative story emerges.

JOHN BUETTNER-JANUSCH

Durham, North Carolina
September 30, 1965

Acknowledgments

AS AN ANTHROPOLOGIST writing about his special field, I would like to acknowledge a debt to my teachers and colleagues who have influenced my intellectual development. I am deeply indebted to S. L. Washburn, James N. Spuhler, F. P. Thieme, James V. Neel, W. J. Schull, H. Eldon Sutton, and Leslie A. White. At various times and in various places they have provided stimulation, intellectual and other, that has been significant and important to me. Needless to say, I no more wish to blame them for what is to be found in this book than I wish to claim sole credit for whatever of value exists in it.

Stanley M. Garn read the entire manuscript. His witty good-humored assistance and suggestions have helped me immeasurably. Irving Rouse, R. J. Andrew, Elwyn L. Simons, Peter Nute, David Pilbeam, and Harold P. Klinger read important segments of the manuscript. Each provided many useful suggestions which helped improve the final draft. Their generosity and good will is gratefully acknowledged.

My wife read every draft of every page of the manuscript from first hand-scrawled versions through final galleys. She read proof of at least five stages of the manuscript. I could not have completed the book without her encouragement and assistance.

A considerable amount of help was given by individuals who provided photographs, drawings, or sketches. I especially wish to thank J. Biegert, S. N. Boyer, Harold P. Klinger, John R. Napier, Adolph H. Schultz, Mr. and Mrs. J. Sorby, and H. Eldon Sutton who made certain photographs available to me. I also wish to thank Margaret L. Estey, Jo Nicholson, and David Dunlap who so ably executed many of the original drawings. Paul Ray drew a number of the sketches of living primates which appear at the chapter headings. Acknowledgment is also due the individuals and publishers who so kindly gave permission to reproduce illustrations.

I also wish to express my thanks to members of the editorial and production departments of John Wiley and Sons for their able assistance and considerable patience and forbearance.

J. B.-J.

Contents

List of Tables

List of Figures

Part One

Introduction

1

THE ORIGINS OF MAN are sought in many places and with many tools. Geologists and archeologists use picks, shovels, and bulldozers. Geneticists and other biologists use test tubes, centrifuges, and calculating machines. Chemists and biochemists also use test tubes and centrifuges along with colorimeters and fancy electronic equipment. Many of the tools used are abstract—the theories and methodologies of science which are themselves products of man's long evolutionary history. Physical anthropologists use all of these tools, and physical anthropology has a theme—the evolution of man and his relatives, the evolution of the Primates. Physical anthropologists study the evolution of man and the relation between man and the living forms closest to him, the Primates. Physical anthropology may be divided into two parts—paleontology and neontology, the study of fossils and the study of living forms, respectively—and these two parts are not mutually exclusive. In the chapters that follow, in which we define and describe man and his relatives, both extinct and extant forms are discussed. One might say that physical anthropologists, being no more or no less egocentric than other scientists, are continually asking an egocentric and anthropocentric question—what is man?

The Question

What is man? This question has engaged philosophers, poets, scholars, scientists, theologians, dramatists, and others for a very long time. The question was put in highly poetic and metaphorical language.

> What would this Man? Now upward will he soar,
> And little less than Angel, would be more;
> Now looking downwards, just as griev'd appears
> To want the strength of bulls, the fur of bears.
> Made for his use all creatures if he call,
> Say what their use, had he the pow'rs of all?
>
> (Alexander Pope, "An Essay on Man," 1732).

The question was phrased and deliberated by austere medieval theologians who chopped away at its logic in their obscure and stately manner. Aristotle, as might be expected, had something to say. His contribution was not particularly valuable or interesting, although his definition of man, "an erect featherless biped," has amused university students for generations.

The unique characteristics of man's mental abilities, symboling and the so-called "power to reason," have elicited mystical, theological, and supernatural answers to anthropocentric questions. Other animals cannot reason, cannot remember complex systems of symbols, nor can they write or produce bodies of knowledge. Man is the only animal that passes any significant knowledge from generation to generation. This very great difference from other living organisms has sometimes overwhelmed man's ability to consider his mental capacities in a naturalistic and evolutionary context. This was one of the first problems argued by the founders of sociology and anthropology in the nineteenth century. It was a novel idea, indeed, that historical events and social systems could be understood and explained as natural phenomena. The concept of man himself as wholly a natural event, a product of the world of matter and energy, did not begin to penetrate the mental fog induced by an excessive dose of spiritualism and supernaturalism until Darwin's time. Man's ideas about himself, about his relationship to the planet on which he lives and to the other living forms on it are part of understanding evolution and evolutionary theory. There is a most important sociocultural chapter in the story of the evolution of man, but the fundamental parts of the story are strictly biological. What is biological man?

The Answer

The answers which many contemporary writers give to this question range from echoes of the pre-Darwinian past to a variety of amorphous and vacillating versions of "Man is an animal." The fact that man builds roads, constructs skyscrapers, composes string quartets, writes poetry, plays games, invents kinship terminologies, lives by codes remembered and handed down from generation to generation, and organizes compassionate systems to aid weak and sick members of his own species does not exclude him from the category animal. Nor do the observations that man commits matricide, patricide, genocide, and suicide exclude him.

Man *is* an animal. He is an interesting, variable, complex animal. The total range of his variability and all the details of his complexity cannot be comprehended by a single scholar. Neither, for that matter, can the total complexity of amoebae.

Man is an animal with a number of unique characteristics. It has been said that to speak so of man is to imply that man is nothing but an animal. Of course he is nothing but an animal! This book is about what it means to be "nothing but an animal," that is, what kind of animal man is.

We begin to answer this question by specifying the zoological categories into which man is put.

KINGDOM:	Animalia
PHYLUM:	Chordata (lampreys, frogs, snakes, birds, wallabies, opossums, bats, rats, cats, moles, whales, elephants, hares, man)
CLASS:	Mammalia (wallabies, opossums, bats, rats, cats, moles, whales, elephants, hares, man)

INFRACLASS: Eutheria (bats, rats, cats, moles, whales, hares, elephants, man)

COHORT: Unguiculata (moles, bats, primates, man)

ORDER: Primates (tree shrews, lemurs, lorises, tarsiers, monkeys, apes, man)

SUBORDER: Anthropoidea (monkeys, apes, man)

SUPERFAMILY: Hominoidea (gorillas, orangutans, gibbons, australopithecines, man)

FAMILY: Hominidae (*Ramapithecus*, australopithecines, man)

GENUS: *Homo* (pithecanthropines, Neandertal, modern man)

SPECIES: *H. sapiens* (Neandertal, modern man)

These categories specify man in progressively more exclusive groups, until he finds himself in the genus *Homo*, monotypic and alone (Fig. 1.1).

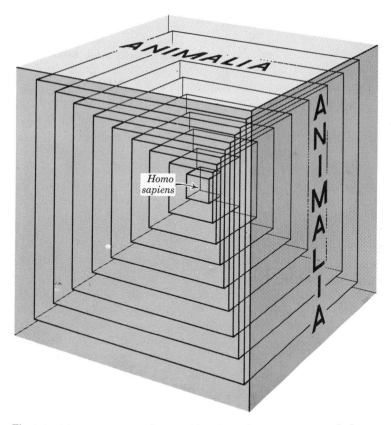

Fig. 1.1.   Man is a species in the animal kingdom. The categories into which man is classified may be viewed as smaller and smaller boxes, one within another, until man is found occupying the box labelled *Homo sapiens*, monotypic and alone.

All living organisms are divided into two major kingdoms—plant and animal. (A third kingdom, the Protista, has been defined to include certain organisms such as Bacteria, Flagellata, and Foraminifera which are neither clearly plant nor clearly animal.) Plants have a fundamental distinguishing characteristic; with almost no exceptions, they use radiant energy to synthesize organic materials from inorganic compounds. This is the process called photosynthesis. Animals are unable to perform photosynthesis. They require organic compounds for food. There are, of course, other distinctions. The phylum Chordata includes animals that have a dorsal tubular nerve cord, a notochord, and gill slits at some stage of the life cycle. The class Mammalia includes those animals that produce milk, are homoiothermal (maintain their body temperature independent of the temperature of the environment), have hair, and have a single bone (the dentary) in the lower jaw. The mammals are divided into several groups called infraclasses, one of which, the Eutheria, includes man. The Eutheria are often called the placental mammals, for the placenta (the structure that nourishes the embryo and fetus) is one of their principal distinguishing features. The order Primates is the group of eutherian mammals that developed a high degree of manual dexterity and a large brain. The suborder Anthropoidea includes those primates whose placental membranes are deciduate and hemochorial (attached to the uterine wall and in contact with maternal blood), and there are special features in the facial skeleton that distinguish the Anthropoidea from the other suborder—the Prosimii. The superfamily Hominoidea is the group of Anthropoidea that lost their tails and developed two modes of locomotion to a high degree of specialization—arboreal brachiation and terrestrial bipedalism. The family Hominidae includes those anthropoid hominoid primates that developed specializations for erect posture and bipedal locomotion. The genus *Homo* is the group of Hominidae that make tools and are symboling animals. The species *Homo sapiens* includes all members of the genus *Homo* that are indistinguishable from present-day man. This probably includes most fossil men of the middle and upper Pleistocene epoch.

The story of man's evolution begins with the origin of life. We shall begin later, at the point at which the order Primates became a distinguishable group of animals, in the Paleocene epoch about 75 million years ago. It is in the Paleocene epoch that the story of the Primates becomes readable in the fossil record. There are many terms, concepts, theories, and techniques which a student must understand in order to follow details of the story. These important points about the Primates in particular and about phylogenetic and systematic concepts in general will be defined and discussed as parts of the story of the evolution of man are told in the ensuing chapters.

Evolution—
Terms and
Concepts

2

CHARLES DARWIN founded modern biology with publication in 1859 of *On the Origin of Species by Means of Natural Selection.* During the century since the appearance of that publication, evolution has become accepted as fact. The validity of the theory of evolution no longer needs defense as the explanation for the diversity of organisms. Discussions of whether a theory is more certainly established than a hypothesis, or whether a fact is more certain than a theory, will take us far from primate evolution and into semantics. The theory of evolution is accepted as established, demonstrated, and valid, as well established as planetary rotation, gravitation, and the atom. We accept the proposition that the earth is, more or less, round and that the present-day living forms on it are the descendants of self-reproducing organic fragments with which the evolutionary process began eons ago.

Evolution has been defined succinctly: It is descent with modification or change. Today, because of the development of modern genetic theory, evolution is viewed as changes in gene frequencies between ancestral and descendant populations. The processes by which these changes come about are natural selection, mutation, sampling error, and migration.

It is now, and probably always will be, virtually impossible to measure directly the changes in gene frequencies between ancestral and descendant populations. There are many difficulties in interpreting the way the evolutionary processes affected the primate populations whose fragmentary bones make up the fossil record of primate evolution. Therefore evolution is also necessarily viewed as changes in the morphology of organisms through time. Morphology is the study of the form of the entire organism or a part of it—size, shape, color, etc. Analyses of degrees of morphological similarities and differences have been and always will be a basic part of evolutionary studies.

Besides being represented as changes in gene frequencies and changes in morphology between ancestral and descendant populations, evolution may also be represented as changes in numbers and kinds of animals within major lineages. The diverse kinds of animals are classified into formal groups called taxa. Taxonomic evolution is the analysis and description of the increasing or decreasing numbers of taxa in the major animal groups. Taxonomic evolution will not be considered in much detail in this book, but it is an important way of considering evolution. The diversity of kinds of living animals is one of the principal facts which the theory of evolution seeks to explain.

The amount of attention given by researchers to major problems in evolutionary biology changed as the theory of evolution became established. The fact of evolution was the first thing that had to be established. Then, given the theory, the phylogenies of various animal and plant groups had to be worked out. After that, the origin of discontinuities between animal populations, that is, speciation, was and is today a major concern of biologists. After

9

the establishment of genetics and the development of genetic concepts, research on rates of evolution and on the evolution of adaptation went forward
rapidly. Today we are the possessors of what is called the synthetic theory of
evolution. Briefly, the synthetic theory states that evolution led to the functional adaptation of the diverse and variable forms of life through the continuous production of variation and the action of natural selection.

Natural Selection

Darwin wrote that evolution occurs through natural selection acting on
organisms so that survival of the fittest results. The phrase "survival of the
fittest" has always aroused much discussion. We now prefer to speak of
this as Darwinian fitness, adaptive value, or selective value. Darwinian
fitness of a population of organisms is measured as the reproductive capacity
of the population. A population is "fit" relative to natural selection if it
can maintain or increase its numbers from generation to generation.

Natural selection, or simply selection, is the name we give to whatever
process, agent, or situation leads to the continuation of one lineage of
organisms and the elimination of another from the evolutionary record. Sir
Ronald Fisher recast the Darwinian theory of evolution by natural selection
in genetic terms. He once said natural selection operates so that an event
that is inherently highly unlikely to occur comes to be the most probable
event after many generations. This event is brought about by a mutation.

Selection is not simply a process that eliminates traits, genes, or organisms
from the record. As stated above, selection (or survival) of the fittest is
positive differential reproduction. The reproductive capacity of one set of
genotypes in a population is higher than the capacity of another set. But
the problem of determining how selection operates involves analysis of
many complex interacting factors. It is often necessary to consider a single
gene or a single characteristic. However selection acts on all the organisms
in a population and on the whole organism, not merely on a single trait.

Mutation

A mutation is a change in the material that carries the genetic information such that an inherited variation in a genetically controlled trait
results. New traits enter a population by mutation. It is the only way a
wholly new characteristic can begin to develop in an evolutionary line. The
way in which mutations occur will be discussed at length in Chapter 24.

Sampling Error

Sampling error may, in the absence of selection or mutation, lead to
changes in the frequencies of genes in a population. This is often called

genetic drift. Genetic drift is a random statistical effect and not a biological process. If a small population lives in isolation, the probability is low that any particular inherited trait will be passed on from one generation to the next. Certain traits may be eliminated purely by chance, and their alternatives may be fixed in a population. Such fixation or elimination of traits is not related to any biological effect they may have. Detailed analysis of sampling error will be presented in Chapter 24.

Migration

Changes in gene combinations in a population may occur by the introduction of genetic material from other populations of the same species. This is usually called migration and mixing. No new traits, such as are introduced by mutation, are produced by migration, but the amount of genetic variation on which selection may work is increased. Changes in human groups due to the mixing of genes of different populations usually occur through the migration of these groups. Such changes undoubtedly occur among nonhuman primates, too.

Extinction

Extinction is the name we give to the disappearance of an animal group, such as a species, from the evolutionary record. There are at least two ways in which a species may become extinct. First, the species may develop a way of life such that a change in the environment would prevent its persistence. This is the negative role of environmental selection in evolution. Second, one species may become extinct when it is transformed into another. A species may be a segment of a continuous, progressive evolutionary lineage. The species of one time period in which this lineage exists is the ancestor of the succeeding species in the next time period. The ancestral species becomes extinct through the processes by which it is transformed into its descendants. The early Pleistocene hominids, the australopithecines, are extinct, yet it is likely that some direct descendants of australopithecine genetic material exist in modern *Homo sapiens*.

Irreversibility

Evolution is irrevocable. Once an animal lineage has passed through a number of stages, a reversion, stage by stage, to the original ancestral condition does not occur. A structure that changes its form in evolution will not revert to its earlier form. This is often called Dollo's principle, after a nineteenth century Belgian paleontologist.

In studies of primate evolution this is an important principle, and applications of it will be cited. The dentition of a given form is often crucial evidence of its ancestral or descendant status with respect to another form. Once a tooth of a particular series (incisor, premolar, or molar) is lost, it does not recur again in the same series in the same form. Changes in the dentition are irreversible. This does not mean that similar structures or even the same adaptive patterns will not be repeated a second time in the evolutionary record. As another example, consider the flying reptiles. After these reptiles became extinct, wings and the adaptation to an airborne way of life recurred in two other distinct lineages—birds and mammals.

Irreversibility is a descriptive generalization, it is *not* a law of nature. The student must realize that it is not a property of living organisms. Although often called Dollo's law, it has no logical or substantive similarity to the law of gravitation or to the second law of thermodynamics. There are a large number of similar "historical laws" which students of evolutionary biology have enunciated, probably in an attempt to appropriate some of the prestige of the more exact sciences of chemistry and physics. Biological laws are not natural laws as are the laws of physics. They are generalizations which apply to the living organisms on this planet. Until such time as exobiologists, now concerned with ways to study living matter on other planets, discover and describe extraterrestrial life, the universal generality of biological laws cannot be substantiated.

The following four principles are sometimes believed to be evolutionary biological laws. All are descriptive generalizations which are demonstrably not inherent in the phenomena they describe.

1. Cope's law—The survival of unspecialized animals.
2. Cope's law—Animals get larger in the course of evolution.
3. Dacqué's principle—Different groups of animals evolve in the same way at the same time period.
4. Williston's law—In the course of evolution, the morphology of animals becomes simplified and specialized.

None of these laws is at all generally true but seems to apply to certain groups of animals studied by Cope, Dacqué, and Williston. One of Cope's laws merely generalizes his observation that the animals which seem to be less specialized are found for a much longer time in the fossil record. The other law points out that the fossil record appears to show that animals in the same lineage get larger as they evolve. The earliest fossil primates are very tiny; the later ones are larger. Dacqué's principle asserts that parallel evolution occurs in related lineages. Williston's law comes from his observation that among certain groups of animals there is a reduction in the number of structures and an increase in the special functions of those that remain, for example, legs of crustaceans. None of these four laws is a natural law in the sense that it is true of all animals at all times; each is a valid generalization from a specific set of observations.

Parallelism and Convergence

Similar structures, similar adaptive relationships, or similar behaviors occur in different groups of animals as the result of similar evolutionary opportunities. A fundamental principle of evolutionary biology is that, if there is a close similarity in the total morphological pattern of two organisms, there is a reasonably close phylogenetic relationship between them. The problem which this phenomenon of similarity brings up is whether such similarities are examples of parallelism or convergence or whether they are evidence of evolutionary affinity between the organisms. Parallelism and convergence imply that a close phylogenetic relationship does not exist. We must emphasize that parallelism should not be indiscriminately invoked to explain similarities. If we constantly refer similarities to parallelism, the whole concept of evolution is rendered meaningless. Simpson has pointed out that parallelism, evident in detailed similarities in two groups, is based both upon initial similarity of structure and adaptive type *and* upon subsequent recurrent homologous mutations. The term parallelism is usually restricted to the development of similar adaptive features in animals that are related, such as animals belonging to the same order. The parallel resemblances are, most likely, the realization of a genetic potential that is present in the entire group.

When two animal species or major groups that are not closely related develop similarities in adaptive relationships or structures, the two are said to converge. The exploitation of a volant way of life through the development of the wings of birds and of bats is clearly an example of convergence. The streamlined shape of marine mammals such as whales and dolphins is convergent to that of fish.

Not all cases of similarity are easy to classify as convergent or parallel. Many parts of the anatomy, many metabolic processes, even protein structures, are very much alike in different animals. If we insist that all similar cases are parallel developments, we are emphasizing the differences rather than the relationships between animals. It is always simpler to find differences between two animals than it is to demonstrate affinities. The examples presented in the following paragraphs were chosen to show some of the difficulties we encounter in attempting to categorize similarities.

There are two kinds of photoreceptors in the vertebrate eye, rods and cones. The rods are highly sensitive and function when there is very little light, but they have little power of discrimination. The cones are sensitive to higher intensities of light and make possible a high degree of discrimination of spatial relationships, colors, and textures. Rods have been found in the eyes of many nocturnal vertebrates—owls, bats, lorises—and of those that must live in dim light—whales, cats, and some fish. The question is: In which of these animals are the rods convergent and in which are they parallel developments? The answer depends upon the definitions we give

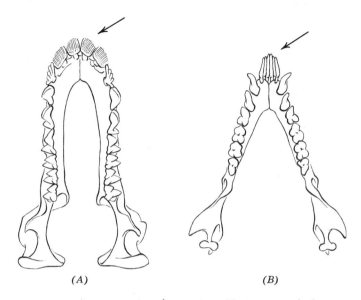

(A)   (B)

Fig. 2.1.   Toothcombs or tooth scrapers, indicated by arrows, in the lower jaws of (A) *Cynocephalus,* the "flying lemur," order Dermoptera, and (B) *Lemur catta,* the ring-tailed lemur, order Primates. In *Cynocephalus,* each front tooth is a comb. In *Lemur,* the toothcomb is formed by the lower incisors and canines.

parallel and convergent *and* upon a detailed study of the rods in these various eyes. The rods in each of the various animals noted may have evolutionary derivations from very different structures (convergence). Or they may be derived from the same parts of the basic vertebrate eye (parallelism). In this latter case the term parallelism refers to similarities that exist among animals separated at a much higher taxonomic level than that of the order.

The "song" or territorial call of *Indri indri,* a large, almost tailless, arboreal, diurnal lemur of Madagascar, has some remarkable similarities to the hoot of the gibbon of southeast Asia. The similarities in the calls of these two primates include the social context in which they are used and the pattern made by the song when it is analyzed on a sound spectrogram. Because the gibbon and the indri are members of the same order, these similarities in vocalization are parallelisms. It is difficult to make a case for close evolutionary affinity between the indri and the gibbon, for little is known of the fossil lineage of the indri. At the same time a close affinity cannot be ruled out.

The incisor toothcomb of the "flying lemur" (order Dermoptera) and of the various prosimian primates is clearly a case of convergent evolution (Fig. 2.1). All the living species, except two, of the prosimian lineages have toothcombs formed by the lower incisors and canines. Brachiation, locomotion by swinging arm over arm through the trees, is highly developed in some monkeys of both the Old World and the New World and in certain apes.

These examples can be, and have been, interpreted either as convergences or parallelisms. The closeness of the postulated relationship between the animals is usually the deciding factor. What is important is the way these situations illustrate the opportunism of evolution. Similar environmental opportunities are exploited by different organisms for their long-range evolutionary advantage. The problems presented by the environmental pressure on the organisms lead to similar solutions. Argument about specifying these as parallel or convergent may obscure this important point.

The terms homologous and analogous are often used to describe particular structures in animals. Homologous structures are those that are related by evolutionary descent and divergence. The wing of a bat and the forelimb of a monkey are homologous—they are descended from the same ancestral structure. The wing of a bat and the wing of a butterfly are analogous—they have similar functions and similar forms, but they are not related by descent from the same ancestral structure. Perhaps thinking of parallelism as homologous evolution and convergence as analogous would help us distinguish the two processes.

Adaptive Radiation

Adaptive radiation is the name we give to the rapid increase in numbers and kinds of any evolving group of animals. A group of animals—a species, a genus, a superfamily—may take advantage of environmental changes and exploit a number of new places in the planetary living space. These places in the environment are called niches or econiches.

The mammals radiated into all parts of the earth as the climate grew cooler during the Paleocene. The reptiles which had been the most successful group were at a selective disadvantage when compared to animals that could maintain a constant internal body temperature. Reptiles have a body temperature that is approximately that of the environment. They are poikilothermal (cold-blooded) animals. Thus, metabolism and all physiological processes that are a function of temperature slow down when the external temperature drops. Mammals have a complex system for maintaining constant body temperature. Their metabolic processes do not slow down when the temperature drops. They are homoiothermal (warm-blooded) animals. Most paleontologists agree that the great reptiles died out and the warm-blooded mammals spread into almost every part of the living space on the planet as the worldwide climate grew colder. But it is not at all certain that it was the change in climate alone that limited further evolution of the reptiles. The Darwinian fitness achieved by mammals, relative to reptiles, may have been decisive.

An adaptive radiation need not be a planet-wide event such as the example just cited. The spread of arboreal primates, the Old World monkeys, into trees of the tropical forests is an example of a more limited but no less important radiation.

An adaptive radiation is said to have occurred when a group of organisms fits into a part of the planetary living space into which it could not have moved earlier, as a result of changes in the group's relationship to the environment. We deduce these changed relationships from two lines of evidence: morphology of the fossils we find and comparative studies of the living forms which are the most likely descendants of these fossils.

One group of organisms may develop a relationship to the environment so that it has a reproductive advantage over another group or species which occupies the same environmental niche. This relative reproductive fitness was once called the "struggle for existence." The replacement of one species by another is the result of a reproductive advantage one has over the other. We usually are unable to reconstruct a very detailed account of the way in which a new adaptive radiation was achieved by a group of animals. However inferences from the morphology of a fossil primate, for example, about the way the living animal functioned, behaved, lived, and adapted, are possible and provide us with some basis for the evolutionary history of a group.

SUGGESTED READINGS AND SOURCES

Darwin, C., *On the Origin of Species by Means of Natural Selection.* John Murray, London (1859).

Dobzhansky, T., *Evolution, Genetics, and Man.* Wiley, New York (1955).

Fisher, R. A., Retrospect of the criticisms of the theory of natural selection. *In* J. Huxley, A. C. Hardy, and E. B. Ford (Ed.), *Evolution as a Process,* p. 84. George Allen and Unwin, London (1954).

Simpson, G. G., *This View of Life.* Harcourt, Brace and World, New York (1964).

Evolutionary Rates

3

THE TIMING OF EVENTS in evolution is fundamental to understanding the evolutionary history of any group of animals. We must be able to determine the sequence of events in the past. For example, our view of the functional course of human evolution depends upon which came first, the pelvic girdle and the foot that made efficient bipedal locomotion possible or the large skull in which man's large brain is housed. The relative dates or ages of the fossil structures involved are of crucial importance when the interpretations of the last stages of human evolution are made. We are now certain that the pelvis adapted for bipedalism came first and the brain grew larger later.

The rate of evolution is also of importance. We should like to know, for instance, how long it took for an erect bipedal primate to become the symboling cultured species it is today. It is not enough to know that *Ramapithecus* is older than the australopithecines and that the latter are older than the pithecanthropines. We must know, or be able to estimate within reasonable, assignable limits of error, the number of years that separate these three groups of fossil hominids. The methods used for assigning or determining the age in years of a specimen are called chronometric dating or absolute dating methods; the term chronometric is preferred.

Methods of Dating

There are a number of chronometric techniques in use for determining the age of a specimen. Two are of special importance to anthropologists concerned with primate evolution. Each method is based on precise measurements of the decay products of radioactive isotopes of a fairly common element. In the one case it is potassium and in the other carbon. Some attempt will be made to explain what is measured and why the dates that result should not be taken as unequivocally established.

The potassium-argon (K-A) technique was developed by geologists in order to date very old deposits of rock and mineral. The age of a rock is estimated by measuring the amount of potassium that has been transformed into argon. The method is based on certain properties of isotopes of potassium (K). Potassium is a mixture of isotopes, K^{39}, K^{41}, and K^{40}. K^{40}, which is present as 0.0118 per cent of the mixture, is radioactive and decays to form calcium-40 (Ca^{40}) and argon-40 (A^{40}). Calcium is a common element in the earth's crust, and argon is one of the rare "noble" gases of the atmosphere. The half-life of K^{40} is 1.30×10^9 years, which is within the range of years estimated for the age of the earth. Half-life of a radioactive element is the period of time in which one-half of the amount of the element disappears through radioactive decay. Thus, in 1.30×10^9 years, one-half the radioactive K^{40} will have been transformed into Ca^{40} and A^{40}.

The K-A method of dating depends in part upon the following: For every 100 K^{40} atoms that decay, approximately 89 become Ca^{40} and 11 become A^{40}. The Ca^{40} is formed by what is known as beta decay; K^{40} emits beta radiation which leads to the transformation of potassium to calcium. A^{40} is formed by what is known as K-capture. (The K here refers to an electron shell and is not to be confused with K, the symbol for potassium.) An electron of the K-shell is captured by the nucleus of the potassium-40 atom, and an atom of A^{40} is formed. The ratio of electron capture from the K-shell to beta decay is called the branching ratio of K^{40}; it has been calculated to be 0.123 ± 0.004.

Since 1937 when it was discovered that A^{40} is a decay product of K^{40}, the K-A method of dating has been developed and applied to many geochronological problems. There are several reasons for this wide applicability. First, most minerals contain some potassium, since it is one of the most abundant elements in the earth's crust. Second, sufficient amounts of A^{40} for accurate determinations are formed within time periods that are of interest and concern to anthropologists and geologists. Third, the amount of A^{40} in a sample can be measured very easily, much more easily than many other rare elements. The quantity of potassium can also be measured easily.

The methods for measuring argon and potassium are not simple. However with adequate precautions and proper equipment accurate results can be obtained readily. The interested reader should consult the reference by

Carr and Kulp, cited at the end of this chapter, for the details of the methods employed for accurate and precise determinations of argon and potassium and for the derivation of the mathematical formula by means of which the age of a sample, in years, can be calculated. Suffice to say here that the age of a sample can be calculated by use of an exponential equation (logarithmic mathematical formula) expressing time as a function of the amounts of A^{40} and K^{40} determined in the sample.

There are at least three assumptions made when a K-A date is assigned to a sample. First, at time zero the crystalline mineral being tested had the same ratio of argon isotopes as exists in the atmosphere. Second, the ratio of potassium to argon has not been changed by any chemical or physical process other than the radioactive decay of K^{40}. Third, the period during which the mineral was formed has been shorter than the age of the mineral. Unfortunately these are not always valid assumptions. Two of the biggest difficulties are that significant amounts of argon and potassium are lost by diffusion and by metamorphic changes and that the mineral tested contained radioactive argon at time zero.

If a mineral containing potassium happens to be buried under proper conditions, the A^{40} is trapped as it is formed. Samples of such buried rock must be very carefully chosen. When a crystalline mineral that contains potassium is heated to very high temperatures, the A^{40} that has already been trapped within it is driven off. If this heated mineral is then cooled and buried, the A^{40} atoms formed by the decay of K^{40} are collected and trapped anew as of the date of cooling. Therefore rocks of volcanic origin are excellent sources for dating by this method, since they have been heated to the extremely high temperatures necessary to drive off all the A^{40} trapped before heating. We may then have a high degree of confidence that the A^{40} in such a mineral has been formed since the date at which the lava or other mineral was formed.

The following are suitable for potassium-argon dating: igneous minerals, such as muscovite, biotite, phlogopite, orthoclase, sanidine, microcline, lucite; volcanic glass; sedimentary materials, such as glauconite, illite, carnallite, sylvite. Most K-A laboratories prefer to date samples of biotite, muscovite, and sanidine. Since it is now possible to date rocks that have only traces of potassium, minerals such as anorthoclase, oligoclase, augite, calcite, and hornblende are also useful.

The assumptions on which the K-A method of dating rocks is based are of critical importance. The nature of the results obtained, especially for lower Pleistocene strata, will be discussed when we turn our attention to Pleistocene hominid fossils (Chapter 9).

The carbon-14 (C^{14}) method was developed by Libby, and it is based on the fact that living organisms maintain a constant proportion of carbon-14 to carbon-12 (C^{12}) no matter where on the planet they live. C^{12} is the normal, nonradioactive, stable form of carbon; C^{14} is a radioactive isotope of C^{12}. C^{14} is formed high in the atmosphere when nitrogen (N^{14}) is bombarded by cosmic rays. Each time an atom of N^{14} is struck by a high energy cosmic ray, one C^{14} atom and one proton is produced. The C^{14} formed at this time combines with oxygen in the atmosphere to make radioactive carbon dioxide ($C^{14}O_2$). Carbon dioxide, containing $C^{14}O_2$ and $C^{12}O_2$ in constant ratio, is absorbed by plants during their normal metabolic processes. Animals absorb C^{14} from plants they eat and from other parts of the worldwide carbon reservoir. When an organism dies, plant or animal, the metabolic exchange of carbon with C^{14} and C^{12} of the environment stops. The radioactive C^{14} present at the time of death decays at a constant rate. If a specimen for dating is properly prepared, the measurement of the ratio of C^{14} to C^{12} will give a good estimate of the time since the organism from which the specimen was taken died. An older sample contains less C^{14} than a younger one. The mathematical formula used in this technique expresses the age of a sample as a logarithmic function of the amount of C^{14} in the sample.

The half-life of C^{14} has been determined by a number of laboratories, and at present two estimates, either 5568 ± 30 years or 5720 years, are quoted widely. The difference between the two has a negligible effect on published dates. When a specimen is 50,000 to 70,000 years old, the ratio of C^{14} to C^{12} is so low that it is not, at present, possible to measure accurately the amount of radioactive C^{14}. Thus the C^{14} method is useful for dating those specimens that are no older than about 50,000 years.

The radiocarbon method of dating depends upon at least three fundamental assumptions, any one or all of which may not be wholly valid. First, the specific activity of carbon (the ratio of C^{14} to C^{12}) in living organic material has been constant for a very long period, and assays of C^{14} in contemporary specimens are universally valid. Second, the biological materials analyzed for their radiocarbon content have not been contaminated. That is, no carbon has been introduced to the specimen by such things as plant rootlets. Third, it is assumed that the half-life of C^{14} has been determined accurately. This last assumption, already discussed, is now being subjected to careful scrutiny.

The constancy of specific activity of carbon depends upon the rate of formation of the radioactive isotope. It is assumed that this rate has been and is constant and that the ratio of the isotopes of carbon to each other maintains itself in equilibrium in the carbon reservoir on the planet. Today there are suggestions that the rate of C^{14} formation and the isotope equilibrium ratio are not constant, as was originally assumed by Libby.

The second assumption is the one that is most likely to cause difficulty. Specimens submitted for dating may have been contaminated in some manner, either while the organism or fragments of it were lying in the earth or during handling by scientists. It is also true that fission products produced by the large number of high atmosphere nuclear tests have introduced technical difficulties in the determination of C^{14}.

These two methods for chronometric or absolute dating have made possible many important contributions to evolutionary studies, archeology, and geology. We have discussed some of the difficulties and uncertainties inherent in the techniques, because students should realize that the dates determined by physical and chemical methods as elegant and sophisticated as these may have a spurious accuracy.

Relative dates, sequences of events—geological, paleontological, and climatic—have been established by application of the very simple principle of superposition. Geologists long ago devised the use of superposition for establishing relative sequences of strata and events contained in them. In a succession of undisturbed sedimentary rocks, superimposed on each other, the older layers or strata are on the bottom, the younger ones on top. Fossils which are found in the deepest strata are older than those found in the shallowest. Yet things are not that simple in reality. The earth's crust has been subjected for millions of years to movements and rearrangements called diastrophisms. At one time geologists thought that the various diastrophic movements were worldwide and could be used to correlate strata. This is not true, but geologists have been able to make excellent correlations between strata in various parts of the world, so that it is possible to put fossils from all over the world into relative sequences. These correlations are based on the synchronous occurrence of geological *and* biological events. Similarity in several species of fossil mammals contained in deposits in two parts of the world is taken as evidence that the deposits were laid down within the same time period. Stratigraphic correlations in two parts of the world imply that the fossils contained in the strata are of about the same age.

There are special techniques available that make it possible to determine whether a fossil actually came from a particular stratum. The relative age of a fossil may be determined by comparing its chemical composition with that of other fossils of known age from the same site or same area, if all have been preserved under the same conditions. Once bones are buried, they are subject to chemical changes. Some of these changes are rapid, others slow. Generally, it is not good practice to depend on the change in the organic matter in bones, most of which is fat and protein (collagen), to determine relative age. It is also not wise to rely on the texture or the apparent degree of "fossilization" of the bone. Bones buried in certain places may lose their organic content and become fossilized very rapidly. In other cases fossilization occurs very slowly. Fossilization consists of the introduction of new mineral matter, particularly lime or iron oxide, to the pores of the bones and alterations in the hydroxy-apatite matrix of the bones. Measuring the change in hydroxy-apatite, the phosphate, calcium, and carbonate material of which bones are mainly composed, is the most reliable means of relative dating. It is the slow constant change through the irreversible substitution of one element for the other in hydroxy-apatite that is used in various methods of relative dating such as the fluorine method. Fluoride ions in the ground water replace hydroxyl ions in the crystals of hydroxy-apatite. Once fluorine is fixed in the bone, it is not readily dissolved. When a paleontologist takes a fossil from the ground, it should have the same amount of fluorine in it as other fossils in the same deposit. If it does not, it is highly probable that the fossil was not laid down with the other fossils in that stratum, but is intrusive. Carnot demonstrated, as long ago as 1892, that in a given place the oldest fossilized material contains the most fluorine. Although the fluorine content of soil and ground water varies enormously, the fluorine method is a good one for determining if several specimens came from the same level in a site.

The most famous example of the use of the fluorine method is contained in Oakley's exposure of the Piltdown skull as a hoax. Oakley demonstrated that the fluorine content of the various fragments of Piltdown man was different from the fluorine content of other mammalian bones found in the same site.

The age of a specimen, then, may be determined in several ways. A relative age may be assigned by dating it in reference to the deposit in which it was found (fluorine method). It may be assigned a place in a relative sequence on the basis of local or worldwide stratigraphic correlations. It may also be put in a relative morphological sequence. The absolute age in years may be determined by a test of the specimen itself (C^{14} method), or the age of the deposit in which it was found may be determined (K-A method). The degree of confidence one has in the date of a specimen depends on the method used to determine the age. It also depends on the temporal and spatial relationship between the specimen itself and the sample (rock, plant, bone) on which laboratory tests were made.

Sequences

Time is represented in paleontology by sequences of geological events. Two kinds of geological sequences are extremely useful in studies of primate evolution—geochronologic sequences and stratigraphic sequences. Geochronologic sequences, geochronologies, are often spoken of as time terms. These are names for abstract, conceptual categories—eras, periods, and epochs such as Cenozoic, Tertiary, Paleocene, and Pleistocene. Stratigraphic sequences or categories are sometimes called rock terms or time-rock terms. Time is represented as closely as possible by particular sequences of rocks or strata. For example in the Pleistocene epoch there is a sequence of glaciations in Europe named Günz, Mindel, Riss, and Würm. The corresponding sequence in the midwest of the United States is called Nebraskan, Kansan, Illinoian, and Wisconsin. Each of these sequences was derived from the stratigraphy determined by geologists interested in analyzing the evidence left by the glaciers of the Pleistocene in the particular parts of the world in which they were working. Stratigraphy is the description and the study of sequences of stratified rocks and of the relationships particular stratified rocks or deposits have to each other. Stratigraphers provide the data on which the more abstract general categories of geochronology are based. Figure 3.1 is an example of the use of geochronologic as well as stratigraphic terms in describing time sequences in certain parts of the world. As we come closer to our own day, it is possible to subdivide the periods into more and more divisions. For example, Fig. 3.2 shows a summary of the correlation of Pleistocene climatic phases in certain parts of Europe and North America.

Certain sequences of paleontological and geological events are basic to understanding primate evolution. Graphic representations of these sequences are shown in Fig. 3.1 and 3.2 and Table 3.1.

TABLE 3.1
Geologic Record

Era	Period	Epoch	Time		Important Fauna
			Years B.P.[a]	Duration (years)	
Cenozoic	Quaternary	Pleistocene	0–2,000,000	2,000,000	Man, large mammals, modern marine invertebrates
Cenozoic	Tertiary	Pliocene	2,000,000–12,000,000	10,000,000	Hominoids, camels, giraffes, antilocaprids, bovines, dogs, hyenas
Cenozoic	Tertiary	Miocene	12,000,000–28,000,000	16,000,000	Grazing mammals, mastodonts
Cenozoic	Tertiary	Oligocene	28,000,000–40,000,000	12,000,000	Modern families of mammals, primitive Anthropoidea, cats, oreodonts
Cenozoic	Tertiary	Eocene	40,000,000–60,000,000	20,000,000	Modern orders of mammals, tarsiers, lemurs, horses, whales
Cenozoic	Tertiary	Paleocene	60,000,000–75,000,000	15,000,000	Archaic mammals, tarsiers, lemurs, modern birds, marine invertebrates
Mesozoic	Cretaceous		75,000,000–145,000,000	70,000,000	Toothed birds, pouched and placental mammals, modern insects

[a] B.P. = before present.

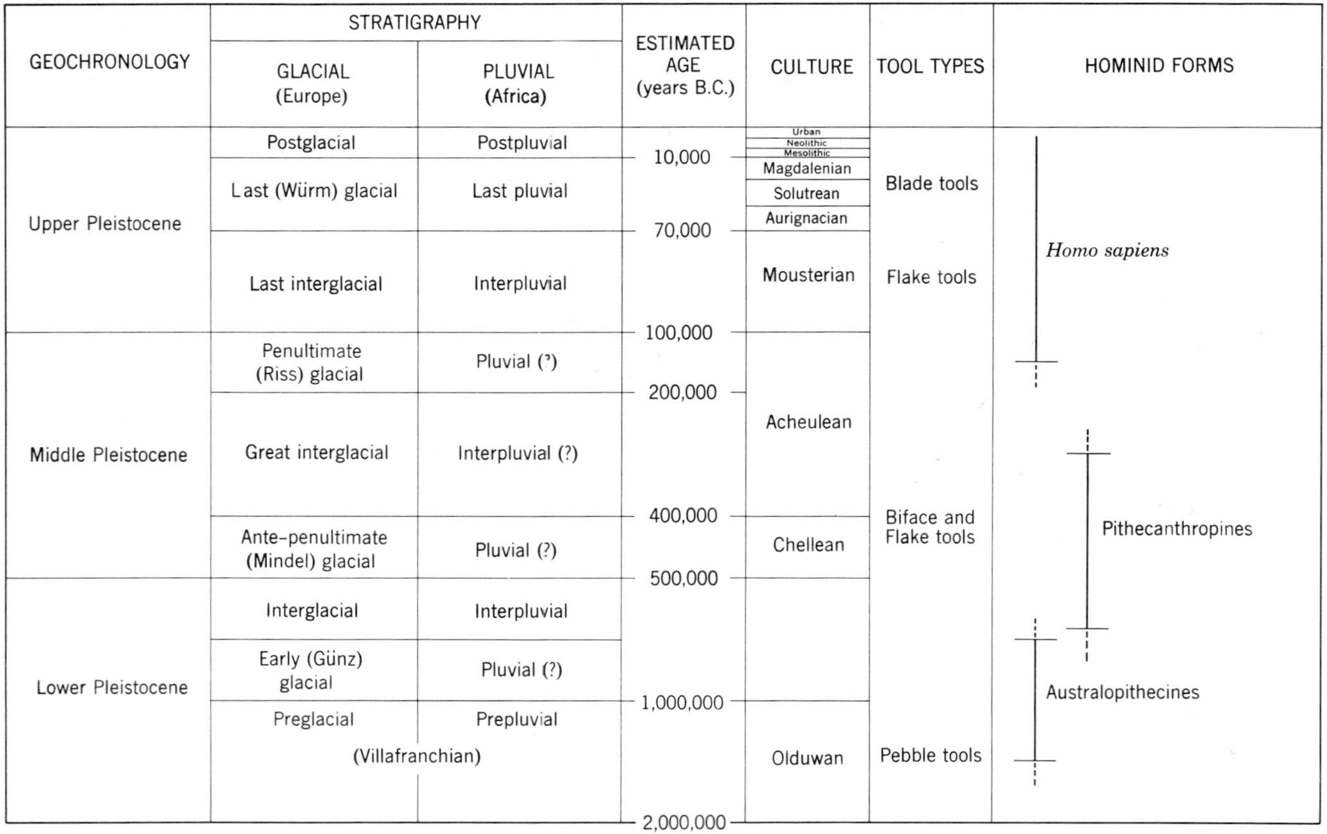

GEOCHRONOLOGY	STRATIGRAPHY		ESTIMATED AGE (years B.C.)	CULTURE	TOOL TYPES	HOMINID FORMS
	GLACIAL (Europe)	PLUVIAL (Africa)				
Upper Pleistocene	Postglacial	Postpluvial	10,000	Urban / Neolithic / Mesolithic / Magdalenian / Solutrean / Aurignacian	Blade tools	*Homo sapiens*
	Last (Würm) glacial	Last pluvial	70,000			
	Last interglacial	Interpluvial	100,000	Mousterian	Flake tools	
	Penultimate (Riss) glacial	Pluvial (?)	200,000			
Middle Pleistocene	Great interglacial	Interpluvial (?)	400,000	Acheulean		
	Ante-penultimate (Mindel) glacial	Pluvial (?)	500,000	Chellean	Biface and Flake tools	Pithecanthropines
Lower Pleistocene	Interglacial	Interpluvial				
	Early (Günz) glacial	Pluvial (?)	1,000,000			
	Preglacial (Villafranchian)	Prepluvial	2,000,000	Olduwan	Pebble tools	Australopithecines

Fig. 3.1.   The Pleistocene epoch and man's place in it. Durations of each period are not drawn to scale.

EPOCH	STRATIGRAPHY	
	EUROPE	NORTH AMERICA
Pleistocene	Würm glacial	Wisconsin
	Last interglacial	Sangamon
	Riss glacial	Illinoian
	Great interglacial	Yarmouth
	Mindel glacial	Kansan
	Interglacial	Aftonian
	Günz glacial	Nebraskan
	Villafranchian	Blancan
Pliocene	Plaisancian	Hemiphilian
	Pontian	Clarendonian
Miocene	Sarmatian	Barstovian
	Tortonian	Hemingfordian
	Helvetian	
	Burdigalian	Arikareean
	Aquitanian	
Oligocene	Chattian	Whitneyan
	Stampian	Orellan
	Sannoisian	Chadronian
Eocene	Ludian	Duchesnian
	Bartonian	Uintan
	Lutenian	Bridgerian
	Guisian	Wasatchian
	Sparnacian	
Paleocene	Thanetian	Clarkforkian
		Tiffanian
	Montian	Torrejonian
		Dragonian
		Puercan

Fig. 3.2.   Geological deposits in Europe and North America during the Quaternary and Tertiary periods.

Rates

The rate at which evolutionary changes occur cannot be determined directly. The sequences of geological and paleontological events enumerated give us a chronological framework into which evolutionary events fit. Determining the rates of evolution of animals is not a simple problem of comparing and correlating data with geochronological sequences. Indeed so many different notions of what is meant by evolutionary rates have been used that there is, even today, a need for some definitions. Three definitions of evolutionary change were given earlier. First, evolution may be viewed as the change in genetic composition of populations. Second, evolution is the morphological differentiation exhibited by a set of animals. Third, evolution is the progressive diversification of taxa in a larger taxonomic set. Each of these views implies a different criterion or method for determining the rate at which evolution takes place.

Genetic rates will probably always be beyond any but the most indirect evaluation. Nonetheless there are several kinds of genetic studies which have important implications. Determination of the rate at which a new gene, either a mutant or one introduced from outside, spreads in a population is one. The determination of fluctuation of gene frequencies in laboratory and natural populations over short periods of time is another. The comparison of the genetic similarities and differences of two populations that derive from a common stock is a third. It was once considered unlikely that genetic rates as such would be important in our present or future determination of actual evolutionary rates. But it is possible, today, to estimate the rate of change in the genetic material of an organism by examining, for example, the ordering of amino acids in certain proteins in closely related living animals. The number of differences in amino acid sequences of a specific protein of two organisms can be used as an indirect measure of the probable rate at which mutations were fixed in the genetic material (Chapter 30).

Changes in morphology are readily determined and are the principal parameters by which evolutionary rates in fossil forms are described. Changes in morphology are studied in several ways—change in a single trait (the length of the skull or of the humerus), change in a set of related traits (the size and positions of the orbits, the whole dentition), change in the whole animal. The choice of which kind of morphological study to make is often dictated by the nature of how much fossil material has been found.

Simpson has made the most cogent analysis of the ways in which morphologic changes may be converted into evolutionary rates, and the discussion here is based on his writings. There are temporal and relative morphologic rates. The rate of morphologic change per absolute unit of time (years, millions of years, etc.) is the most precise evolutionary rate obtainable. Unfortunately the data needed for such calculations are rare. Among the Primates, absolute rates of morphologic change have been calculated only for the Pleistocene hominids, and these rates are tentative. The rate of change in morphology per linear unit of deposited sediment or stratum is often useful and can, in some cases, be determined very precisely. The rate of morphological change per geochronological unit—era, epoch, or period—is occasionally used. None of these three absolute morphological rates has been calculated for the order Primates as a whole.

The relative rates of morphological change are of much more importance in studies of primate evolution because these rates can be determined, albeit without the precision of absolute rates. Relative rates do not require that we measure the actual elapsed time between events. The time element is actually eliminated from a consideration of relative rates. Intragroup and intergroup relative rates are determinable when we interpret the data of primate evolution. The rate of change in one character may be measured against the rate of change in another character in the same group of animals. This intragroup rate of change may radically alter our views of the nature of a particular group's evolution. The rate of change in a trait in one group relative to the rate of change of the same trait in another group, an intergroup relative rate, enables us to compare groups of animals.

The calculation of evolutionary rates based on any changes in morphology is subject to serious error due to the phenomenon of mosaic evolution. Not all parts of the organism change at the same rate in the course of evolution. Nor do all parts of the organism change in the same time period. For example, the foot and pelvis of the Hominidae were clearly transformed from quadrupedal to bipedal types in a relatively short time. The skull, particularly the braincase, of the hominids changed relatively little until the erect bipedal structure had been perfected. Then the braincase changed rapidly relative to further changes in the pelvis and foot.

Taxonomic rates are sometimes important in describing primate evolution. The two most useful and important are, to use Simpson's terms, phyletic rates and rates of taxonomic diversification. Phyletic rates measure, or estimate, the time required for the origin of a new taxonomic unit (taxon) from an ancestral taxon or measure the time a taxon persists. Taxonomic diversification rates estimate the number of taxa that develop within a higher taxon in a specified period of time. The number of genera that existed in a family in the Eocene or in the Pliocene compared to the number that exist in the same family today may be a measure of the rate of taxonomic diversification.

There are many ways in which rates and ages can be determined. The salient feature of all the methods we have discussed can be summed up by the statement that, as a general rule, older materials are found at the bottom of the pile and younger ones at the top. At the present time, calculations of relative and absolute dates are much more feasible than calculations of evolutionary rates. Rates of evolution depend on the chronological framework constructed by use of various dating methods and on the quantity of available fossil material to hang on the framework.

SUGGESTED READINGS AND SOURCES

Campbell, B., Quantitative taxonomy and human evolution. *In* S. L. Washburn (Ed.), *Classification and Human Evolution*, p. 50. Aldine, Chicago (1963).

Carr, D. R., and J. L. Kulp, Potassium-argon method of geochronometry. *Bull. Geol. Soc. Am.* **68**, 763 (1957).

Gentner, W., and H. J. Lippolt, The potassium-argon dating of upper Tertiary and Pleistocene deposits. *In* D. Brothwell and E. Higgs (Ed.), *Science in Archaeology*, p. 72. Basic Books, New York (1963).

Kulp, J. L., Geologic time scale. *Science* **133**, 1105 (1961).

Libby, W. F., *Radiocarbon Dating.* Univ. of Chicago Press, Chicago (1955).

Oakley, K. P., Analytical methods of dating bones. *In* D. Brothwell and E. Higgs (Ed.), *Science in Archaeology*, p. 24. Basic Books, New York (1963).

Romer, A. W., Time series and trends in animal evolution. *In* G. L. Jepsen, G. G. Simpson, and E. Mayr (Ed.), *Genetics, Paleontology and Evolution*, p. 103. Princeton Univ. Press, Princeton (1949).

Simpson, G. G., Rates of evolution in animals. *In* G. L. Jepsen, G. G. Simpson, and E. Mayr (Ed.), *Genetics, Paleontology and Evolution*, p. 205. Princeton Univ. Press, Princeton (1949).

Stirton, R. A., *Time, Life and Man.* Wiley, New York (1959).

Weiner, J. S., K. P. Oakley, and W. E. Le G. Clark, The solution of the Piltdown problem. *Bull. Brit. Museum (Nat. Hist.)* **2**, 139 (1953).

Willis, E. H., Radiocarbon dating. *In* D. Brothwell and E. Higgs (Ed.), *Science in Archaeology*, p. 35. Basic Books, New York (1963).

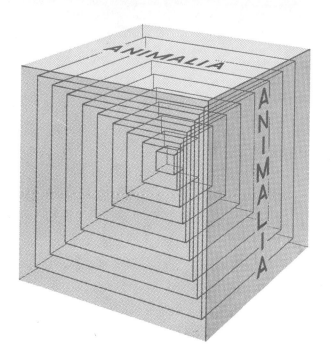

Systematics

4

THE THEORY OF EVOLUTION was the greatest unifying event in the history of the biological sciences. The vast numbers of diverse plants and animals, their forms, behaviors, geographical distributions, and their special adaptations and relationships to each other and to the environment were brought from chaos into order by the Darwinian revolution. An appreciation of how we order newly discovered fossils and the living primates is necessary. Systematics, taxonomy, classification, and nomenclature of the living primates should come to be understood and appreciated by anthropologists and primatologists and should not be thought of as merely a special and esoteric branch of biology.

Systematics in General

Systematics is "the scientific study of the kinds and diversity of organisms and of any and all relationships among them" (Simpson, 1961). Systematics deals with any and all aspects of organisms with the view that they may be put into some kind of order. Science cannot discuss or handle organisms, or anything else for that matter, unless the organisms are ordered. Systematics thus becomes a very general and inclusive activity. Data from any and all of the sciences dealing with organisms may be used. All the comparative sciences (or subdivisions) of biology affect systematics. Once comparative studies have provided data about the traits by which relations between organisms are determined, a formal system of relating the organisms to each other is required by systematics. This formal system is a classification, and classification is also a kind of study or activity.

Formerly, comparative anatomy and comparative osteology were two of the most important sciences used in establishing the systematics of the Primates. But today, comparative studies using the methods of biochemists, cytologists, physiologists, psychologists, and ethologists are becoming equally important. Many of these newer aspects of systematics will be discussed in later chapters.

Taxonomy and Classification

The terms systematics and taxonomy have often been used interchangeably or in such a way as to imply that they are the same thing. Simpson's excellent discussions distinguish systematics and taxonomy clearly. Systematics is the study of the diversity of animals and of all possible relationships among and between them. Taxonomy is the theoretical study of how classifications are made. Classifications put animals into groups on the basis of the relationships they have with each other. The subjects of classifications are organisms, the subjects of taxonomies are classifications, and, one might say, the subjects of systematics are everything that is relevant to the study of organisms.

The rules for constructing classifications, the technical procedures used, and the theoretical foundations on which classifications are based are taxonomy. We should know why classifications are made, how they are made, and something of the history of the one we choose to use. The principal reasons for constructing a classification are to provide as simple a system of groupings *and* names as possible and to give zoologists a practical way to understand what they are talking about. It is, of course, extremely important that the classification be genuinely consistent with animal affinities, i.e., with phylogeny. The problems of making classifications are immediate and among the most troublesome in physical anthropology and primatology. The act of classifying groups of things, in our case certain animals—the Primates, is the act of putting them into categories and into systems of categories which are defined according to the characteristics of the animals. We then name the categories, that is, indulge in the art or science of nomenclature.

The present hierarchical (sequentially stratified) classification of animals grew out of the original work of the great Swedish naturalist, Carolus Linnaeus, in the eighteenth century. The categories, or taxa (singular: taxon), of the current classification are listed in Table 4.1. Each of the categories, each taxon, is a group of real organisms. These groups are recognized or defined as formal units at each level of the classification, for example, at the level of species, genus, or family. A taxon is a population, and a taxon at one level includes those populations in taxa at lower levels. For example a genus is a population of species; it includes all of the populations that make up the species contained in it.

It is important to understand that classifications are constructed by scientists for particular purposes. The zoological classification in current use was made to provide a convenient and universally intelligible organization of living and fossil animals. Since Darwin's day the zoological classification has been slowly overhauled to bring it into accord with the view that animal groups are related phylogenetically. But the classification is an historical product, too. It was first proposed by Linnaeus who believed that animal diversity was the result of the special creation of each species. The species existing in his day were considered permanent and immutable. There are difficulties in using this classification in evolutionary biology, but none so great that they cannot be overcome.

We make the classifications; we do not discover them in nature as Columbus discovered America. We are working with real objects, organisms, and we attempt to organize data about them in order to understand and specify the phylogenetic relationships among them. Therefore we attempt to construct categories at all levels of the hierarchy that are consistent with such phylogenetic relationships. Although it may be argued that the categories are arbitrary, and indeed it often is, they are not chosen or defined capri-

TABLE 4.1
Categories in the Current Classification
(The Linnaean Hierarchy)

Kingdom
Phylum
Subphylum
Superclass
Class
Subclass
Infraclass
Cohort
Superorder
Order
Suborder
Infraorder
Superfamily
Family
Subfamily
Tribe
Subtribe
Genus
Subgenus
Species
Subspecies

ciously. At least one category in the classification, the species, is nonarbitrary, and others such as the genus are not always used in a wholly arbitrary fashion. When we say that the species is a natural unit in the classification and is nonarbitrary, we mean that the definition of species depends upon real biological relationships among its members.

From the time of Linnaeus to the time of Darwin, the typological or archetypal concept was the basis for biological classification. Since the time of Darwin, the typological concept has gradually been replaced by the phylogenetic concept, the basis of the present-day classification. The type or archetype of a category is the ideal or typical member. Animals are compared to this ideal and put into or removed from a category on the basis of the extent to which they resemble the ideal. The phylogenetic approach to classification is based on a statistical analysis of the members of a category. The breeding population to which a specimen belongs is the category, an individual specimen is not. The typological classification was based upon the morphology of the animals and depended upon the concept of the type. Technically the type in zoological nomenclature is a zoological object, an organism—it is a particular specimen which bears the name for the taxon. Philosophically the type or archetype came to mean a supernatural, ideal model for the taxon (the *eidos* of Plato). The living animals in the taxon were imperfect copies. The definition of a type is part of the formal operation of the currently accepted system of zoological nomenclature. The student must bear in mind that the type specimen for a taxon is often not at all typical of the population of organisms whose name it bears.

The present, generally accepted, classification of animals is based upon the assumption that the categories are evolutionary entities. At least we strive to base modern classifications on evolutionary theory. The actual differences in two classifications, one phylogenetic, the other typological, may very well be small, but the theoretical gap between them is enormous. Typological classifications are based on the concept of type. Phylogenetic classifications are based on the concept of variable populations. A species consists of many individual animals which exhibit variations of each of the many traits which are used to characterize it. What relates all these individuals into one population or a series of populations, which are then categorized into a species, is the degree of genetic homogeneity which they may be shown to have when compared to other populations.

Proponents of the typological basis for classification and proponents of the phylogenetic basis for classification have disputed these concepts for many years. It is quite true that phylogeny is an inference, for we cannot observe it. The best data we have for phylogeny come from the study of the genetics of organisms. It is unfortunate that the evidence from genetics is rather indirect and usually comes only from living organisms. It is of course difficult to produce a large body of data that supports the view that the genetic study of organisms is the best base from which to develop sound, phylogenetic relationships. The repertory of traits, characteristics, features, and systems described by comparative morphology of organisms, particularly comparative anatomy and osteology, is infinitely greater. This fact has long supported the typological view as the basis for classification.

Phylogeny is a dynamic concept; morphology is a static concept, and although it is useful as a tool, it is not the final criterion. It is a serious mistake to look at a morphological trait or system, described from one or two specimens, as representative of all the criteria for defining this system throughout the group of organisms with which one is working. We must always remember that this system—a jaw, a tooth, a facial bone, a pelvis, an occiput—in a living, interbreeding population of animals is highly variable, and this variability in the characteristic is basic. The trait is but one item from one individual member of a population subject to variations.

The classifications we are interested in are those which represent the phylogeny of contemporary animals. There are scholars who believe that the purpose of a classification is to express phylogenetic relationships. But, as Simpson showed, no method of classification, particularly one based upon the Linnaean hierarchy, symbolizes phylogeny in any consistent way. This or that segment of the system expresses phylogenetic relationships, but, in general, the system of classification does not present a true picture of phylogeny. It is also quite clear when one analyzes the history of the present Linnaean hierarchy that it was never meant to express phylogenetic relationships. This does not mean, however, that phylogeny should not be recognized in the classificatory system. The groups constructed for classification should be valid phylogenetic entities, for, as we have seen, the basis for the modern system of zoological nomenclature is phylogenetic.

We o ten hear the following questions asked by students as well as by senior scientists—Do the categories in the classifications correspond to or reflect some supraindividual, real biological level of integration, or are the taxa at various levels arbitrary analytical devices invented by scientists? These questions rephrase in the context of biology the ancient philosophical controversy between nominalism and realism. Nominalism is the doctrine that universals or abstract concepts are merely names. Realism is the doctrine that universals or abstract concepts have objective existence. The nominalists would consider the individual animals as the only realities and the classification with its levels of taxa a convenient way to sort data. The realist considers the various groupings of animals as real natural phenomena and speaks of species or genera as real evolutionary entities. For a realist the species is a biological phenomenon independent of the construction of a taxonomic system. However the classification itself is not a real biological phenomenon but a set of analytical categories which scientists use to order data. The reader must bear in mind that the objective existence of the species as a natural phenomenon is one thing and the construction of the analytical system, the zoological classification, is another. It is *probable* that biological species and genera and families are real natural phenomena. It is quite *certain* that the taxa called species, genus, and family are not. They are analytical devices created by scientists.

Nomenclature

Nomenclature is the act of assigning names to the groups which are recognized in the classification we construct. The most important names are those given species. Linnaeus devised a system of Latin binomials and a concept of the species which together created classification and nomenclature as it is known today. The binomial *Homo sapiens* names the genus (*Homo*) and the species (*sapiens*) to which man belongs. This two-term name, the binomial, is unique to this species and cannot be applied to another. Each binomial is unique to the species it names.

It sometimes seems that giving names and disputing about them become major activities in certain branches of science. The name given an animal *is* important, and because it is important an *International Code of Zoological Nomenclature* has been drawn up. This Code lists rules which must be followed if clear and unambiguous communication is to occur among scientists who work with organisms. Unfortunately, men are fallible and the Code is not always applied correctly. Nomenclature is a distinct branch of zoology and should be better understood by anthropologists than it is. The rules of nomenclature and the processes of assigning valid names to taxa are not based on phylogenetic principles or principles of classification except indirectly. The entire body of the current rules and their interpretations and modifications cannot be listed in this book. At least two of the rules have special importance for anthropologists—the designation of types and the law of priority.

A type specimen must be designated when a new species is described. The type of a species is a particular specimen, and it is an object. This requirement of a type specimen must not be confused with a typological concept of species. The type of a genus is a particular species; it includes all the individuals of the designated type species. The type of a family is a genus. For taxa of higher rank than the family there are no requirements for a type, nor does the law of priority apply to more than species, genus, and family. A diagnosis or definition of a species, a genus, or a family specifies the characteristics which define the taxon. Examples of such diagnoses are given in Chapter 10.

The law of priority applies to names used as of 1 January 1758. This is the date, by convention, of the publication of the authoritative edition of Linnaeus' *Systema Naturae.* The first name validly used after this date has priority. Priority applies if the author of the original description of a species, genus, or family interpreted his material correctly and proposed and published the name(s) according to rules set forth in the *International Code of Zoological Nomenclature.* All later names given the same taxon are known as synonyms. This law was formulated to bring some order out of the chaos which resulted from the unruly and individualistic use of the formal Latin binomials for species. Various modifications and suspensions of the rules have been proposed in particular instances and occasionally accepted by the International Commission on Zoological Nomenclature. One example will show why adherence to the rules is important. The name *Simia* was used as the name for the genera to which both macaques *and* orangutans belonged. The confusion and ambiguity created by such indiscriminate use of *Simia* finally ended in the suppression of *Simia* as a valid name. Although this decision was protested by many, and some said they would refuse to abide by it, *Simia* passed out of use. *Macaca* and *Pongo* are currently the correct generic names for macaques and orangutans, respectively. Another example will show that the rules are not always strictly adhered to. The source for the name of the genus *Pan,* to which chimpanzees and gorillas belong, is a work by Oken published in 1816. The Commission suppressed Oken's work as a source for Linnaean binomials. The Commission has not yet put *Pan* on the list of suppressed generic names. *Pan* has been used without ambiguity, and we continue to use it in this book as the generic name for chimpanzees and gorillas. We shall abide by the decision of the Commission, should *Pan* be suppressed. If *Pan* is suppressed, the next available generic name for chimpanzees and gorillas is *Chimpansee* Voigt 1831.

Pan troglodytes (chimpanzees) would become *Chimpansee troglodytes,* and *Pan gorilla* (gorillas) would become *Chimpansee gorilla.*

It is well to point out that, even if there is a recognized, valid name for a taxon, this does not imply that such a taxon is valid on biological or genetic grounds. Also the etymology of the word that names a taxon does not necessarily imply a description of the population included in the taxon. Examples will make this clear. Carnivora names a particular order of mammals which includes fruit-eating as well as flesh-eating members! Although the derivation of the word primates implies "first among animals," Primates is simply the name of a particular taxon.

Certain conventions have been devised for the purpose of forming names of some of the higher categories in the Linnaean hierarchy. Some of these conventions, "rules of thumb" for forming names, are presented in Table 4.2. Often, when we discuss the primates, the usage of names becomes more colloquial. For example, man (genus *Homo*) is referred to as a hominoid, a member of the superfamily Hominoidea; he is also a hominid, a member of the family Hominidae. An orangutan (genus *Pongo*) is also a hominoid; and it is a pongid, a member of the family Pongidae. A guenon (genus *Cercopithecus*) is a cercopithecoid, superfamily Cercopithecoidea; it is a cercopithecid, a member of the family Cercopithecidae; and it is a cercopithecine, a member of the subfamily Cercopithecinae. A colobus monkey (genus *Colobus*) is a cercopithecoid, a cercopithecid, and a colobine, a member of the subfamily Colobinae. In the chapters on living primates, some more examples will be discussed with the hope that the problems of nomenclature, as distinct from problems of classification or phylogeny, may be better understood by the student.

TABLE 4.2
Names of Higher Categories

Category	Suffix	Examples		Name of higher category
		Genus	Stem[a]	
Infraorder	−IFORMES	*Lemur*	LEMUR−	LEMURIFORMES
		Tarsius	TARSI−	TARSIIFORMES
Superfamily	−OIDEA	*Lemur*	LEMUR−	LEMUROIDEA
		Daubentonia	DAUBENTONI−	DAUBENTONIOIDEA
Family	−IDAE	*Lemur*	LEMUR−	LEMURIDAE
		Cercopithecus	CERCOPITHEC−	CERCOPITHECIDAE
Subfamily	−INAE	*Lemur*	LEMUR−	LEMURINAE
		Alouatta	ALOUATT−	ALOUATTINAE

[a]The stem is usually taken from the name of a genus within the higher category.

What Are Species?

The species concept, one of the most important in modern evolutionary biology, is fundamental to any discussions of systematics and taxonomy. A full appreciation of the biological properties of species is of overwhelming importance to students of human evolution. And an understanding of the species is necessary to many biomedical sciences, such as entomology, parasitology, or virology. Vaguely and poorly conceived notions of species have led, among other things, to confusion and proliferation of names and to irrelevancy and redundancy in discussions of interpretations of the fossil record.

The definition of species which will be used throughout this book is based on the view that a species is an objective, nonarbitrary grouping of animals. This view is championed by Mayr and is the prevailing view today. The definition of species which this leads to is called the genetical species or the biospecies. A species is a population or group of populations of actually or potentially interbreeding animals that are reproductively isolated from other such groups. There are scholars who use other definitions of species that suit particular problems. Among these are definitions based on the typological concept of species.

The typological concept of species is a consequence of too much Platonic philosophy. In Platonic terms, the type is considered to have an immutable, eternal, ideal existence. Individual animals are thought of as copies or representations of the type. The variation that exists from individual to individual is the result of imperfections in the copies of the ideal type. Today we consider this view supernatural. The definition of species which this leads to is the morphological species or the morphospecies. Morphospecies are considered to be established only on the basis of morphological traits. Since the species concept in modern biology applies to populations of whole organisms, this definition, if taken literally, implies a category of traits, not of organisms.

Many populations of animals in a particular place at a particular time are easily distinguishable, as many a naturalist has observed. There are no intermediates between any two populations, and there is a reproductive gap between them. Such populations or species which occur in the same place are called sympatric. There are many populations whose geographical ranges do not overlap, yet these populations are members of the same species. Such populations are called allopatric. Some populations of fossils are put into the same species as living animals. These groups whose ranges in time do not overlap are called allochronic species.

There are two practical ways of delineating species: first, by showing that they are distinct from other species that live in the same place; second, by contrasting them with populations that have geographical ranges that are mutually exclusive. Those that meet the first criterion are called sympatric species; these are species whose ranges coincide or overlap. The second criterion defines allopatric populations or species that occupy separate geographical areas or ranges which do not overlap. The gaps or distinctions between sympatric species are complete and absolute, whereas the distinctions between allopatric species are not.

Sibling species are sympatric species which are very similar morphologically, indeed, quite often indistinguishable. Despite close similarities, each possesses specific traits which are unique, and each is reproductively isolated. Sibling species may be confused with biological races, but they are not races, and the difference is clear in theory. Races are not reproductively isolated (Chapter 32). Among the Primates, race is probably a good term for any and all subgroups of animals included in a single species. For other organisms the term subspecies is preferred by many authors.

To speak about human subspecies, as some recent authorities have done, is quite misleading. Few will question the need to define subgroups of certain populations within a primate species such as man or lemur. It is the name we give these populations within a well-defined species that leads to disagreements. The emotional connotations surrounding the word race are well known when human varieties are discussed.

The story of primate evolution includes a major concern with fossils. We often speak of an evolutionary species when discussing fossils. Following Simpson, we consider an evolutionary species to be a lineage (a line), an ancestral-descendant sequence of populations, that is evolving separately from others. It has its own evolutionary role and tendencies, and these must be inferred from rather scanty paleontological evidence.

Other names are those used for species present in nonoverlapping time periods, the allochronic species, called by some paleospecies. Since the fossil record is incomplete, it is possible to delimit such species. If the fossil record were complete, it would be difficult, if not impossible, to delimit allochronic species. A species changes very slowly through time. The evidence would show slight, gradual changes in the organisms, such that distinctions would be difficult to make. When a paleontologist has an intergrading series of fossils with no discontinuities, he breaks up the series for convenience. Although the fossil record of the Primates is among the best known for mammals, it contains few lineages in which the paleontologist must arbitrarily introduce breaks.

To summarize then, there are several kinds, or at least names, of species that concern us. First, there are genetical species, sometimes called biospecies—groups of actually or potentially interbreeding populations. Genetical species may be sympatric, distinct populations with overlapping geographical ranges or allopatric populations with nonoverlapping ranges. Some genetical species may be characterized as sibling species. These are sympatric species that are particularly difficult to distinguish. They are good genetical species that differ hardly at all in anatomy. Second, there are evolutionary species, defined as lineages of ancestral-descendant populations that have their own evolutionary tendencies. Allochronic species probably should be referred to this definition. Basically, the concept of the species is the same for both genetical and evolutionary species. It is the nature of the evidence that differs.

The definitions of species in use today stress the *distinctness* of species rather than their differences. They stress that species are *populations* of variable animals, not types, and that they are *reproductively isolated* from each other. These definitions of species are based on a nonarbitrary, multidimensional, biological concept of species. The species is considered to be a real, natural phenomenon and not an arbitrary, analytical category. This is a valid view. The contribution of the biological concept of species to science and the importance and usefulness of the genetical definition of the species are independent of any metaphysical and logical discussions of the species as a natural phenomenon.

Recognition of a species is another matter. The student should not confuse the theoretical problem just discussed with the more practical problem faced by the scholars who must classify actual specimens, particularly those in museum collections. Nonetheless, many of the difficulties researchers have in recognizing valid species would disappear if they had a sound grasp of the theoretical basis for the species concept.

The "good species" of the local naturalist, many a museum curator, and the zoologist of a generation ago were usually based upon *differences* in the pelage, skeleton, and body proportions of the specimens. The existence of differences does not necessarily imply genetic distinction. It must be remembered that individual variation within a species is great. Species imply relationships; a population is a species relative to other populations. A species is a closed genetic, behavioral, and ecological system. Unless these ideas are kept in mind, the confusion present in the systematic biology of the order Primates will be present forever.

Speciation

The several definitions of the species proposed in the preceding section lead us to examine how a species is formed. Implicit in definitions of the various species or species concepts are functional assumptions. That is, we assume that the processes of species formation are known. There are probably two points of view about species formation. One is that of the systematist or taxonomist who must catalog and classify animals by categories. The cataloger's approach will tend to emphasize the existence of the differences in characteristics of populations and individual specimens. The other approach is that of the biologist who studies functional mechanisms which lie behind the existence of the diversity of animals found in nature.

Reproductive isolation is the principal criterion used in defining the genetical species. The development of reproductive isolation between diverging populations, for example subspecies or races, is the process by which species are formed. When reproductive isolation is complete a population has been transformed from an open genetic system (a race or a subspecies) into a closed genetic system (a species).

Reproductive isolation is most often achieved by geographical isolation. Geographical isolation is believed to be the most important way in which speciation occurs. A population that radiates to any accessible parts of the planet where it can live successfully covers a considerable geographical range. It will live in various local environments which differ from each other in climate, soil, vegetation, predators, and even parasites. Each local population will tend to adapt to these different local conditions. Adaptations will be molded by natural selection. A population responds adaptively by developing genetic differences in various local areas. The more remote two segments of the population are from each other, both in actual distance and in differences of environmental conditions, the greater is the probability that they will differentiate into distinct subpopulations. Eventually the distinctions between the two subpopulations will become sufficiently great so that two species will be formed from a single parent population. These are allopatric species. The geographical isolation between them may eventually break down, and they may enter into each other's territory. They will then be called sympatric populations or sympatric species. This kind of speciation is illustrated in Fig. 4.1.

It is impossible to observe speciation directly, for it is a slow, historical process. The method most suitable for reconstructing this continuous process is by arranging fixed stages in correct chronological sequence. We do this when we construct paleospecies. However, there is a functional aspect to the problem of defining and constructing species which suggests that we should be able to describe the processes of speciation. We should be able to find natural populations in all stages of speciation and from them reconstruct the processes. Most of the interesting examples that bear upon the process of speciation, particularly geographical speciation, have been found in lower vertebrates and insects. Generally, in a group that is actively evolving, there are some populations that are extremely similar to each other, others that are sufficiently different that they may be classed as subspecies, others that have almost speciated, and still others that have reached specific status. Often these populations are still allopatric. Occasionally their ranges begin to overlap, and thus they reach the stage of sympatric speciation. Formerly, continuous fossil sequences were the best evidence for speciation. As Mayr has pointed out, one of the most convincing proofs of geographical speciation is the existence of a complete set of successive levels of speciation in living populations of, for example, fish, birds, and fruit flies. The Cercopithecinae (Old World monkeys) provide, as far as we know, the best example of speciation among higher mammals.

Almost all carefully studied species that have wide geographical ranges have representative populations that differ from each other in the details of their ecology. That is, the specific and detailed relationship the animals in

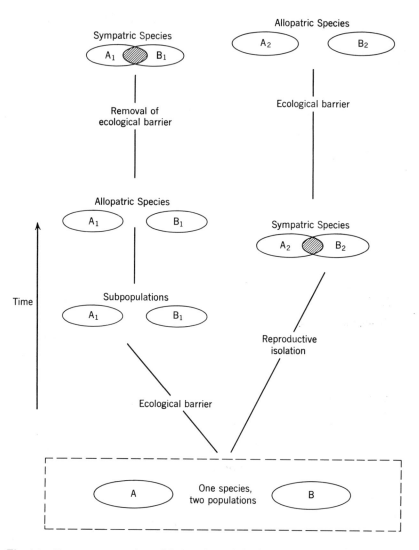

Fig. 4.1. Diagrammatic and simplified analysis of the formation of sympatric and allopatric species.

the population have to their environment differs in various parts of the range of the species. The differences in ecological niches may lead to a kind of pre-adaptation for future speciation. Still other isolating mechanisms may be important, particularly ethological (behavioral) mechanisms. Although relatively little is known about the nature of ethological isolation in many species, there is good evidence that presumed ethological barriers may be more easily bypassed among primates than expected. In effect, any mechanisms, either ecological or ethological, that lead to isolation and that impede the flow of segments of populations will lead to reproductive isolation and, hence, to speciation. The geographical dispersion of a population into different ecological zones or environments will accelerate and facilitate this process.

Small isolates or isolated segments of larger populations, which have some of the characteristics of a new species but not others, exist at the periphery of the parental species. We might expect some segments of a species which have been isolated geographically to fail to develop efficient isolating mechanisms and, eventually, to rejoin the original parental species or become extinct.

Geographical dispersion and speciation permit genetic reshuffling within a population, leading to the development of isolating mechanisms. Geographical dispersion is a prerequisite for speciation, but it is what happens to the genotype of the population that is fundamental in speciation. Mayr has stated that speciation may be looked upon as a process of rejuvenation in evolution, an escape from a rigid system of genetic homeostasis (maintenance of genetic stability). The genetic reconstruction of a population disrupts the balanced equilibrium of the population gene pool and forces the population either into a different environment or into a different relationship with its environment. The greater the genetic change in a population, the greater is the probability that the new species or pair of species can radiate into new ecological zones and successfully adapt. It is probable that most incipient species, populations in the process of speciation, are not successful and die out, but those that do complete the process and succeed have entered a new adaptive zone and may establish the basis for a new adaptive radiation. Speciation is a kind of evolutionary experiment. It is the process by which new evolutionary units are created.

Higher Categories

A definition of a higher category based upon phylogenetic theory has been given by Simpson (1961). "A higher category is one such that a member taxon includes either two or more separate (specific) lineages or a segment of a lineage long enough to run through two or more successional species." A higher category includes taxa from a lower level of the Linnaean hierarchy. A genus is a taxon that consists of species. A family is a taxon that consists of genera. It happens that a higher taxon may be monotypic at a particular time, i.e., contain a single member taxon. Therefore we also define a higher category as a segment of a single lineage, long enough to run through two or more successional species.

The genus is the next category after the species in the hierarchical classification. Part of the formal Linnaean binomial for a species contains its generic designation, i.e., *Homo* in *Homo sapiens* is the name for the genus to which man belongs. A genus consists of a group of related species; few genera consist of, or should consist of, a single species. In light of that, it is, perhaps, ironical that the genus *Homo* is monotypic.

The genus is a different sort of category from the species; it should contain several related species. Definition or construction of a genus is believed by some to be the recognition of a natural phenomenon. It is one of a number of higher categories—higher since they are more inclusive than the category species. The basis for recognition of such higher taxa has been a matter for considerable discussion and argument. The genus is considered to be a natural unit of groups (species) that are closely related by descent.

There are, of course, different conceptions of the genus commonly used by mammalogists and paleontologists. Some consider a genus a wholly subjective and arbitrary grouping of as many species as are convenient to put in a single category. Others, particularly paleontologists, consider the genus as a stage arbitrarily separated from other stages in an evolving sequence. Most useful in primatology is the third view of genus. In this concept, the genus is considered a definite evolutionary entity, hardly more arbitrary than the species. This view of genus reflects adaptive relationships which are not obscured by local speciation, once the adaptive relationships have been achieved. We consider the genus a natural unit, in the same sense that we consider the species a natural unit. The line between genera and species has often been obscured both by the practice of using generic names when fossils are discussed and by confusing morphological differences or variations with taxonomic distinctions.

Genera of living animals can be established in many cases on the basis of genetic evidence. When animals that are nominally placed in two different genera produce living hybrids, even if the hybrids are sterile, the two nominal genera should be made one genus. There are several famous examples that are worth mentioning. Crosses between polar bears and ordinary North American bears or grizzly bears (genera *Thalarctos* × *Ursus*) have occurred in zoos. Baboons and macaques (*Papio* × *Macaca*) have produced hybrid offspring. Perhaps the most famous are mules produced by mating horses and asses (*Equus* × *Asinus*) and the crosses between cattle and buffalo (*Bos* × *Bison*). The generic names of both parents should be, respectively, *Ursus, Papio, Equus,* and *Bos.*

The occurrence of living hybrids demonstrates that the genotypes of the two animals have not diverged enough to warrant categorical, generic, distinctness. Certainly the animals should be classed as different species, particularly if the hybrids are sterile or are less fertile than either parent. But the production of living hybrids demonstrates that the two populations (species) involved are reasonably close genetically. Interfertility is, in a sense, a negative criterion. It may be used to put into a single genus animals now in separate genera. On the other hand it is not always possible to use this criterion. There are strains of inbred laboratory mice, clearly members of the same species, which are sterile when mated. A single gene difference between the two strains is all that prevents successful mating. Thus making the criterion of interfertility a *necessary* part of our conception of the genus will lead to difficulties, even if the difficulties are largely semantic. But, interfertility with production of a viable hybrid offspring is a *sufficient* criterion to put two populations now in different nominal genera into the same genus.

Many of the hybrids between members of populations now included in separate genera occurred in captivity. This disturbs a number of scholars. It should not. The important point here is the demonstration that the genotypes of the two populations have not diverged sufficiently to completely prevent exchange of genetic material. If such hybrid offspring are fertile, particularly if they can be backcrossed with either parent, the potential exchange of genetic material is demonstrated.

If individuals that are members of two different species mate and produce fertile offspring, the measurement of the fertility of the offspring is of critical importance. A reduction in the fertility of the offspring may be taken as evidence that the two species *are* correctly defined. But how do we measure the fertility of such hybrid offspring? It is measured relative to that of the parents or the parental populations or relative to the average fertility of the two species. How much must fertility be reduced before we confirm that the two parental populations are separate species? A 2 per cent reduction is considered by some authorities as a sufficient reduction to confirm the existence of the two species. It is no easy research problem to determine the fertility of primate populations or individuals.

Higher categories at various levels of the taxonomic hierarchy are considered by some authorities to represent natural phenomena, the evolutionary origin of which may be understood when sufficient information is available. Lengthy discussion of this subject cannot be undertaken here, but a summary of some general principles will guide us in attempting to understand those phenomena which the higher categories represent.

The origin of higher categories usually depends on the development of a general and basic adaptive complex which remains virtually unchanged throughout the evolutionary history of a group. Among mammals and especially among Primates a general improvement occurs in many groups of traits, in the total organization of the animals, and in their relationship to the environment. The patterns of phylogeny of higher categories may be based upon a single lineage, splitting of a lineage at successive time levels, or multiple splitting and radiation at one time level. The Primates as a whole provides examples of all three kinds of phylogenetic patterns, at different taxonomic levels. The Tarsiiformes (an infraorder) appears to be a single lineage; the Hominidae, Pongidae, and Oreopithecidae (families) are examples of a lineage splitting at successive time levels; modern Lemuriformes is apparently a case of multiple splitting and radiation at one time level. Some groups developed allopatrically and allochronically, such as the superfamilies Ceboidea and Hominoidea.

The subject of higher categories is complex and is discussed in several of the books listed at the end of this chapter. It is essential for the reader to understand that our views of higher taxonomic categories, such as genera, families, orders, and classes, are based upon phylogenetic theory. The definition of the family Hominidae, for example, is made with an evolutionary concept of the family in mind.

SUGGESTED READINGS AND SOURCES

Darlington, P. J., Jr., *Zoogeography: The Geographical Distribution of Animals.* Wiley, New York (1957).

International Code of Zoological Nomenclature. N. R. Stoll, R. Ph. Dollfus, J. Forest, N. D. Riley, C. W. Sabrowsky, C. W. Wright, and R. V. Melville (Ed.), International Trust for Zoological Nomenclature, London (1964).

Lack, D., *Darwin's Finches.* Cambridge Univ. Press, London (1947).

Mayr, E., *Animal Species and Evolution.* Belknap, Cambridge, Mass. (1963).

Mayr, E., E. G. Linsley, and R. L. Usinger, *Methods and Principles of Systematic Zoology.* McGraw-Hill, New York (1953).

Rensch, B., *Evolution above the Species Level.* Methuen, London (1959).

Simpson, G. G., The principles of classification and a classification of mammals. *Bull. Am. Museum Nat. Hist.* **85**, 1 (1945).

Simpson, G. G., *Principles of Animal Taxonomy.* Columbia Univ. Press, New York (1961).

The Order Primates

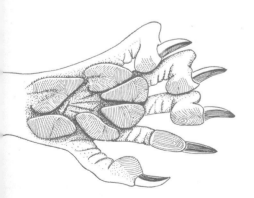

5

PRIMATES ARE: Unguiculate, claviculate placental mammals, with orbits encircled by bone; three kinds of teeth, at least at one time of life; brain always with a posterior lobe and calcarine fissure; the innermost digits of at least one pair of extremities opposable; hallux with a flat nail or none; a well-developed caecum; penis pendulous; testes scrotal; always two pectoral mammae (Mivart, 1873).

This definition has withstood the test of time. Primates have nails (unguiculate) instead of claws, well-developed clavicles, and orbits enclosed with bone (except among plesiadapids). Primates have incisor, canine, premolar, and molar teeth. There is at least one pair of grasping extremities (prehensile hands or feet), and the thumb or big toe is at least partly opposable. The rest of the definition applies only to the living primates. It is difficult enough to apply the first part to the fragments of fossil primates usually recovered. (Definitions of many of the terms used by Mivart are given in the glossary.)

Characteristics

The Primates are a highly gregarious, lively, clever, and successful group of mammals. They are probably some of the most successful mammals that have ever developed during the course of evolution. They have managed to

become one of the most widespread and diversified groups of mammals because of their special characteristics, or perhaps in spite of them. Some groups of animals produce many young in a litter, have short periods of gestation, breed several times a year, and have young that mature quickly. The Primates, on the other hand, seldom have more than a single offspring at a time; they all have a rather long gestation period. Many species, perhaps most, breed less often than once a year, and the young are dependent on their mothers for a long time. The shortest period of infant dependency among Primates appears to be about 5 months and the longest about 15 years. In a metaphorical sense we may say that other groups of animals use the shotgun approach during evolution—they produce many offspring in short spurts, and they gamble on the chance that some will survive. The Primates use a more refined and conservative approach—they produce a small number of offspring and have developed mechanisms which ensure a high rate of survival. The Primates eventually produced a species that can, at will, manipulate the environment in such a way that survival of the offspring becomes highly probable. This species can remake its environment or destroy it or, possibly, move to another planet. All of the Primates, extinct and extant, are part of the evolutionary developments that led to this species.

All Primates share certain general characteristics, and there are certain trends evident in the record of their evolution which are unique to the order. The major characteristic of the order *as a whole* is probably the development of an increasing ability to adapt to varying environments or ecological opportunities rather than a progressive or increasingly detailed adaptation to specific environments or niches. It is for this reason that the Primates are often considered more generalized and more primitive than other animals. It is the trend of increasing ability to adapt to or to manipulate the environment that is so characteristic of our own mammalian order. A consequence of this is the probable elimination of forms that develop a very close relationship to a specific set of environmental circumstances. The kind of restriction imposed upon the koala bears of Australia by their efficient, if highly special, adaptation to eucalyptus trees is not common among the Primates. Where such an adaptation develops, as is likely among the Malagasy lemurs, the consequences lead the animal group into what is likely to be an evolutionary dead end. The principal features and trends that are distinguishable among the Primates include the following rather broadly conceived characteristics.

1. Increasing refinement of the hands and feet for grasping objects. This trend may be characterized as the development of a high degree of manual dexterity. It includes the development of flat nails on the digits of the hands and feet, in place of the sharp claws of primitive, earlier mammals; the development of very sensitive tactile pads on the fingers and toes; the retention of the primitive, early mammalian pentadactylism—5 fingers or toes on each extremity; the development of mobile digits, particularly emphasized by the thumb and big toe. This gradually increasing mobility of the digits and the manual dexterity that is its consequence are the basis for man's

efficient grasping hand with its opposable thumb. The generalized limb structure of early mammals also is retained, as is the clavicle.

2. Reorganization of the special senses and certain skeletal structures associated with them. The sense of smell has been reduced in importance, functionally and structurally. The olfactory center of the brain (the rhinencephalon) has decreased proportionally in size, and the length of the snout and the degree of protrusion of the face have gradually and progressively been reduced. The visual sense and the anatomical apparatus for vision have been emphasized throughout primate evolution, leading to very efficient binocular color vision, such as man has. It is important to note here that all members of the order have binocular vision. Color vision has also very likely developed to some degree in all diurnal primates, although the evidence to support this is not clear cut.

3. Continuous development of the brain with special elaboration and differentiation of the cerebral cortex. A high brain-to-body ratio is characteristic of the members of the order. The reduction in the snout and the increased size of the cranium are related to this neurological development as well as to the reorganization of emphasis on vision over olfaction. The increase in manual dexterity and in eye-hand coordination that is so prominent a feature of primate evolution also required an enlargement of the coordinating centers in the cerebral cortex. These developments, of course, had profound consequences on the morphology of the skull.

4. Apparent increase and elaboration of the development of the processes of gestation and of the uterine and placental membranes. The period of gestation has lengthened, probably as a consequence of this. Among the most primitive primates (*Tupaia*) the gestation period is 43 to 46 days; among the more primitive Prosimii (*Microcebus, Galago*) the gestation period is at least 4 months, and it increases in length to 9 months among *Homo sapiens*.

5. Greatly increased time during which infant primates are dependent upon their mothers or upon other adults. This is related to the preceding trend; perhaps it is a consequence of it. Of course, this is not recognizable in the fossil record. But the fact that the length of this period of infant dependency increases from lower to higher living primates suggests that it is a real evolutionary trend. The length of the period of postnatal growth and development to sexual maturity increases markedly from lower to higher primates. This period ranges from less than 1 year among the nocturnal prosimian primates to almost 15 years in man.

6. Marked increase in the "natural" life span for all members of the order. Tiny mouse lemurs (*Microcebus murinus*) have lived for 8 years in a laboratory colony and are still alive. These are animals whose average weight is about 80 grams. It is likely that longevity is related to the relatively few offspring normally produced by the primates. Multiple births are rare in the majority of primate species, and many species normally have less than one birth per female per year. The female sexual cycle varies from a single ovula-

tion per year, in many prosimians, to as many as thirteen per year in higher primates. Perhaps we could categorize (4), (5), and (6) as part of an extremely complex adaptation to ensure survival of sufficient offspring to maintain the species.

7. Increasing complexity and quantity of social behavior. This fact also is not interpreted from the fossil record and is based on the study of the living primates. The vocalizations of primates, their displays, and their special social behavior, such as grooming, infant care, and vigilance, are extremely varied and complex. Primates are the most social of mammals, and the higher primates are more social than the lower.

8. Retention of a simple cusp pattern on the molar teeth. Certain elements of primitive mammalian dentitions have been lost, but this simple cusp pattern has been retained.

Certain terms used in the list are usually taken for granted but should be explained—*higher* and *lower, simple* and *complex, generalized* and *specialized, progressive, primitive*. Some may be taken as value judgments. They are often taken to be symbols of approval or praise. Here they mean something quite simple. The more advanced or higher members of the order Primates are usually the most recent, the newest, evolutionary developments and the most progressive. They are also probably the most complex.

The criteria for determining simple and complex traits or animals must always be explicit. The list of characteristics of the order Primates just enumerated provides such a set of criteria. The simplest of the Primates are those that have the smallest number or the least development of these characteristics. Those with the largest number or greatest development may be considered complex.

Higher and lower refer to time periods in which particular lineages differentiated. The living lower primates are those whose ancestors made a particular adaptive radiation in the Eocene (Chapter 7); the higher primates differentiated later. The ecological adaptations were successful and the environments were reasonably constant so that the lineages remained fairly stable. It is possible, in a graphic way, to symbolize this by making a family tree for the order and then placing the various taxa on higher or lower branches of the tree (Fig. 5.1).

It is also necessary for us to understand what generalized and specialized mean in an evolutionary context. They mean something simple and straightforward. A generalized trait or structure is one that permits future evolution; a generalized animal is one that is clearly capable of further adaptive radiations. A specialized trait is one that is unlikely to or actually does not permit further evolutionary development except along one restricted line. A specialized animal is one that will become extinct when the environment to which it is specially adapted changes. As Le Gros Clark has shown, pentadactylism is a very generalized, primitive characteristic of mammals. Specializations have occurred in various lineages in which one or another digit has been lost. Specialized characteristics indicate trends of evolution that diverge

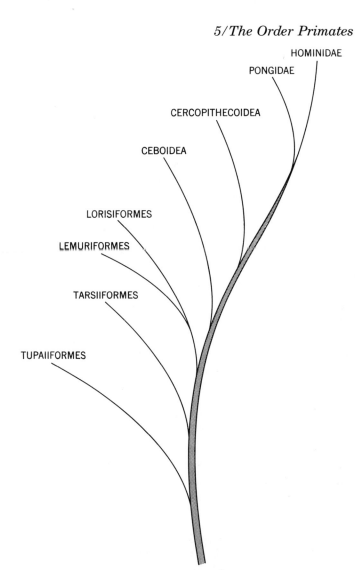

Fig. 5.1. A family tree of the order Primates. This tree is in no way to be considered as an authoritative opinion or accurate diagrammatic representation of primate phylogeny. It is simply a convenient device to symbolize, approximately, the times at which each of the eight major lineages diverged.

from the main line and generally give rise to aberrant organisms, organisms that cannot have an ancestral relationship to living forms.

A single trait, a complex of traits, or even a whole group of animals may be called progressive if the trait or complex clearly leads to further evolution. A trait that is the basis for an adaptive radiation is a progressive trait. Progressive, in evolutionary biology, is a word generally applied to any situation, event, morphology, or adaptation which permits or enables an animal lineage to continue to develop and change. It is a retrospective concept, for we can only judge whether a trait led to evolutionary development of an animal lineage by the record of its fossils.

Traits, usually anatomical or skeletal, which most closely resemble those found in early mammals, are called primitive. Another way to think of primitive is to look at traits which seem to be relatively unchanged throughout an evolutionary lineage. The fact that a trait is unchanged and relatively the same throughout a lineage implies that it has reached a kind of equilibrium with the pressures of natural selection. It is therefore a primitive trait in an evolutionary sense and also a generalized trait in the sense that it is found throughout the group of animals we are studying. In other words primitive, generalized, and progressive are terms denoting the opposite of specialized.

Classification

There are many ways one may classify the animals grouped in the order Primates. The system of classification used here (Table 5.1) is more or less generally accepted. The eight taxa—four infraorders, two superfamilies, and two families—categorize the living primates. The fossil primates are, of course, the ancestors of these eight monophyletic groups, but the exact relationships between known fossil and living primates has not yet been worked out. The eight taxa are at various levels in the Linnaean hierarchy, and the names used here for these taxa are not the same as those Simpson used in the past, but the taxa encompass the same animals (Table 5.1). Each taxon is considered to be monophyletic; that is, the members of each of the eight major taxa are considered to be more closely related to each other by evolu-

TABLE 5.1

The Eight Major Primate Taxa

The positions in the Linnaean hierarchy of the eight major taxa are shown. Each major taxon is indicated by an asterisk (°). Simpson uses an alternative set of names for these major taxa: Tupaiidae, Tarsiidae, Lorisiformes, Lemuroidea, Ceboidea, Cercopithecoidea, Pongidae, and Hominidae.

Suborder	Infraorder	Superfamily	Family
Prosimii	Tupaiiformes°		
	Tarsiiformes°		
	Lorisiformes°		
	Lemuriformes°		
Anthropoidea		Ceboidea°	
		Cercopithecoidea°	
		Hominoidea	Pongidae°
			Hominidae°

tionary descent from a unitary lineage than they are to any members of the other taxa.

It should be noted that the suborder Anthropoidea is divided into three superfamilies—Ceboidea, Cercopithecoidea, and Hominoidea. There appear to be well-marked distinctions among the three. An alternative division of this suborder into two distinct and coordinate groups—Catarrhini and Platyrrhini, the Old World and New World primates, respectively—was once common. However it has never been fully accepted that the Cercopithecoidea and Hominoidea together (Catarrhini) form a unit at the same level as the Ceboidea (Platyrrhini). There is much to be said about primate systematics, and there are many problems involved. As each major group is presented in later chapters, some of these problems will be discussed in greater detail.

Some Trends in Evolution

Today it is well established that the more highly organized forms of life, the more complex organisms, have developed through time from simple organisms. Indeed no other hypothesis for the evidence that has been gathered and is still accumulating since the days of Darwin, Huxley, and Wallace is at all convincing or even plausible. It is no longer necessary to fight this battle again.

Those organisms in which we ourselves are interested, or which we know well, are apt to present such complex problems to our inquiring minds that we lose some of our ability to judge just what is complex or elaborate, simple or advanced, on an evolutionary scale. A biologist who specializes in *Paramecium* may be heard to declare that even man has no more complicated, subtle, or interesting biology. If we use relatively explicit standards for assessing levels of complexity or elaboration of organization, we have fewer difficulties in our discussions. The fact that we end with our own species as the most elaborate, most complex, and most advanced is, we hope, only partly an egocentric phenomenon. *Homo sapiens* is probably the latest primate species to have developed. It is the most successful primate known; it has the widest geographical distribution of any primate; it has developed new modes of behavior by which to adapt to the planet, and, indeed, to adapt the planet to its presumed needs. But this does not mean that *Homo sapiens* is the goal or the objective of primate evolution.

The fossil record has often been interpreted as if there was an inevitable trend that produced our species. And, if we take this view in a strictly retrospective manner, there *is* a kind of inevitability about it. However, there is no evidence at all that each stage was planned to grow into the next stage. As we shall see, there has been no steady, continual progression. Rather, many primate lineages differentiated, radiated, persisted, and then vanished. This can be seen in the fossil record. It is only by very careful use of the retrospectroscope that a straight line of development, often presented in texts, may be discerned. The evolution of life is full of false starts, dead ends, and animals specifically adapted to very narrow econiches. In this book we cannot discuss all parts of evolutionary thought or theory. But it will be possible to discuss some of these concepts with reference to the Primates and related animals.

The view that evolution proceeded in a straight, inevitable progression of species culminating in a particular animal such as *Homo sapiens* is called orthogenesis. It is only by ignoring much of the fossil evidence, by biased focusing of the retrospectroscope, that evidence of orthogenesis can be found in the fossil record. When we relate the events of human evolution, our desire to present a clear and coherent story based on the small number of relevant fossils may often result in an impression of orthogenesis. It is also possible that our understandable wish to organize the available data—fossil material in particular—into reasonable evolutionary lineages leads to an overemphasis on what appear to be linear progressions. On the other hand, too rigorous an attempt to avoid the implication of orthogenesis may lead to the kind of situation that existed until recently in the interpretaion of hominid fossils. Practically all of the fossils that belong in the hominid evolutionary lineage had been put into side branches, leaving almost no fossils for the direct ancestral line. In this book we shall try to avoid each of these extreme positions.

Simpson has shown that orthoselection may occur in a limited way, and the result, as read from the fossil record, may lend weight to orthogenetic postulates. An interesting example is found in the evolutionary record of the increase in the body size of animals. There seems to have been a general increase in body size in many mammalian groups throughout the early Pleistocene. There was also an apparent increase in the size of reptiles from the small lizard-like creatures of the Pennsylvanian and Permian to the gigantic reptiles of the Jurassic and Cretaceous. But there is no law internal to the species which requires that animals evolve into larger forms as a function of evolutionary time. Rather, we are observing a short-term trend, short term from the point of view of the evolution of life, during which natural selection favored larger animals. When the planetary environment changed, selection no longer favored large animals, and they became extinct.

Several examples of straight-line evolution among the vertebrates are quite striking and well known. The elephant-like herbivores progressively increased in size from rather small dog-like animals to the huge woolly mammoth of the Pleistocene. With the passing of this ice age the smaller, contemporary elephants took their place. There are also several examples among the Primates. The Malagasy lemurs produced one of the largest primates known, donkey-sized *Megaladapis*, which became extinct when the swampy rain forest in which it lived was invaded and destroyed by the competing primate, man. Contemporary lemurs range in size from *Microcebus*, the size of a small mouse, to *Indri,* the size of a police dog. The general increase in size of African primates may be seen to culminate in the massive gorilla, yet his more successful relative and successor, man, is considerably smaller. In these examples the complex requirements of adaptation to the environment led natural selection to emphasize increased body size until environmental changes reversed the trend. This is properly termed orthoselection, since an adaptive trend, initiated by natural selection, was continued in a kind of straight-line fashion for a long time. As long as the selective pressures exerted by the environment on the organism were constant, the trend continued in the same direction. But the trend was reversed, modified, or stopped when the environment changed. There is nothing orthogenetic about the evolutionary record in the cases cited here.

The living primates have a characteristic which makes them, and has always made them, of exceptional interest to students of the evolutionary process. Huxley, in 1876, was one of the first to refer explicitly to the Primates as an *échelle des êtres* (scale of being) in miniature, in which almost every development—every major adaptive radiation that occurred during primate evolution—is represented by a living group of the order. The surviving groups of Primates all retain some primitive anatomical features. The retention of such features makes it possible for paleontologists to relate various living primates to their fossil ancestors and to better interpret many of the fossils. To a degree the Primates have members who are "living fossils," and the order as a whole is a kind of living family tree (Fig. 5.1). There is some danger in taking this notion too literally. Each of the contemporary primates has behind it a long and complex path of adaptation and change. They, as we, are present-day representatives of evolutionary processes that have been underway for a very long time. The evolutionary history of the Malagasy lemurs, for example, is much longer than ours in one sense, since the lemurs are the living descendants of prosimians that have been on Madagascar since at least Eocene times. They did not remain stationary as far as evolutionary adaptive developments go. The present-day lemurs are very different from the known lemuroid fossils, yet there are enough similarities to enable us to see the relationships between them.

The reasons for the survival of "living fossils," such as the lemurs, are many, but we must infer most of them. The lemurs probably survived because they were living on an island. They were isolated from other primates and from other large mammals. On the continent of Africa and elsewhere, other primates, such as monkeys, very likely had a competitive advantage for the ecological niche which the lemurs inhabited. Large carnivores never appeared on Madagascar, and hence the lemurs were under little or no predator pressure until man arrived. Lemurs diversified, adapted, and radiated into most of the arboreal econiches in which a primate might be expected to be found. They have come far from the early fossil lemuroids such as *Pronothodectes.* Valid interpretations of possible life ways of fossil primates based upon such living forms as *Lemur* or *Tarsius* must be made cautiously. Nonetheless, fruitful interpretations and reconstructions may be made using living *and* fossil primates.

The relationships between living primates are determined to a large extent by the way we interpret the fossil record. Phylogenetic relationships between living primates and the assignment of primates to levels in the Linnaean hierarchy depend upon phyletic branching and upon all the adaptive, mutational, and functional events which occurred after branching. Phyletic branching is the splitting of a single lineage into two or more lineages. The fossil record is the only body of data that can document this event with any degree of precision. If we assess the evidence of the fossils correctly we may be able to determine when phyletic branching occurred and, sometimes, how many new lineages differentiated from the line that branched. The time that has elapsed since two evolutionary lineages were one is of some importance in determining how we view the relationship between two primate groups. But length of time is not the only factor and, in some cases, may not even be the essential factor. All the adaptive, mutational, and functional events which have taken place since phyletic branching occurred are part of what we must evaluate when deciding to what extent two or more groups of primates are related. All the events, all the traits, all the information must become a part of the evidence we use when we construct our phylogenies and when we make our classifications. Our conception of the relationships among Primates and, as a consequence, the classification of the Primates depend upon many lines of evidence. We should not radically change our notions every time a new trait is added to the evidence. Neither should we hold too rigidly to established ideas. But we should attempt to understand the evolutionary processes that phylogenies describe. The relationships between species, genera, and families can be assessed after we understand as fully as we can the entire adaptive complex achieved by each group.

SUGGESTED READINGS AND SOURCES

Barnett, S. A. (Ed.), *A Century of Darwin*. Harvard Univ. Press, Cambridge, Mass. (1958).

Carter, G. S., *A Hundred Years of Evolution*. Sidgwick and Jackson, London (1957).

Clark, W. E. Le G., *The Antecedents of Man* (Second edition). Edinburgh Univ. Press, Edinburgh (1962).

deBeer, G., *Charles Darwin*. Doubleday, New York (1964).

Dobzhansky, T., *Genetics and the Origin of Species*. Columbia Univ. Press, New York (1951).

Genetics and Twentieth Century Darwinism. Cold Spring Harbor Symposia Quant. Biol., **24** (1959).

Huxley, J., *Evolution the Modern Synthesis*. George Allen and Unwin, London (1942).

Mayr, E., The new systematics. *In* C. A. Leone (Ed.), *Taxonomic Biochemistry and Serology,* p. 13. Ronald Press, New York (1964).

Mivart, St. G., On *Lepilemur* and *Cheirogaleus,* and on the zoological rank of the Lemuroidea. *Proc. Zool. Soc. London,* 484 (1873).

Simpson, G. G., The principles of classification and a classification of mammals. *Bull. Am. Museum Nat. Hist.* **85,** 1 (1945).

Simpson, G. G., *The Meaning of Evolution.* Yale Univ. Press, New Haven (1949).

Simpson, G. G., *The Major Features of Evolution.* Columbia Univ. Press, New York (1953).

Simpson, G. G., Primate taxonomy and recent studies of nonhuman primates. *Ann. N. Y. Acad. Sci.* **102,** 497 (1962).

Simpson, G. G., The meaning of taxonomic statements. *In* S. L. Washburn (Ed.), *Classification and Human Evolution,* p. 1. Aldine, Chicago (1963).

Simpson, G. G., *This View of Life.* Harcourt, Brace and World, New York (1964).

Teeth, Bones, and Muscles

6

ATTENTION READER! Before we present the story of primate evolution and discuss various interpretations of it, we present a somewhat technical excursion into anatomy. We shall emphasize some facts of the anatomy of the teeth, bones, and muscles of man. The reader, we hope, will use this chapter primarily as a reference. Its contents will enable him to understand some of the expressions used in later chapters and should help him in reading the literature of physical anthropology. For some readers, the treatment here will appear to be sketchy; for others, it will seem far too detailed. What we have tried to do is to present some discussion of anatomical features of importance in physical anthropology, to mention some topics only briefly, and to omit others entirely.

The fossil remains of our ancestors are bones and teeth. If we know a little about our own anatomy and the anatomy of some of our living relatives, the fossil record can become something more than a catalog of bones and their sizes and shapes. By use of our scientific imagination, we can extrapolate from bones to living, leaping, jumping, climbing, behaving animals. The end to which we direct the reader is not the retention in his memory of each term, each bone, and each muscle, but the interpretation of the bones and fragments of bones found by paleontologists.

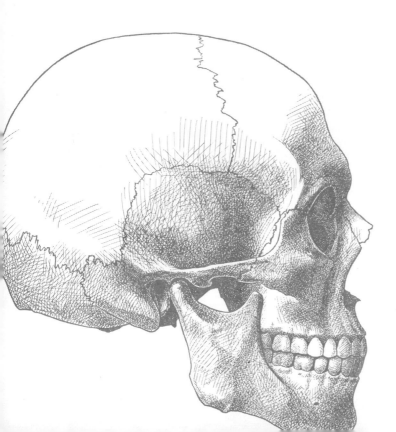

Some Anatomical Terms

Most terminology in anatomy is purely descriptive, and there are often many synonyms for each term. The surfaces of the body are designated dorsal (back), ventral (belly), and lateral (the two sides). Cranial or cephalic means toward the head, and caudal means toward the tail or, in man, where the tail would be if he had one. Superior means upward and implies a relationship to gravity. For this reason its exact meaning may differ depending upon the position in which the body is placed. The anatomical position is fixed by convention. It is the one to which all such terms as superior, anterior, and posterior refer. The anatomical position is, for man, the erect, bipedal position, heels together, feet pointing outward, arms by the sides with palms facing forward. Therefore, superior means toward the head and may be used interchangeably with cephalic or cranial. Inferior is toward the feet and is generally the same as caudal. Anterior is the part of the body carried forward when one walks and is the same as ventral; posterior and dorsal are synonyms. The different referents for these terms in orthograde and pronograde animals are shown in Fig. 6.1. There are other words used to refer to relative positions: for example, medial and lateral, that is, toward the midline of the body and toward the side, respectively. Proximal means toward the point at which the limb is attached to the trunk. Distal is away from the point of attachment. Palmar refers to the palm of the hand, plantar to the sole of the foot.

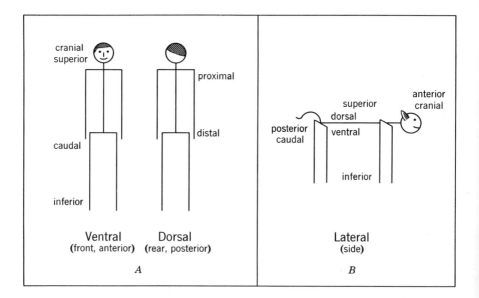

Fig. 6.1.   Some anatomical directions in (*A*) an orthograde animal and (*B*) a pronograde animal.

The words we use to describe movements are also important, and the reader should understand that whenever these terms are used they should be used precisely. Bending of the trunk or limbs occurs very freely toward the ventral surfaces of these members. The term flexion means bending, specifically bending in the ventral direction. Dorsiflexion is bending in the dorsal direction. Extension means the straightening of a bent part of the body or a limb and is a movement that opposes flexion. Abduction means that the limb is moving away from the midline or that two limbs are being moved apart. Abductor muscles are those that move limbs away from the midline of the body. Adduction is the reverse movement; it is movement of limbs or parts of the body toward the midline. Rotation is the movement of a limb or part of the body as it is twisted around its own longitudinal axis. Rotation is lateral or external if the anterior surface of the limb is turned laterally. It is medial or internal rotation if the limb is turned medially. Circumduction combines a number of movements, so that the distal end of the limb being moved goes in a circle.

The skeletal or striated muscles are the organs of voluntary action. When muscles contract, they move various parts of the body. The TENDONS, APONEUROSES, and FASCIAE secure the ends of the muscles, control the direction of their action, and make the energy of muscle contraction effective. The attachments of a muscle, one at each end, are called the ORIGIN and the INSERTION. The origin is the fixed proximal end and the insertion is the movable, distal end. The origins and insertions of muscles have been designated arbitrarily and are a matter of convention.

Many anatomical terms refer to the shape, size, location, or function of a muscle or to its apparent similarity to some structure or object outside of anatomy. Muscles have often been given fanciful names. For example a BICEPS is a muscle with two heads of origin, and there are two biceps in the human body, the BICEPS BRACHII, which is the two-headed muscle of the arm, and the BICEPS FEMORIS, the two-headed muscle of the thigh. SARTORIUS (Latin: tailor) is a muscle of the hip that allows one to sit tailor-fashion.

Almost all muscles that act at joints are paired; these pairs are known as antagonists. For example BICEPS BRACHII is a flexor, and TRICEPS BRACHII is an extensor; their action is at the elbow.

Fossil fragments of pelvis, scapula, skull, and humerus are not mere lumps of rock with a specific shape but were once parts of a living animal and participated in the movements this animal made every day of its life. We may infer much about the behavior of a long-dead primate from the shape, the morphology, of such fragments. The muscles which make it possible to eat, sit, climb, walk, run, jump, squat, vocalize, etc., are all attached to the skeleton in living animals. On fossilized bones, the points of origin and/or insertion of muscles can often be located. The marks left by muscles now lost forever tell us much about the repertory of movements and behaviors in extinct forms.

Teeth

The teeth of mammals are differentiated in various parts of the mouth, probably for special purposes. In the front of the jaw flat, sharp-edged INCISOR teeth serve as cutting instruments. Just beyond these a sharp projecting tooth, the CANINE, grasps and tears and in some animals is used in threatening displays. The next several square, broad teeth, PREMOLARS and MOLARS, are rather complicated in form and are the chewing teeth.

Various groups of mammals have characteristic dental formulae. The dental formula is the number of teeth of each kind to be found in one-half of the upper jaw, the MAXILLA, and the number of each kind to be found in one-half of the lower jaw, the MANDIBLE. Counting starts from the midline of each jaw, so that the total number of teeth is twice the number given by the formula. The dental formula of the primitive ancestral eutherian mammals is believed to be $I\frac{123}{123} C\frac{1}{1} P\frac{1234}{1234} M\frac{123}{123}$. This may be simplified to $\frac{3.1.4.3.}{3.1.4.3.}$. This means that there were three incisors, one canine, four premolars, and three molars in each half of both the upper and lower jaw of such a generalized primitive mammal. Among the Primates, the trend in evolution has been toward reduction in the number of teeth. Thus, two incisor teeth are the rule, and some reduction in the number of premolars is generally found. Table 6.1 lists the dental formulae of a number of extinct and living primate forms. There are several genera, for example, *Daubentonia* (Daubentonioidea), with very specialized dentitions. Nonetheless, teeth seem to be a very conservative trait in evolution, especially in primate evolution, and it is this conservatism that makes them so useful for paleontologists.

TABLE 6.1
Dentition of Primates

	Dental formula	Total number of teeth
PRIMITIVE MAMMAL	$\frac{3.1.4.3.}{3.1.4.3.}$	44
FOSSIL PRIMATES		
Adapidae		
Pelycodus	$\overline{2.1.4.3.}$	
Anaptomorphidae	$\frac{3.1.2 \text{ or } 3.3.}{2.1. \quad 2.3.}$	34 or 36
Tetonius	$\overline{1.0 \text{ or } 1.3 \text{ or } 2.3.}$	
Carpolestidae		
Carpolestes	$\frac{?. \quad ?. \quad ?3.3.}{1.1 \text{ or } 0.3 \text{ or } 4.3.}$	

TABLE 6.1 (*Continued*)
Dentition of Primates

	Dental formula	Total number of teeth
Omomyidae	$\frac{2.1.3.3.}{2.1.3.3.}$	36
Oreopithecidae		
Apidium	$\overline{2.1.3.3.}$	
Parapithecidae		
Parapithecus	$\overline{1 \text{ or } 2.1.3 \text{ or } 2.3.}$	
Phenacolemuridae		
Palaecthon	$\frac{?\,?\,?.3.}{1.1.3.3.}$	
Paromomys	$\frac{?.1.3.3.}{1.1.3.3.}$	
Phenacolemur	$\frac{?.1.3.3.}{1.0.1.3.}$	
Plesiadapidae		
Pronothodectes	$\frac{?.?.?.3.}{2 \text{ or } 1.1.3.3.}$	
Tarsiidae		
Necrolemur	$\frac{2.1.3.3.}{1.1.3.3.}$	34
LIVING PRIMATES		
Tupaiiformes	$\frac{2.1.3.3.}{3.1.3.3.}$	38
Tarsiiformes	$\frac{2.1.3.3.}{1.1.3.3.}$	34
Lorisiformes	$\frac{2.1.3.3.}{2.1.3.3.}$	36
Lemuriformes		
Lemuridae	$\frac{2.1.3.3.}{2.1.3.3.}$	36
Lepilemur	$\frac{0.1.3.3.}{2.1.3.3.}$	32
Indriidae	$\frac{2.1.2.3.}{1.1.2.3.}$	30
Daubentonioidea	$\frac{1.0.1.3.}{1.0.0.3.}$	18
Ceboidea		
Cebidae	$\frac{2.1.3.3.}{2.1.3.3.}$	36
Callithricidae	$\frac{2.1.3.2.}{2.1.3.2.}$	32
Cercopithecoidea	$\frac{2.1.2.3.}{2.1.2.3.}$	32
Hominoidea	$\frac{2.1.2.3.}{2.1.2.3.}$	32

The dental formula is important because of the following generalization: if a tooth is lost during evolution, it cannot be regained as the same tooth. If a species is believed to be ancestral to or closely related to the ancestors of a particular primate, it must have the same or a larger number of each kind of tooth than the descendant form. This is a specific example of the irreversibility of evolution, of Dollo's principle (Chapter 2).

The teeth of early primitive mammals (Fig. 6.2) have certain character- istic structures that are the roots from which the dentition of the Primates developed. The incisor teeth are relatively small, rather flattened, resembling small spatulas. The canine is large, pointed, rather sharp, slightly curved, and projected beyond the other teeth. The premolars in this generalized mammalian dentition are relatively simple and cone shaped. They have a CUSP, the major elevation on the crown of a tooth. The base of the crown of the premolar is thickened to form a ring of enamel around the base of the tooth. This ring is called the CINGULUM. It is an important feature for it is believed that the cingulum developed later in evolution into additional cusps as the premolars and molars developed into the forms they now have. Premolar crowns in living primates are considerably more elaborate than the crowns of the simpler ancestral mammals (Fig. 6.2).

The upper molars of the primitive mammal have three large, or major, cusps. These cusps are called the PROTOCONE, PARACONE, and METACONE. These three main cusps of each upper molar form a triangle on the crown, called the TRIGONE. This three-cusp or tritubercular pattern is believed to be the source of all the more complicated dentitions of later, living mam- mals. This trituberculy theory of the evolution of mammalian dentition was first developed by Cope and Osborn.

The upper molars are fastened to the jaw by three roots, two lateral and one medial. The medial root is the strongest. Enamel crests extend laterally from the protocone along the margins of the crown. Subsidiary CUSPULES may develop on these enamel crests, the one in front is called the PROTOCO- NULE, the one in back the METACONULE. The cuspules are morphological features that may be absent in some species or have special characteristics in others. The enamel itself may be crinkled or crenulated in various ways in different individuals of a species. Such individual variation has often been given undue significance in assessing similarities and differences between two specimens.

The crown of the lower molar of the generalized mammalian dentition is divided into two segments. The front part is called the TRIGONID and has three main cusps; the back part is called the TALONID. The trigonid is higher than the talonid in this generalized dentition. The three cusps—the PROTO- CONID, the PARACONID, and the METACONID—are also set in a triangular for- mation. The trigones of the upper molars alternate with the trigonids of the lower molars to produce a shearing force during mastication. When the teeth of the upper and lower jaws are occluded, the protocones of the upper molars fit into the talonid basin of the lower molars. The term tribosphenic

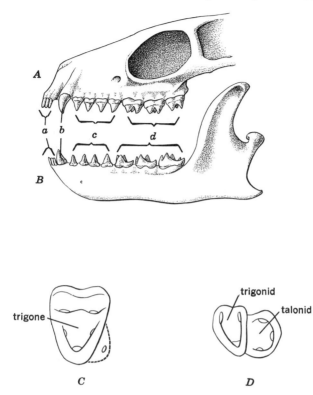

Fig. 6.2.   Mammalian dentition. (A) Generalized dentition of the upper jaw, (B) generalized dentition of the lower jaw, (C) cusp pattern on occlusal surface of a left upper molar, and (D) cusp pattern on occlusal surface of a left lower molar. The generalized dentition in one half of each jaw contains (a) three incisors, (b) one canine, (c) four premolars, and (d) three molars (Redrawn after Le Gros Clark, 1962.)

is used to describe the generalized molar dentition of mammals. It refers to the shearing force of trigones and trigonids and the grinding action of the protocones in the talonid basin. The complex crown patterns of the molar teeth of Primates and other Recent mammals are elaborations of this simple primitive pattern.

The general trend of dental evolution in the Primates has been the retention of a fairly primitive molar pattern and specializations of the other teeth. The incisor teeth are reduced in number, usually to two. Remember, we are speaking of the dental formula—of one half of the dentition counting from the midline of the jaw. In the higher primates the incisors resemble the early mammalian form. The upper incisors are reduced in size in some prosimians and the lower incisors are specialized to form the tooth scraper or toothcomb (Fig. 2.1*B* and 6.3). Throughout the lower primates (the Prosimii), the incisors are not part of a single, uniform evolutionary trend.

The canines tend to become large and very sharp. The canines of higher primates are usually much larger in males than in females. They function in displays, in defense, and as tools for tearing at bark and leaves. The premolar teeth of higher primates are reduced in number to only two. The last two in the premolar series (P_3 and P_4 of the primitive, generalized dentition) grow larger and cusps grow up from the cingulum. This trend is called the molarization of the premolars and becomes very pronounced among certain of the Prosimii.

The upper molars become quadritubercular in form, specializing away

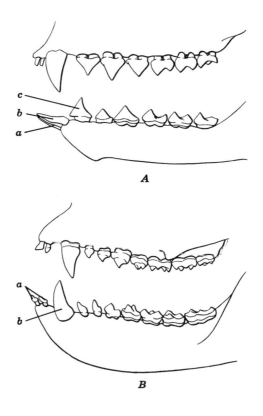

Fig. 6.3.   Dentition of (A) *Lemur*, a living prosimian, and (B) *Notharctus*, a fossil prosimian. The toothcomb, formed by the lower incisors (*a*) and canines (*b*), is present in *Lemur* and absent in *Notharctus*. The first lower premolar (*c*) in *Lemur* is caniniform. (Redrawn after Le Gros Clark, 1962.)

from the tritubercular primitive morphology. The lower molars also develop a quadritubercular crown. Since the upper and lower molars must function together in shearing and grinding, we would expect a change in the upper series to be matched by a change in the lower.

The teeth of primates most often survive the passage of time during which the rest of the body is slowly reduced to soluble compounds which are dissipated in the soil. Therefore the teeth of an ancient primate are the feature most likely to be found by paleontologists. Bones of the skull, particularly of the jaw, are also often found. But it is the recovery of teeth of ancient and extinct primates that has made possible so many of the evolutionary generalizations about them.

Skulls

The primate skull is a complex organ. The description that follows is based on the skull of man, and the terms used are applicable to the skulls of other primates.

The human skull (Fig. 6.4–6.8) consists of two separate parts, the CRANIUM and the MANDIBLE, which are articulated when the CONDYLES of the mandible fit into the GLENOID FOSSA on the base of the skull (Fig. 6.5 and 6.6). The cranium is divided into two principal parts, the CALVARIUM which contains the brain and the FACE. The calvarial part of the skull is usually described as having a vault (the roof of the calvarium), a base, side walls, and a rounded posterior wall. The technical terms given various bones and landmarks are indicated in the various figures. The bones that form the calvarium are joined to each other by SUTURES or seams (Fig. 6.7) which do not become completely fused until after growth is completed. Estimates of the age of an individual at time of death are often based upon the degree to which the sutures have been obliterated or fused. During the course of human evolution the calvarium has become a large globular case for the brain. This characteristic shape of modern man's braincase is a relatively recent development.

The facial skeleton has undergone some remarkable changes during primate evolution. The ORBITS, the eye sockets, have rotated to the front of the skull (Fig. 6.4 and 6.5). This is part of the development of efficient stereoscopic vision. The brow ridge over each orbit, the SUPRAORBITAL TORUS, has almost completely disappeared. These structures, massive in some other primates and in the remains of Pleistocene men, are a complex bit of anatomy. They are part of the frontal bone which in part forms the orbits. The sinuses (passages inside the bone) are involved in forming this structure, as are the MALAR bone and the complex of muscles of the face and jaw.

The largest facial bones are the MAXILLAE, the bones of the upper jaw. The two maxillary bones are united to form the bony part of the PALATE (Fig. 6.6). The maxillae are of special interest to us, since bits and pieces of fossil primate maxillae are often found with teeth in them. If a proper degree of

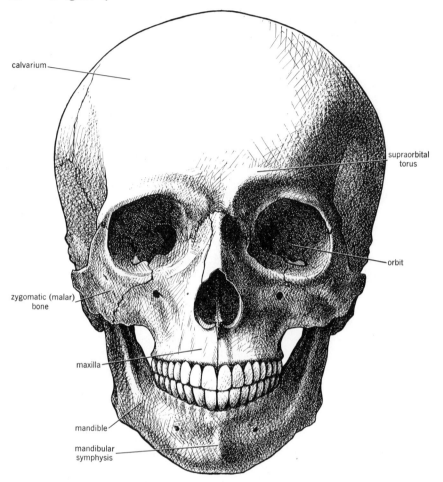

Fig. 6.4. Human skull, front view.

caution is exercised, a reconstruction of important parts of the facial skeleton of a long extinct primate may be based on such fragments of maxillae (Chapters 8 and 9).

The base of the cranium has one specially notable feature as far as studies of primate evolution go—a hole, the FORAMEN MAGNUM (Fig. 6.6). It is through this hole that the spinal cord enters the brain. The two OCCIPITAL CONDYLES on either side of this hole are the points of articulation with the vertebral column. The positions of this great hole and of the occipital condyles on the base of the skull may indicate the position in which the head was carried by a fossil primate. The expansion of the braincase has also influenced the position of the foramen magnum.

The mandible (Fig. 6.8) consists of a pair of bones, fused at the MANDIB-ULAR SYMPHYSIS (Fig. 6.4 and 6.5). This point of juncture, like the maxillary symphysis, divides the dentition into two parts. The lower jaw of a primate is often separated from the skull after death. The chance that saved a lower jaw for the paleontologist to recover is responsible for many a primate genus. It is unfortunate that maxillae and mandibles are not always found together. In one case, the maxillae and mandibles of members of the same fossil primate species were once placed in different genera (Chapter 8).

The muscles of the head (Fig. 6.9) are conveniently divided into two major groups—those of mastication and those of the face. The muscles of mastication are quite significant in certain parts of the story of human evolution. They have an important effect upon the morphology of the skull. Massive, powerful chewing muscles, such as the TEMPORALIS and the MASSETER, require large, strong, bony supports for their origins and insertions. Such features of the skull as a prominent ZYGOMATIC ARCH or a large SAGITTAL CREST are partly the result of the tension exerted upon growing bone by the action of these muscles. The other muscles of mastication are the INTERNAL and

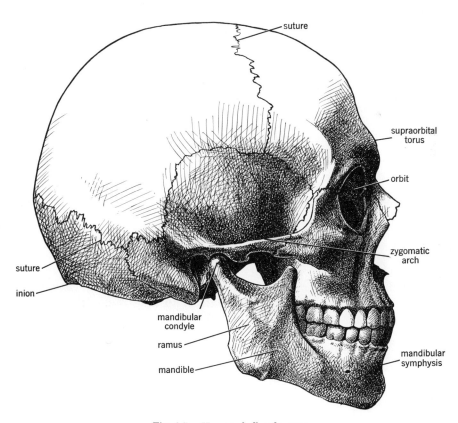

Fig. 6.5.   Human skull, side view.

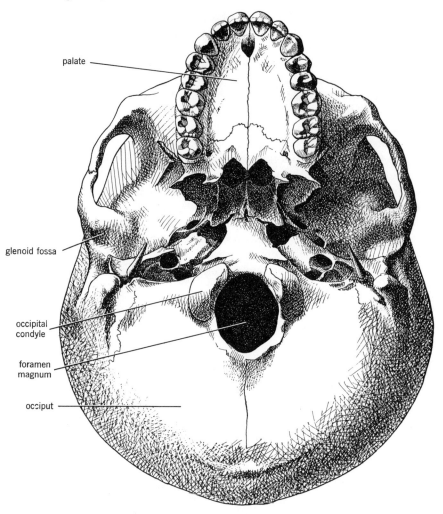

palate

glenoid fossa

occipital
condyle

foramen
magnum

occiput

Fig. 6.6.   Base of the human skull.

EXTERNAL PTERYGOIDS. Anatomists describe the morphology of the internal pterygoid and masseter muscles as the MANDIBULAR SLING. These two muscles suspend the angle of the mandible in a sling. When the mouth is opened and closed the mandible moves around a center of rotation made by this sling and a ligament.

The muscles of the face (Fig. 6.9) are often called the muscles of expression. They lie in the superficial layers of the subcutaneous fasciae. During primate evolution several of these muscles grew increasingly mobile and became vital parts of the facial expressions and displays, so important a part of the evolution of primate behavior.

There are four other groups of muscles of the head and neck that must

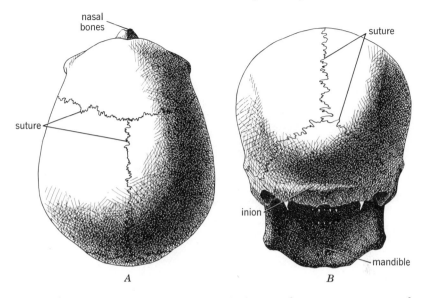

Fig. 6.7. Human skull, (A) top view and (B) back view. The sutures are exaggerated.

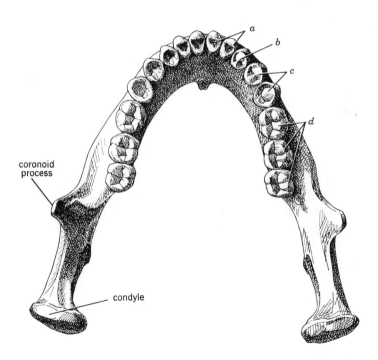

Fig. 6.8. Human mandible (lower jaw), occlusal view. The dental formula is $\frac{2.1.2.3.}{2.1.2.3.}$. (a) incisors, (b) canine, (c) premolars, and (d) molars.

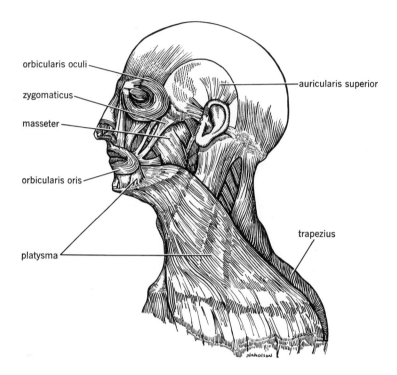

Fig. 6.9.   The arrangement of muscles in the human head and neck. Only part of the masseter is shown. The temporalis is not shown; it lies in a superficial layer above the auricularis muscles. The pterygoids lie in a layer beneath the masseter.

have changed significantly during primate evolution: the muscles of the eye, auditory ossicles, tongue, and pharynx. We shall not discuss these in detail. Table 6.2 lists the important muscles of the head and neck and their actions.

Whole skulls and bones of the skull, like teeth, are a most important part of the record left by the Primates. The characteristics which distinguish primate skulls from those of other mammals are the result of the following functional changes that occurred during primate evolution.

1.   The development of forelimbs to take over the grasping functions of the teeth;
2.   The development of various degrees of upright posture;
3.   The high degree of development of the visual apparatus and the correlative reduction in the sense of smell;
4.   The enlargement of the brain.

These four major functional, adaptive transformations in the Primates had profound effects on the form of the skull. Actually the changes in the skull

TABLE 6.2
Muscles of the Face, Head, and Neck

Muscle	Principal Actions
MASTICATION	
Temporalis	Closes jaw, retracts mandible
Masseter	Closes jaw
Pterygoideus internus	Closes jaw
Pterygoideus externus	Opens jaw, protrudes mandible, and moves it from side to side
FACIAL EXPRESSION	
Levator labii superioris	Elevates upper lip
Levator labii superioris alaeque nasi	Dilates nasal aperture
Levator anguli oris	Forms furrows in face
Zygomaticus major	Draws angle of mouth upward
Zygomaticus minor	Forms furrow between nose and lip
Risorius	Retracts angle of mouth
Depressor labii inferioris	Draws lower lip down
Depressor anguli oris	Depresses angle of mouth
Mentalis	Raises and protrudes lower lip, wrinkles chin
Orbicularis oris	Closes lips, brings lips together
Buccinator	Compresses cheek
NOSE	
Procerus	Wrinkles nose, draws down medial angle of eyelid
Nasalis	Depresses cartilaginous part of nose
Depressor septi	Constricts nasal aperture
Dilator naris posterior	Enlarges nasal aperture
Dilator naris anterior	Enlarges nasal aperture
EYELIDS	
Levator palpebrae superioris	Opens eyelid
Orbicularis oculi	Closes eyelid, sphincter muscle
Corrugator	Draws eyebrow downward, wrinkles forehead
SCALP	
Occipitofrontalis	Draws scalp back, raises eyebrows, wrinkles forehead
Temporoparietalis	Moves scalp, in some persons wiggles ears

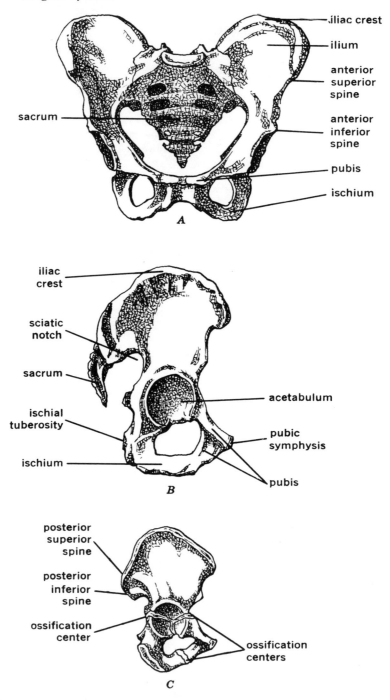

Fig. 6.10. The human pelvis, (A) whole pelvis, ventral view, (B) right os innomina-tum, lateral view, and (C) right os innominatum, immature, lateral view. Three ossi-fication centers are indicated; these are the places at which the bones fuse when maturity is reached.

plus our knowledge of living primates led to the deduction that the four major functional, adaptive changes occurred. When the forelimbs took over the grasping functions of the teeth and jaws, a reduction in the size of the jaw resulted. The development of upright (orthograde) posture is one of the reasons for the movement of the foramen magnum forward on the base of the skull (Fig. 6.6). The enormous development of the visual sense led to enlargement of the eye sockets, rotation of the sockets to the front of the skull, and development of a bony ring around the sockets (Fig. 6.4). The enlargement of the brain led to increase in the size of the braincase and to the rounded shape of the skull of man. The forward movement of the foramen magnum is partly a consequence of the enlargement and rounding off of the braincase. These are the simplest and most generalized statements we can make about the relationship between the changing form of the skull and the changes in way of life of the evolving populations. One general rule should be kept in mind. Gross similarities in structure may be brought about by similar or identical functional requirements. The detailed morphology of the eye sockets of *Tarsius* (Chapter 7) and of other primates is one example in which the same general kind of structure, filling a similar function, is formed of rather different morphological elements in the. two groups of animals.

Pelvic Girdle

The pelvic girdle consists of a right and left os INNOMINATUM (Latin: unnamed bone) connected dorsally by the SACRUM. The two OSSA INNOMINATA are usually called the right and left halves of the pelvis. Each half is constructed of three bones—ILIUM, ISCHIUM, and PUBIS. When adulthood is reached they fuse in a round cup, the ACETABULUM. The head of the FEMUR (thigh bone) fits into this cup, making a ball and socket joint (Fig. 6.10).

The whole pelvis forms a bony tube through which, in living females, the fetus must pass as it is born (Fig. 6.11). In the living organism the upper cavity formed by the iliac blades of the pelvis is filled by the lower part of the abdominal peritoneum, and the walls of the cavity support the intestines. The lower cavity has bony walls that are more complete than the upper cavity. It is defined as that part of the pelvic cavity that lies below the pelvic brim. This lower or lesser

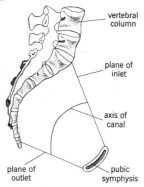

Fig. 6.11. Diagram of medial section of the birth canal in the human female.

cavity contains part of the large intestines—the pelvic colon and rectum—the bladder, and some of the reproductive organs. In females the uterus and vagina are contained within this part of the pelvic cavity. We must remember that considerable rearrangements in these organs must have taken place as natural selection changed the morphology of the pelvis in the course of primate evolution.

The pelvis also links the legs to the vertebral column. It is fixed rigidly to the axial skeleton by short, strong ligaments, and the points of attachment for the muscles that move the legs are on the pelvis. The shape and positioning of the pelvis is of fundamental importance in locomotion (Fig. 6.12). The mode of locomotion of a fossil primate, for example, may be deduced if the pelvis is found. Few fossil primate pelves have been recovered. As we shall see, a certain critical fossil pelvis completely changed the common view about the nature of the course of human evolution.

A number of morphological features of the pelvis (Fig. 6.10 and 6.12) have come to be, in a sense, landmarks in the evolution of man. The CREST of the human ilium is sinuously curved and convex in outline. It ends in two pairs of spines, the ANTERIOR and POSTERIOR SUPERIOR ILIAC SPINES and the ANTERIOR and POSTERIOR INFERIOR ILIAC SPINES. The anterior superior iliac spine serves for the insertion of the ILIACUS muscle and for the origin of the SARTORIUS muscle. The anterior inferior iliac spine is the point of attachment for the RECTUS FEMORIS muscle and for the ILIOFEMORAL LIGAMENT. The posterior superior and inferior iliac spines are two projections separated by a notch; to the former is attached part of the SACROILIAC LIGAMENTS and the MULTIFIDUS muscles. The great SCIATIC NOTCH lies below the posterior inferior iliac spine.

The ischium is the lower part and the back of the hip bone. The ISCHIAL SPINE is prominent and forms the lower edge of the sciatic notch. In man, the ISCHIAL TUBEROSITY is placed in close proximity to the acetabulum. The following muscles, important in locomotion, insert or originate on the ischial tuberosity: BICEPS FEMORIS (long head), SEMIMEMBRANOSUS, SEMITENDINOSUS, and ADDUCTOR MAGNUS (Fig. 6.13). The position of the ischial tuberosity in relation to the acetabulum in orthograde animals is different from that in pronograde animals. The position is related to the differences in posture and locomotion of the two.

The muscles of locomotion and posture associated with the pelvis and thigh are numerous and important. Table 6.3 summarizes the names and actions of these muscles. There are important differences in the actions and mass of many of these muscles between erect bipeds, man, and the more pronograde animals, great apes. These are discussed in Chapter 20 where the major functional changes that occurred in primate evolution are analyzed.

The muscles of this critical functional complex of hip and thigh are most easily understood if they are grouped according to their principal actions: flexion, extension, adduction, abduction, and rotation. The major flexor muscles of thigh and hip are the ILIOPSOAS, SARTORIUS, PECTINEUS, and TENSOR FASCIAE LATAE (Fig. 6.12 and 6.13). The iliopsoas is a complex of two muscles, PSOAS MAJOR and ILIACUS. Its origins are the entire iliac fossa and anterior lateral aspects of the lumbar vertebrae and their transverse processes.

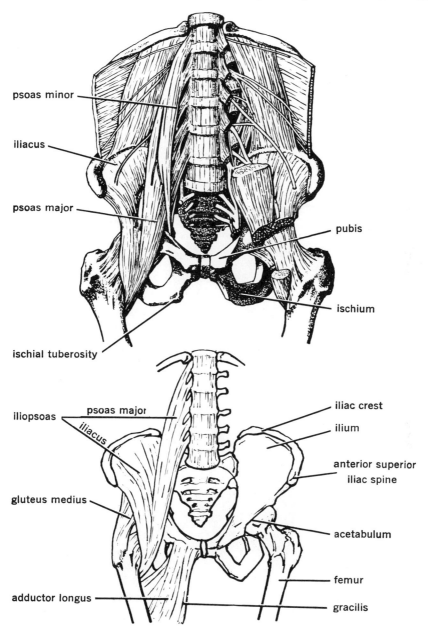

Fig. 6.12. The pelvic girdle and the muscles important in maintaining erect posture. The iliopsoas muscles (psoas major and iliacus) flex the trunk at the hip and exert a pull upon the lumbar region of the vertebral column. In errect posture, this pull prevents the trunk from falling over backwards. In addition, the iliopsoas muscles act to rotate and adduct the femur. Also important in erect posture is the position of the acetabulum, the cup into which the proximal end of the femur fits.

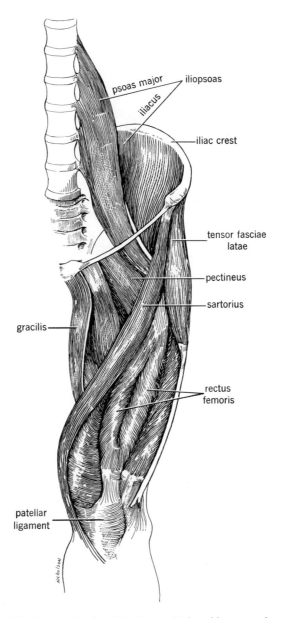

Fig. 6.13.   Muscles of the human thigh and hip, ventral view.

TABLE 6.3

Muscles of the Pelvic Girdle, Leg, and Foot

Muscle	Principal actions
MULTIPLE ACTION	
Sartorius	Flexion, rotation of hip and knee
Tibialis anterior and *T. posterior*	Flexion, supination of foot
Extensor hallucis longus	Extension of proximal phalanx of hallux, flexion, supination of foot
Extensor digitorum longus	Extension of proximal phalanges of toes 2–5, flexion, pronation of foot
Extensor digitorum brevis	Extension of hallux and digits 2–4
Peroneus tertius	Dorsiflexion, pronation of foot
Peroneus longus and *P. brevis*	Flexion, pronation of foot
Gastrocnemius	Flexion of foot and leg, supination of foot
Popliteus	Flexion, medial rotation of leg
Flexor hallucis longus	Flexion of hallux, flexion, supination of foot
Flexor digitorum longus	Flexion of toes 2–5, flexion and supination of foot
Lumbricales (4)	Flexion of proximal phalanges, extension of distal two phalanges of toes 2–5
Articularis genu	Pulling upward of synovial membranes of knee joint
FLEXORS	
Iliopsoas	Flexion at hip
Pectineus	Flexion, adduction of hip
Tensor fasciae latae	Flexion, medial rotation of thigh
Biceps femoris (short head)	Flexion, lateral rotation of leg
Plantaris	Flexion of leg and foot
Soleus	Plantar flexion of foot
Flexor digitorum brevis	Flexion of second phalanges of toes 2–5
Quadratus plantae	Flexion of terminal phalanges of toes 2–5
Flexor hallucis brevis	Flexion of proximal phalanx of hallux
Flexor digiti quinti brevis	Flexion of proximal phalanx of toe 5
EXTENSORS	
Quadriceps femoris	Extension of thigh
Gluteus maximus	Extension, internal rotation of thigh
Semitendinosus[a]	Extension of thigh, flexion of leg
Semimembranosus[a]	Extension of thigh, flexion of leg
Biceps femoris (long head)[a]	Extension of thigh
Adductor magnus (posterior portion)	Extension of thigh

TABLE 6.3 (*Continued*)
Muscles of the Pelvic Girdle, Leg, and Foot

Muscle	Principal actions
ADDUCTORS	
Adductor longus and *A. brevis*	Adduction, flexion, medial rotation of thigh
Adductor magnus (anterior portion)	Adduction, medial rotation of thigh
Gracilis	Adduction of thigh
Adductor hallucis	Adduction of hallux
Interossei plantares	Adduction of toes toward imaginary longitudinal axis through toe 2
ABDUCTORS	
Gluteus medius	Abduction, medial rotation of thigh
Gluteus minimus	Abduction, medial rotation of thigh
Abductor hallucis	Abduction of hallux
Abductor digiti quinti	Abduction of toe 5
Interossei dorsales	Abduction of toes from imaginary longitudinal axis through toe 2
ROTATORS	
Piriformis	Lateral rotation of thigh
Obturator internus and *O. externus*	Lateral rotation of thigh
Superior and *inferior gemellus*	Lateral rotation of thigh
Quadratus femoris	Lateral rotation of thigh

[a] *Semitendinosus, semimembranosus,* and *biceps femoris* (long head) together are the HAMSTRING muscles.

The complex has its insertion on the femur, at the LESSER TROCHANTER and below. It is a powerful flexor at the hip and may act as a slight external rotator of the femur. The iliopsoas is one of the most powerful muscles in the human body. It is attached to the pelvis, vertebral column, and femur. It not only is able to move the trunk, pelvis, and femur, but also stabilizes the trunk. It has an important effect upon the maintenance of erect posture. The strength and the length of the iliopsoas muscle increased as upright posture and bipedal locomotion developed in the primate lineage. The distance between the origin and the insertion of the iliopsoas muscle is greater in *Homo sapiens* than it is in the great apes and in Neandertal fossils. We deduce that the muscle gradually became thicker and stronger as this distance increased so that the erect trunk could be stabilized in an upright position.

The sartorius muscle is a long muscle whose origin is the anterior superior iliac spine and whose insertion is just medial to the tuberosity of the tibia. Although it is principally a flexor of both the hip joint and the knee, it is also a weak abductor and an external rotator of the thigh. These four actions are

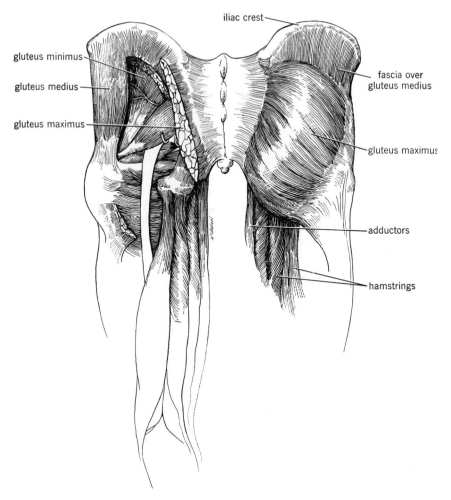

Fig. 6.14.   Muscles of the human thigh and hip, dorsal view.

needed to sit tailor-fashion. The pectineus has its origin on the superior ramus of the pubic bone and its insertion on the femur close to the insertion of the iliopsoas. The tensor fasciae latae arises on the iliac crest behind the anterior superior spine and inserts on the iliotibial tract, a band of fascia that itself inserts on the tibia.

The major extensor muscles of the hip (Fig. 6.13 and 6.14) are the QUADRI-CEPS FEMORIS and the GLUTEUS MAXIMUS. The quadriceps is a four part (four-headed) muscle. It is the muscular mass that covers the anterior and medio-lateral aspects of the femur. The four parts of the head of the quadriceps are called RECTUS FEMORIS (origin—anterior inferior iliac spine and the ilium above the acetabulum), VASTUS MEDIALIS (origin—posterior aspect of the femur), VASTUS LATERALIS (origin—posterior aspect of the femur), and VASTUS INTERMEDIUS (origin—all over the femur). All four heads of the quadriceps insert on the PATELLA (kneecap), and, through the patellar ligament, their action is exerted as a pull on the tibia. The quadriceps is a powerful extensor at the knee.

Gluteus maximus, the muscle mass upon which we sit, is a powerful extensor of the hip. It is important in walking up inclines and stairs, in straightening up after bending, in running and jumping, and in bipedal locomotion generally. Its major origin is the sacrum, and a small part of it originates in the region of the posterior superior iliac spine. It inserts in the iliotibial tract and upon the gluteal tuberosity of the femur.

The adductor and abductor muscles of hip and thigh (Table 6.3) are numerous and are responsible for most of the refined movements of the leg. The adductor muscles are anterior and medial in position, and they also function as flexors and rotators of the hip. The abductor muscles, GLUTEUS MEDIUS and GLUTEUS MINIMUS, abduct the femur. Some fibers of these muscles assist in rotation of the femur as well as in its flexion and extension.

Man's pelvic girdle is efficient, and it is a structure molded by natural selection to make erect bipedal locomotion possible. But the pelvic girdle is not mechanically perfect. Anyone who doubts this need only listen to someone who has "pulled" a back muscle or who suffers from "low back pain." Man has not achieved a perfect and painless adaptation for erect bipedalism.

Shoulder Girdle

The shoulder girdle connects the arms to the skeleton. The two bones of the shoulder girdle are the SCAPULA (shoulder blade) and the CLAVICLE (collarbone). Figure 6.15 shows how these two bones connect the arm to the axial skeleton. The broad scapula and the short, stout clavicle are the supporting structures of man's powerful and mobile arm. The differences in these bones among living primates are related to differences in the animals' modes of locomotion. It has been suggested that the arboreal mode of life of many primates, and possibly of man's ancestors, required the kind of shoulder which made the development of the specifically human type of shoulder girdle a relatively simple matter. The anatomy of the shoulder girdle is related to the complex question of brachiation, arm-swinging locomotion through the trees, as a functional stage in human evolution. We shall discuss this in Chapter 20.

The muscles of the shoulder and arm are divided into six general groups: (1) muscles connecting the arm to the vertebral column, (2) muscles connecting the arm to the anterior and lateral walls of the thorax (chest), (3) muscles

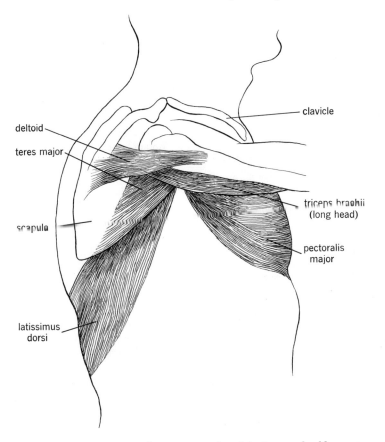

Fig. 6.15. Bones and extensor muscles of the human shoulder.

of the shoulder, (4) muscles of the arm, (5) muscles of the forearm, and (6) muscles of the hand. Table 6.4 lists the various muscles and their principal actions.

Although all the shoulder and arm muscles are important in understanding the various shifts in locomotion during primate evolution, several are of particular significance (Fig. 6.15 and 6.16). PECTORALIS and LATISSIMUS DORSI muscles are used for propulsion by quadrupedal primates, and in man they are flexors, extensors, adductors, and rotators of the arm. SERRATUS ANTERIOR and TRAPEZIUS raise the arm. They are rotators and adductors of the scapula when the arm is being raised (flexed) and abducted. The TRICEPS BRACHII is an extensor of the forearm. Its mass and orientation must have changed during primate evolution. In the primates that brachiate, for example, the muscles that raise the arm are relatively more massive than they are in wholly quadrupedal primates.

TABLE 6.4
Muscles of the Shoulder Girdle, Arm, and Hand

Muscle	Principal actions
MULTIPLE ACTION	
Sternocleidomastoideus	Lateral flexion, rotation of head, flexion and bending of vertebral column
Trapezius	Rotation, adduction of scapula, elevation of shoulder tip
Levator scapulae	Elevation, rotation of scapula
Rhomboideus major and *R. minor*	Adduction, retraction of scapula
Subclavius	Depression of scapula
Teres major	Extension, adduction, medial rotation of arm
Teres minor	Lateral rotation, weak adduction of arm
Pectoralis major (anterior portion)	Adduction, flexion, medial rotation of arm
Pectoralis major (posterior portion)	Adduction, extension, medial rotation of arm
Pectoralis minor	Depression of shoulder
Latissimus dorsi	Extension, adduction, medial rotation of arm
Coracobrachialis	Adduction, flexion of arm
Biceps brachii	Flexion of arm, supination of hand
Supinator	Supination of forearm
Pronator teres and *P. quadratus*	Pronation of forearm
Opponens pollicis	Opposition of thumb
Opponens digiti minimi	Cupping of hand
Palmaris brevis	Holding hypothenar pad in place, wrinkling skin of palm, increasing hypothenar eminence as in making a fist
Lumbricales (4)	Flexion of metacarpophalangeal joints, extension of two distal phalanges
FLEXORS	
Brachialis	Flexion of forearm
Brachioradialis	Flexion of forearm
Flexor carpi radialis	Flexion of hand, abduction of hand
Palmaris longus	Flexion of hand
Flexor carpi ulnaris	Flexion, adduction of hand
Flexor digitorum superficialis	Flexion of second phalanges of fingers
Flexor digitorum profundis	Flexion of terminal phalanges of fingers

TABLE 6.4 (*Continued*)
Muscles of the Shoulder Girdle, Arm, and Hand

Muscle	Principal actions
Flexor pollicis longus	Flexion of second phalanx of thumb, flexion of first phalanx and metacarpal by continued action
Flexor pollicis brevis	Flexion at metacarpophalangeal joint of thumb, flexion, adduction of thumb
Flexor digiti minimi	Flexion of little finger

EXTENSORS

Triceps brachii	Extension of forearm
Anconeus	Extension of forearm
Extensor carpi radialis longus	Extension of hand
Extensor carpi radialis brevis	Extension of hand
Extensor carpi ulnaris	Extension, adduction of hand
Extensor digitorum	Extension of phalanges and wrist
Extensor digiti minimi	Extension of little finger
Extensor indicis	Extension of index finger
Extensor pollicis longus	Extension of second phalanx of thumb, abduction of hand
Extensor pollicis brevis	Extension of first phalanx of thumb, abduction of hand

ADDUCTORS

Adductor pollicis	Adduction of thumb—moving thumb toward palm
Interossei volares	Adduction of fingers toward imaginary longitudinal axis through middle finger

ABDUCTORS

Deltoideus	Abduction of arm
Supraspinatus	Abduction of arm
Abductor pollicis longus	Abduction of thumb and wrist
Abductor pollicis brevis	Abduction of first phalanx of thumb
Abductor digiti minimi	Abduction of little finger
Interossei dorsales	Abduction of fingers from imaginary longitudinal axis through middle finger

ROTATORS

Subscapularis	Internal, medial rotation of arm
Infraspinatus	External, lateral rotation of arm
Serratus anterior	Upward rotation of scapula

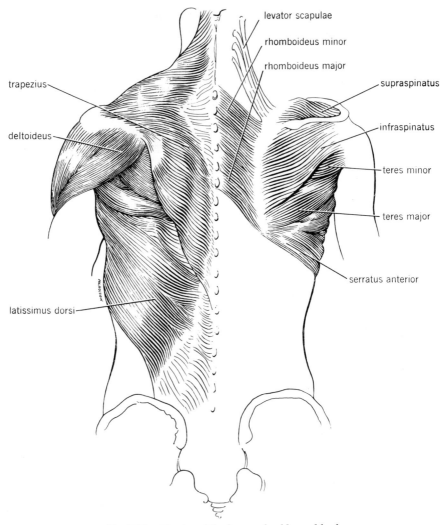

Fig. 6.16.   Muscles of the human shoulder and back.

Long Bones

The bones of the arms and legs are usually referred to as the long bones, which they are! The large bone of the upper arm is the HUMERUS, and the large thigh bone is the FEMUR. These two bones, particularly their proximal ends, are most useful in deducing the posture and the mode of locomotion of an animal. The RADIUS and ULNA, the more slender bones of the arm, connect the humerus to the hand. The TIBIA and FIBULA are the analogous bones of the leg. (See Fig. 6.17 and 6.18.)

When long bones of fossil primates are found, they provide considerable information about the animal's mode of locomotion. The intermembral in-

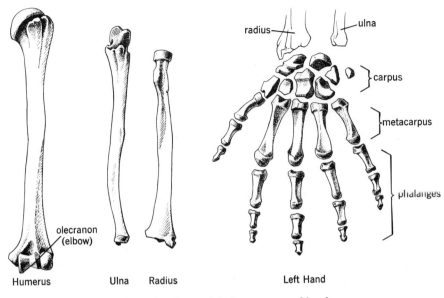

Fig. 6.17.   Bones of the human arm and hand.

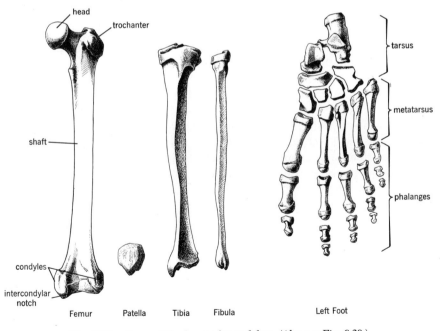

Fig. 6.18.   Bones of the human leg and foot. (Also see Fig. 6.20.)

dex, the ratio of the length of the long bones of the arm to those of the leg, is used as evidence in arguing whether an animal was a brachiator or not. The details of the structure of the head of the humerus, the ridges on which muscles are inserted, and the design of the OLECRANON PROCESS of the ulna are all part of a functional muscle-bone complex. The very short, stout olecranon process of man is part of the complex that powers the strong flexor muscles of man's forearm. Quadrupeds have strong extensor muscles in this complex. The stability of the forelimb, in true quadrupeds, is maintained by tension on a large tendon (the TRICEPS TENDON) which keeps the partially flexed forelimbs extended when the animal is standing. A long olecranon process is important in such a complex. Similar analysis of the role of the head of the femur in relation to the pelvis and the muscles of the hind limb in extension, flexion, and other movements is possible. When a piece of fossilized humerus or femur is found, inferences are made about the bone-muscle complex of the once-living animal.

Vertebral Column

The vertebral column, the skull, the ribs, and the sternum are referred to as the AXIAL SKELETON. The vertebral column is a chain of five groups of VERTEBRAE—the CERVICAL (neck), THORACIC (trunk), LUMBAR, SACRAL, and COCCYGEAL or CAUDAL (tail). There are always seven cervical vertebrae in mammals, although the number of vertebrae in the other regions is variable. Figure 6.19 illustrates the segments of a typical vertebra. The neural spines of the vertebrae are most important in a study of primate evolution. The number of vertebrae in the thoracic and lumbar region that have spines pointed toward the head (cranially), at right angles to the column, or toward the rear (caudally) is significant. Many muscles important in locomotion and in support of the back are attached to these spines. The direction in which the neural spines point is an indication of how to define, in functional terms, the relation between muscles of locomotion and mode of locomotion.

Comparative studies of living primates suggest that there is a constellation of morphological features of the rib cage which are related to various types of locomotion and posture. These features need not be discussed here, for the ribs and sternum of extinct primates are seldom found.

Hands and Feet

The MANUS, the hand (Fig. 6.17), consists of three groups of bones: the PHALANGES or finger bones, the METACARPALS or bones of the hand, and the CARPALS or bones of the wrist. The bones of the wrist provide a firm but elastic link between the bones of the arm and those of the hand. The metacarpals articulate with the phalangeal bones in a way that is closely related to the efficiency and mobility of the fingers. In man, the first metacarpal (thumb or pollex) is a short, stout bone. The articulation of the thumb with

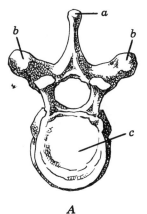

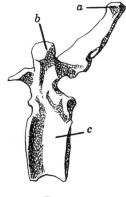

A

B

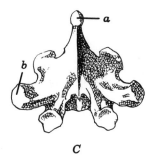

C

Fig. 6.19. A typical human thoracic vertebra, (A) cranial view, (B) left lateral view, and (C) dorsal view. (a) spinous process or spine, (b) transverse process, and (c) body of vertebra.

the carpal bone (CARPUS TRAPEZIUM) allows it greater mobility than the other metacarpals. It articulates with the carpus trapezium in a different plane than do the other metacarpals. The palmar surface of the thumb faces across the palm, making the thumb opposable to the fingers. This is the basis for the mobile, efficient, dextrous, grasping hand of man. The shape of the finger bones, the phalanges, the nature of the surfaces that articulate with the metacarpals, and the arrangements of tendons and muscles in the hand are also part of the structural basis for the power and precision grips of man. The hand is a wonderful and complex example of the relationship of structure and function. There is little information about the evolution of the hand in the fossil record. Most of our knowledge comes from comparative studies of the hands of living primates (Chapter 20).

The PES or foot (Fig. 6.18 and 6.20) is made up of three groups of bones,

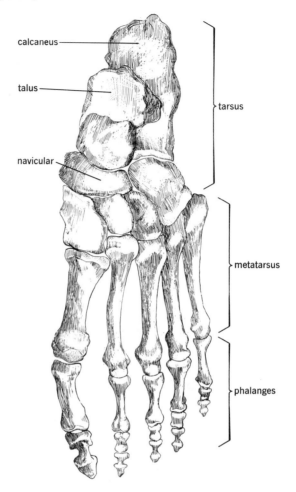

Fig. 6.20.   Human left foot, dorsal view. (Also see Fig. 6.18.)

analogous to those of the hand—the PHALANGES, the METATARSALS, and the TARSALS. The phalanges are the toes; the metatarsals and tarsals are bones of the arch and the heel. The TALUS, or ankle bone, connects the leg bones, tibia and fibula, to the rest of the foot. Its shape and size are related to the mode of locomotion of the animal (Chapter 20). The CALCANEUS is the largest of the tarsal bones and transmits the weight of the body to the ground. It forms the heel of the foot. Man's foot is designed to provide a firm platform to support the body. This firm support is particularly important in the erect bipedal locomotion which is the morphological-functional basis for the adaptive radiation of the genus *Homo*. The size of the first digit (the big toe or hallux) and its position and robustness relative to the other toes are important in determining the way the foot functions. This digit is different in the great apes, for instance, where the foot may be used to grasp objects or to cling to tree branches.

The preceding brief description serves to introduce the rudiments of the bony structure and some important muscle-bone complexes of man. The bony structure is the contemporary end-product of a very long evolutionary process. The bones by themselves can be made to reveal much about the behavior of the living organism from which they came. The reader must understand that certain muscles appear to be extremely important in anthropological discussions of human and primate anatomy for special reasons. First, only portions of skeletons are recovered by paleontologists, and deductions about the way muscles were attached to the bones must be made from the fragmentary material available. Second, the evolutionary advance of the Primates is, in major part, the result of changes in posture, locomotion, and manipulative abilities. The bone-muscle complexes relevant to these functional changes will naturally be emphasized.

Man's locomotion, posture, grasping abilities, and arm movements are the result of a very precise relationship between his muscles and bones. The relationships of homologous muscles and bones of other living primates are not the same. We must infer what these relationships were in the living forms whose fossil remains we find. The inferences we make are based upon human anatomy and on the comparative anatomy of the living primates and other mammals.

Suggested Readings and Sources

Brothwell, D. H. (Ed.), *Dental Anthropology*. Pergamon, Macmillan, New York (1963).

Clark, W. E. Le G., *The Antecedents of Man* (Second edition). Edinburgh Univ. Press, Edinburgh (1962).

Erikson, G. E., Brachiation in the New World monkeys and in anthropoid apes. *Symp. Zool. Soc. London* **10**, 135 (1963).

Gray, H., *Anatomy of the Human Body*. C. M. Goss (Ed., Twenty-seventh edition). Lea and Febiger, Philadelphia (1959).

Hartman, C. G., and W. L. Straus, Jr. (Ed.), *The Anatomy of the Rhesus Monkey*. Hafner, New York (1933).

Hollinshead, W. H., *Functional Anatomy of the Limbs and Back* (Second edition). Saunders, Philadelphia (1960).

James, W. W., *The Jaws and Teeth of Primates*. Pitman Medical, London (1960).

McLean, F. C., and M. R. Urist, *Bone*. Univ. of Chicago Press, Chicago (1955).

Murray, P. D. F., *Bones*. Cambridge Univ. Press, Cambridge (1936).

Napier, J. R., The prehensile movements of the human hand. *J. Bone and Joint Surg.* **38B**, 902 (1956).

Oxnard, C. E., Locomotor adaptations in the primate forelimb. *Symp. Zool. Soc. London* **10**, 165 (1963).

Schultz, A. H., Relations between the lengths of the main parts of the foot skeleton in primates. *Folia Primat.* **1**, 150 (1963).

Scott, J., Factors determining skull form in primates. *Symp. Zool. Soc. London* **10**, 127 (1963).

Thieme, F. P., Lumbar breakdown caused by erect posture in man. *Anthropological Papers, Museum of Anthropology, Univ. of Mich.* 4 (1950).

Paleocene, Eocene, and Oligocene Fossil Primates

7

Primates are "Unguiculate, claviculate placental mammals, with orbits encircled by bone . . ." (see p. 51). When did the first primates appear? In geological deposits of the Paleocene, a period that began about 75 million years ago (Table 3.1). But how can we distinguish the first primates from other early mammals? The answer is that we cannot, easily. By the time fossils that are recognizable as Primates are present, they are already clearly differentiated in the primate direction, away from the insectivores. The question is often asked whether primate-like forms of the Paleocene-Eocene epochs are "really" primates. This question, as most paleontologists know, makes no sense when asked that way. Many mammalian orders have key characteristics or fundamental adaptive patterns which are distinctive. The origin of an order is taken to coincide with the group of animals, usually fossils, in which such distinctive characters or adaptive patterns are fixed. Such characters, when they exist, provide definite, clear-cut criteria for deciding whether a newly found fossil belongs to one group or another. The early primates do not carry the marks of their special adaptive patterns on their bones until they are clearly and distinctly Primates. The transitional forms have not yet been recognized.

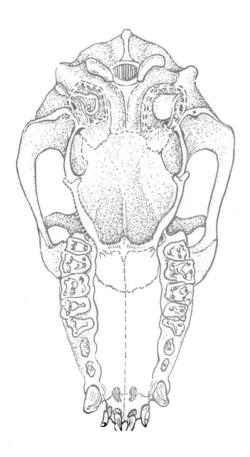

An interesting example of this difficulty is the problem of deciding whether *Zanycteris* is a bat or not. As Simpson says, if it had wings, it is a bat; if it did not, it is not. Unfortunately, there are no such definite criteria to make all the Primates distinct from the Insectivora. The evolutionary development and differentiation of primates from insectivores was gradual. It occurred in many lineages, not all of which were segments of the "main line." There is a continuous range or spectrum of similarities and differences in various characters. In the fossil record there is no threshold over which the insectivore lineages stepped to become primate lineages. There is no missing link. The concept of missing link is self-contradictory and is suitable only in oversimplified, sensationalized writing.

The decision whether a group of mammals belongs in the Primates depends upon how we evaluate a whole range of similarities and differences. Interpretations change with time. To cite only one example, 30 years ago an eminent authority broadened the range of the order Primates and included the tree shrews of southeast Asia, the Tupaiiformes (Chapter 12).

A very logical way to define the Primates would be to include only the Anthropoidea (Table 5.1). The Prosimii are certainly distinct from the Anthropoidea, but they are not quite as distinct from the early Insectivora—which is the crux of the problem of defining the Primates. The prosimians are a genetic and adaptive group of animals noticeably different but not completely distinct from shrews, moles, and hedgehogs. They are definitely related to the anthropoids. The "real" answer to the problem is that it is not reasonable to expect a clear distinction between the taxa Insectivora and Primates when no sharp distinction between them has ever existed in nature.

Ancestors are both similar to and different from their descendants. Related forms do show many differences from each other. The number of differences or the degree of resemblance is often related to the duration in time that separates the ancestor from his descendant or that separates two forms from the ancestor they have in common. Individual variation between members of the same population is great. Many of the traits considered important in anthropology, such as the dimensions of the skull, often vary more between members of the same population than the averages of such dimensions vary between two different populations. It is, therefore, most important to remember that there is no such things as *the* chimpanzee jaw or *the* orangutan facial skeleton. It is, perhaps, difficult to follow this exhortation when only a small part of one fossil jaw exists. But the evaluation of the material on hand must be tempered by the knowledge that it is but one part of one individual animal from what is most likely a highly variable population. As we pointed out earlier (Chapter 4) and as we shall discuss later (Chapter 19), variability is a characteristic of living animals.

When we evaluate primate fossil remains it is important that we understand how various traits are related to the evolutionary course taken by the order. One way to interpret the course of primate evolution is to view it as several major adaptive advances with extensive radiations into various eco-logical zones as a consequence. Most of these advances were based upon advantageous changes in locomotion, manual dexterity, and the special senses. The primary changes in anatomy, usually known to us only from the skeleton, are those that made radiation into new ecological zones possible. These primary traits are subject to much change and refinement during evo-lution. They are the basis for evolutionary change, and we should expect to find that they vary considerably from one time period to another. The changes in the primate ilium, to be described later, were the basis for the adaptive radiation that we now call *Homo sapiens.* This adaptive radiation is based upon erect posture and bipedal walking, behavior made possible by the transformed ilium and pelvic muscles. Secondary traits are those which are consequences of the development of the primary traits, or correlates of them. The development of the large braincase and the small face are second-ary to the development of erect posture and bipedality. Conservative traits are characters that are neither the basis for an evolutionary advance nor a consequence of the reorganization of the body plan as the result of a primary change. Particularly among the Primates, teeth are a conservative trait when considered from the long-range evolutionary view. We expect greater simi-larities in the teeth of ancestors and descendants separated by a long period of time than we expect in the hands or in the shape of the face. Teeth may be of very great selective and adaptive significance in certain lineages. By infer-ence, they may provide much information about the environmental changes to which the lineage had to adapt.

This discussion should not be confused with older arguments about adap-tive and nonadaptive traits. Today there are few who would argue that non-adaptive traits exist. The fact that some parts of the organism seem to change very little over long periods of time merely indicates that such traits have reached a high degree of stability. They are in equilibrium with natural selection. The term mosaic evolution has been used to describe the different rates at which separate anatomical systems evolve.

The taxonomic status of a fossil and the assessment of its phylogenetic affinities must be based upon a consideration of the total pattern of the various morphological characters taken in combination rather than individ-ually. Total morphological pattern is a concept that must be kept in mind when we study fossils. This concept is of added significance when we attempt to characterize traits as primary, secondary, and conservative. Examples will be cited during the discussion of various primate fossils, particularly the Hominidae of the Pleistocene epoch.

Primates is one of the thirty-three orders of Mammalia that are recognized in a widely accepted classification of mammals. Very few, perhaps only six, of these orders have a larger number of fossil genera than the Primates. The popular view that primates are rare and not often recovered with other fossil animals is not true. Primate fossils are not particularly rare. Simons, for example, points out that from the Paleocene strata of the Big Horn Basin, Wyoming, only four uintatheres (large, extinct ungulates) are known. Yet several hundred primates have been recovered. McKenna has reported that primates make up from 10 to 20 per cent of the mammals recovered from small-mammal quarries in the Eocene strata of northwestern Colorado.

Another common assumption asserts that primate fossils are not apt to be abundant because the Primates are tropical or subtropical arboreal animals. We gather this is supposed to mean that the dead primate will not be preserved and fossilized. Yet since they are and were arboreal animals, those that were fossilized perhaps fell out of the trees into streams and were preserved in riverine deposits.

Primate fossils are quite abundant in areas that are certainly not tropical now—Wyoming, Colorado, and northern Europe. The deduction is that these areas were much warmer in the past than at present and were probably covered by forests. At least one fossil primate species is found in both Europe and North America, implying that the subtropical or tropical forest extended from what is now one continent to the other. This primate is a member of the genus *Plesiadapis,* found both in France (near Quercy) in Paleocene deposits and in Clarkforkian deposits in North America. Since the Primates evolve quite rapidly compared with other mammals, it is probable that the French and American deposits are of similar age.

Fossil primates are useful as evidence for the existence of land bridges between continents in the remote past, for the kinds of climates and ecologies that existed in certain geological periods, and for making correlations in time between geological deposits in different continents. The *similarity* in the fossil primate genera of Europe and America is considerable during the early Eocene. The *divergence* between fossil primate genera of the two continents was great by the end of the Eocene, a fact which implies that land bridges between the continents no longer existed.

The early primates undoubtedly lived in warm climates, although whether they were restricted to such climates cannot be proven. Some contemporary primates, langurs and macaques, for example, live in temperate and even cold areas. The most primitive living primates, the prosimians, occur only in tropical or subtropical zones. Their particular econiche seems to be in the forest canopy of the tropics. Le Gros Clark, among others, has suggested that the forests of Africa and Asia contain a stratified primate fauna that represents the various successive evolutionary grades produced by the order. The distribution of living primates supports the view that the earliest primates were adapted to and lived in warm climates.

The various families, or subfamilies, of fossil primates and the probable time span during which they existed are shown in Fig. 7.1 and 7.2. These figures summarize, in schematic form, much of the data presented in the following sections devoted to descriptions of the fossil primates. These figures are subject to revision as new data are collected.

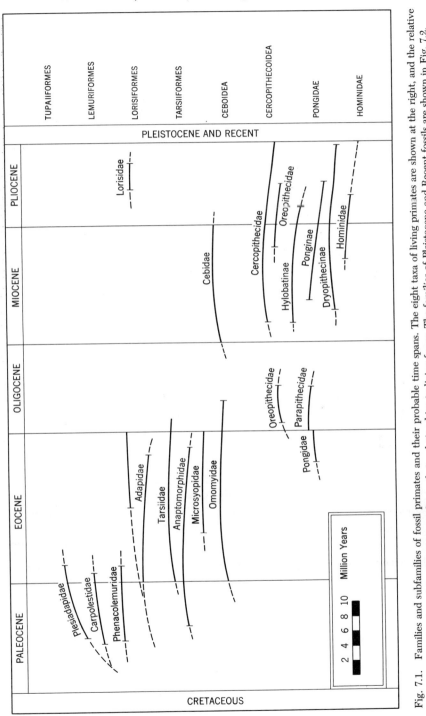

Fig. 7.1. Families and subfamilies of fossil primates and their probable time spans. The eight taxa of living primates are shown at the right, and the relative positions of the families of fossil primates indicate their relationships to living forms. The families of Pleistocene and Recent fossils are shown in Fig. 7.2.

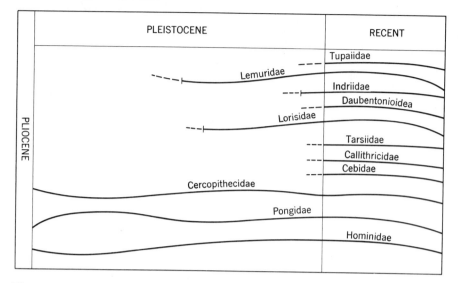

Fig. 7.2.   The major families and superfamilies of Pleistocene and Recent primates. The time span of the Recent epoch relative to the Pleistocene is not drawn to scale.

Paleocene Forms

The earliest primate fossils come from deposits laid down about the middle of the Paleocene epoch (Tables 7.1 and 7.2). These are the earliest *known* members of our own mammalian order, although obviously not the first. The

TABLE 7.1
Paleocene Fossil Primates

Family	Subfamily	Genus
Carpolestidae		*Carpodaptes*
		Carpolestes
		Elphidotarsius
Phenacolemuridae	Paromomyinae	*Navajovius*
		Palaechthon
		Palenochtha
		Paromomys
		Plesiolestes
	Phenacolemurinae	*Phenacolemur*
Plesiadapidae		*Chiromyoides*
		Platychaerops
		Plesiadapis
		Pronothodectes

TABLE 7.2
Paleocene Fossil Primates

Genus	Continent
Middle Paleocene forms	
Elphidotarsius	North America
Palaechthon	North America
Palenochtha	North America
Paromomys	North America
Plesiolestes	North America
Pronothodectes	North America
Late Paleocene forms	
Carpodaptes	North America
Navajovius	North America
Chiromyoides	Europe
Late Paleocene to early Eocene forms	
Carpolestes	North America
Phenacolemur	North America
Platychaerops	Europe
Plesiadapis	North America, Europe

earliest, distinctly primate mammals, as we said earlier, will never be "recognized" in the fossil record. The primates of the middle Paleocene were already distinctly different from insectivores. It follows that the original members of these fossil lineages must have become differentiated from the insectivore stem before or during the early Paleocene.

The earliest primates known are six genera from the middle Paleocene, all found in deposits and sites of the same age (Table 7.2). These genera occur in North America—in Colorado, Montana, New Mexico, and Wyoming. Several authorities consider that the known Paleocene primate fossils do not make completely convincing ancestors for later, Eocene and Recent, members of the order.

The differences between Paleocene primates and the primates of the next epoch, the Eocene, are interesting and important. The front teeth of all the Paleocene primates are large and procumbent. The known and described skulls of all Eocene to Recent primates have a postorbital bar that is continuous (Fig. 7.3). Neither of the two known Paleocene primate skulls has such a continuous postorbital bar. The Paleocene primates probably had claws on their digits instead of flattened nails. The evidence for this consists of the recovered foot bones of one Paleocene form, *Plesiadapis*.

The known fossil primates from the Paleocene are listed in Table 7.1. The thirteen genera are grouped into three families, Carpolestidae, Phenacolemuridae, and Plesiadapidae.

1. Carpolestidae (*Carpodaptes, Carpolestes, Elphidotarsius*). This group is

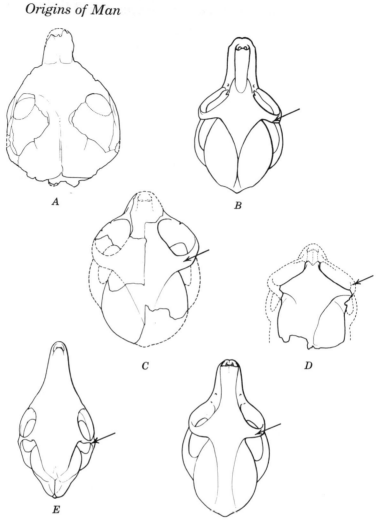

Fig. 7.3.   Prosimian and cercopithecoid skulls, top views. The postorbital bars are indicated by arrows. It should be noted that the orientation of the orbits differs among the skulls diagramed here and that the length of the snout varies. (A) *Plesiadapis*, Paleocene prosimian, (B) *Notharctus*, Eocene prosimian, (C) *Tetonius*, Eocene prosimian, (D) Fayum cercopithecoid, (E) *Tupaia*, living prosimian, and (F) *Lemur*, living prosimian. (Adapted from Simons, 1962.)

known only from fossilized jaws—both mandibles and maxillae—and teeth. The name of the family means fruit stealers. This case is similar to one noted earlier for the Carnivora, where the name means the taxon and is not descriptive. The name apparently refers to characteristics of the teeth. Presumably such teeth were adapted to splitting open seeds and hard, woody stems. The last lower premolar (P_4) is greatly enlarged and serrated to form a longitudinal cutting edge. This trait, along with others in the upper teeth, is convergent to, or at least resembles, the dentition of the Multituberculata, a long extinct order of mammals. There are similar dental elements in the rat kangaroos of Australia, and it is from this animal that the function of the teeth of the Carpolestidae is inferred.

2. Phenacolemuridae (*Navajovius, Palaechton, Palenoctha, Paromomys, Plesiolestes, Phenacolemur*). These are the earliest of the Primates and are probably the least known and understood. The name of the family means deceptive ghosts, which can have only a metaphorical significance. The family is divided into two subfamilies, Phenacolemurinae and Paromomyinae. The former are believed to be more specialized adaptively than the latter. The Phenacolemurinae are considered prosimian primates because they have a close common ancestry with other lineages that are prosimian. There does not seem to be any progressive trend in the known fossil specimens. Dental peculiarities and specializations are found in specimens that are the youngest members of the group. All fundamentals of a specialized dental adaptation seem to have been acquired by *Phenacolemur* by the late Paleocene. Specifically, the root of the lower incisor is long; the molar trigonids are very much compressed, and the paraconids are minute; the metacone on the fourth premolar is distinct; the talonid basin on the third molar projects far to the rear. The Paromomyinae are more primitive and generalized in every respect than the Phenacolemurinae. They are more apt to be on the main line that leads to later primates.

3. Plesiadapidae (*Chiromyoides, Platychaerops, Plesiadapis, Pronothodectes*). Among Paleocene primates, this family is the best known group. There are two partial skeletons known, one from France, the other from Colorado. An entire skull was found in the French site. The claws are flattened somewhat, as are the claws of the living *Tupaia* (the tree shrew of southeast Asia). The front teeth are large and procumbent in a way that suggested to many a close resemblance to the living *Daubentonia*(the aye-aye of Madagascar, Chapter 15). Simpson believes that the resemblance does not indicate an ancestral connection between *Plesiadapis* and the aye-aye, and this is the opinion generally accepted today.

Paleocene primates are put in the order Primates on the basis of the close morphological resemblance of certain cheek teeth (premolars and molars) to the cheek teeth of later fossils which are clearly primates. If only the unusual front teeth, enlarged and procumbent, had been available, it is likely that primate jaws from the Paleocene would have been placed in a category with the Insectivora or with the earliest of the Rodentia.

Eocene Forms

Some primates that first appear in the Paleocene persisted into the next epoch, the Eocene. Nevertheless, the onset of the Eocene coincides with the

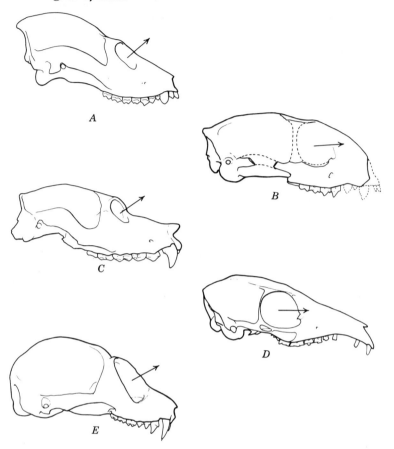

Fig. 7.4. Prosimian skulls, side views. The orientation of the orbits, indicated by arrows, and the relative size of the orbits differ in the forms pictured here. (A) *Adapis,* Eocene form, (B) *Pronycticebus,* Eocene form, (C) *Palaeopropithecus,* Pleistocene form, (D) *Tupaia,* living prosimian, and (E) *Nycticebus,* living prosimian. (Adapted from Simons, 1962.)

appearance of primate species that are more modern in total morphological pattern. Eocene primates had larger brains and bigger eyes than their Paleocene forebears. They are the earliest fossils which are unmistakably and unarguably Primates.

The major evolutionary trends in Eocene primate fossils are found in features of the skull (Fig. 7.4 and 7.5). The fossil remains of several large-brained prosimians, with enlarged orbits rotated toward the front of the skull for the eyes, imply that several important traits found in man and other higher primates began to develop over 50 million years ago. The snout is already foreshortened in these Eocene forms, compared with that of Paleocene primates or other mammals of the same periods. The foramen magnum (Fig.

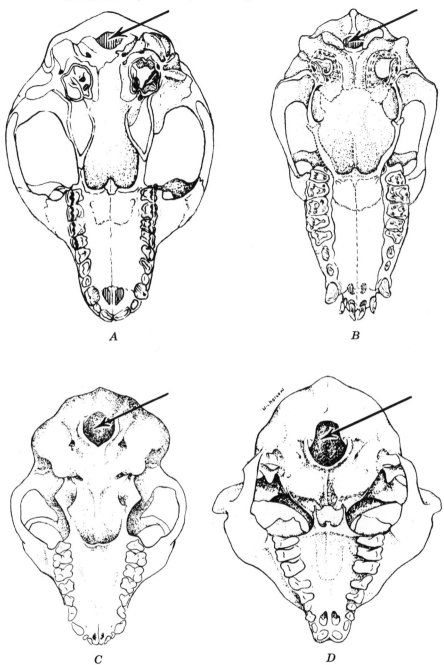

Fig. 7.5.   Base of the skull of Eocene and living prosimians, (A) *Adapis,* (B) *Notharctus,* (C) *Lemur,* and (D) *Tarsius.* The foramen magnum in each skull is indicated by an arrow. These skulls have not been drawn to the same scale.

7.5) is shifted forward toward the front of the base of the skull. This provides evidence about the way in which the spine and skull are attached to each other and implies that these primates were more in the habit of holding their bodies erect than forms in which the foramen magnum is farther back on the base of the skull. They were erect while hopping and sitting, most likely much in the manner of living lemurs, galagos, and tarsiers. Hopping and climbing primates sit and hop with trunk erect. These hoppers and climbers have fore- and hindlimbs of unequal length; the forelimbs are shorter. This is a most important point, which only a few present-day students have emphasized sufficiently. The forward movement of the foramen magnum, correlated with development of modes of locomotion that grew out of a tendency to hold the trunk erect, indicates that primate locomotion began to develop away from true quadrupedality at an early time. Although many primates are classed as quadrupeds, they are, at the least, very specialized quadrupeds.

Enlargement of the brain is another trait which is evident in the Eocene primates. Simons estimates the brain-to-body ratio of *Necrolemur antiquus* as 1:35, a very large ratio. For example, uintatheres (rhinoceros-like animals), contemporaries of *Necrolemur,* have a ratio of 1:2000.

We know that the Eocene was the epoch of maximum radiation of Prosimii, for the largest number of prosimian genera are found in deposits of Eocene age. Forty-three genera in five families have been recognized (Table 7.3). Most of them are considered valid genera today. The geographical distribution of these forms and their probable positions in Eocene horizons are presented in Table 7.4. In the Eocene there are also two fossil members of the Anthropoidea, possibly Pongidae (Tables 7.3 and 7.4). A brief description of each of the families of prosimians and a note on the possible Eocene pongids follow.

TABLE 7.3
Eocene Fossil Primates

PROSIMII

Family	Subfamily	Genus
Adapidae	Adapinae	*Adapis*
		Anchomomys
		Caenopithecus
		Gesneropithex
		Pronycticebus
		Protoadapis
	Notharctinae	*Notharctus*
		Pelycodus
		Smilodectes
Anaptomorphidae		*Absarokius*
		Anaptomorphus
		Anemorhysis
		Tetonius

TABLE 7.3 (*Continued*)

Family	Subfamily	Genus
		Tetonoides
		Trogolemur
		Uintalacus
		Uintanius
		Uintasorex
Microsyopidae		*Alsaticopithecus*
		Craseops
		Cynodontomys
		Microsyops
Omomyidae		*Cantius*
		Chlororhysis
		Chumashius
		Dyseolemur
		Hemiacodon
		Hoanghonius
		Loveina
		Lushius
		Niptomomys
		Omomys
		Ourayia
		Periconodon
		Shoshonius
		Stockia
		Teilhardina
		Utahia
		Washakius
Tarsiidae	Necrolemurinae	*Microchoerus*
		Nannopithex
		Necrolemur
		Pseudoloris

ANTHROPOIDEA

Pongidae		*Amphipithecus*
		Pondaungia

TABLE 7.4
Eocene Fossil Primates

Genus	Continent or country
Early Eocene forms	
Notharctus	North America
Pelycodus	North America
Absarokius	North America

TABLE 7.4 (*Continued*)

Genus	Continent or country
Anemorhysis	North America
Tetonius	North America
Tetonoides	North America
Uintalacus	North America
Cynodontomys	North America
Cantius	Europe
Chlororhysis	North America
Loveina	North America
Niptomomys	North America
Shoshonius	North America
Middle Eocene forms	
Adapis	Europe
Anchomomys	Europe
Caenopithecus	Europe
Protoadapis	Europe
Smilodectes	North America
Anaptomorphus	North America
Trogolemur	North America
Uintanius	North America
Uintasorex	North America
Alsaticopithecus	Europe
Microsyops	North America
Hemiacodon	North America
Lushius	China
Omomys	North America
Periconodon	Europe
Teilhardina	North America
Utahia	North America
Washakius	North America
Nannopithex	Europe
Necrolemur	Europe
Late Eocene forms	
Gesneropithex	Europe
Pronycticebus	Europe
Craseops	North America
Chumashius	North America
Dyseolemur	North America
Hoanghonius	China
Ourayia	North America
Stockia	North America
Microchoerus	Europe
Pseudoloris	Europe
Amphipithecus	Burma
Pondaungia	Burma

1. Adapidae. The best known and described members of this family are *Notharctus* and *Adapis*. These forms do not have elongated tarsal bones; they appear to have an opposable thumb and big toe, and the limbs have a generalized structure. The snout is long and resembles that of modern lemurs, although it is not as long as that of many terrestrial mammals. The orbits are rotated to the front of the face, increasing the likelihood that the fields of vision of each eye overlap at least as much as they do in modern living prosimians. There is a free tympanic ring in the middle ear and numerous other detailed features of the skulls that are similar to those found in living Malagasy lemurs. (The tympanic ring is a thin, bony ring which supports the tympanic membrane [the eardrum].) Many authorities consider that *Adapis*, an Old World form, is closer to modern lemurs and lorises than *Notharctus*, a New World form. Whether the adapines are lineal ancestors of the Malagasy lemurs or African lorises is a question that cannot be answered yet.

2. Anaptomorphidae. This group of fossil primates is wholly North American and is found in the early and middle Eocene. The snout has been shortened, and this shortening may be correlated with, or the result of, the reduction in the size of the front teeth. The reduction in tooth size may well have been a response to a special diet, the nature of which we can hardly speculate about. The Anaptomorphidae are a wholly New World radiation. Although they originated and finally became extinct there, it is possible to exclude them from the direct ancestry of living New World monkeys on the basis of dental formula. The anaptomorph formula is $\dfrac{3.1.2 \text{ or } 3.3.}{2.1. \qquad 2.3.}$; that of New World monkeys is $\dfrac{2.1.3.3.}{2.1.3.3.}$. One skull of an anaptomorphid, *Tetonius homunculus*, is known in which there are features that some take to imply relationship to tarsioid fossils. There may well be a common ancestor for *Tetonius* and the living *Tarsius*, as a recent authority suggests. But to call *Tetonius* tarsioid implies the existence of many more precise morphological similarities between the two than there actually are.

3. Microsyopidae. Until very recently this was a little known group of fossil prosimians. The major character of note was the molarization of the premolar teeth (Chapter 6). The Microsyopidae were relatively large, about the size of the living Indriidae of Madagascar. Their relationship to other fossil primates and to living primates is now under study.

4. Omomyidae. At least eighteen genera of this family have been found in Asia, Europe, and North America (Tables 7.3 and 7.4). They were the most widely distributed geographically, and they radiated more extensively than any other prosimian group. There are bits of fossil material from the Oligocene suggesting that the Anthropoidea may have developed from some African omomyid. If this material is supported by further discoveries, the lineage of the higher primates may be extended back to a specific Eocene form.

However, the Omomyidae have one specialization that may eventually be used to rule out all the discoveries in this group from the direct lineage of the higher primates—the incisor teeth are larger than the canine teeth in the upper and lower jaws. They are believed to show no other specializations that might eliminate the anthropoids as descendants. It is possible to argue that the omomyids are on the direct ancestral line of each of the major taxa of higher primates—the Ceboidea, the Cercopithecoidea, and the Hominoidea.

Omomys and its close relatives were smaller than living South American monkeys, with unspecialized molar tooth-crown patterns and small third molars. They share the dental formula of South American monkeys, and they have at least one unique and very special dental trait in common with these monkeys—a pericone cusp. Thus, some members of the fossil omomyid group may prove to be on the direct lineage of South American monkeys. Unfortunately, there are few post-Eocene candidates for this lineage because there are few fossil primates known from the Oligocene and a gap in the fossil sequence exists.

Hemiacodon has a postcranial skeleton which may resemble that of a lemuroid or a more generalized grade of primate. If this or another omomyid should prove to be ancestral to the Anthropoidea, the argument that a tarsioid grade occurred on the lineage would have little force. *Hemiacodon* has a tooth pattern considered to be typical of a primate near or on the lineage of the ancestral groups of later primates. Nonetheless, the eighteen genera of Omomyidae occur in the right places and, based on quite fragmentary data, have the anatomy expected of a prosimian ancestor of each of the three major stocks of higher primates. The evidence is based almost wholly on the teeth and bits of jaws which, up to now, are all that have been recovered.

5. Tarsiidae. There is one fossil subfamily in this group, the Necrolemurinae, which has many features in the skull that imply close relationship with living tarsiers. Hence they are placed in the larger taxon which includes the living form *Tarsius*. The specific Eocene ancestor of *Tarsius* will probably be found, eventually, among fossils of this group.

Necrolemur itself can be ruled off the direct ancestral line of *Tarsius* because of the greatly reduced size of the paraconids of the lower molars and the absence of front teeth from the lower jaw. Since there are no known fossils intermediate between modern *Tarsius* and *Necrolemur,* the distance between them in space and time must also be given some weight in assessing their relationship. Almost all the bones of the skull of *Necrolemur* have been recovered. There is enough material for good comparisons with other primates, fossil and living.

The total morphology of Necrolemurinae raises an important question—what is a lemuroid and what is a tarsioid Eocene fossil? A much discussed problem among students of fossil primates is whether to put a Paleocene-Eocene fossil in a lemuroid or tarsioid group. To call early Tertiary primate fossils either lemuroid or tarsioid depends upon the clear understanding or definition of these two categories. The tarsioid-lemuroid problem involves using data from living primates to interpret the fossils. It depends on understanding what are specialized and unspecialized morphological features in the skull. It requires understanding the evolutionary developments and specializations of a living primate, *Tarsius*. Finally, the fossil fragments that have been recovered consist of teeth and some bones of the skull. Therefore, comparisons between these fossils and between fossil and living forms will have to be made on extrapolations from skulls and teeth.

The assignment of Eocene and Paleocene primate fossils to a lemuroid or tarsioid assemblage is difficult. *Pronycticebus gaudryi* was first discovered in large Eocene deposits in France and was put with the lorisoids by its discoverer. About 20 years later, two renowned and competent paleontologists decided this was a tarsioid fossil. Several years later, a most eminent anatomist and paleoanthropologist demonstrated that *Pronycticebus gaudryi* was lemuroid rather than tarsioid. Very recently a student of the latter authority suggested that the affinity of this fossil is with certain lorisoids rather than with lemuroids. Thus, in half a century, a number of eminent and competent students decided that a single prosimian fossil was an ancestor or belonged to an ancestral stock of each of the three major taxa of living prosimians. This story could be repeated for a number of Eocene prosimian fossil discoveries. It illustrates the care with which an assessment of a fossil primate must be made. Table 7.5 lists the characters that may be useful in distinguishing the tarsioids from the lemuroids.

The orbits and the probable manner by which postorbital closure is achieved in *Necrolemur*, presumably an Eocene relative of *Tarsius*, is of great interest in this connection. Postorbital closure occurs in *Tarsius* in at least three stages. First, the posterior midregion of the orbital plate of the maxilla grows upward. Second, the middle section of the postorbital bar spreads from front to rear of the orbit. Third, a flange from the frontal bone grows laterally. Ossification occurs downward, as this latter flange or plate grows along the postorbital bar and eventually fuses with it.

In *Necrolemur* the frontal element of postorbital closure is present in an incipient form. The large flanges of bone that surround the orbit of an adult tarsier are not present in *Necrolemur*. These flanges were developed in *Tarsius* to support the enormous eyes. Since they are unique to tarsiers, we would not necessarily expect to find them, fully developed, in a presumed Eocene ancestor of *Tarsius*, and we do not (Fig. 7.6). As we said above, although the specializations of *Necrolemur* rule it out as a direct ancestor of *Tarsius*, it is definitely a tarsioid form.

6. Pongidae. There are two fossils sometimes assigned to the Anthropoidea, possibly to the family Pongidae—*Pondaungia* and *Amphipithecus*—which come from Eocene deposits in Burma (Tables 7.3 and 7.4). They are of late Eocene or of early Oligocene age. *Pondaungia* consists of very fragmentary bits of fossil primate—a piece of a single left maxilla with the first

TABLE 7.5

A Comparison of Lemuroid and Tarsioid Characters

The lemuroid characters are based on living Lemuridae and fossil Adapidae. The tarsioid characters are based on *Tarsius* and *Necrolemur*. The characters of Lorisiformes, which are intermediate in some cases, are presented here for comparison and are described from living and fossil forms. (From Simons, 1961.)

Character	Lemuroid Lemuriformes	Lorisiformes	Tarsioid Tarsiiformes
Lower canines	shorter than upper or incisiform	incisiform	longer than upper
Zygomatic arches	typically stout (flaring)	variable (flaring)	slender (close to skull)
Postorbital opening, area	large	rather large	almost closed or small
Brain case, transverse width	very narrow to expanded	expanded	much expanded
Muzzle length	long	long to medium	very short
Muzzle width	typically broad	variable	very narrow
Contact between jugal and lacrimal	typically occurs	often occurs	does not occur
Ectotympanic position	in bulla	at margin of bulla	extends out of bulla
Ectotympanic shape	annular	annular but broad	tubular
Elongation of calcaneus and talus	none	some	some
Inflation of bullae	typically inflated	little inflation	much inflation
Bony canal of promontory artery	small, in bulla	not ossified, not in bulla	large, in bulla
Median lacerate foramen	variably present	present	not present
Carotid foramen, position	at posteriolateral angle of bulla	not present	on ventromedial face of bulla
Septum between tympanic cavity and hypotympanic sinus	incomplete	complete	incomplete
Tibiofibular fusion	does not occur	does not occur	may occur
Palate	broad anteriorly	broad anteriorly	narrow anteriorly
Tooth rows	parallel (U-shaped)	parallel (U-shaped)	convergent (V-shaped)
Upper canines	much larger than anterior incisors	much larger than anterior incisors	smaller than anterior incisors
Molar hypocone	variable, but often large	variable, but often large	often small
Posterior nares, shape	broad	broad	narrow
Posterior nares, position	anterior to M_3	anterior to M_3	posterior to M_3
Pterygoid alae	long anterio-posteriorly	long anterio-posteriorly	short anterio-posteriorly
Bullae, position	well separated	well separated	not well separated anteriorly
Mastoid region	not prominent	not prominent	prominent
Foramen magnum, orientation	largely backward	largely backward	largely downward
Posterior palatine torus	absent	absent	present
Contact of external pterygoid alae with bulla	touching	touching	overlapping
Interfrontal suture	typically open in adults	typically open in adults	fused in adults

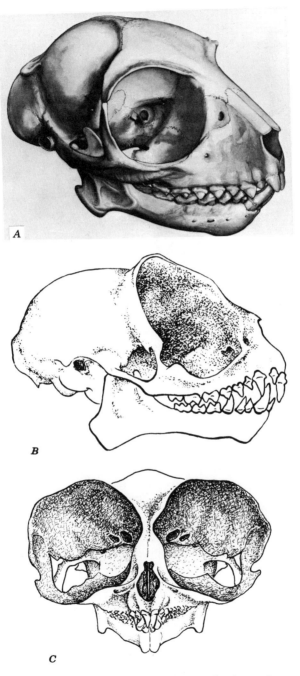

Fig. 7.6.   Skulls of tarsioids. (A) Photograph of *Necrolemur*. (From Simons and Russell, 1960.) (B) and (C) Drawings of *Tarsius*. The prominent flanges of bone which surround the orbits in *Tarsius* are noteworthy.

two upper molars and a partial mandible with both rami and corroded second and third molars. Careful microscopic examination of the two upper molars showed that they have the four primary cusps placed in the manner characteristic of early Hominoidea. The molar crown structure can be interpreted as transitional, morphologically, between the Omomyidae and the Pongidae. *Amphipithecus* is a fragment of the left mandible with two premolars and one molar. Since the roots of a third premolar are present in the fragment, the dental formula is not that of a hominoid $\left(\dfrac{2.1.2.3.}{2.1.2.3.}\right)$. Nevertheless *Amphipithecus* resembles Oligocene fossils which are definitely hominoid, and it, too, may be a transitional form. If pongids were distinct as early as the late Eocene, it is reasonable to suggest that they differentiated directly from some prosimian lineage and not from some ancestor resembling a monkey. The three major anthropoid taxa—Ceboidea, Cercopithecoidea, and Hominoidea—may have been coordinate developments from distinct prosimian ancestors.

Oligocene Forms

Primates from this epoch are rare; none have been found in Europe. There is one dubious discovery in Asia; two have been found in the New World; and several have been found in Africa (Table 7.6). All the African Oligocene primates have been discovered in Egypt, most of them in the Fayum. The Oligocene leaves a gap in the sequence of fossils which provide the data for our knowledge of primate evolution. The primate fossils currently being recovered from Oligocene deposits in Egypt—the Qatrani formation in the Fayum—are beginning to fill the gap. These fossils are the only ones that provide evidence about the differentiation of the higher primates of the Old World.

TABLE 7.6
Oligocene Fossil Primates

Family or superfamily	Genus	Continent or country
Omomyidae	*Macrotarsius*	North America
Oreopithecidae	*Apidium*	Egypt
Oreopithecidae	*Parapithecus*	Egypt
Hominoidea	*Aeolopithecus*	Egypt
Pongidae	*Aegyptopithecus*	Egypt
Hominoidea	*Oligopithecus*	Egypt
Hominoidea	*Propliopithecus*	Egypt
?	*Moeropithecus*	Egypt
Cebidae	*Dolichocebus*[a]	Argentina

[a]Either late Oligocene or early Miocene.

Parapithecus. One specimen of this genus has been described. Both segments of the lower jaw are known. *Parapithecus* is probably a transitional primate, intermediate in many characteristics between prosimians and anthropoids. One of the major problems in interpreting *Parapithecus* is the evaluation of the dentition. All the lower teeth are known, but several prominent authorities are not certain what kinds of teeth are represented. One German paleontologist believed that the canine of *Parapithecus* was a transformed second premolar, and the original canine was shifted to the incisor series and became the second incisor. Thus the dental formula of this fossil may be, for the lower jaw, either $\overline{1.1.3.3.}$ or 2.1.2.3. . The former suggests that it is related to prosimians and the latter to anthropoids. As noted earlier, it is not necessary that the dental formulae of all species of a higher category, such as Hominoidea, be identical. Thus the $\overline{1.1.3.3.}$ formula would not rule *Parapithecus* out of this superfamily, but it would rule it off the direct ancestral lineage. Descriptions of recent discoveries of *Parapithecus* may allow us to decide its phylogenetic affinities.

Apidium. *Apidium* is another of the primate fossils found in the Fayum of Egypt. *Parapithecus* and *Propliopithecus* are often considered closely related to this genus, primarily because the conformation of the major cusps on the molar teeth are similar in all three. The similarities between *Oreopithecus* and *Apidium* have led to the suggestion that these two genera represent a major lineage of higher primates of the Old World distinct from cercopithecoids and hominoids.

Moeropithecus. A fragment of lower jaw with two molar teeth is all that exists of this form. *Moeropithecus* is a fossil whose immortal remains are shorter than its name. It is believed that these teeth resemble the teeth of other primates found in deposits of the same age in the Fayum. These teeth may prove to have the proper molar morphology to place *Moeropithecus* close to the ancestors of modern monkeys.

Propliopithecus. The teeth and mandible given this name were once thought to closely resemble those of later primate fossils, *Limnopithecus* and *Pliopithecus* of the Miocene. If true, this would put *Propliopithecus* on or close to the ancestry of the gibbons. But, the discovery of *Aeolopithecus* demonstrated that a lineage much more like modern gibbons than *Propliopithecus* was distinct from *Propliopithecus* at the same time and in the same place.

It can be argued that *Propliopithecus* is most reasonably the exact ancestor of the other Hominoidea. The lower molars are rather like those of the pongids. This may be a primitive hominoid pattern. It can also be argued that *Propliopithecus* is already too specialized for a generalized hominoid— it may be too specialized to be ancestral to both hominids and pongids.

Oligopithecus. *Oligopithecus* is known from a small piece of mandible with five teeth in it. It resembles *Propliopithecus* more than other Fayum forms and is clearly a member of the Anthropoidea. The teeth resemble those of the late Eocene fragment *Amphipithecus*. If this resemblance means that *Amphipithecus* is an anthropoid, we can put the origin of the anthropoids back to at least late Eocene times.

Aegyptopithecus. Aegyptopithecus consists of three mandibular fragments, which contain enough teeth, among them, so that the dental formula of the lower jaw has been determined as 2?.1.2.3. . *Aegyptopithecus* seems to be much more closely related to Miocene-Pliocene dryopithecines than it is to *Propliopithecus*. This would make it an ancestor or put it close to the ancestral lineage of the East African Miocene ape, *"Proconsul."*

Aeolopithecus. Aeolopithecus is a complete mandible, with dental formula 2.1.2.3. . It was named after Aeolus, god of the winds, for it was uncovered in the Qatrani formation of the Fayum by a windstorm rather than by a paleontologist. The large simian shelf and the reduction in size of the third molar are among the features of this mandible leading to the consideration of *Aeolopithecus* as a front-running candidate for the ancestor of the gibbons.

The Oligocene, in the Old World, is the period in which the Hominoidea became a distinct and, probably, a prominent group of animals. We find dentitions that are much like those of modern pongids and some that suggest our own.

The known Oligocene fossil primates are all from a small region in northern Africa. There are none clearly prosimian, although *Parapithecus* may be intermediate between prosimians and anthropoids. All are now assigned to the Anthropoidea. Some are definitely closely related to the ancestors of the various living Hominoidea; some may indeed be the direct ancestors.

SUGGESTED READINGS AND SOURCES

Clark, W. E. Le G., On the skull structure of *Pronycticebus gaudryi. Proc. Zool. Soc. London,* 19 (1934).

Gazin, C. L., A review of the middle and upper Eocene primates of North America. *Smithsonian Misc. Collections* 136, 1 (1958).

Gregory, W. K., On the structure and relations of *Notharctus*, an American Eocene primate. *Am. Museum Nat. Hist. Mem.* 3, 49 (1920).

Gregory, W. K., *The Origin and Evolution of the Human Dentition.* Williams and Wilkins, Baltimore (1922).

Hürzeler, J., Zur Stammesgeschichte der Necrolemuriden. *Schweiz. Paläont. Abhandl.* 66, 1 (1948).

Jones, F. Wood, Some landmarks in the phylogeny of Primates. *Human Biol.* 1, 214 (1929).

Kälin, J., Sur les primates de l'Oligocène inférieur d'Egypte. *Ann. Paleont.* 74, 1 (1961).

McKenna, M. C., Fossil mammalia from the early Wasatchian four mile fauna, Eocene of northwest Colorado. *Univ. Calif. (Berkeley) Publ. Geol. Sci.* 37, 1 (1960).

Russell, D. E., Le crâne de *Plesiadapis,* note préliminaire. *Bull. Soc. Géol. France,* 7, 312 (1959).

Simons, E. L., Notes on Eocene tarsioids and a revision of some Necrolemurinae. *Bull. Brit. Museum (Nat. Hist.) Geol.* 5, 43 (1961).

Simons, E. L., Fossil evidence relating to the early evolution of primate behavior. *Ann. N. Y. Acad. Sci.* 102, 282 (1962).

Simons, E. L., Two new primate species from the African Oligocene. *Postilla, Yale Peabody Museum* No. 57 (1962).

Simons, E. L., A critical reappraisal of Tertiary primates. *In* J. Buettner-Janusch (Ed.), *Evolutionary and Genetic Biology of Primates*, Vol. I, p. 65. Academic Press, New York (1963).

Simons, E. L., New fossil apes from Egypt and the initial differentiation of Hominoidea. *Nature* **205**, 135 (1965).

Simons, E. L., and D. E. Russell, The cranial anatomy of *Necrolemur*. *Brev. Museum Comp. Zool. Harvard* No. 127, 1 (1960).

Simpson, G. G., The Fort Union of the Crazy Mountain Field, Montana, and its mammalian faunas. *Bull. Smithsonian Inst. U. S. Natl. Museum* **169**, 1 (1937).

Simpson, G. G., Studies on the earliest primates. *Bull. Am. Museum Nat. Hist.* **77**, 185 (1941).

Simpson, G. G., The Phenacolemuridae, new family of early primates. *Bull. Am. Museum Nat. Hist.* **105**, 411 (1955).

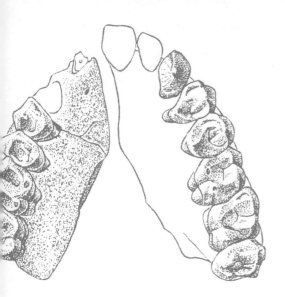

Miocene and Pliocene Fossil Primates

8

DURING THE MIOCENE and the epoch which followed, the Pliocene, the Anthropoidea continued to radiate (Table 8.1). The adaptive radiation of the Pongidae probably reached its peak during the Pliocene. We consider the two epochs together, for there is no clear and complete break in the fossil history of certain lineages of Anthropoidea during the Miocene and Pliocene (Fig. 7.1). These lineages expanded their geographic range to the limits to which anthropoids have radiated. Fossil primates found in Miocene deposits of South America are very distinctly members of the superfamily Ceboidea (*Cebupithecia, Dolichocebus, Homunculus*). Hominoid fossils have been found in Miocene deposits throughout Eurasia and Africa—Spain, Central Europe, India, China, Kenya. Except for one or two teeth and several jaws found in Kenya, there are few fossil remains from the Miocene of the Old World assignable to the Cercopithecoidea. This is puzzling, for the Old World monkeys, anatomically and in other ways, are "primitive." The assumption has always been that the living monkeys represent a stage of evolution between prosimians and hominoids.

TABLE 8.1
Miocene-Pliocene Fossil Primates

Family	Genus	Continent or country	Epoch
Lorisidae	*Indraloris*	India	Pliocene
Cercopithecidae	*Dolichopithecus*	Europe	Pliocene
	Libypithecus	Egypt	Pliocene
	Mesopithecus	Kenya, Europe	Miocene-Pliocene
Cebidae	*Cebupithecia*	Colombia	Miocene
	Homunculus	Colombia	Miocene
	Neosaimiri	Colombia	Miocene
Oreopithecidae	*Oreopithecus*	Italy	Pliocene
Hylobatidae	*Pliopithecus*	Europe	Miocene-Pliocene
Pongidae	*Dryopithecus*	Europe, Asia, Africa	Miocene-Pliocene
	Gigantopithecus	China	Pliocene
Hominidae	*Ramapithecus*	Asia, Africa	Miocene-Pliocene

The evidence is growing that the Old World monkeys (Cercopithecoidea) diverged at about the same time as the Ceboidea and the Hominoidea. We would expect to find numerous fossil cercopithecoids at least as early as the Miocene. Since we do not, the large number of genera and species of living Old World monkeys may well be the result of a relatively recent radiation, one that occurred after the adaptive radiation of the Hominoidea. The differentiation of the Cercopithecoidea as an evolutionary lineage occurred during the Oligocene, or earlier. The adaptive radiation, the maximum spread, of this group of primates may well have been a much later event.

Pliopithecus

Pliopithecus is a gibbon-like fossil that occurs in European deposits that range from late Miocene to early Pliocene. The morphological similarities of *Limnopithecus* and *Epipliopithecus* to *Pliopithecus* led one paleontologist to suggest recently they are members of a single genus. If two or more fossils are not sufficiently different morphologically to be placed in separate genera, the fact that they are separated in geological time should not be used to do so. The dentition of *Pliopithecus* is much like that of the modern gibbons. But the skeletal remains show that the characteristic great length of the forelimb of living gibbons was not present in these Miocene apes. *Prohylobates*, once thought to be similar to living gibbons and closely related to *Pliopithecus*, is a cercopithecoid.

The *Dryopithecus* Problem

We now come to one of the most interesting and controversial parts of our story of primate evolution—the problem of the dryopithecines of the Miocene-Pliocene epochs. Among the many fossil remains placed in this group are some direct lineal ancestors of man and of the great apes. There are many reasons why the dryopithecines have created problems for students of human evolution. This group of ape-like fossils from the Miocene and Pliocene is ancestral to modern apes and modern man. Therefore to an anthropocentric species, determination of exactly what or who diverged when and where into a lineage leading to the great apes and into another lineage leading to itself raises many important questions that are not wholly scientific. But there are also a number of difficulties of a scientific sort which affect the way these fossils are interpreted.

First, the fossils are incomplete; almost all the remains are teeth with fragments of upper and lower jaws. Second, until 1963 no single paleontologist or other authority had a chance to look at and study most of the 500 to 600 individual fossils. They have been found in many parts of the Old World—Europe, Africa, India, China. The discoveries began to be described in print as long ago as 1856. No single point of view was ever brought to the examination of these finds. Third, almost every new dryopithecine-like fragment was assigned to a new genus or a new species. Fourth, the relative age of the various fossil remains or types is often in doubt. It is not always easy to decide if two forms are reasonably contemporaneous or belong to different time periods. New methods of dating will undoubtedly help to solve some of these problems. Fifth, since many of the fossils in the dryopithecine group are found in widely separated regions, it has been assumed that geographical and ecological barriers existed between them. Sixth, the persistence of the belief among anthropologists that man had to originate in one place has made it difficult to have reasonable discussions of what and where the Miocene and Pliocene ancestors of man are. This belief is based on the assumption that the species ancestral to the Pleistocene Hominidae was restricted to a very small geographical area. One of the consequences of this point of view is the so-called "failure" to find the pre-Pleistocene ancestors of man. As we shall see, the Miocene-Pliocene ancestors *have been found.* They have been around in various museum collections for over a hundred years. Yet they were not generally recognized. Many prominent anthropologists, anatomists, and others, during these hundred years, have been busy eliminating each and every one of them from man's ancestral line.

A few highlights from the story of the revision of the dryopithecines are worth telling. The following discussion is based on the work of Simons and Pilbeam who have reorganized our views of the phylogeny of Miocene-Pliocene hominids. The multitude of genera to which dryopithecine primate fossil specimens have been assigned (Table 8.2) have recently been reduced to three—*Dryopithecus, Ramapithecus,* and *Gigantopithecus. Ramapithecus* includes *Bramapithecus* and *Kenyapithecus* and is a member of the family Hominidae. The others have been assigned to the family Pongidae: *Gigantopithecus* is now a genus of the subfamily Dryopithecinae; *Dryopithecus* (subfamily Dryopithecinae) probably includes the rest of the genera listed in Table 8.2.

The generic definition of *"Proconsul"* was based on a comparison with an upper jaw fragment which was mistakenly assigned to the genus *Dryopithecus,* and species *punjabicus.* But *"Dryopithecus" punjabicus* is a member of the genus *Ramapithecus,* not of the genus *Dryopithecus.* The original taxonomic description of *"Proconsul"* simply demonstrated that it was not *Ramapithecus* but did not distinguish it from *Dryopithecus,* the genus to which it is now assigned.

The genus *Bramapithecus* consisted only of lower jaw fragments. The genus *Ramapithecus* consisted of upper jaw fragments. This interpretation was a result of the fragmentary nature of the recovered fossils and of the fact that most who had described them had seen and studied very few of the available specimens. Recently an authoritative study demonstrated that the lower jaw fragments called *Bramapithecus* went with the upper jaw fragments called *Ramapithecus.* Thus two genera became one! It is clear that assigning specimens to particular taxa must be done with care.

The dryopithecine problem illustrates what ignorance or violation of some

TABLE 8.2
Names, Names, and More Names

This list contains the multitude of names now assigned to three genera—*Dryopithecus,*
Gigantopithecus, and *Ramapithecus.*

Genus	Continent or country
Adaetontherium	India
Ankarapithecus	Middle East
Anthropodus	Europe
Austriacopithecus	Europe
Bramapithecus	India
Dryopithecus	Europe, Africa
Gigantopithecus	China
Griphopithecus	Europe
Hispanopithecus	Europe
Hylopithecus	India
Indopithecus	India
Kansupithecus	China
Kenyapithecus	East Africa
Neopithecus	Europe
Paidopithex	Europe
Paleopithecus	India
Paleosimia	India
Proconsul	East Africa
Proconsuloides	East Africa
Rahonapithecus	Europe
Ramapithecus	India
Rhenopithecus	Europe
Sivapithecus	Europe, India
Sugrivapithecus	India
Xenopithecus	East Africa
Udabnopithecus	Georgia (U.S.S.R.)

general principles may lead to. The fossil dryopithecine material, although abundant, consists almost entirely of fragments of upper and lower dentitions. Paleontologists experienced with primate material are convinced that teeth provide data for excellent taxonomic criteria which distinguish higher categories such as infraorders, superfamilies, and subfamilies. But teeth are not quite so useful in distinguishing between genera and even less so in defining species. It must be clearly restated that variability is a basic feature of single morphological traits, such as teeth, as well as a basic feature of the total morphology of living animals. The revision of dryopithecine taxonomy was made by taking full account of the dental variability that occurs among the living Hominoidea. Before referring specimens of primate fossil jaws and teeth to several distinct genera, it should be shown that the differences between any two fossil specimens are at least as great as the differences between two living genera of the same group of animals.

We expect ancestors to be both similar to and different from their descendants. The differences may be very great, but the presumed ancestors should not have any traits which are so specialized that reversible evolution must be postulated to get them, so to speak, back on the family tree. The continual elimination of dryopithecines from the direct lineage of man brings to mind a quotation from William King Gregory. He wrote that those who insisted that dryopithecines were not direct ancestors of man ". . . would fail to recognize a direct ancestor of man of Miocene age even if it were represented by a complete skeleton, since they would expect to find it abounding in the diagnostic characters of recent Hominidae. . . ." In other words, what do they expect the ancestors to look like? The large number of genera and species to which the various dryopithecine finds have been referred are also the consequence of inadequate taxonomic procedures. Single teeth have been the basis for describing a new dryopithecine taxon. As many paleontologists know, a single fossil tooth may be assigned to a genus and a species under certain special circumstances. But it is usually considered unwise to define a taxon on the basis of a single tooth, particularly one belonging to such a well-known group as the dryopithecines.

Since research on this Miocene-Pliocene group of hominoid fossils is in such a stage of change, the conclusions which follow, generalized from recent publications, are certainly subject to future qualifications. There are two major groups represented within what has been called the dryopithecine group. One is a subfamily of the Pongidae, the Dryopithecinae, which includes the genera *Gigantopithecus* and *Dryopithecus*. The other, a smaller group, may be assigned to a single genus, *Ramapithecus*. *Ramapithecus* is a member of the family Hominidae and is most likely ancestral to the Pleistocene hominids.

The subfamily Dryopithecinae. These fossils are found in deposits that span the period from the early Miocene to the Pleistocene (Fig. 7.1). They were probably quite ape-like. They ranged in size from animals as big as the larger gibbons to animals larger than the modern gorilla. Since most of the remains are jaws and teeth, the characters which distinguish Dryopithecinae from Hominidae are based on the dentition. The canines of Dryopithecinae are larger than those of the Hominidae. The lower third premolar teeth are sectorial. The molars are elongated from front to back, and they usually increase in size, progressively within the series—$M_1 < M_2 < M_3$. The incisors are smaller and more vertical than those of Ponginae. There seem to be no definite and consistent features of the mandibles distinguishing Dryopithecinae from Hominidae. The limbs seem very generalized, in that their structure gives no indication of specializations in the direction of either brachiation or bipedalism. The most authoritative recent study of this group of fossils leads to the conclusion that they are the best candidates for ancestors of the modern Ponginae (chimpanzees and gorillas).

The genus *Ramapithecus*. *Ramapithecus* probably split from the dryo-pithecine lineage in the Miocene and was fully differentiated by the Plio-cene. It is now assigned to the family Hominidae and is considered the ancestral form of the Pleistocene hominids—the australopithecines and *Homo*. The present fossil remains indicate that it was about the size of a modern gibbon, a smaller animal than the australopithecines and most of the Dryopithecinae. The crenulation patterns of the teeth are less complex than those of later hominids or dryopithecines. The lower jaw is shallower than in either of the latter groups, and the face is shorter. The incisors and the canines are smaller in relation to molars than is the case for *Dryo-pithecus*. The molar cusp patterns are markedly different from those of the teeth assigned to *Dryopithecus,* and there are several details of the dentition that mark off *Ramapithecus* from *Dryopithecus*. The palate of *Ramapithe-cus* is arched, the tooth row is arcuate, and there is a shorter muzzle (Fig. 8.1). *Ramapithecus* was probably an erect biped. From the reduction in size of the anterior teeth, the incisors and the canines, we may infer that the forelimbs had taken over certain tasks the front teeth perform among the Pongidae—the grasping and tearing up of vegetation. We must point out that it is quite an imaginative jump from data to inference—from the size of a few teeth to a statement that an animal was an erect biped. We shall justify making this jump in the next few pages.

At this stage, it is appropriate to consider briefly the origin of the family Hominidae. The Hominidae and the Pongidae, as noted earlier, are believed to have a special relationship. Recent authoritative studies show that the

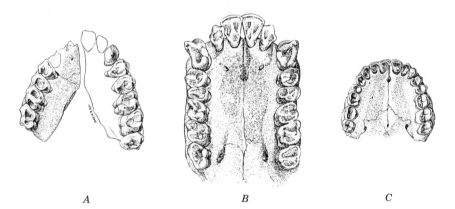

A   B   C

Fig. 8.1.   Comparison of upper jaws of (A) *Ramapithecus*, a Miocene fossil form; (B) *Pan*, a living chimpanzee, family Pongidae; and (C) *Homo*, contemporary man, family Hominidae. The reconstructed palate and dentition of *Ramapithecus* is more like man's than it is like a chimpanzee's. The arch of the dental arcade (the curve made by the teeth) of *Ramapithecus* is more like man's. The canine tooth of *Ramapithecus* is small. The anterior teeth (incisors and canines) are small relative to the cheek teeth (premolars and molars). The canine is spatulate like man's. (Redrawn after Simons, 1964.)

two families were distinct from each other by the late Miocene. The probabilities are highest that they are descended from one distinct lineage, which includes the Dryopithecinae. The lineage of Dryopithecinae, which is now classified as a subfamily of the Pongidae, split into two segments in the Miocene or earlier. One segment continued as the pongid lineage, ancestral to modern gorillas and chimpanzees. The other became the Hominidae and is represented in the late Miocene by *Ramapithecus*. The Hominidae include all those species which show distinct evolutionary trends toward *Homo*. The Hominidae are unified by a major shift in their adaptive zone. Reduction in size of incisor teeth and canines implies that the hands were used for grasping and tearing vegetation. It also implies that the development of bipedal locomotion was well underway and that the forelimbs were no longer a major part of the animal's locomotor system. The discovery of a pelvic bone adapted for bipedal locomotion and belonging to a true hominid of the early Pleistocene (Chapter 9) suggests that some Pliocene hominid already had a pelvis well on the way to becoming reorganized for habitual erect posture and a bipedal gait.

Oreopithecus

Oreopithecus was found in brown coal or lignite deposits in Tuscany about 1870. It was first described in 1872 by Gervais and since that time has been interpreted as related to baboons, to the great apes, and to man. In 1958 the recovery of an almost complete skeleton of this animal produced a newspaper sensation. The bones of the pelvis and the leg are those of an animal capable of bipedal locomotion. The anterior inferior iliac spine is fairly well developed and is hominid rather than cercopithecid or pongid. The presence of this iliac spine is believed to be particularly closely related to the erect posture and bipedal locomotion of man. This spine is the point of attachment of the iliofemoral ligament. The ligament plus this pelvic spine maintain stability in erect posture. The dimensions of the skeleton of the foot suggest that *Oreopithecus* developed some degree of adaptation to terrestrial life. Morphological analysis of certain foot bones shows that there are marked similarities between *Oreopithecus* and man, between *Oreopithecus* and the great apes, and there are also some resemblances between *Oreopithecus* and cercopithecoids.

To which taxon should *Oreopithecus* be referred? It is definitely not a cercopithecoid, and it is just as clearly a hominoid. Saying that *Oreopithecus* belongs in the Hominidae may be justified. Some would prefer to erect a special taxon, the family Oreopithecidae. The arguments for putting it in either the Hominidae or the Oreopithecidae are equally sound, but we prefer Oreopithecidae. The most recent authoritative study, by Straus, points out that *Oreopithecus*, if it is a hominid, is a very specialized hominid. It is also significant that the molar teeth of *Apidium*, an Oligocene primate discussed earlier, closely resemble those of *Oreopithecus*. If this similarity in dentition indicates close biological relationship, then the *Apidium–Oreopithecus* lineage was diverging in Oligocene times, and *Oreopithecus* must be considered a specialized and aberrant hominid. Since it is clearly a member of the Hominoidea, it seems best to emphasize its distinctness by placing it in a separate family, the Oreopithecidae. The great significance of *Oreopithecus* should not be overlooked in the process of assigning it to a taxonomic category. This fossil demonstrates that characters which are part of the adaptive complex of the Hominidae appeared at least 12 million years ago within the Hominoidea. But *Oreopithecus* probably occurs too late in time for the dental features to represent the direct ancestral hominid types.

Other Miocene and Pliocene Forms

Ceboidea, South American monkeys, are not represented among Pliocene fossils; however, they are reported from the Miocene of South America. There is, apparently, a single prosimian, *Indraloris lulli*, from the Siwalik Hills of India. A number of fossil lorisoid-like specimens have been found in East Africa. These are perhaps of early Pliocene age, but they have not yet been described.

Some Old World monkey lineages are found as fossils in Pliocene deposits (Table 8.1). (Specimens that can be assigned with confidence to the Cercopithecoidea have not yet been reported from earlier epochs. *Prohylobates* is an exception.) As we said earlier, it is puzzling that the fossil ancestors of a living group which is considered "primitive" in anatomy should appear so much later than the more advanced hominoids. Those fossils which are definitely cercopithecoid do not provide information about the ancestry of the group. *Libypithecus* is a well-preserved skull found in middle to late Pliocene deposits in Egypt. It has a pronounced sagittal crest and an elongated muzzle. These traits suggest relationships to baboons and macaques. *Mesopithecus* is a fossil monkey found in Europe and the Near East—Czechoslovakia, Romania, Hungary, southern Russia, Greece, and Iran. It has also been found in Kenya. Although it has been interpreted as being a leaf-eating monkey, such as modern *Presbytis* or *Colobus*, the paleoecology of the regions where it was found is that of open tundra or steppes. *Dolichopithecus*, a French cercopithecoid fossil of the late Pliocene, has similarities with modern baboons (an elongated muzzle) and macaques (short limbs). Fossil primate fragments from southern France and Tuscany have been assigned to *Presbytis* (a living genus). The Pliocene deposits of the Siwalik Hills in India have produced a number of fossil primates that may belong to the living genera *Pygathrix*, *Macaca*, and *Cercopithecus*. During the Pleistocene, the cercopithecines in particular radiated very widely. Fossils of this group are recovered from Pleistocene deposits in Europe, Asia, and Africa.

Suggested readings and sources

Clark, W. E. Le G., The Miocene Hominoidea of East Africa. *Fossil Mammals of Africa*, No. 1. Brit. Museum (Nat. Hist.), London (1951).

Gervais, P., Sur un singe fossile, d'espèce non encore décrite, qui a été découvert au Monte Bamboli. *Comptes Rendus Acad. Sci.*, 74 (1872).

Gregory, W. K., *The Origin and Evolution of the Human Dentition.* Williams and Wilkins, Baltimore (1922).

Hopwood, A. T., Miocene primates from Kenya. *J. Linnean Soc. Zool. London* 38, 437 (1933).

Hürzeler, J., *Oreopithecus bambolii* Gervais. A preliminary report. *Verhandl. Naturforsch. Ges. Basel* 69, 1 (1958).

Leakey, L. S. B., A new lower Pliocene fossil primate from Kenya. *Ann. Mag. Nat. Hist.* 4, 689 (1962).

Napier, J. R., and P. R. Davis, The fore-limb skeleton and associated remains of *Proconsul africanus. Fossil Mammals of Africa*, No. 16. Brit. Museum (Nat. Hist.), London (1959).

Pilgrim, G. E., New Siwalik primates and their bearing on the question of the evolution of man and the Anthropoidea. *Records Geol. Surv. India* 45, 1 (1915).

Simons, E. L., *Apidium* and *Oreopithecus. Nature* 186, 824 (1960).

Simons, E. L., Some fallacies in the study of hominid phylogeny. *Science* 141, 879 (1963).

Simons, E. L., A critical reappraisal of Tertiary primates. *In* J. Buettner-Janusch (Ed.), *Evolutionary and Genetic Biology of Primates*, Vol. I, p. 65. Academic Press, New York (1963).

Simons, E. L., On the mandible of *Ramapithecus. Proc. Natl. Acad. Sci. U. S.* 51, 528 (1964).

Simons, E. L., and D. R. Pilbeam, Preliminary revision of the Dryopithecinae (Pongidae, Anthropoidea). *Folia Primat.* 3, 81 (1965).

Straus, W. L., Jr., The classification of *Oreopithecus. In* S. L. Washburn (Ed.), *Classification and Human Evolution*, p. 146. Aldine, Chicago (1963).

Pleistocene Fossil Primates

9

THE MAJOR EVENTS of human evolution occurred during the late Pliocene and early Pleistocene epochs. *Homo sapiens* clearly differentiated from the hominid stock sometime during the Pleistocene. One of the more interesting questions for students of primate evolution is: How rapidly did man evolve? For a long time, anthropologists believed that the Pleistocene epoch began about 600,000 years ago. Since all fossils referable to the genus *Homo* came from Pleistocene deposits, the differentiation of man from the hominid stock was thought to have occurred during this relatively brief span of time. Recently, potassium-argon (K-A) dates for the beginning of the Pleistocene have been determined, and it appears that the Pleistocene began 1,750,000 ($\pm$300,000) years ago. The change in dates, critical in the evolution of our species, gives a much longer span of time in which to fit the hominid fossil ancestors of man. The new dates were obtained by the application of the K-A method to materials from a very famous African fossil-man site, Olduvai Gorge in Tanganyika.

Pliocene-Pleistocene Border

The appearance of the Villafranchian faunal assemblage is believed to be coincidental with the terminal phases of the Pliocene epoch and the beginning of the Pleistocene. Despite the straight lines we draw on our

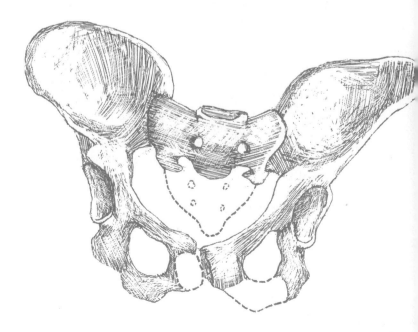

charts and diagrams of the sequences of epochs, it is, in reality, quite difficult to define boundaries between epochs such as the Pliocene and Pleistocene. In what is now Europe and North America, the Pleistocene was an
epoch during which several major glaciers advanced and retreated. The
climate grew colder and a characteristic group of plants and animals developed. The sequence of Pleistocene geological events in Europe and in much
of North America has been worked out by geologists and paleontologists
(Fig. 3.2). The Pliocene-Pleistocene border is drawn where a cold-adapted
fauna first appears among fauna typical of warm climates in terminal Pliocene deposits.

Two successive mammalian fauna mark the transition from the Pliocene
to the Pleistocene—the Villafranchian fauna and the later typically Pleistocene assemblage which implies a cooler, more continental climate than
existed in Pliocene times. The Villafranchian fauna is a kind of worldwide
reference point or marker. Its presence makes it possible for us to correlate
and compare widely separated horizons, geological or archeological, that
cannot be correlated by other means. The Villafranchian fauna suggest a
moist, warm to temperate climate, primarily in a forest environment, with
some steppe and plains regions. There is marked variation from region to
region, and this suggests that the continents were being divided into topo-
graphic and climatic segments. This process of continental climatic segmen-
tation was continued during the colder Pleistocene.

The change in fauna is known from Europe, specifically from Italy, where
true elephants (*Elephas planifrons*), zebra-like horses (*Equus stenonis*), and
cattle (*Leptobos*) appear in the transitional Villafranchian zone between
Pliocene and Pleistocene. These three genera define the Villafranchian
horizons. In Africa the Villafranchian fauna and the lower Pleistocene mammals include *Elephas africanavus* (at Olduvai), *Mammuthus planifrons*,
Lycyaena and *Hyaenictis* (hyena-like animals), *Bos* (true cattle), various
zebra-like horses, *Megantereon* and *Machairodus* (cats), a variety of wolves,
camels, antelopes, and other now extinct genera. Some of these are also
found in Asiatic and European Villafranchian horizons.

The correlations between Europe and Africa are based on faunal assem-
blages, on a presumed correlation between African pluvial periods and
European glaciations, and on the correlation of North African and Italian
littoral (shoreline) deposits. The fauna of the African lower Pleistocene is
Villafranchian in character. This fauna persisted because of the continuation
of the warm Pliocene climate, particularly in South Africa, whereas in
Europe a cold, continental, glacial climate developed. Correlation of European and African fossil fauna of the late Pliocene and lower Pleistocene is
not easy. Animals that were extinct in Europe by the lower Pleistocene were
still a part of the fauna of East and South Africa. The post-Villafranchian
Pleistocene sequences of mammals in East and South Africa are indicated
in Fig. 9.1.

| EPOCH | Climatic Zone | EAST AFRICA | | SOUTH AFRICA |
		Known Fauna Now Extinct (per cent)	Time Span of Leading Mammals	Fauna
Upper Pleistocene	Post-Gamblian and Gamblian pluvial	≤1	Recent forms	Very similar to that of present day.
	Dry		*Homoiceros*	
Middle Pleistocene	Kanjeran pluvial	41	*Loxodonta antiqua recki* *Bularchus* *Pelorovis*	Buffalo, wart hog, horse, giraffe-like forms, mammoth, mastodon, hippopotamus.
	Dry	40	*Mammuthus oxoptatus*	
	Kamasian pluvial			
Lower Pleistocene	Dry (?)	57	*Stegodon* *Stegolophodon* *Hippopotamus kaisensis* *Dinotherium bozaei* *Metachyatherium* *Mammuthus planifrons*	Baboons, rodents, hippopotamus, hoofed mammals, cats, hyena, mongoose, wolf.
	Kageran pluvial (?)			

Fig. 9.1.   The Pleistocene epoch in Africa.

There is some agreement about the most general correlation between European and African Pleistocene subdivisions. However, the age in years of each stage and whether African horizons are completely synchronous with those of Europe are not known with certainty. The date in years assigned to the earliest phase in Africa is based upon Leakey's samples from Olduvai Gorge. These samples were taken from the bottom of Bed I at Olduvai. This stratum contains the early hominid forms, Villafranchian fauna, and crude stone tools called the Olduwan culture. The K-A date—1,750,000 years ago—in East Africa places the beginning of the Pleistocene almost three times as far back as the previous estimates held it to be. At the very least, these African hominids, Villafranchian fauna, and tools are 1,750,000 (±300,000) years old. Whether the African dates apply to the earliest Pleistocene horizons in the rest of the world is a matter awaiting further research.

Leakey, whose work at Olduvai for the past 35 years is world renowned, submitted a large number of carefully chosen samples of volcanic rock (anorthoclase, oligoclase, and biotite) for K-A dating, from beds containing early examples of the Villafranchian fauna in Africa (Fig. 3.1 and 3.2). Early hominid fossils and crude stone tools also were found in these beds. As we pointed out in Chapter 3, the method of dating and the material dated are critical when we decide how much confidence to have in the dates assigned certain events in evolution. The problems raised by the dates of these strata in Olduvai Gorge are worth investigation, for they touch upon the use of fauna in dating, the worldwide correlations of strata, and the kinds of implications one may draw from a K-A date.

If these dates for Olduvai are correct, the hominid lineage was transformed into *Homo sapiens* at a rate considerably slower than anthropologists believed when the length of the Pleistocene was estimated at 600,000 years. If the faunal horizon at Olduvai is indeed coincidental with the beginning of the Pleistocene, this dates the beginning of that epoch in Africa. But the Pleistocene in Africa might not have been synchronous with that in Europe, and the conventional age for the European Pleistocene may still be valid. In a way the K-A dates at Olduvai liberate us from dependence upon European Pleistocene sequences. We no longer need to chain the sequences of fossil hominids to estimates for the age of Pleistocene geological strata which must be fitted into less than a million years. Of course, these statements all depend upon the validity of the K-A dates. These K-A dates were not determined on the fossils but on rocks which came from the strata on which the fossil-bearing stratum lay, from rocks in the fossil-bearing stratum, and from rocks lying above the fossil-bearing stratum. The results obtained from the various samples were consistent; that is, the dates for the rocks from the lowest level were earlier than those from the highest level. The minerals used in the tests were well suited to the K-A method. The results enable us to accept the validity of 1,750,000 (±300,000) years as the minimum age of certain fossils assignable to the genus *Homo*.

Australopithecines

There are several major groups of Pleistocene hominid fossils that have been discovered in several parts of the Old World. They will first be discussed under their older names and in a sequence corresponding to their geological ages. First, the australopithecines; second, the pithecanthropines; third, the sapiens group which includes the so-called Neandertal man. In Chapter 10 we will discuss the systematics, particularly the classification and nomenclature, of these fossils.

The first of the australopithecine group was discovered in South Africa by Raymond Dart, who reported it to the world in 1925. Table 9.1 is a list of most of the australopithecine discoveries reported up to July, 1964. Australopithecines have been found in many parts of Africa, and we may infer from dentitions that several fossil discoveries in Java and China belong to this group. This geographical distribution suggests that the australopithecines were widespread. We may expect future reports of fossil australopithecines in other parts of the Old World.

TABLE 9.1
Reported Discoveries of Australopithecines

Region	Site	Name	Number of individual fossils[a]
South Africa	Taung	AUSTRALOPITHECUS AFRICANUS	1
	Kromdraai	PARANTHROPUS ROBUSTUS	3
	Makapansgat	AUSTRALOPITHECUS PROMETHEUS	5
	Sterkfontein	PLESIANTHROPUS TRANSVAALENSIS	20
	Swartkrans	PARANTHROPUS CRASSIDENS	35
	Swartkrans	TELANTHROPUS CAPENSIS	2
East Africa	Olduvai	ZINJANTHROPUS BOISEI	3
	Olduvai	HOMO HABILIS	6
	Kanam	HOMO KANAMENSIS?	1
	Garusi	AUSTRALOPITHECUS	1
West Africa	Tchad (Sahara)	AUSTRALOPITHECUS	1
Palestine	Tell Ubeidiya	AUSTRALOPITHECUS	1
Java	Djetis	MEGANTHROPUS PALEOJAVANICUS?	2
China	Drugstore	AUSTRALOPITHECUS	1?

[a]The exact numbers of different *individuals* are difficult to determine for several of the species. The numbers reported here are based upon summaries in the literature.

Australopithecine fossils have been found in at least four places in South Africa—Taung, the spot from which the original fossil described by Dart came, Sterkfontein, Kromdraai, and Swartkrans. Leakey has found fossils which belong to the same genus in East Africa, at Olduvai Gorge. The fact that he has called his finds *"Zinjanthropus boisei"* and *"Homo habilis"*

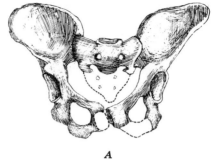

A

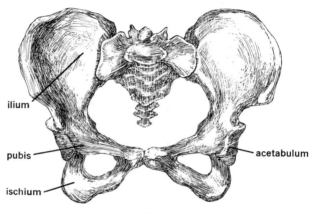

ilium

pubis   acetabulum

ischium

B

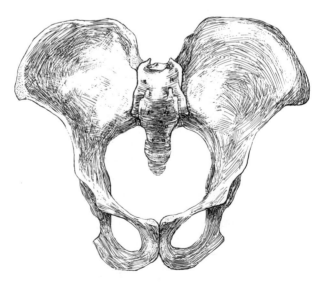

C

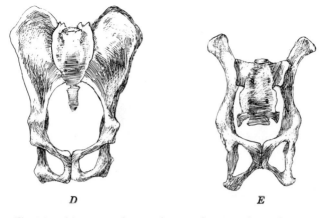

Fig. 9.2.   Pelves, ventral view, of Hominidae, Pongidae, and Lemuridae. (A) *Australopithecus*, (B) *Homo*, (C) *Pan gorilla*, (D) *Hylobates*, and (E) *Lemur*. The relative size of the ilium and the shape of the os innominatum (see Fig. 9.3) in *Australopithecus* resemble those features in *Homo* and contrast with the pelves of the other orthograde forms (C, D) and with the pelvis of the pronograde form (E).

makes no difference. As we shall show in the section on the nomenclature of Pleistocene hominids, they are probably all members of the taxon to which australopithecine fossils are referred. Other australopithecine fragments have come from West Africa, North Africa, and Palestine.

The fauna associated with the original South African discoveries include *Equus*, and such extinct mammals as *Lycyaena, Hyaenictis, Dinopithecus, Griquatherium, Notochoerus, Parapapio, Machairodus*. The fact that many of these, such as *Lycyaena*, are represented only in the Pliocene in Europe led many to suppose that the South African man-apes were of Pliocene age. Cooke, a South African geologist and paleontologist, and Wallace before him, pointed out that the southern tip of Africa was a cul-de-sac in which many archaic mammals survived long after they had vanished from Europe with its rigorous quaternary climatic conditions.

The most important morphological feature of the australopithecines, the primary trait upon which the adaptive radiation of the hominids is based, is the modification of pelvis and long bones for habitual erect posture and bipedal locomotion. The evolutionary segregation of the Hominidae from the other primates occurred when they achieved efficient bipedal locomotion and erect posture. In Chapter 8 it was suggested that *Ramapithecus*, a Pliocene hominid, was a bipedal, erect animal. But this suggestion was entirely inferential. It was based on two bits of evidence—the dentition of *Ramapithecus* (Fig. 8.1) and the pelvis of *Australopithecus* (Fig. 9.2). It seems reasonable to suppose that a less perfectly bipedal primate preceded the australopithecines. The appearance of three pelvic bones that belong

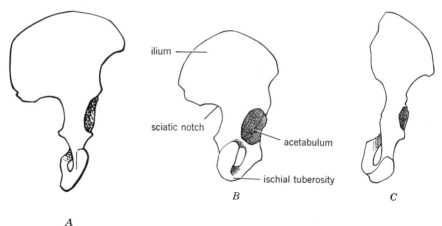

Fig. 9.3.   Shape of lateral surface of the pelves of (A) *Australopithecus*, (B) *Homo*, and (C) *Pan gorilla*. The sharper sciatic notch in (A) and (B) is especially noteworthy. See Fig. 9.2 for relative sizes of these pelves.

to the australopithecines in South Africa is direct evidence that this group had achieved efficient bipedal locomotion.

The total morphological pattern of the pelvis of the australopithecines is distinctively hominid (Fig. 9.2 and 9.3). The following features from the australopithecine fossil pelvic bones are the evidence that these are hominid pelves. The anterior inferior iliac spine is strongly developed. The iliofemoral ligament attaches to this spine. This powerful ligament braces the front of the hip joint when it is extended as the animal stands. The posterior extremity of the iliac crest is extended backwards and the ilium is relatively much broader and somewhat shorter than it is in apes (Fig. 9.3). The broad ilium lengthens the attachment of the gluteus muscles, important in maintaining balance of the trunk on the legs. The backward extension of the posterior extremity of the iliac crest brings the gluteus maximus posterior to the hip joint. The gluteus maximus becomes a powerful extensor muscle, part of the complex needed for erect bipedal walking. In monkeys and apes, gluteus maximus is lateral to the hip joint and is an abductor (Fig. 9.4). Gluteus medius and gluteus minimus become abductors of the thigh in man. They are extensors in monkeys and apes. As Washburn has emphasized, changes in the gluteus maximus muscle were critical in the evolution of erect posture and bipedal locomotion, as critical as changes in the iliopsoas muscle. Normal human locomotion requires extreme extension of the thigh. This is not possible unless gluteus maximus is a powerful extensor. But the ilium is bent back and broadened for another very important reason. The pelvis forms the bony structure around the birth canal (Fig. 6.11), and the ilium is bent back for obstetrical reasons. If, through natural selection for erect bipedalism, the ilium is made shorter, a large angle with the ischium must result if the birth canal is to be of adequate dimensions. The sciatic notch (Fig. 9.3) is more acute in the pelves of modern man and the australopithecines than it is in the pelves of the great apes, a result, in part, of the development of a prominent ischial spine. A strong ligament is attached to the ischial spine, binding the pelvic bone to the sacrum. The articulation of the sacrum is brought close to the acetabulum. This provides greater stability for transmission of the weight of the trunk to the hip joint in the erect stance. The ischial tuberosity is relatively high which makes the action of the hamstring muscles more efficient. These muscles maintain full extension of the hip. The position of the ischial tuberosity in man and australopithecines (Fig. 9.3) brings the hamstring muscles to a position in back of the hip joint rather than below. The tuberosity and hip joint are much farther apart in great apes and pronograde animals. This arrangement gives the hamstrings (Fig. 9.4) great power in the lever action of the hind limb. In man, and by implication in the australopithecines, the hind limb is a propulsive strut rather than a propulsive lever.

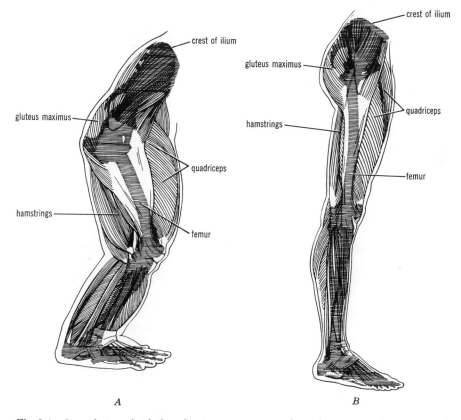

Fig. 9.4.   Lateral view of right legs showing arrangements of muscles necessary for extension of leg in (A) gorilla and (B) man.

On the australopithecine ilium, there is a deep, well-marked groove on the ventral surface which is the origin of part of the iliopsoas muscle. This muscle must turn backwards at a pronounced angle in order to attach to the femur when the thigh is extended. Since the thigh is habitually extended in erect posture, this is a most important feature of the pelves in question. The pelves of the australopithecines are not identical with those of *Homo sapiens*, but they are clearly hominid in total morphological pattern. They are functionally well designed for erect posture and bipedal locomotion, but they are not perfectly designed for a bipedal gait. Australopithecines could run bipedally but were probably clumsy bipedal walkers. They would have made excellent soccer players but very poor place kickers, and as ballet dancers they would have been ludicrous.

Further corroboration of the functional interpretation of the australopithecine pelvis comes from parts of the lower limbs which have been found. Two specimens of the distal end of the femur have been described. These show a combination of characters with a total morphological pattern that is very similar to the human femur and represents a mechanical adaptation for erect bipedalism. The combination of morphological characters which support this view include obliquity and robustness of the shaft, the particular alignment of the condyles, the contour of the patellar surfaces, and the forward lengthening of the intercondylar notch.

Some bone fragments of the hands and feet were among the South African australopithecine discoveries. A piece of the talus (ankle bone) was found at Kromdraai with some pongid as well as hominid features combined. It was interpreted as indicating a stable ankle joint, to bear the weight of the body, and a very mobile foot. Although perhaps this bone must bear the weight of excessive interpretation, such a combination is clearly to be expected in an animal that has developed efficient erect posture recently. The first metatarsal bone is also significant. It is robust, as is ours, and suggests that the australopithecines' habitual gait was bipedal.

Fortunately the discoveries at Olduvai Gorge include many bones of the hands and feet of the East African australopithecines. Twelve bones from a foot were found. Napier studied these bones and was able to articulate them (Fig. 9.5). Therefore, he got an impression of the foot skeleton as a whole. His analysis indicates that this Olduvai foot skeleton belongs to a bipedal hominid. Many of the proportions and articulations between the bones are typically hominid, or closer to *Homo sapiens* than to living pongids such as the gorilla. The foot of modern man is specialized in order that he may walk bipedally. Among the specializations that permit this are the shape of the arch and the position and robustness of the big toe. Also the proportions of the bones associated with the lever action of the foot (p. 315) are such that they support the body weight adequately when man walks. The walk of man is unique—it is a stride. In striding, the sequence in which various parts of the foot bear the weight of the body is heel, lateral edge of the foot, big toe. The big toe bears all the weight at the end of the step and

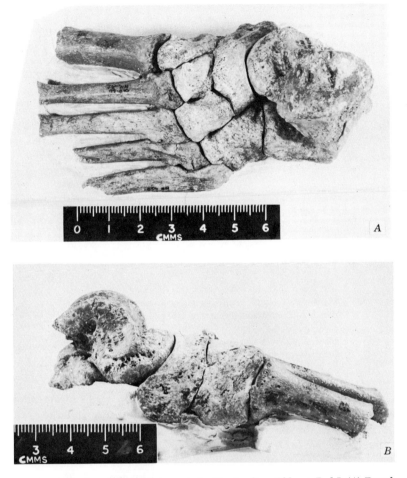

Fig. 9.5. The foot skeleton of the australopithecine from Olduvai, Bed I. (A) Dorsal view and (B) lateral view. (Photographs courtesy of J. R. Napier, (A) from Day and Napier, 1964.)

at the beginning of the next. The metatarsal and phalanges of man's big toe are more robust than the other metatarsals and phalanges. The first metatarsal of the Olduvai australopithecine is robust and resembles that of contemporary man, although it is not as robust relative to the other metatarsals as is that of *Homo sapiens*. But then the Olduvai hominid was a much smaller animal. The Olduvai hominid had a foot which enabled him to run well bipedally. But he probably could not walk bipedally as efficiently as we do.

Napier has also analyzed the bones of the hand found at Olduvai. They have a greater similarity to those of juvenile gorillas and adult *Homo sapiens* than to those of adult great apes.

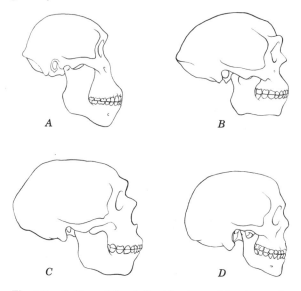

Fig. 9.6.   Outlines of the skulls, side views, of four hominids. (A) An australopithecine, (B) a pithecanthropine, (C) a Neandertal, and (D) a modern man. It is clear that there has been a reduction in the size of the jaw (the mandible), an increase in the size of the braincase, and a progressive rounding of the braincase since early Pleistocene times.

One of the most interesting characteristics of the skulls of australopithecines is the combination of a relatively small braincase with large jaws (Fig. 9.6). This gives the skull a most ape-like appearance. It is this primitive characteristic that contrasts most strongly with modern man. Nevertheless there are a number of characteristics of the skull of the australopithecines that contrast with the pongids and are clearly hominid (Fig. 9.7). First, cranial height—the height of the vault of the cranium above the levels of the orbits—is outside the range of variation found in modern apes and is within the range of modern hominid skulls. Second, the occipital torus and the inion are low, as is the case in *Homo sapiens* and in other hominids that occur later in time than the australopithecines. Third, the occipital condyles are forward in position relative to the total length of the skull and relative to the transverse level of the auditory apertures. Among the Pongidae the occipital condyles are behind the midpoint of the cranial length and also behind the auditory apertures. In other words the position of the occipital condyles implies that the australopithecines held their heads much the way erect bipedal *Homo sapiens* does. As Le Gros Clark has pointed out, these three hominid features of the australopithecine skull are related to each other. They are part of the complex that is associated with the way the head is held in relation to the vertebral column. A statistical analysis has shown that the three indices relevant to the way the head is held—the supraorbital height, nuchal area height, and condylar position, taken together (Fig. 9.8)—place the australopithecine skull outside the limits of variation of the great apes and within the range found in the skulls of the true hominids.

The jaws, palate, and teeth of the australopithecines have been studied in great detail. The upper jaw and the palate have the marked reduction and recession of the incisor region, typical of the hominid skull. There is no simian shelf, and, in general, the jaws have many hominid features.

The dentition of the australopithecines is important in the assessment of the phylogenetic affinities of these fossils. The total morphological pattern of the dentition of australopithecines conforms to the pattern of the dentition of the Hominidae (Fig. 9.9). The dental arcade is an even curve. There are no diastematic intervals. If the canine of the upper jaw is large, it is usually separated from the lateral incisor by a space, the diastema. If there is a diastematic interval, the lower canine will fit closely into this space in the upper dentition when the jaws are occluded. The upper incisors of australopithecines are small and retracted to the level of the canines. The canines are small in size, smaller relative to those of the great apes, and they are spatulate in form. There is no obvious sexual dimorphism in the size of the canines of known specimens. The anterior upper premolars have two roots. The anterior lower premolars are nonsectorial and bicuspid and have one root, the surface of which is marked by longitudinal grooves. (The anterior lower premolars of pongids are sectorial and unicuspid and have two roots.) The morphology of the cusps of the

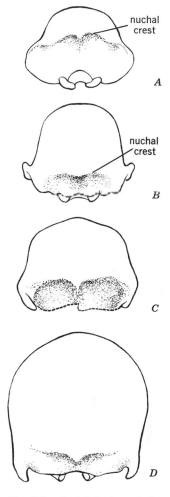

Fig. 9.7. Occipital view of the skulls of four hominids. (A) A chimpanzee, (B) an australopithecine, (C) a pithecanthropine, and (D) a modern man. The position of the nuchal crest is relatively high in the chimpanzee skull. This is a typical pongid character. The position of the nuchal crest is much lower in the three hominid skulls. (Redrawn after Le Gros Clark, 1964.)

molar teeth is similar in general form and in many details to that of the teeth of the pithecanthropine forms of Asia.

One important consequence of the discovery of the australopithecines

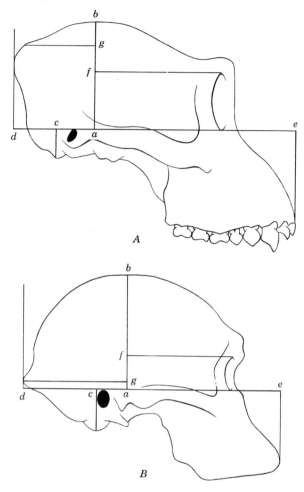

Fig. 9.8. Skulls of (A) a female gorilla and (B) an australopithe-cine from Sterkfontein. The various distances for calculating indices to distinguish two skulls are: $\dfrac{fb}{ab} \times 100 =$ index of supraorbital height; $\dfrac{ag}{ab} \times 100 =$ index of nuchal area height; $\dfrac{dc}{ce} \times 100 =$ index of condylar position. (Redrawn after Le Gros Clark, 1964.)

is the demonstration that a primitive hominid with a very small cranial capacity, indeed not much larger than that of the anthropoid apes, had a pelvis and limb structure that enabled it to walk bipedally and hold itself erect. The fact that the australopithecine pelvis is not as perfected for erect posture and bipedal locomotion as that of modern man is raised as an argument by those who believe that the australopithecines are aberrant apes rather than hominids. The rapid evolution of the pelvic girdle of the Hominidae relative to the evolution of the skull and brain during the late Pliocene and early Pleistocene is an excellent example of the principle of mosaic evolution (Chapter 4). Since we believe *Homo* evolved from the australopithecines, the features which the two share are of importance and must be carefully evaluated. It is quite legitimate to suggest that the characters peculiar to *Homo* are specializations, that is, special features developed after phyletic branching. We ought to ask how like the australopithecines is *Homo*, not vice versa.

The belief has been current that brain size is one of the most important criteria for defining the Hominidae. Cranial volume is a trait that has been much overemphasized by anthropologists in their discussions of hominid evolution. As noted earlier, the brain-to-body ratio is very high among all the Primates relative to most other animals. We are still not certain what relationship exists between amount of brain and so-called higher functions. Furthermore, the brain consists of many structures, only some of which are concerned with such high-level activities as speech, intellect, toolmaking, and anthropology. The range in cranial volume found in modern man is large, from 850 to 1700 cc. The smaller end of the range is *not* correlated with mental insufficiency. A relatively small brain, weight 1017 grams (volume estimated at 1000 cc), belonged to Anatole France, a distinguished and celebrated French literary artist of the nineteenth century.

Methods of measuring the brain volume are quite inexact. One way to measure cranial capacity is to fill the cranium with lead shot, millet seeds, or sand and then measure the volume of the lead shot, millet seeds, or sand in a calibrated vessel. Or the volume may be measured by dropping the

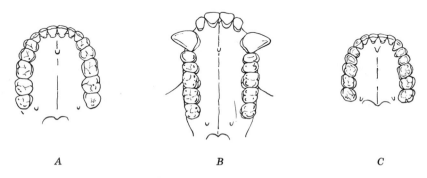

A   B   C

Fig. 9.9.   Upper jaws of (A) *Australopithecus,* (B) *Pan,* and (C) *Homo.* The total morphological pattern of the dentition of *Australopithecus* is a hominid pattern, as illustrated by *Homo.* It contrasts with the pattern of the Pongidae, as illustrated by *Pan.* See text for details. (Redrawn after Le Gros Clark, 1964.)

lead shot, for example, into a flask of water and the volume of water displaced then measured.

Most fossil skulls are not preserved well enough for such a technique to be successful. Estimates are made by reconstructing the skull from recovered fragments. Such estimates may be extremely misleading if only a very small mistake is made in fitting together bones of the cranial vault—or in reconstructing missing pieces of them. A number of estimates of cranial volumes of fossil Pleistocene hominids is listed in Table 9.2. Some are quite reliable

TABLE 9.2
Cranial Volumes of Hominids

Hominid	Volume (cc)	Range (cc)
Australopithecines		
Taung	494	
Makapansgat	600[a]	
Sterkfontein	435	
Sterkfontein	480	
Kromdraai	650[a]	
Swartkrans		550–700[b]
Olduvai		
"*Zinjanthropus*"	600[c]	
"*Homo habilis*"	673[d]	
Homo erectus		900–1225
Neandertal		1300–1640
Modern man		850–1700

[a]Rough estimates from fragmentary remains.
[b]Upper and lower estimates based on differing opinions of reconstruction of cranium.
[c]Lower limit as postulated by Leakey.
[d]Based on reconstruction of cranium.

estimates, others are not. We find a trend—the hominid brain grew larger during the Pleistocene. However, the range of cranial capacities of the australopithecines (435–700 cc) is about the same as the range of cranial capacities of modern gorillas (420–752 cc). The functional meaning of the trend to a larger cranial capacity is not as well understood as is sometimes thought.

Probably more significant than absolute cranial capacity is the relationship between brain size and body weight. This relationship is expressed as the brain:body ratio. A living gorilla weighs about 250 kg and its brain:body ratio is about 1:420. *Homo sapiens* has a brain:body ratio of about 1:47 (see Chapter 20). If we equate the cranial capacity estimated for the australopithecines with brain weight and use reasonable estimates for body weight based upon the dimensions of various skeletal fragments, we can calculate a brain:body ratio for these early Pleistocene hominids. Brain weight of australopithecines is about 600 g and body weight is about 25 kg (25,000 g). The brain:body ratio is 1:42, a ratio close to that of *Homo sapiens*.

Pithecanthropines

The first of the pithecanthropine group of fossils was discovered by Dubois in 1891 at Trinil in central Java. Dubois found them in alluvial deposits on the banks of the Solo river, and he gave them the name which had been suggested by Haeckel for the hypothetical missing link. Haeckel had proposed that the name for the extinct hominid stage between apes and modern man be *Pithecanthropus* (Greek: ape-man). He even gave a specific name *P. alalus* which means speechless ape-man. Haeckel was a little premature with his name, for no fossil to which it might be applied was discovered for more than 20 years. The fauna associated with these hominid remains were post-Villafranchian, and today these Trinil beds are considered to be middle Pleistocene in age. Subsequent discoveries in other

TABLE 9.3
Reported Discoveries of Pithecanthropines

Region	Site	Name	Number of individual fossils
Java	Ngangdong	HOMO SOLOENSIS[a]	10
	Trinil	PITHECANTHROPUS ERECTUS	4
	Sangiran	PITHECANTHROPUS ERECTUS	2
	Kedoeng Broeboes	PITHECANTHROPUS ERECTUS	1
	Sondé	PITHECANTHROPUS	1
North Africa	Ternifine	ATLANTHROPUS	1
	Abd er-Rahman	ATLANTHROPUS	1
East Africa	Olduvai	HOMO ERECTUS	1
Northern Rhodesia	Broken Hill	HOMO ERECTUS[a]	2
South Africa	Saldanha Bay	HOMO ERECTUS[a]	1
China	Choukoutien	SINANTHROPUS PEKINENSIS	40

[a]May also be considered Neandertal (Table 9.4).

parts of Java (Table 9.3) added to the number of individuals that belong to this group. Fluorine dating has confirmed that Dubois' original finds are of the same age as the fauna and alluvial beds with which they are associated.

The sinanthropine group of fossils was discovered in China, at Choukoutien, in 1927. The first find was a single tooth, a lower molar. In 1929 an almost complete skull was discovered, and within a few years remains of forty individuals had been recovered. The sinanthropine fossils are from deposits that are believed to be of middle Pleistocene age.

The total morphological pattern of these fossils is distinctly hominid. Massive brow ridges are characteristic, and there is a distinct postorbital constriction. The foramen magnum is set farther forward on the base of the skull than that of the australopithecines. A very heavy occipital torus suggests that the neck muscles were massive. The mastoid process of the skull is well developed. Several femurs have been found and all are characteristic of erect bipeds. Even though there were no pelvic bones among the remains of this group, the group were undoubtedly erect bipeds. All are considered to be in the taxon *Homo erectus,* although we and other anthropologists would prefer to put them in the taxon *Homo sapiens* (Chapter 10).

Sapiens

Fossils of *Homo sapiens* are found in middle and upper Pleistocene horizons. A great amount of energy and ink has gone into the discussion of how to identify and define what or who is the first "true" *Homo sapiens* fossil. As we have said so often, ancestors should not be expected to look exactly like descendants. Therefore we should not rule fossils out of the taxon *Homo sapiens* because they differ from contemporary man. Indeed if we take the theory of evolution seriously, we expect them to be different. The problem of assigning Pleistocene fossil hominids to properly defined taxa is a problem in delimiting an allochronic species. The numbers of relevant fossils are now sufficiently large to make the systematics of *Homo sapiens* a problem when deciding where to divide what is essentially a continuum.

The first fossil *man* ever taken seriously was a skull cap found near Düsseldorf, Germany, in 1856. It was found in the Neander river valley (in German, valley is *tal*), hence Neandertal man. (The old German spelling of *tal* was *thal*, hence the spelling Neanderthal is seen in older writings.) Anthropologists have worried each other and their students for at least a generation with analyses and, sad to say, virulent polemics, about the "Neandertal Problem." There are a number of reasons for the existence of a "Neandertal Problem." Before we can analyze these reasons, we must describe the problem itself. Neandertal fossils of the middle and upper Pleistocene (Table 9.4) have been divided into two groups by many, if not most, who have worked with them. They have been grouped as Classic and Progressive Neandertals. The Classic Neandertals are believed to be a specialized evolutionary lineage that developed in northwestern Europe.

TABLE 9.4
Some Hominids of the Middle and Upper Pleistocene

Country	Locality	Remains
Germany	Neandertal	parts of skull and skeleton
	Ehringsdorf	cranium, two mandibles, and child's skeleton
	Steinheim	cranium
	Heidelberg (Mauer)	mandible
	Oberkassel	parts of skulls and skeletons of two individuals

TABLE 9.4 (*Continued*)

France	La Chapelle-aux-Saints	skull and nearly complete skeleton
	La Ferrassie	bones from six individuals
	Le Moustier	parts of skull
	La Quina	parts of skull
	Montmaurin	mandible and teeth
	Fontéchevade	frontal bone of skull
	Chancelade	skull and part of skeleton
	Cro-Magnon	parts of skulls and skeletons of five individuals
	Combe-Capelle	skull and part of skeleton
Belgium	Engis	skull fragments of infant
	La Naulette	mandible
	Spy	parts of two skulls
Spain	Cova Negra	parts of skull
Gibraltar	Forbes' Quarry	parts of skull
	Devil's Tower	parts of child's skull
Italy	Saccopastore	parts of two skulls
	Monte Circeo	parts of skull and one mandible
	Grimaldi (Grottes des Enfants)	parts of skulls and skeletons of two individuals
Czechoslovakia	Gánovce	brain cast
	Brno	parts of skull
	Predmost	parts of skulls and skeletons from twenty individuals
Yugoslavia	Krapina	fragments of thirteen individuals
Great Britain	Swanscombe	part of skull
	Bury St. Edmunds?	
	Galley Hill?	
Palestine	Tabun, Mount Carmel	skull and adult skeleton
	Skūhl, Mount Carmel	nine adult and one child's skeletons
	Galilee	cranial fragment
	Jebel Qafza	five adult and one child's skeletons
	Shukbah	one adult and six children's skeletons
Uzbekistan (U.S.S.R.)	Teshik-Tash	child's skeleton
Crimea (U.S.S.R.)	Starosel'e	infant's skull
Iraq	Shanidar	parts of skulls and skeletons of seven individuals
Algeria	Afalou-bou-rhummel	fragments of fifteen individuals
	Mechta-el-arbi	skulls and skeletons of thirty-two individuals
Zambia	Broken Hill	skull and skeletal fragments
East Africa	Gamble's Cave	fragments of four individuals
South Africa	Saldanha Bay	skull
Java	Ngangdong	fragments of ten skulls

These Classic Neandertals became extinct when *Homo sapiens* invaded Europe after the retreat of the last ice sheet. The Progressive Neandertals are believed to have developed more or less apart from the Classic Neandertals. These were the direct ancestors of *Homo sapiens*.

Classic Neandertals are middle and upper Pleistocene hominid fossils with certain features that are believed to make them distinct from modern man. The noteworthy features of the skull, which are said to rule them off the direct human lineage, are: a very heavy supraorbital torus making an uninterrupted shelf of bone above the eye sockets, no vertical forehead, an occipital bone that is "bun-shaped," and a face that is long, prognathous, and narrow. The fossils attributed to the Classic Neandertal group and named after the localities in which they were found include: Spy (Belgium), La Chapelle-aux-Saints, La Ferrassie, Le Moustier, La Quina (France), Neandertal (Germany), Gibraltar, Monte Circeo (Italy), Shanidar (Iraq), and possibly Montmaurin (France).

Progressive Neandertals, although possessing the neandertaloid characters listed, are said to be more heterogeneous in morphology. Comparisons of some of them with some Classic Neandertals show that they have shorter and narrower braincases and more highly arched foreheads; the supraorbital torus does not form a complete uninterrupted bony shelf above the eye sockets; the occiput is expanded and has a poorly developed occipital torus. The Progressive Neandertals include fossils found at: Ehringsdorf, Steinheim (Germany), Saccopastore (Italy), Krapina (Yugoslavia), Swanscombe (England), Teshik-Tash (Uzbekistan), Galilee, Shukbah (Palestine), and the caves of Skūhl and Tabun (Mount Carmel, Palestine).

The fact that the Classic Neandertal skulls seem to be clustered in Europe, and a part of Europe at that, suggested to some that these fossils represented an isolated, specialized population which eventually became extinct. The crucial evidence which enabled many to argue that all Neandertals were a specialized development, a subspecies which became extinct, was the existence of hominid skulls of the early middle Pleistocene with the features of essentially modern *Homo sapiens*. The famous Piltdown skull was one of these early sapiens-like fossils, but it proved to have been planted in the stratum where it was presumably found. It was a fake. Piltdown was a group of prehistoric, but not early Pleistocene, human cranial bones and mandibular fragments of a modern ape, probably an orangutan. The hominid fossils, believed to be like modern man in morphology, but of early middle Pleistocene age, have been found at Swanscombe, Galley Hill, Bury St. Edmunds (England), Fontéchevade (France), Ehringsdorf, Steinheim (Germany), Krapina (Yugoslavia), and Tabun (Mount Carmel, Palestine). Some of these have been included in Progressive Neandertals; others have been considered true sapiens; others have been removed from consideration because of uncertainties in the dates assigned to them. If the entire face and occiput of each of these early Pleistocene skulls had been recovered, many doubts about the resemblance to modern man would be allayed. The best documented of these fossils from the early Pleistocene do not have all the features upon which an unequivocal diagnosis of sapiens-like or Neandertal-like may be based. The skull cap alone is not sufficient, and much of the discussion of whether these forms are sapiens-like depends upon how one reconstructs the area around the brow ridges. Except for the Mount Carmel fossils, there is no unambiguous evidence that these forms must be ruled out of the population which produced the more extreme Neandertal faces.

The "Neandertal Problem" is essentially a dilemma of interpretation. The evolutionary interpretation of the morphology of these fossils depends upon geographical, chronological, and ideological factors as well as upon anatomy. Perhaps the crucial question really is, can Neandertal faces such as are found in the Classic Neandertals be assigned to the population that is our own direct ancestor? We believe the answer is yes. Many of those who believe the answer is no are adherents of catastrophism as an evolutionary process. The australopithecines, pithecanthropines, and Neandertals are believed to be the remains of populations whose extinctions were the result of a catastrophic meeting with *Homo sapiens*. The meeting was a catastrophe for each at separate times and in separate places. The explicit expression of this view was made by the French paleontologist, Teilhard de Chardin. He stated that neither Peking man (sinanthropines) nor Neandertal man had any genetic connection with *Homo sapiens*. *H. sapiens* swept them away! The replacement of Neandertals in Europe by invading hordes of *Homo sapiens* was, indeed, a catastrophe for the Neandertals, if such an invasion took place. This interpretation of the fossil record is conditioned by a highly anthropocentric attitude toward human evolution and is no longer widely held among physical anthropologists.

We reject the notion that the species *Homo sapiens* met a separate Neandertal species and wiped it out, but we do not reject the notion that populations moved from the Near East and Mediterranean into northwestern Europe. The archeological and geological evidence suggests that populations from the Middle East and the Mediterranean regions moved into Europe at various times during the late middle and upper Pleistocene. There was probably considerable mixture of genes as the result of migrations. The morphological variability of the late middle and upper Pleistocene hominids of Europe reflects such movements. The Neandertals may have been absorbed and dominated culturally much as the American Indians were, beginning in the sixteenth century, by European settlers and conquerors. When examining the cultural artifacts found with Neandertal skeletons, archeologists discern a cultural shift from a Mousteroid (early upper Pleistocene) complex to a later Perigordoid complex. If this shift is correlated with consistent changes in the morphology of the skeleton, then perhaps a kind of cultural catastrophe did overcome the early human inhabitants of northwestern Europe. But it is not necessary to interpret the appearance of new kinds of cultural artifacts, tool types, as coincident with the appearance of a new species.

The geographical area through which the Neandertal populations were distributed is better known today than it was in the past. Instead of being a narrowly circumscribed area of northwestern Europe, the range of middle and upper Pleistocene hominid fossils includes Asia, the Near East, and Africa. A large number of hominid fossil specimens comes from middle and upper Pleistocene beds in Europe (Table 9.4). The number of individual European fossil hominid specimens is much larger than the number of specimens from any other region. A large amount of variability in the skull and skeleton, such as is found in all the Primates, is found among these specimens. They are, in one sense, a biased sample of once-living Pleistocene hominid populations. A great number of scholars concentrated on the geological, stone-tool, and fossil sequences of the Pleistocene of Europe. Granted that it is pleasanter, easier, and more convenient to work in western Europe than in many other regions, Europe is not the world. Europe was not the center of the major events, particularly the major transitional events, of primate and specifically human evolution. Europe is a periphery, a cul-de-sac with respect to human evolution, as was clearly suggested by paleontologists of the late nineteenth century and shown by the finds in East Africa, South Africa, and China. The fossils found in Europe do not represent the entire middle and upper Pleistocene population of the genus *Homo* (probably species *sapiens*). These European fossils probably represent one of the allopatric populations of which the Pleistocene species of the genus *Homo* consisted. And to make it more difficult for us, the European Neandertals are an allochronic population as well.

Each Neandertal fossil, relative to every other, must be placed in a temporal sequence, for the fossils occur within a very short time span (from 400,000 to 50,000 years ago). A probable error of 50,000 or 100,000 years is a significant percentage of the total time available. Our notion of the duration of the Pleistocene and the time available during which the evolutionary changes from pithecanthropines to modern *H. sapiens* occurred has been radically altered by chronometric dating methods. The Pleistocene probably lasted for a much longer time than was once thought—between 1.5 and 2 million years. Until the past decade, evolutionary events had to be crammed into a total Pleistocene that spanned less than 600,000 years. Pithecanthropines, Neandertals, and sapiens-like fossils were believed to have evolved in that short span of time. Now we think perhaps as much as a million years were available for these developments. We need no longer assume an unusually rapid rate for hominid evolution. The hominid fossils of the middle and upper Pleistocene are samples of an allochronic and allopatric species, probably all properly classified as *Homo sapiens*. Morphological variation must be interpreted with regard to their range in time and in space.

The concepts of subspecies and race have been applied somewhat recklessly to the fossils of the European Pleistocene. The racial characters of living men are not reflected in the skeletons of all the individual members of the various racial groups. Variations among individuals in a population are great, and individual skulls are often mistakenly identified. Some traits of the skulls of Australian aborigines and Eskimos are, in certain individuals, quite distinctive. Yet it is not possible for most skulls from African or European populations to be identified positively as belonging to one or the other population. Certain Neandertal skulls (the Grimaldi specimens from Italy) have been identified by several eminent paleoanthropologists as resembling modern African skulls. Yet other equally eminent anthropologists have declared them to be indistinguishable from skulls of populations that are lumped in the so-called Mediterranean race. A famous skull, Chancelade, has been diagnosed as Eskimo, yet other opinions hold that it is but a variant of the Cro-Magnon skeletal population of western Europe. Le Gros Clark summed up the problem when he pointed out that unquestionably there were racial or population differences related to geographical isolation and separation among Pleistocene populations. The skull may be diagnostic of species differences, but not of differences at lower taxonomic levels. The anatomical evidence available is not sufficient, nor is it diagnostic, for relating differences among living human populations to differences found among Pleistocene fossil skulls.

It is difficult to find out from those concerned what constitutes the morphological complex that makes Neandertal distinct from other members of the genus *Homo*. At one time, Neandertal fossils were thought to be a special group, a distinct species distantly related to *Homo sapiens* because of certain special features. These include the very marked supraorbital torus that forms an uninterrupted shelf of bone above the eye sockets, a flattened cranial vault, no vertical forehead as in modern man, and absence of a chin eminence. A very small number of Neandertal fossils display taurodontism (teeth with enlarged pulp cavities and reduced roots). Fossil hominids of Pleistocene age not having these characteristics were considered non-Neandertal.

The discovery, at Mount Carmel in Palestine, of a group of Neandertal skulls and skeletons that included some individuals similar to modern man led to several new interpretations of the relationship of Neandertal to *Homo sapiens*. The Mount Carmel group was thought by some to be a population that showed a mixture between two racial types. Another interpretation stated that here was Neandertal man being transformed, in orderly sequence, to modern man. The most plausible interpretation, in light of the genetical concept of species, is that this group shows the variation that existed within the species *Homo sapiens* during the last stages of the Pleistocene.

Fossil hominids are not obvious candidates for genetic analyses. Yet the attitude of mind that produced the genetical concept of species would be a valuable addition to the points of view which have been brought to studies of Neandertal and other Pleistocene hominids. Instead of explaining them away as aberrant isolated populations that became extinct, some attention should be given an alternative hypothesis. The Neandertal fossils are representatives of an allopatric, allochronic species—*Homo sapiens*. We have a reasonably large sample in the specimens of middle and upper Pleistocene hominid fossils. The total sample comes from a wide area of the Old World and from a variety of Pleistocene strata; thus temporal and geographical as well as morphological variation may be studied. The question to be asked is: Does this sample represent a population of once-living animals actually or potentially capable of interbreeding? We have not been convinced that the answer is no. If we accept the implications of our discussion and accept this last interpretation, the "Neandertal Problem" disappears.

During the Pleistocene the evolutionary divergence of the Hominidae was completed. The adaptive radiation of the genus *Homo* began and, presumably, still continues. The small, erect australopithecines, awkwardly bipedal, were the first population of an evolutionary adaptive radiation that is now preparing to radiate into outer space in search of new econiches.

SUGGESTED READINGS AND SOURCES

Broom, R., and J. T. Robinson, Thumb of the Swartkrans ape-man. *Nature* **164,** 841 (1949).
Brace, C. L., The fate of the "classic" Neanderthals: a consideration of hominid catastrophism. *Current Anthropol.* **5,** 3 (1964).
Clark, W. E. Le G., The Rhodesian man. *Man* **28,** 206 (1928).
Clark, W. E. Le G., General features of the Swanscombe skull bones and endocranial cast. *J. Roy. Anthropol. Inst.* **68,** 58 (1938).
Clark, W. E. Le G., *The Antecedents of Man* (Second edition). Edinburgh Univ. Press, Edinburgh (1962).
Clark, W. E. Le G., *The Fossil Evidence for Human Evolution* (Second edition). Univ. of Chicago Press, Chicago (1964).
Cooke, H. B. S., Mammals, ape-men, and Stone Age men in southern Africa. *S. African Archeol. Bull.* **7,** 59 (1952).
Dart, R., *Australopithecus africanus:* the man-ape of South Africa. *Nature* **115,** 195 (1925).
Dart, R., The Makapansgat proto-human *Australopithecus prometheus. Am. J. Phys. Anthropol.* **6,** 259 (1948).
Day, M. H., and J. R. Napier, Hominid fossils from Bed I, Olduvai Gorge, Tanganyika. *Nature* **201,** 967 (1963).
Dubois, E., *Pithecanthropus erectus, eine Menschenähnliche Übergangsform von Java.* Landes Druckerei, Batavia (1894).
Dubois, E., The distinct organization of *Pithecanthropus erectus* now confirmed from other individuals of the described species. *Proc. Koninkl. Akad. Wetenschap. Amsterdam* **35,** 716 (1932).
Flint, R. F., *Glacial and Pleistocene Geology.* Wiley, New York (1957).
Haeckel, E., *Natürliche Schöpfungsgeschichte.* Georg Reimer, Berlin (1873).
Hooton, E. A., *Up from the Ape* (Revised edition). Macmillan, New York (1947).
Howell, F. C., The place of Neanderthal man in human evolution. *Am. J. Phys. Anthropol.* **9,** 379 (1951).
Howell, F. C., Pleistocene glacial ecology and the evolution of "classic Neanderthal" man. *Southwestern J. Anthropol.* **8,** 377 (1952).
Koenigswald, G. H. R. von, *Hundert Jahre Neanderthaler.* Wenner-Gren, New York (1958).
Leakey, L. S. B., A new fossil skull from Olduvai. *Nature* **184,** 491 (1959).
Leakey, L. S. B., *Olduvai Gorge 1951–1961.* Cambridge Univ. Press, London (1965).
Leakey, L. S. B., J. F. Evernden, and G. H. Curtis, Age of Bed I, Olduvai Gorge, Tanganyika. *Nature* **191,** 478 (1961).

Leakey, L. S. B., and M. D. Leakey, Recent discoveries of fossil hominids in Tanganyika: at Olduvai and near Lake Natron. *Nature* **202**, 5 (1964).

Leakey, L. S. B., P. V. Tobias, and J. R. Napier, A new species of the genus *Homo* from Olduvai Gorge. *Nature* **202**, 7 (1964).

Napier, J. R., Fossil metacarpals from Swartkrans. *Fossil Mammals of Africa*, No. 17. Brit. Museum (Nat. Hist.), London (1959).

Pilbeam, D. R., and E. L. Simons, Some problems of hominid classification. *Am. Sci.* **53**, 237 (1965).

Robinson, J. T., The australopithecine-bearing deposits of the Sterkfontein area. *Ann. Transvaal Museum* **22**, 1 (1952).

Robinson, J. T., The nature of *Telanthropus*. *Nature* **171**, 33 (1953).

Robinson, J. T., *Homo "habilis"* and the australopithecines. *Nature* **205**, 121 (1965).

Schultz, A. H., Die Schädelkapazität männlicher Gorillas und ihr Höchswert. *Anthrop. Anz.* **25**, 197 (1962).

Teilhard de Chardin, P., *Fossil Man: Recent Discoveries and Present Problems*. Henri Vetch, Peking (1943).

Tobias, P. V., The Olduvai Bed I hominine with special reference to its cranial capacity. *Nature* **202**, 3 (1964).

Wallace, A. R., *The Geographical Distribution of Animals*, Vol. 1–2. Macmillan, London (1876).

Weidenreich, F., The *Sinanthropus* population of Chou Kou Tien. *Bull. Geol. Soc. China* **14**, 427 (1935).

Weidenreich, F., The mandibles of *Sinanthropus pekinensis:* a comparative study. *Paleont. Sinica*, Ser. D, **7**, Fasc. 3, 1 (1936).

Weidenreich, F., Six lectures on *Sinanthropus pekinensis* and related problems. *Bull. Geol. Soc. China* **19**, 1 (1939).

Weiner, J. S., K. P. Oakley, and W. E. Le G. Clark, The solution of the Piltdown problem. *Bull. Brit. Museum (Nat. Hist.)* **2**, 139 (1953).

These two references are, respectively, a popular account of the evolution of human nature and a fictional account of Neandertal catastrophism.

Ardrey, R., *African Genesis*. Dell, New York (1961).

Golding, W., *The Inheritors*. Harcourt, Brace and World, New York (1962).

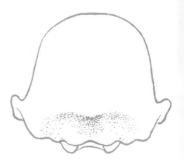

Systematics of the Hominoidea and the Hominidae

10

THE CLASSIFICATION OF THE HOMINOIDEA and the selection of valid names for the categories in the classification have always been subjects fraught with confusion, error, and passion. We believe it is possible to make a sensible classification of fossil man and other fossil hominoids—one that will reflect phylogenetic affinities, at least to some degree, and one that will follow the rules in the *International Code of Zoological Nomenclature*.

We have shown how many names were given to the fossil fragments of what are essentially a single group—the dryopithecines (Table 8.2). The multitude of names completely confused those who wished to understand the phylogenetic affinities of this group of animals. A similar confusion, to put it mildly, has existed among students of the hominids. The reader will find the various names given to the three groups of Pleistocene hominids—australopithecines, pithecanthropines, and sapiens—in Tables 9.1 and 9.3 and on pages 173–176. In some cases, *"Homo habilis"* and *Homo erectus*, the names imply the degree of genetic distinction that exists between two species. In other cases, *"Zinjanthropus boisei"* and *Australopithecus robustus*, the names imply the degree of genetic isolation that exists between two living genera.

Most anthropologists have not realized what the biological implications of assigning these names are. Before we erect new species and genera for fossil

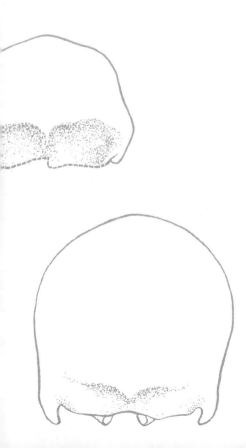

primates, it is incumbent upon us to demonstrate that proposed taxa are significantly distinct from already accepted taxa. The distinction must be demonstrated in a number of characteristics. We must take account of the variability in the same anatomical systems of related living species and genera when making a diagnosis for a new taxon. When new and different specific and generic names are bestowed on fossil hominids, when new taxa are constructed, genetic distinctions are inferred. When two individual fossils are given two different specific names, he who names them infers that two populations once existed and these could not have been members of one genetical species. The formal Linnaean binomials imply that the two newly named fossils are samples from two populations. If different generic names are bestowed on two groups of fossil remains, the inference is that the populations were genetically incompatible. A long period of isolation, probably a consequence of geographical separation, would have been necessary for development of complete genetic distinctness. As we have said and as others have pointed out, it is highly improbable that isolation of the required magnitude occurred. So large a number of geographically isolated hominid populations, as the large number of names given in the past implies, is quite improbable.

Another of the difficulties in making classifications of the Hominidae is the variety of opinions about the diagnosis or definition of the genus *Homo* and the family Hominidae. According to the International Code a diagnosis or definition accompanies the naming of a taxon. During the past 15 years several definitions have been published of the hominoid taxa with which we are concerned. These appear in Table 10.1. Some revisions are necessary. The revised definition of the genus *Homo* will be presented after we discuss some of the oustanding problems of hominid systematics.

TABLE 10.1
Definitions of Members of the Hominoidea

Family Pongidae (Le Gros Clark, 1964)	Family Hominidae (Le Gros Clark, 1964)
A subsidiary radiation of the Hominoidea distinguished from the Hominidae by the following evolutionary trends: progressive skeletal modifications in adaptation to arboreal brachiation, shown particularly in a proportionate lengthening of the upper extremity as a whole and of its different segments; acquisition of a strong opposable hallux and modification of morphological details of limb bones for increased mobility and for the muscular developments related to brachiation; tendency to relative reduction of pollex; pelvis retaining	A subsidiary radiation of the Hominoidea distinguished from the Pongidae by the following evolutionary trends: progressive skeletal modifications in adaptation to erect bipedalism, shown particularly in a proportionate lengthening of the lower extremity, and changes in the proportions and morphological details of the pelvis, femur, and pedal skeleton related to mechanical requirements of erect posture and gait and to the muscular development associated therewith; preservation of well-developed pollex; ultimate loss of op-

TABLE 10.1 (*Continued*)

the main proportions characteristic of quadrupedal mammals; marked prognathism, with late retention of facial component of premaxilla and sloping symphysis; development (in larger species) of massive jaws associated with strong muscular ridges on the skull; nuchal area of the occiput becoming extensive, with relatively high position of the inion; occipital condyles retaining a backward position well behind the level of the auditory apertures; only a limited degree of flexion of basicranial axis associated with maintenance of low cranial height; cranial capacity showing no marked tendency to expansion; progressive hypertrophy of incisors with widening of symphysial region of mandible and ultimate formation of "simian shelf"; enlargement of strong conical canines interlocking in diastemata and showing distinct sexual dimorphism; accentuated sectorialization of first lower premolar with development of strong anterior root; postcanine teeth preserving a parallel or slightly forward divergent alignment in relatively straight rows; first deciduous molar retaining a predominantly unicuspid form; no acceleration in eruption of permanent canine.

posability of hallux; increasing flexion of basicranial axis associated with increasing cranial height; relative displacement forward of the occipital condyles; restriction of nuchal area of occipital squama, associated with low position of inion; consistent and early ontogenetic development of a pyramidal mastoid process; reduction of subnasal prognathism, with ultimate early disappearance (by fusion) of facial component of premaxilla; diminution of canines to a spatulate form, interlocking slightly or not at all and showing no pronounced sexual dimorphism; disappearance of diastemata; replacement of sectorial first lower premolars by bicuspid teeth (with later secondary reduction of lingual cusp); alteration in occlusal relationships, so that all the teeth tend to become worn down to a relatively flat even surface at an early stage of attrition; development of an evenly rounded dental arcade; marked tendency in later stages of evolution to a reduction in size of the molar teeth; progressive acceleration in the replacement of deciduous teeth in relation to the eruption of permanent molars; progressive "molarization" of first deciduous molar; marked and rapid expansion (in some of the terminal products of the hominid sequence of evolution) of the cranial capacity, associated with reduction in size of jaws and area of attachment of masticatory muscles and the development of a mental eminence.

Australopithecus (Le Gros Clark, 1964)	*Pithecanthropus* (Le Gros Clark, 1955)
A genus of the Hominidae distinguished by the following characters: relatively small cranial capacity, ranging from about 450 to well over 600 cc.; strongly built supra-orbital ridges; a tendency in individuals of larger varie-	A genus of the Hominidae characterized by a cranial capacity with a mean value of about 1,000 cc.; marked platycephaly, with little frontal convexity; massive supra-orbital tori; pronounced postorbital constriction; opisthocranion

TABLE 10.1 (*Continued*)

ties for the formation of a low sagittal crest in the frontoparietal region of the vertex of the skull (but not associated with a high nuchal crest); occipital condyles well behind the mid-point of the cranial length but on a transverse level with the auditory apertures; nuchal area of occiput restricted, as in *Homo;* consistent development (in immature as well as mature skulls) of a pyramidal mastoid process of typical hominid form and relationships; mandibular fossa constructed on the hominid pattern but in some individuals showing a pronounced development of the postglenoid process; massive jaws, showing considerable individual variation in respect of absolute size; mental eminence absent or slightly indicated; symphysial surface relatively straight and approaching the vertical; dental arcade parabolic in form with no diastema; spatulate canines wearing down flat from the tip only; relatively large premolars and molars; anterior lower premolar bicuspid with subequal cusps; pronounced molarization of first deciduous molar; progressive increase in size of permanent lower molars from first to third; the limb skeleton (so far as it is known) conforming in its main features to the hominid type but differing from *Homo* in a number of details, such as the forward prolongation of the region of the anterior superior spine of the ilium and a relatively small sacroiliac surface, the relatively low position (in some individuals) of the ischial tuberosity, and the marked forward prolongation of the intercondylar notch of the femur.

coincident with the inion; vertex of skull marked by sagittal ridge; mastoid process variable, but usually small; thick cranial wall; tympanic plate thickened and tending toward a horizontal disposition; broad, flat nasal bones; heavily constructed mandible, lacking a mental eminence; teeth large, with well-developed basal cingulum; canines sometimes projecting and slightly interlocking, with small diastema in upper dentition; first lower premolar bicuspid with subequal cusps; molars with well-differentiated cusps complicated by secondary wrinkling of the enamel; second upper molar may be larger than the first, and the third lower molar may exceed the second in length; limb bones not distinguishable from those of *H. sapiens.*

Homo (Le Gros Clark, 1964)	*Homo* (Leakey, Tobias, and Napier, 1964)
A genus of the family Hominidae, distinguished mainly by a large cranial capacity with a mean value of more than 1,100 cc. but with a range of vari-	A genus of the Hominidae with the following characters: the structure of the pelvic girdle and of the hind-limb skeleton is adapted to habitual erect

TABLE 10.1 (*Continued*)

ation from about 900 cc. to almost 2,000 cc.; supra-orbital ridges variably developed, becoming secondarily much enlarged to form a massive torus in the species *H. erectus* and *H. neanderthalensis,* and showing considerable reduction in *H. sapiens;* facial skeleton orthognathous or moderately prognathous; occipital condyles situated approximately at the middle of the cranial length; temporal ridges variable in their height on the cranial wall, but never reaching the mid-line to form a sagittal crest; mental eminence well marked in *H. sapiens* but absent in *H. erectus* and feeble or absent in *H. neanderthalensis;* dental arcade evenly rounded, usually with no diastema; first lower premolar bicuspid with a much reduced lingual cusp; molar teeth rather variable in size, with a relative reduction of the last molar; canines relatively small, with no overlapping after the initial stages of wear, limb skeleton adapted for a fully erect posture and gait.

posture and bipedal gait; the fore-limb is shorter than the hind-limb; the pollex is well developed and fully opposable and the hand is capable not only of a power grip but of, at the least, a simple and usually well developed precision grip; the cranial capacity is very variable but is, on the average, larger than the range of capacities of members of the genus *Australopithecus,* although the lower part of the range of capacities in the genus *Homo* overlaps with the upper part of the range in *Australopithecus;* the capacity is (on the average) large relative to body-size and ranges from about 600 cc. in earlier forms to more than 1,600 cc.; the muscular ridges on the cranium range from very strongly marked to virtually imperceptible, but the temporal crests or lines never reach the midline; the frontal region of the cranium is without undue post-orbital constriction (such as is common in members of the genus *Australopithecus*); the supra-orbital region of the frontal bone is very variable, ranging from a massive and very salient supra-orbital torus to a complete lack of any supra-orbital projection and a smooth brow region; the facial skeleton varies from moderately prognathous to orthognathous, but it is not concave (or dished) as is common in members of the Australopithecinae; the anterior symphyseal contour varies from a marked retreat to a forward slope, while the bony chin may be entirely lacking, or may vary from a slight to a very strongly developed mental trigone; the dental arcade is evenly rounded with no diastema in most members of the genus; the first lower premolar is clearly bicuspid with a variably developed lingual cusp; the molar teeth are variable in size, but in general are small relative to the size of these teeth in the genus *Australopithecus;* the size of the last upper molar is highly variable, but

TABLE 10.1 *(Continued)*

it is generally smaller than the second upper molar and commonly also smaller than the first upper molar; the lower third molar is sometimes appreciably larger than the second; in relation to the position seen in the Hominoidea as a whole, the canines are small, with little or no overlapping after the initial stages of wear, but when compared with those members of the genus *Australopithecus*, the incisors and canines are not very small relative to the molars and premolars; the teeth in general, and particularly the molars and premolars, are not enlarged bucco-lingually as they are in the genus *Australopithecus;* the first deciduous lower molar shows a variable degree of molarization.

Homo neanderthalensis (Le Gros Clark, 1964)	*Homo sapiens* (Le Gros Clark, 1964)
The skull is distinguished by an exaggerated development of a massive supra-orbital torus, forming an uninterrupted shelf of bone overhanging the orbits (with complete fusion of the ciliary and orbital elements); absence of a vertical forehead; marked flattening of the cranial vault (platycephaly); relatively high position of the external occipital protuberance and the development (usually) of a strong occipital torus; a massive development of the naso-maxillary region of the facial skeleton, with an inflated appearance of the maxillary wall; a heavy mandible, lacking a chin eminence; a pronounced tendency of the molar teeth to taurodontism (that is, enlargement of the pulp cavity with fusion of the roots); a relatively wide sphenoidal angle of the cranial base (about 130° . . .); angular contour of the occiput; certain morphological details of the ear region of the skull (including the rounded or trans-	A species of the genus *Homo* characterized by a mean cranial capacity of about 1,350 cc.; muscular ridges on the cranium not strongly marked; a rounded and approximately vertical forehead; supra-orbital ridges usually moderately developed and in any case not forming a continuous and uninterrupted torus; rounded occipital region with a nuchal area of relatively small extent; foramen magnum facing directly downward; the consistent presence of a prominent mastoid process of pyramidal shape (in juveniles as well as adults), associated with a well-marked digastric fossa and occipital groove; maximum width of the calvaria usually in the parietal region and axis of glabello-maximal length well above the level of the external occipital protuberance; marked flexion of the sphenoidal angle, with a mean value of about 110°; jaws and teeth of relatively small size, with retrogressive features in the last molars; maxilla having a con-

TABLE 10.1 (*Continued*)

versely elliptical shape of the auditory aperture, the conformation of the mastoid process, and of the mandibular fossa); a slightly backward disposition of the foramen magnum; and a large cranial capacity (1,300–1,600 cc.). The limb skeleton is characterized by the coarse build of the long bones (which show pronounced curvatures and relatively large extremities), the morphological features of the pubic bone, and by certain of the morphological details of the talus and calcaneus bones of the ankle, which are said to be somewhat "simian" in character. In addition, the vertebrae of the cervical region of the spine in some cases show a striking development of the spinous processes, which, however, though somewhat simian in appearance, does not exceed the extreme limits of variation in *H. sapiens*.

cave facial surface, including a canine fossa; distinct mental eminence; eruption of permanent canine commonly preceding that of the second molar; spines of cervical vertebrae (with the exception of the seventh) usually rudimentary; appendicular skeleton adapted for a fully upright posture and gait; limb bones relatively slender and straight.

First come the Hominoidea. The nomenclature of the taxa in this superfamily reflects or should reflect their systematic relationships. The phylogenetic affinities and distinctions of these animals are implied by the names they are given. The evolutionary segregation of the gibbons (*Hylobates*) occurred a very long time ago. It is probable that this occurred in the late Eocene or early Oligocene (Chapter 7). Because modern gibbons have many characteristics that make them quite different and distinct from the other apes, they have been placed in a separate subfamily, the Hylobatinae, of the family Pongidae. There are fossil primates referable to the same subfamily from Oligocene times. The Miocene fossil *Pliopithecus* has a gibbonlike dentition, although not the elongated forelimbs of modern *Hylobates*. The separation of the gibbons from the rest of the pongids is probably quite ancient.

The other apes, the orangutan (*Pongo*) and the gorilla and the chimpanzee (*Pan*), are placed in a subfamily, the Ponginae, a second distinct evolutionary lineage of the Pongidae. The Ponginae most likely became a fully differentiated and distinct group during the early Miocene. The discovery and description of *Aegyptopithecus* suggest that the differentiation of Ponginae may have occurred even earlier. The dryopithecines (Chapter 8) are placed in another subfamily of the Pongidae, the Dryopithecinae.

The lineage leading directly to man is placed in a separate family, the Hominidae. The Hominidae achieved evolutionary segregation from the Pongidae sometime in the Miocene or possibly in the Oligocene. Man's lineage (Hominidae) is distinct from that of the other Hominoidea at a higher taxonomic level than the level at which the separate lineages of the Pongidae are distinct from each other.

How do we decide at what taxonomic level to make the divisions we are discussing? Strange as it may seem to students and others who believe that there are no ambiguous answers in science, or ought not to be, this decision is largely a matter of opinion formed from admittedly fragmentary materials. Some people prefer to emphasize distinctions, others continuities. It is quite clear that *Oreopithecus*, for example, should be separated at a high level from the other subgroups of the Hominoidea. The orangutan, on the other hand, is still considered a member of the subfamily Ponginae. The distinctness of the orangutan from the African great apes has long been recognized. But the orangutan, unlike the gibbon, has no obvious fossil ancestor. It is not at all unlikely that the morphological specializations that distinguish orangutans from gorillas and chimpanzees are relatively recent developments.

The classification of the superfamily Hominoidea follows. It is believed to reflect the evolutionary affinities within this group as they are understood today.

SUPERFAMILY: Hominoidea
    FAMILY: Oreopithecidae
        GENERA: *Apidium, Oreopithecus, Parapithecus*
    FAMILY: Pongidae
        SUBFAMILY: Ponginae
            GENERA: *Pan, Pongo*
        SUBFAMILY: Hylobatinae
            GENERA: *Pliopithecus, Aeolopithecus, Hylobates*
        SUBFAMILY: Dryopithecinae
            GENERA: *Dryopithecus, Gigantopithecus, Aegyptopithecus*
    FAMILY: Hominidae
        GENERA: *Ramapithecus, Homo* [=*Australopithecus*]

An alternative classification is possible.

SUPERFAMILY: Hominoidea
    FAMILY: Oreopithecidae
        GENERA: *Apidium, Oreopithecus, Parapithecus*
    FAMILY: Hylobatidae
        GENERA: *Pliopithecus, Aeolopithecus, Hylobates*
    FAMILY: Pongidae
        SUBFAMILY: Ponginae
            GENERA: *Pan, Pongo*
        SUBFAMILY: Dryopithecinae
            GENERA: *Dryopithecus, Gigantopithecus, Aegyptopithecus*
    FAMILY: Hominidae
        GENERA: *Ramapithecus, Australopithecus, Homo*

There is no essential difference between the two classifications except that the distinctness of the gibbons is emphasized in the latter by placing them in a separate family.

It is well to note here that the siamang of southeast Asia (formerly the genus *Symphalangus*) is now considered to belong to the genus *Hylobates*, and the genus *Gorilla* is no longer considered distinct from the genus *Pan*. To use the expressive words of taxonomists, *Gorilla* and *Symphalangus* have been sunk in *Pan* and *Hylobates*, respectively.

It has been suggested in recent years that *Pan* should be included in the Hominidae because of certain biochemical and cytogenetic similarities which indicate there is greater affinity between the two species of the genus *Pan* with man than there is between these species and the orangutan or the gibbon, *Pongo* and *Hylobates*. This is not exactly a new idea. For at least 80 years the African great apes have been considered to have a greater evolutionary affinity with each other and man than with the Asiatic apes and greater than man has with the Asiatic apes. The principal reason for excluding *Pan* from the Hominidae is that *Homo* is the most distinctive of all the hominoids. It is distinct anatomically, adaptively, chemically, and neurologically. Primatologists in the past have almost unanimously considered this distinction sufficiently great to separate the Hominidae at the family level from the Pongidae.

There are a number of other reasons for separating *Pan* from *Homo* at the family level. *Pan* is not directly ancestral to *Homo*. The common ancestor of the two was unquestionably much more like modern members of the genus *Pan* than like the genus *Homo*. This does not suggest that *Pan* is a member of the Hominidae but that the common ancestor should be in a family distinct from the Hominidae. There are more theoretical reasons that argue against putting any of the pongids in the same family with hominids. Arguments for putting them in the same family seem to be based upon the fact that they appear to have had a relatively recent ancestor in common. Such lumping in a phylogenetically oriented classification cannot be made. If we adhered strictly to such a principle in classification, we would put more and more animals together in the same family so that the whole animal kingdom, down to the primordial unicellular organism, would be in the family Hominidae.

It must be noted that both ways of handling the classification of *Pan* (placing it either in the Pongidae or the Hominidae) are equally consistent with what we can deduce about hominid phylogeny from fossils. Nevertheless, the animals included in the genus *Homo* and in the family Hominidae have diverged radically from the pongids, and this should be reflected in the classification and nomenclature. By putting *Ramapithecus* into the Hominidae (Chapter 8), the argument about how recently pongids and hominids diverged may be resolved. *Ramapithecus* is evidence that anatomical divergence between pongids and hominids is actually not such a very recent thing at all.

Now comes the problem of handling the nomenclature and classification of the genus *Homo*. With whom does the genus *Homo* begin? As we pointed out in discussing the "Neandertal Problem," the evidence from hominid fossils and our concept of the way the sequences of hominid fossils relate to primate and hominid phylogeny suggest that we are viewing an allochronic lineage. In an allochronic lineage, the division into genera and families is difficult, but the division into species is even more so. The first group of fossils that might qualify for membership in the genus *Homo* are the south African man-apes, the australopithecines. The australopithecines have been placed in different genera at various times by various authors (Table 9.1). The three genera (*"Paranthropus," "Plesianthropus,"* and *Australopithecus*) proposed by the original finders are two too many. The fossils are better placed in a single genus, either *Australopithecus*, with a single species *A. africanus*, or *Homo*, with a single species *H. africanus*. Recently the discovery of other australopithecines has led to the creation of new species and genera. *"Telanthropus," "Zinjanthropus,"* and *"Homo habilis"* have been suggested as valid taxa. Many problems are involved here: most of them concern the interpretation of the morphology of the various recently discovered fossils and their relative ages. Leakey and his co-workers gave much weight to the fact that the teeth of the fossil called *"H. habilis"* were much smaller than those of the australopithecines found in South Africa. The relative proportions of these teeth, too, were considered significantly smaller than the teeth of australopithecines. If significant, this would make the teeth closer to those of modern man than to those of the South African man-apes. Le Gros Clark noted that the differences in the size of these teeth are not significant from a taxonomic point of view and the differences in relative proportions seem to be trivial. It is significant that such competent anatomists and taxonomists as Le Gros Clark and Campbell have pointed out that most if not all of the traits supposed to make *"H. habilis"* generically distinct from the australopithecines are in fact not distinct traits.

The hand of the *"H. habilis"* fossil is supposed to be different. However predictions were made (from a bone of the thumb, a metacarpal found in South Africa) about the appearance of the whole hand of *Australopithecus*, and the hand of *"H. habilis"* fits these predictions remarkably well. The foot skeleton of the *"H. habilis"* find was considered advanced and man-like by Leakey and his colleagues. Yet Napier's studies of the talus of the australopithecine foot indicated it is very similar to the talus of *"H. habilis."* Studies of the leg bones (tibia and fibula) of *"Zinjanthropus"* suggest that there were only minor differences in the way those bones were arranged in the ankle region from the arrangement of the same bones in modern man. This indicates that the foot skeleton of *"Zinjanthropus"* is that of an australopithecine and was already quite as advanced as it was in *"H. habilis."* Thus *"Zinjanthropus"* is also very similar to the South African man-apes.

The supraorbital torus and the postorbital constriction of the skull of *"H. habilis"* are similar to those present in australopithecine skulls. Thus the

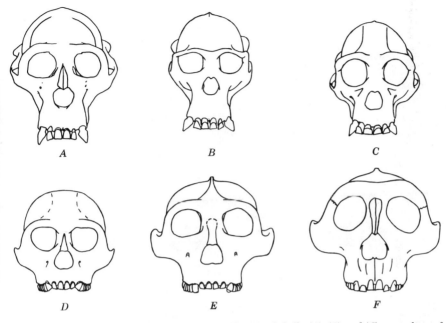

Fig. 10.1.   Variations in relative size and shape of hominoid skulls. (A), (B), and (C) are outlines of chimpanzee skulls. (D), (E), and (F) are outlines of australopithecine skulls, *Homo* [= *Australopithecus*] *africanus*, *Australopithecus* [= *Paranthropus*] *robustus*, and "*Zinjanthropus boisei*," respectively. (Drawings courtesy of E. L. Simons.)

traits supposed to make the "*H. habilis*" fossils distinct from the australopithecines are interpreted by many competent paleoanthropologists, anatomists, and taxonomists as not distinct at all. The "*H. habilis*" discoveries are not identical with the australopithecine finds of South Africa, but they are taxonomically the same. The differences are too small to justify the erection of a separate taxon for "*Homo habilis.*"

Some have suggested that there are two separate species or even two genera of australopithecines, the groups we have called *Homo* [=*Australopithecus*] *africanus* and *Australopithecus* [=*Paranthropus*] *robustus*. Some anthropologists argue that there is morphological and ecological evidence that implies the existence of two distinct hominid species. The robustus forms, with larger molar and premolar teeth, are presumably larger than the africanus forms. Although variability is apparent, the same kind of variability is evident in chimpanzee skulls (Fig. 10.1 and 10.2). No one would use the three chimpanzee skulls in the figures as the basis for erecting three taxa. (The skulls are from three chimpanzees of the same local population.) Yet three genera were proposed on the basis of the fossil australopithecine skulls.

One of the crucial points distinguishing the robustus from the africanus group is supposed to be different wear patterns on the cheek teeth. The

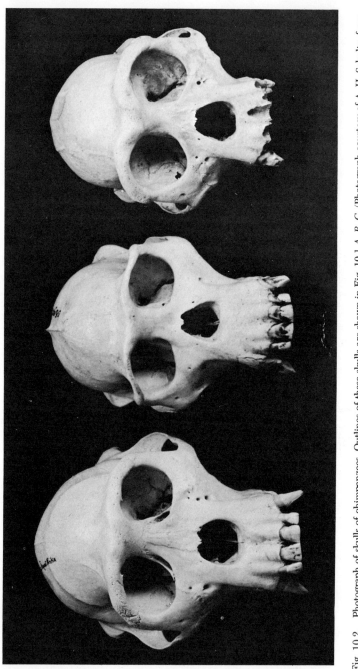

Fig. 10.2.   Photograph of skulls of chimpanzees. Outlines of these skulls are shown in Fig. 10.1 A, B, C. (Photograph courtesy of A. H. Schultz, from Schultz, 1963.)

wear patterns imply that the robustus group ate sandy grass stems and roots and the africanus forms were carnivorous. As the number of specimens grows larger, the distinctness of the wear patterns of the robust and not-so-robust forms grows less clear. The ecological differences between the two rest upon ambiguously interpretable data. After all, a diet of sandy meat may lead to wear patterns on molars and premolars similar to those produced by sandy grass roots. Specimens from Olduvai and other sites in East Africa have been assigned to the robustus group partly on the basis of deep wear patterns on upper and lower molars. These are similar to wear patterns on the cheek teeth of certain Australian aboriginals whose diet includes sandy unwashed grass and roots. The fossil jaws assigned to the taxon variously named *Australopithecus africanus,* *"Homo habilis,"* and *Homo africanus* have upper and lower molars with wear patterns that resemble those found among the Masai of East Africa who are almost exclusively meat eaters. No one has suggested that the Australian aboriginals and the Masai are two distinct species, let alone genera.

All of the australopithecines, those forms given various names such as *Australopithecus robustus, Australopithecus africanus,* and *"Homo habilis,"* have small canines and incisors. The australopithecine anterior dentition is smaller than dryopithecine and pongid anterior dentition. We infer that the anterior dentition was not used for tearing up vegetation. We further infer that the australopithecines used tools and probably used them in preparing their food.

The contention that the robustus forms are found only in geological deposits typical of wet climatic periods and the australopithecines only in strata typical of dry climatic periods does not hold up. The *"Zinjanthropus"* fossil of East Africa is a robustus form not found in a "wet" stratum.

On the whole, the case for two distinct species (or genera) of Hominidae in the early Pleistocene is very weak. Given two specimens of an evolving lineage that represent two populations living at different times, we should ask: Can the observed differences have arisen in the span of time that separates the two specimens? We can rephrase this question: Is there sufficient morphological space between the specimens? The Hominidae provide a good example of the concept of morphological space (Fig. 10.3). The morphological space between *Ramapithecus* of the Miocene-Pliocene and the pithecanthropines of the middle Pleistocene is rapidly filling. The hominid fossil discoveries of the past decade and the accurate determination of the ages of many of them fit into what is essentially a continuum. The species of the genus *Homo* that lived at the time of the Villafranchian fauna during the lower Pleistocene was undoubtedly more variable than the species living today. It is much more reasonable to infer racial variation or population variation from the australopithecine fossils than it is to infer separate species or genera.

Ecological considerations also argue against more than a single hominid genus or species at any time in the past. *Homo* (whether *H. africanus* or

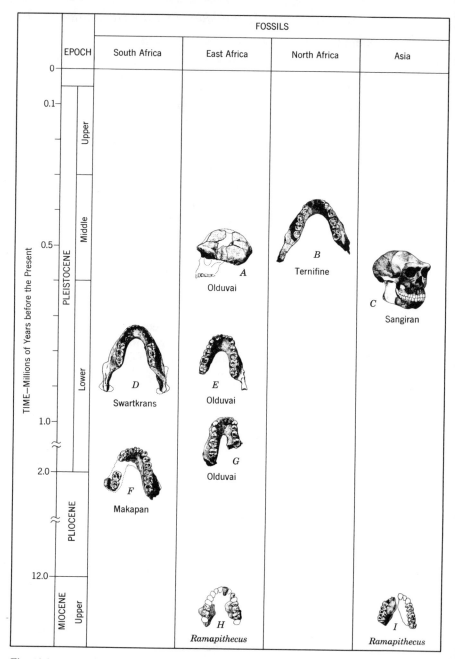

Fig. 10.3. Some fossil hominids of the Miocene, Pliocene, and Pleistocene epochs. (A) *Homo erectus*, Chellean skull, (B) mandible of *H. erectus*, (C) *H. erectus* IV, Djetis horizon, Java, (D) mandible of *H. africanus* [=*Telanthropus capensis* TYPE], (E) mandible of *H. africanus*, Olduvai hominid no. 13, "*H. habilis*," (F) mandible of *H. africanus*, (G) mandible of *H. africanus*, Olduvai hominid no. 7, "*H. habilis*," (H) composite maxilla of *Ramapithecus*, Fort Ternan, and (I) composite maxilla of *Ramapithecus*, Haritalyangar, India. Potassium-argon dates relevant to (A), (C), (E), (G), and (H) have been determined. (Adapted from Pilbeam and Simons, 1965.)

H. sapiens) occupies and is adapted to an extremely broad econiche. The probability is very low that a distinct species of the same genus coexisted in the same econiche. Mobile and adaptable populations of the genus *Homo*, residing in southern and eastern Africa, would have quickly assimilated or eliminated other populations of close relatives. If there were any clear and unequivocal cases of australopithecine and robustus forms occurring at the same place and at the same time, we might infer that two sympatric hominid species coexisted. But there is no unambiguous evidence today to support this inference. The notion that contemporary *Homo sapiens* and the many races into which it is divided developed from distinct species of Pleistocene Hominidae receives, today, no support from the fossil record.

The use of cranial capacity as a significant trait in classification was discussed earlier (Chapter 9). The estimates of cranial capacity of the East African *"H. habilis"* show that its capacity is larger than that of other australopithecines. Yet its capacity is still smaller than the cranial capacity that is supposed to be characteristic of a full-fledged member of the genus *Homo*.

"Telanthropus" is another fossil belonging to the same group as the australopithecines. It is an australopithecine from a more recent stratum than the original discoveries. It also has features that seem to forecast those found in the pithecanthropines. Thus it may also be considered an early pithecanthropine. Leakey and his co-workers see great similarities, indeed possible identities, between *"Telanthropus"* and *"H. habilis."* This makes the question of where *"Telanthropus"* fits of critical importance. Here is an intermediate form. In one sense it has a foot in each taxon. It is a form that links the pithecanthropines and australopithecines. How should we make a division? Should there be two taxa, *Australopithecus* and *Homo*, or should we have one? In 1950 Mayr suggested that there is no reason not to put the South African forms into the genus *Homo*. Leakey and his co-workers give us much evidence, although obscuring its importance with the controversy over the invalid name *"Homo habilis,"* that these forms are best placed in the genus *Homo*.

At the end of this chapter a revised definition for the genus *Homo* is given. If we accept this definition, the australopithecines are certainly members of the genus *Homo*. Since *Ramapithecus* is clearly a hominid, at what point do we draw the line between the genus *Homo* and some earlier genus of Hominidae? We believe it will become more difficult to draw such a line as more discoveries are made. The inclusion of the australopithecines in the genus *Homo* means they must have a name, and the correct name, by priority, is *Homo africanus*.

The diagnoses and definitions listed in Table 10.1 stress certain features of the skeleton. This is quite proper since all we have are skeletal remains. However it is possible to introduce another character into the definition, one that is unique to man. Toolmaking, symboling, and language are all part of an advanced functional development of the primate nervous system.

It is reasonable to construct a definition of the genus *Homo* that includes as one of its criteria the consequences of this significant neurological advance—toolmaking. Goodall's discovery that chimpanzees have a kind of toolmaking ability does not alter the value of adding this activity, deduced from the fossil and tool remains, to the definition of *Homo*. No one in his right mind is going to confuse a chimpanzee with a member of the genus *Homo*. Neither is he going to confuse the incipient toolmaking of the chimpanzee with the kind of toolmaking and other behaviors implied by the assemblage of stone artifacts found with early man.

Next comes the problem of the pithecanthropines. They usually have been considered to be members of a distinct genus, *Pithecanthropus,* or to be a separate species of the genus *Homo, H. erectus.* There is no good reason not to consider them part of the species *H. sapiens.* But most scholars are reluctant to do this, perhaps because the large brow ridges of the Chinese *H. erectus* fossils distress some paleoanthropologists. To have as close lineal relatives such "brutal" looking skeletons is more than some who belong to the species *H. sapiens* can bear. But we must become accustomed to this notion, for they are close relatives.

The Mauer population (deduced from the so-called Heidelberg mandible) and its relationship to Neandertal populations has puzzled a number of anthropologists and paleontologists. The Mauer mandible resembles the mandibles of the pithecanthropines. It is as massive, and it has no chin. The Mauer jaw was found in a stratum in Europe which is about the same age as the strata in which the Asiatic pithecanthropines were found. It is the earliest hominid fossil found in Europe, and it is at one end of the geographical range of the pithecanthropines. The morphological affinities of the Ternifine fossils (Table 9.3) with the pithecanthropines make it somewhat easier to accept the existence of a European pithecanthropine. Since we place the Asiatic pithecanthropines in *Homo sapiens*, the Mauer mandible also must be placed in that group. Alternative opinions are possible. *H. erectus* may be viewed as a successional species in the evolutionary lineage of the genus *Homo.* It may be useful to accord it subspecific status, *H. sapiens erectus,* for convenience in speaking about it. But the weight of the evidence is that all such forms are part of a single polytypic species. Some taxonomists feel that we are obligated to use subspecies to indicate subdivisions of modern man as well as of fossil man in order to handle the observed variations. This seems unnecessary to us.

The species *H. erectus,* if we insist that it is a species, is an example of an evolutionary species. Yet we can find no unassailable reason to consider it so. *H. erectus* can never be a good genetical species (see Chapter 4). If members of the *H. erectus* population were alive today, they would probably be fully interfertile with contemporary man.

Eskimos and Australian aboriginals are both members of the species *Homo sapiens.* The skulls of members of these two populations are quite different, yet none but the most ardent "splitters" would put them in separate taxa.

It is unlikely that there has been any significant genetic interchange between Eskimos and Australian aboriginals for quite a long period of time. The same situation unquestionably existed among many Pleistocene hominid populations. And we must remember that the samples we have of the hominid populations of the Eurasian Pleistocene are from very different times as well as different places. The variability we find in the skeletal samples of Pleistocene hominids is partly because the various individual remains are not necessarily contemporaneous. It is highly likely that thousands of years separate many of them. The fact that we must consider variation as a function of time as well as space does not imply that we must have more than a single taxon for hominid fossils from the middle Pleistocene to the present.

The "Neandertal Problem" has, of course, created a number of nomenclatural and classificatory problems. In our discussion we noted that if we interpret these fossils as part of a genetic continuum the "Neandertal Problem" ceases to exist. If we use the genetical concept of species in assessing these fossils we must consider them as members of *Homo sapiens.* The most acceptable evidence that two fossils do not belong to the same taxon is that a statistical *distinction* exists between the two populations inferred from the fossils. Thus when we apply the genetical concept of species in classifying specific fossils, we must remember that the population inferred from the fossils is what is being classified. Such populations are separate species if they were in separate lineages between which significant interbreeding did not occur. They are separate species if these particular fossils represent successive populations in a single lineage, in which the evolutionary change between the different time stages is sufficiently great so that the two populations differ to the same extent as do two species which live today.

The distribution of hominid fossil discoveries in time and space is an important part of the information used when assessing their phylogeny. Figure 10.4 shows the chronology of the Pliocene–Pleistocene, with the various fossil hominids arranged in the sequence generally accepted today.

There are, we believe, two reasonable classifications for the hominid fossils, each of which is presented here. One is a lumper's classification, the other a splitter's. First, we present the formal taxonomy of the Hominidae as viewed by a lumper. The names in a formal classification are usually followed by a name and a date. The name and date specify the authority and date of publication used to justify priority. These references are seldom given in bibliographies.

FAMILY: Hominidae Gray 1825
  GENUS: *Ramapithecus* Lewis 1934 [includes *Bramapithecus* Lewis 1934]
    SPECIES: *R. punjabicus* Pilgrim 1910 [includes *R. brevirostris* Lewis 1934; *Bramapithecus thorpei* Lewis 1934; *B. sivalensis* Lewis 1934; *Kenyapithecus wickeri* Leakey 1962]
  GENUS: *Homo* Linnaeus 1758 [includes *Australopithecus* Dart 1925; *Plesianthropus* Broom 1936; *Paranthropus* Broom 1938;

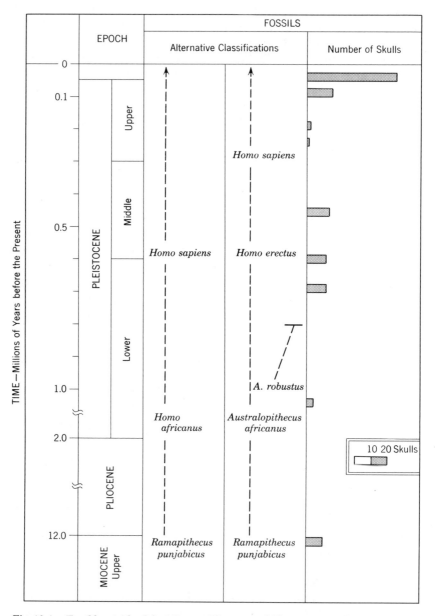

Fig. 10.4. Fossil hominids of the Miocene, Pliocene, and Pleistocene epochs. See text for a discussion of the alternative classifications. For the Miocene forms, the estimate of 10 skulls is a minimum number based on the recovered mandibles or maxillae.

Meganthropus von Koenigswald 1945; *Zinjanthropus* Leakey 1959; *Australanthropus* Heberer 1953; *Telanthropus* Broom and Robinson 1949; *Anthropopithecus* Dubois 1892; *Pithecanthropus* Dubois 1894; *Sinanthropus* Black and Zdansky 1927; *Atlanthropus* Arambourg 1954; *Javanthropus* Oppenoorth 1932; *Cyphanthropus* Woodward 1921; *Protoanthropus* Bonarelli 1909; *Africanthropus* Dreyer 1935]

SPECIES: *H. africanus* Broom 1936 [includes *Australopithecus africanus* Dart 1925; *A. prometheus* Dart 1948; *Plesianthropus transvaalensis* Broom 1936; *Paranthropus crassidens* Broom 1936; *P. robustus* Broom 1938; *Meganthropus paleojavanicus* von Koenigswald 1945; *Zinjanthropus boisei* Leakey 1959; *Telanthropus capensis* Broom and Robinson 1949; *Australanthropus africanus* Heberer 1953; *Homo habilis* Leakey 1964]

H. sapiens Linnaeus 1758 [includes more than 100 binomials and trinomials referring to species and subspecies such as *Pithecanthropus erectus* Dubois 1892; *P. robustus* Weidenreich 1945; *Anthropopithecus erectus* Dubois 1892; *Sinanthropus pekinensis* Black and Zdansky 1927; *Homo heidelbergensis* Schoetensack 1908; *H. modjokertensis* von Koenigswald 1936; *H. neanderthalensis* King 1864; *H. rhodesiensis* Woodward 1921; *H. soloensis* Oppenoorth 1932; *H. krapinensis* Gorganovic-Kramberger 1902; *H. steinheimensis* Berckhemer 1936; *H. sapiens africanus* Blumenbach 1775; *H. sapiens americanus* Linnaeus 1758; *H. sapiens afer* Linnaeus 1758; etc.]

The splitter's classification is presented next. The reader will notice that the synonymies for the genus *Homo* and the species *H. sapiens* have not been repeated from the lumper's classification.

FAMILY: Hominidae Gray 1825
  GENUS: *Ramapithecus* Lewis 1934 [includes *Bramapithecus* Lewis 1934]
    SPECIES: *R. punjabicus* Pilgrim 1910 [includes *R. brevirostris* Lewis 1934; *Bramapithecus thorpei* Lewis 1934; *B. sivalensis* Lewis 1934; *Kenyapithecus wickeri* Leakey 1962]
  GENUS: *Australopithecus* Dart 1925 [includes *Plesianthropus* Broom 1936; *Paranthropus* Broom 1938; *Meganthropus* von Koenigswald 1945; *Zinjanthropus* Leakey 1959; *Australanthropus* Heberer 1953; *Telanthropus* Broom and Robinson 1949]
    SPECIES: *A. africanus* Dart 1925 [includes *A. prometheus* Dart 1948; *Plesianthropus transvaalensis* Broom 1936; *Meganthropus paleojavanicus* von Koenigswald 1945; *Telanthropus capensis* Broom and Robinson 1949; *Australanthropus africanus* Heberer 1953; *Homo habilis* Leakey 1964]

A. *robustus* Broom 1938 [includes *Paranthropus crassidens* Broom 1936; *Zinjanthropus boisei* Leakey 1959]

GENUS: *Homo* Linnaeus 1758

SPECIES: *H. erectus* Dubois 1892 [includes *Pithecanthropus erectus* Dubois 1892; *P. robustus* Weidenreich 1945; *Anthropopithecus erectus* Dubois 1892; *Sinanthropus pekinensis* Black and Zdansky 1927; *Javanthropus soloensis* Oppenoorth 1932; *Homo* (or *Pithecanthropus*) *modjokertensis* von Koenigswald 1936]

H. sapiens Linnaeus 1758

The taxonomic status of *Australopithecus* [= *Homo* [= *Paranthropus*]] *robustus*, the larger and more massive of the two australopithecines, has not yet been determined satisfactorily. We left it either as part of the species *Homo africanus* (lumper's view) or a distinct species of the genus *Australopithecus*, *A. robustus* (splitter's view). The size and morphological relationships between the two australopithecines to some extent resemble those between pygmies and other populations of *Homo sapiens*. If this proves to be reasonable, no taxonomic distinction will be warranted for *robustus*. It is also possible to argue that the fossils originally assigned to the genus "*Paranthropus*" are sufficiently distinct from *Homo africanus* to warrant the erection of a separate taxon, either a species or a genus. The morphology of the teeth and jaw suggest to some that the robustus remains belonged to an animal distinct in behavior and ecology from the australopithecines. Because of these alternative possibilities, no definition of the *Australopithecus* [= *Homo* [= *Paranthropus*]] *robustus* group will be provided.

The confusing multitude of names just presented may be reduced to simpler form (Table 10.2). It is important, nevertheless, for the reader to become acquainted with the synonymy we have just presented. Almost any binomial referring to any of the Hominidae the reader finds in the literature of paleoanthropology may be safely sunk in one of the three binomials of the lumper's taxonomy.

TABLE 10.2
Taxonomy of the Hominidae

	Lumper's version	Splitter's version
Family	Hominidae	Hominidae
Genus	*Ramapithecus*	*Ramapithecus*
Species	*R. punjabicus*	*R. punjabicus*
Genus	*Homo*	*Australopithecus*
Species	*H. africanus*	*A. africanus*
	H. sapiens	*A. robustus*
Genus		*Homo*
Species		*H. erectus*
		H. sapiens

The formal definitions of several important taxa within the Hominoidea are presented in Table 10.1. These definitions are very detailed. This makes it easy for those with the splitter's point of view to justify removing any particular specimen from these various taxa. Simplicity of definition is not necessarily a virtue; neither is it always possible. There are not very many fossil specimens upon which to base the formal definitions. But the definitions should reflect the major changes in adaptive zones which are the basis for the phylogenetic split of the Pongidae and Hominidae. The pongids developed a kind of herbivore way of life. They have a massive anterior dentition as their tool for tearing up the vegetation that forms a major portion of their diet. Their adaptive zone is and was partially arboreal.

The hominids developed erect posture and bipedal locomotion; the anterior dentition grew smaller. Their adaptive zone was terrestrial, and it was successfully exploited because of the development of symbolic communication and, eventually, high intelligence. Intelligence, symbolic communication, toolmaking, and increased cranial capacity are consequences of original adaptive specializations of the Hominidae. They are not useful criteria for definitions of the family. The choice of toolmaking, that is, direct evidence that symbolic communication of a relatively high order had developed, as a criterion for defining the genus *Homo* as distinct from *Ramapithecus* is justified. If tools are found with hominid fossils, toolmaking and the symbolic faculty are sufficiently advanced to be recognized in the taxonomy of these beasts.

We prefer a simple definition of the genus *Homo*. The characteristics include those listed by Le Gros Clark for the genera *Australopithecus* and *Homo*, with one exception. There is no reference to cranial capacity. If toolmaking and other evidence of culture are present, it must be assumed that the cranial capacity was sufficient. Unfortunately too much anthropological writing has focused on cranial volume when there is no evidence that a critical threshold for cranial volume need be exceeded for such "higher" activities as toolmaking and, by implication, culture.

REVISED DEFINITION OF THE GENUS *Homo*—A GENUS OF THE HOMINIDAE. THE STRUCTURE OF THE PELVIC GIRDLE AND THE LEGS ARE ADAPTED TO HABITUAL ERECT POSTURE AND THE BIPEDAL GAIT. THE FORELIMB IS SHORTER THAN THE HINDLIMB. THE HAND HAS A WELL-DEVELOPED THUMB WHICH IS FULLY OPPOSABLE. THE HAND IS CAPABLE OF THE POWER AND THE PRECISION GRIP. THE CAPACITY TO MAKE TOOLS IS DEVELOPED SO SUFFICIENTLY THAT TOOLS ARE MADE TO A PLAN COMMUNICATED SYMBOLICALLY FROM ONE INDIVIDUAL TO ANOTHER AND FROM ONE GENERATION TO THE NEXT. The evidence that this last characteristic is present in early Pleistocene members is inferential. The making of a tool is the result of culturally conditioned behavior, toolmaking is a concept, which is rather difficult to perceive in early Pleistocene strata. The presence of tools in the archeological complex associated with the early Pleistocene hominids is evidence that symbolic communication, education, and a cultural tradition were achieved by the animals whose fossilized bones have been recovered. The presence of stone tools is the best evidence available today. Bone tools, if unequivocally demonstrated, would constitute other evidence.

Changes in the definitions of two species of the genus *Homo* are implied. The criteria used by Le Gros Clark, with the exception of cranial volume, remain valid for the new designations. These two species are part of an allochronic lineage, and they must be arbitrarily separated. As more fossil material becomes available, the distinctions between the two species will be more difficult to discern.

Homo africanus—A SPECIES OF THE GENUS *Homo* WITH THE DISTINGUISH-ING CHARACTERISTICS LISTED FOR *Australopithecus* (Le Gros Clark 1964; Table 10.1).

Homo sapiens—A SPECIES OF THE GENUS *Homo* THAT INCLUDES THE POPULATIONS CHARACTERIZED BY THOSE FEATURES USED TO DEFINE THE GENUS *Pithecanthropus* (Le Gros Clark 1955; Table 10.1) AND THE SPECIES *Homo neanderthalensis* (Le Gros Clark 1964; Table 10.1), DISTINGUISHED FROM *Homo africanus* BY A LARGER AND MORE ROUNDED BRAINCASE, A SMALLER JAW, AND A LESS PROGNATHOUS FACE.

SUGGESTED READINGS AND SOURCES

Campbell, B., Quantitative taxonomy. *In* S. L. Washburn (Ed.), *Classification and Human Evolution*, p. 50. Aldine, Chicago (1963).

Campbell, B., Just another 'man-ape'? *Discovery* 25, No. 6, 37 (1964).

Clark, W. E. Le G., *The Fossil Evidence for Human Evolution* (First edition). Univ. of Chicago Press, Chicago (1955).

Clark, W. E. Le G., *The Fossil Evidence for Human Evolution* (Second edition). Univ. of Chicago Press, Chicago (1964).

Clark, W. E. Le G., The evolution of man. *Discovery* 25, No. 7, 49 (1964).

International Code of Zoological Nomenclature. N. R. Stoll, R. Ph. Dollfus, J. Forest, N. D. Riley, C. W. Sabrosky, C. W. Wright, and R. V. Melville (Ed.), International Trust for Zoological Nomenclature, London (1964).

Mayr, E., Taxonomic categories in fossil hominids. *Cold Spring Harbor Symposia Quant. Biol.* 15, 109 (1950).

Mayr, E., E. G. Linsley, and R. L. Usinger, *Methods and Principles of Systematic Zoology*. McGraw-Hill, New York (1953).

Pilbeam, D. R., and E. L. Simons, Some problems of hominid classification. *Am. Sci.* 53, 237 (1965).

Robinson, J. T., *Homo 'habilis'* and the australopithecines. *Nature* 205, 121 (1965).

Schultz, A. H., Age changes, sex differences, and variability as factors in the classification of primates. *In* S. L. Washburn (Ed.), *Classification and Human Evolution*, p. 85. Aldine, Chicago (1963).

Simons, E. L., Some fallacies in the study of hominid phylogeny. *Science* 141, 879 (1963).

Simpson, G. G., *Principles of Animal Taxonomy*. Columbia Univ. Press, New York (1961).

Simpson, G. G., The meaning of taxonomic statements. *In* S. L. Washburn (Ed.), *Classification and Human Evolution*, p. 1. Aldine, Chicago (1963).

Living Primates

11

WRITING ABOUT THE PRIMATES is fraught with difficulties. The misconceptions current about the Primates are many. Someday it will be interesting to write a book about the role these misconceptions play in our ideology. In this book we shall discuss a few of the misconceptions as they impinge upon essential points. It is difficult not to attribute enormous intelligence to almost all of the Primates, except perhaps the tree shrew. Some anthropologists have written that lemurs, for example, are rather dull witted. Nothing could be farther from the truth to one who has worked with lemurs extensively. But we must be honest and admit it is possible that the absorption and fascination we have with these particular animals may lead us to project fantasies about their intelligence, good nature, etc. Actually almost all members of the Lemuriformes, with the exception of the tiny nocturnal *Microcebus*, seem to be rather gentle, placid, easily tamed creatures who might make excellent pets and whose general intelligence seems to be considerably greater than that of dogs and cats. Even a statement of this sort is suspect, for it is very difficult for us to say precisely what we mean by general intelligence. To us, some primates, particularly those of the New World, seem extraordinarily unpleasant, unfriendly, and difficult to work with. Others insist this is not the case. Perhaps it is the projection of our own attitudes and our own treatment of the animals that result in these difficulties. There is no question that all the Primates are considerably more intelligent and more skillful than most other mammals, perhaps all other mammals. They have acute vision, quick reactions, larger brains, and extraordinarily skillful, manipulative hands. As a result they can do things beyond the capacity of most other mammals solely because of the anatomy involved.

It is not unlikely that much of what has been written about the behavior, temperament, or intelligence of Primates is simply the projection of an author's personality. For example we believe that most primate females are extremely good mothers. That is quite an anthropomorphic statement but conveys an impression from our own observations and those of others which is difficult to phrase any other way. An infant primate is dependent on its mother longer than are most mammalian infants, and the mother does care for the infant. But beyond that there is the problem of projecting our own anthropomorphic fantasies into the animals. It is said that people make pets of dogs that resemble, symbolize, or caricature certain of their own salient personality traits. We usually tell our undergraduate students that the reason we are so interested in baboons and lemurs is that baboons remind us of our colleagues on university faculties—alert, intelligent, quarrelsome, untidy, fickle, disagreeable, intriguing; and lemurs remind us of undergraduates—bright-eyed, bushy-tailed, with facial expressions that reflect incredulity and disbelief that the way of the world is what it is.

Classification, Distribution, Significant Characteristics

The living primates may be divided into eight monophyletic taxa (Table 5.1). These taxa are the Tupaiiformes, Tarsiiformes, Lorisiformes, Lemuriformes, Ceboidea, Cercopithecoidea, Pongidae, and Hominidae. The content of these taxa is identical with the eight taxa proposed by Simpson in 1962. He chose different names and, by this choice, different levels in the taxonomic hierarchy for these groups. We have raised the four taxa of prosimians to the level of infraorders. Our reasons for doing this are essentially two. First, it represents a conservative point of view, one which Schultz used many years ago. Our choice is not a radical change in the organization of the classification of these primates. Second, the difference or the separation between the tree shrews, the tarsiers, the lorises, and the lemurs is great enough that the difference might as well be recognized with infraordinal status. Although the separations are very ancient, length of time since divergence should not be an essential criterion for choosing the level at which to separate these taxa. Separating them at the level of the infraorder recognizes that each is very different from the other and, at the same time, suggests that we are taking some account of their long-standing separation.

As we pointed out earlier, the Anthropoidea, the other major division within the order Primates, is the only group of Primates clearly distinct from all other mammals. This is so for the fossil and the living forms. The line we draw between Insectivora and Prosimii is more arbitrary than the line we draw between Anthropoidea and Prosimii. Since there is, and was, no clear and distinct line in nature between Insectivora and Prosimii, it is difficult to draw one in our classification and definitions. Within the suborder Anthropoidea, we have followed Simpson's lead and do not recognize the usefulness of the classic division of Platyrrhini, the New World monkeys, and Catarrhini, the Old World monkeys, apes, and man. This division was based upon the belief that there was a single, clear-cut subdivision. We now know there are three well-marked subdivisions among the Anthropoidea that ought to be recognized in the classification. These are the New World forms, the Ceboidea; the Old World monkeys, the Cercopithecoidea; and the apes and man, the Hominoidea. These three superfamilies appear to have differentiated in approximately the same major time period. As we said earlier, the length of time living groups are presumed to have been separated should not be taken as an essential criterion for putting them into higher categories. At the very least as Simpson pointed out so clearly, the Cercopithecoidea and the Hominoidea do not form a single and exclusive unit.

One problem in classification is the almost complete lack of understanding, on the part of many who have written about classification of primates, of the nature of individual variation, variation within a population, and variation within a species. Color phases, individuals of different colors, have been proposed as species, if not as genera. Many an aberrant individual primate, we would suspect, has been used by someone to define a taxon. As Simpson has pointed out, almost every genuinely distinct population that should be called a species has been called a genus and a large number of genera have been given higher taxonomic statuses. At this point we refer the reader to our earlier discussion of the nature of species and of the problems faced by those who must classify actual specimens (Chapter 4).

We have an incorrigible lumper's point of view and prefer to avoid what seems to us to be an excessive number of divisions into subspecies, species, genera, subfamilies, families, and superfamilies. The classification of the living primates is reasonably presented in Fig. 11.1. The classification of the fossil ancestors of the living primates is shown in Fig. 7.1 and 7.2. A grand synthesis in which both were placed on the same piece of paper would probably require a few more families and superfamilies and possibly subfamilies, but in general there are no changes on the subordinal or infraordinal level. The classification we have chosen to present is based primarily upon that published by Simpson in 1945. We have made some changes in it because research since 1945 suggests certain appropriate adjustments. As Simpson pointed out in 1962, a rigorous revision of only the nomenclature of the Primates would very likely cause greater confusion than is necessary. Eventually, official recognition of a classification somewhat like the one presented here will be necessary.

ORDER	SUBORDER	INFRAORDER	SUPERFAMILY
PRIMATES	PROSIMII	TUPAIIFORMES	
		TARSIIFORMES	
		LORISIFORMES	LORISOIDEA
		LEMURIFORMES	LEMUROIDEA
			DAUBENTONIOIDEA
	ANTHROPOIDEA		CEBOIDEA
			CERCOPITHECOIDEA
			HOMINOIDEA

FAMILY	SUBFAMILY	GENUS
TUPAIIDAE	TUPAIINAE	*Tupaia* *Anathana* *Dendrogale* *Urogale*
	PTILOCERCINAE	*Ptilocercus*
		Tarsius
LORISIDAE	LORISINAE	*Loris* *Nycticebus* *Perodicticus*
	GALAGINAE	*Galago*
LEMURIDAE	LEMURINAE	*Lemur* *Hapalemur* *Lepilemur*
	CHEIROGALEINAE	*Cheirogaleus* *Microcebus* *Phaner*
INDRIIDAE		*Indri* *Lichanotus* *Propithecus*
		Daubentonia
CEBIDAE	CEBINAE	*Cebus* *Saimiri*
	ALOUATTINAE	*Alouatta*
	AOTINAE	*Aotus* *Callicebus*
	ATELINAE	*Ateles* *Brachyteles* *Lagothrix*
	CALLIMICONINAE	*Callimico*
	PITHECIINAE	*Pithecia* *Cacajao* *Chiropotes*
CALLITHRICIDAE		*Callithrix* *Cebuella* *Leontideus* *Saguinus* *Tamarinus*
CERCOPITHECIDAE	CERCOPITHECINAE	*Cercopithecus* *Cercocebus* *Cynopithecus?* *Macaca?* *Papio*
	COLOBINAE	*Colobus* *Nasalis?* *Presbytis* *Pygathrix* *Rhinopithecus?* *Simias?*
PONGIDAE	PONGINAE	*Pongo* *Pan*
	HYLOBATINAE	*Hylobates*
HOMINIDAE		*Homo*

Fig. 11.1.   Classification of the living primates.

One specific problem that we have had to discuss over and over again with colleagues and acquaintances all over the world is the relationship between speciation and the distribution or range through which a population is found. If two populations are sympatric they are usually, and quite legitimately, placed in separate species. But the crux of the matter is: *Are* the two forms or populations sympatric? The black colobus (once named *Colobus satanas*) and the black and white colobus (*Colobus polykomos*) are believed by some to be two separate species of guerezas because they appear to occupy the same area; their distributions overlap. On the face of it, this is good evidence that they should be separate species. However we have found groups of *Propithecus* in southern Madagascar, once described as sympatric populations, in which there were animals of three different colors and markings—almost pure white, white and brownish red, and almost completely brownish black. All three forms were living together in a single troop. Our observations of their behavior support the simpler view that this is another example of polymorphism and variability exhibited by the Primates. Some insisted that this polymorphic group was the result of hybridization between neighboring populations of even more distinct forms. Only very careful field observations will demonstrate unequivocally whether these are examples of sympatric species, sibling species, or color varieties of a single species.

Throughout the discussion of the living primates there will be other things said about problems of classification, nomenclature, and systematics. To trace the prior and valid name of a species or genus of Primates is often a fascinating exercise in scholarly detection. The rules in the *International Code of Zoological Nomenclature* should be followed as rigidly as possible. But there are times when the point of coherent nomenclature seems lost in a game of who can find the prior name or the correct grammatical form of a name. *Saimiri sciurea* is the name of a South American squirrel monkey, and considerable ink has been spilled over whether it is S. *sciurea* or S. *sciureus*. The grammatical gender of *Saimiri* is said, by some classical scholars, to be feminine, by some, masculine. Here the battle lines are drawn. The name for one of the subfamilies of the Lorisiformes is assumed to be Galaginae, but careful examination of the literature shows that Galagoninae appears to have priority. Does this really matter? Unfortunately, the answer is yes. Every step that enforces uniform nomenclatural practice is a step toward clarity in the system.

The amount of data and material available varies for the seven monophyletic nonhuman primate taxa. Our own interest, for that matter, varies considerably. We are completely fascinated by the lemurs of Madagascar, by baboons, and by African prosimians. We have slightly less interest in research on the other Old World living primates. For a variety of reasons we have still less interest in the living New World primates. It is our personal prejudice as well as the varying quality and amount of data that make the treatment of the various living groups somewhat uneven in the succeeding chapters.

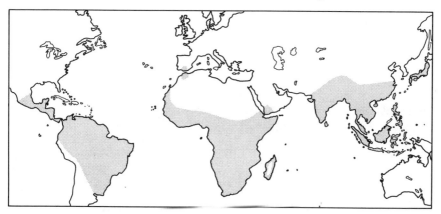

Fig. 11.2.   Worldwide distribution of the nonhuman primates is indicated by the shaded areas.

Many of the trends, discussed in Chapter 5, that we deduce as having occurred during the evolutionary history of the Primates may be seen in the living members of the order. The living members may be considered an evolutionary stratification of a number of the major adaptive radiations within the order. The evolution of the distinguishing features and characteristics of the order Primates can be discerned in the living forms, if one looks carefully.

The living primates, other than man, are found on almost all the major land masses of the world except in North America north of Mexico City. They are found on some of the islands of the East Indian archipelago (Fig. 11.2), but they are not found on Australia and many other Pacific islands. There is no fossil evidence that they ever lived there.

Primates are nocturnal, diurnal, and crepuscular in habit. Some of them seem to be all three at once. Nocturnal primates are those that carry out their cycle of activity during the night; diurnal primates, during daylight hours; and crepuscular primates, mostly in very late afternoon, early evening, and early morning. Most primates considered crepuscular prove, after intensive field work, to be diurnal and spend a large portion of the middle part of the day resting or sleeping.

Most of the primates have adapted to an arboreal way of life. The anatomy of the bones and muscles of the back and limbs have been modified for appropriate kinds of locomotion in the trees. There are other primates who spend a great deal of time on the ground. Only those that have developed anatomical modifications to adapt them for a life on the ground are considered terrestrial. Almost all terrestrial primates spend at least their nights in the trees, and many arboreal primates spend much of the day on the ground. But by definition arboreal primates are those that have adapted to and modified their anatomy for life in the trees, and terrestrial primates are those that have adapted to and modified their anatomy for life on the ground.

Primates come in a variety of sizes, from the tiny *Microcebus* to the giant gorilla. Most primates are vegetarian, although their range of diet is large. Some are so specialized in their requirements that they will die unless they are able to eat a certain amount of particular kinds of vegetable material. Others seem to be completely omnivorous. In general the experience of zoo keepers and those who keep captive primate colonies is that adaptability of diet is much greater among the Primates than among most other animals. As we pointed out earlier, almost all primates live in tropical or semitropical regions today. This has led to the presumption that primates always have been adapted to a tropical or semitropical climate. This is not necessarily true for there are langurs that live in very cold regions near the Himalaya mountains, and the Japanese macaque lives in what is essentially a temperate climate. It is also not necessarily true that extinct primates lived in tropical or semitropical regions, for fossil primates have been found in Europe in deposits that indicate a temperate climate.

All the Primates have prehensile hands, that is, they can grasp things with their hands. Almost all of them have prehensile feet also; man is a notable exception. The prehensile hand is one in which the thumb or some other digit is opposed to the remaining digits, and the animal is able to grasp objects between the thumb and the other digits. A truly opposable thumb, by definition, is not present in all the Primates. Two grips are used, a precision grip and a power grip. All hand positions used by the Primates to grasp objects or to cling to things are variations on these two grips. The power grip is formed by partial flexion of the fingers and palm with counterpressure applied by the thumb. The precision grip is formed when an object is pinched between flexed fingers and the opposing thumb. The details of these are discussed in Chapter 20.

It is not at all unexpected that animals with prehensile extremities should be capable of using tools, and many primates besides man use tools. At least they manipulate objects—sticks, stones, bits of grass, crushed leaves—in what appears to be a purposive manner. The objects are manipulated to get food or to throw at and chase intruders and predators. The tool that distinguishes the genus *Homo* from other primate taxa is not the same tool that the chimpanzee uses. There is no evidence that there is a tool tradition among chimpanzees (a cumulative body of knowledge which chimpanzees pass on by means of symbolic communication from individual to individual). The tools of nonhuman primates are not made according to a plan, they do not conform to a style tradition, they are objects which are picked up and used on the spur of the moment. The manual dexterity achieved by the Primates enables them to manipulate objects with versatility greater than that of other animals. We expect related animals to have similar abilities, but we should not overdo our enthusiasm in describing the tool-using skills of nonhuman primates. Nonhuman primates have an incipient tool-making capacity, they have not achieved the special, peculiar ability of man.

It is well to dispel the common notion that binocular vision is restricted to the Anthropoidea. Almost all prosimians have overlapping fields of vision and have neurological structures which permit us to deduce that they have depth vision. Color vision is also probably more widespread among the Primates than primatologists have allowed us to believe in the past.

Sociality

The Primates are an unusually sociable group of mammals. The nocturnal primates are relatively less gregarious and less social, although this statement is not based on very much evidence. The nocturnal primates, in general, appear to live alone or in pairs and probably do not form tightly structured social groups. But almost all of the diurnal primates form structured social groups, at least as large as a family group and many considerably larger. In discussing primate sociality, terms such as groups, congeries, pairs, family groups, harems, troops, and solitary animals are often used. We try to use the expression group or groupings when we are not certain of the extent to which the social unit is structured. We call an association of primates a congeries if numbers of them are found in some kind of association, even though the animals themselves may be considered solitary or even antisocial. In other words, a congeries of nocturnal primates may be found in one section of a forest, whereas a neighboring section may have no nocturnal primates in it. A pair is a male and a female. In general when we refer to pairs we imply a social relationship, either a consort pair (a male with a female in estrus) or a family (an adult male and an adult female). A family group usually consists of an adult male and an adult female with one or more infants and subadult or immature animals. A troop is a number of primates associated together in their daily cycle of activities. It is usually larger than a family group. The membership in the association does not vary significantly from one part of the year to another. Some troops of primates, such as baboons, are highly structured and extremely well organized. A harem is a male with complete dominance over a number of mature and subadult females in association with him. There may be a few infant and juvenile animals associated with a harem group. Many nocturnal primates are solitary in that they do not associate with other members of their own species except at the breeding season. Some of these may actually associate in pairs, but the evidence is not decisive. Remember that it is difficult to study nocturnal primates in their natural habitat, so we must not expect as many observations of them as of diurnal primates. The term solitary is also applied to those diurnal primates that normally live in social groups of one sort or another but are sometimes found alone.

The sociality and sociability of primates is expressed in many aspects of their daily and annual cycles of behavior. Almost all primates groom each other and themselves. Grooming behavior may be as elaborate as that of baboons wherein with very fine precision they are able to grasp individual hairs and parasites. This behavior keeps the animals healthy and satisfies certain psychological needs. The prosimians also groom each other and themselves, but they use their hands much less precisely. They groom with their procumbent lower incisors and canines, the famous toothcomb or tooth scraper.

Marking behavior, particularly scent marking, is quite common. Some prosimians urine mark and urine wash. When excited or interested in some object or other, they urinate upon their hands and then rub their wet hands on the object. At other times they urine mark in the vicinity of something that disturbs or interests them. Many diurnal lemurs will scent mark with their anal, carpal, or brachial scent glands. This marking behavior is pronounced during estrus, and it is noticed when the animals are disturbed by other members of their own species or by man.

The displays of the Primates make a full-time research field of their own. Vocalizations, facial expressions, and certain gestures are the principal kinds of displays among the Primates. These displays serve to convey information which other primates may or must use as a cue for their behavior. Among some of the most sociable primates, displays are connected with the territorial behavior of the social group or troop.

A primate or a troop of primates does not wander at random during the normal cycle of activities—getting enough to eat, mating, caring for young, etc. A fairly well-defined home range is covered during its normal daily activities. It is likely that all primate species have a home range, the size of which depends upon the size of the social group. A territory is the part of the home range that is exclusively occupied by a particular troop or pair and that may be defended against other members of the same species. Many primates maintain territory with vocal and facial displays. It is rather unusual for primates other than man to fight over territory, despite a number of colorful popular accounts that they do.

The study of the behavior of primates has become of great interest and, indeed, most fashionable among anthropologists in the last few years. There is no question that much can be learned about the possible origin of human behavior by examining the behavior of man's relatives. There are some limitations on this statement which we should be careful to observe. Man, *Homo sapiens,* has diverged radically from the other primates, and, as a result, some false analogies may be constructed. Nevertheless much valuable information which can be used in a metaphorical as well as direct way to make deductions about the behavior of our fossil ancestors has been gathered.

There is one trait or property man has that is not shared with the non-human primates. Man is a symboling creature; that is, he has developed one great neurological advance during primate evolution. It occurred somewhat after the development of erect posture and bipedal locomotion. Whatever the structural change was, the functional consequence is that man is able to construct arbitrary systems of symbols enabling him to pass on information from one individual to another and from one generation to another. This capacity to use symbols led to the origin and differentiation of culture. Primates, other than man, do not develop cultural systems. All animals may be said to live in some kind of society, because some sort of social interaction between individuals occurs in all species of organisms. Man's social interactions are conditioned by the fact that he possesses culture.

SUGGESTED READINGS AND SOURCES

Buettner-Janusch, J. (Ed.), The relatives of man. *Ann. N. Y. Acad. Sci.* **102**, 181 (1962).

Buettner-Janusch, J. (Ed.), *Evolutionary and Genetic Biology of Primates*, Vol. I–II. Academic Press, New York (1963, 1964).

Carpenter, C. R., *Naturalistic Behavior of Nonhuman Primates*. Penn. State Univ. Press, University Park (1964).

Clark, W. E. Le G., *The Antecedents of Man* (Second edition). Edinburgh Univ. Press, Edinburgh (1962).

Elliot, D. G., *A Review of the Primates*, Vol. I–III. Am. Museum Nat. Hist., New York (1913).

Hill, W. C. O., *Primates, Vol. I. Strepsirhini*. Wiley, Interscience, New York (1953).

Hill, W. C. O., *Primates, Vol. II. Haplorhini: Tarsioidea*. Wiley, Interscience, New York (1955).

Hill, W. C. O., *Primates, Vol. III. Pithecoidea: Platyrrhini—Hapalidae*. Wiley, Interscience, New York (1957).

Hill, W. C. O., *Primates, Vol. IV. Cebidae, Part A*. Wiley, Interscience, New York (1960).

Hill, W. C. O., *Primates, Vol. V. Cebidae, Part B*. Wiley, Interscience, New York (1962).

Hofer, H., A. H. Schultz, and D. Starck (Ed.), *Primatologia*. S. Karger, Basel (1956 *et seq.*).

Hofer, H., A. H. Schultz, and D. Starck (Ed.), *Bibliotheca Primatologica*. S. Karger, Basel (1962 *et seq.*).

Hooton, E. A., *Man's Poor Relations*. Doubleday, New York (1942).

Napier, J., and N. A. Barnicot (Ed.), *The Primates. Symp. Zool. Soc. London* **10** (1963).

Zuckerman, S., *The Social Life of Monkeys and Apes*. Kegan Paul, London (1932).

Zuckerman, S., *Functional Affinities of Man, Monkeys, and Apes*. Harcourt, Brace, New York (1933).

Tupaiiformes

12

THE TUPAIIFORMES (TREE SHREWS) represent a link with the insectivore stock from which the Primates came. They are the most primitive living primates. They have a rather primitive mammalian appearance because of their pointed muzzles. They are small, extremely active creatures with long feathery tails. They look much like small squirrels (Fig. 12.1).

Characteristics

In 1879 Doran noted that among Tupaiiformes certain features of the ossicles of the middle ear differ from those of the insectivores and closely resemble those of primates. The incus bone in the middle ear appears to be very similar to that of monkeys and lemurs. The comparative anatomy of the muscles and the skull and the microscopic anatomy of the brain of *Tupaia* and *Ptilocercus* support the suggestions that the Tupaiiformes have more affinities with the Primates. Because the tree shrews are often placed in an order intermediate between Insectivora and Primates—the order Menotyphla—we include comments on both the primate and insectivore affinities of the Tupaiiformes. We accept the proposition that the Tupaiiformes are Primates, and our reasons will be presented here.

Fig. 12.1.   *Tupaia glis*, the Malaysian tree shrew.

The visual centers in the brain and in the eye are differentiated to a high degree, like the Primates; the olfactory centers in the brain and the olfactory apparatus are smaller than in primitive insectivores; the dentition is primate-like; the pollex (thumb) and hallux (big toe) are extremely mobile; the placenta is a hemochorial type. Some indication of the phyletic position of tree shrews may be grasped in a quotation from Meister and Davis who studied the placental membranes of these animals: ". . . the placenta and fetal membranes of *Tupaia* are almost ideally intermediate between the generalized insectivores . . . and those of the more generalized Anthropoidea."

The total morphological pattern is characteristic of the Lemuriformes. For example the elements of bone that form the interior wall of the orbit are very similar. They include a forward extension of a plate-like process of the palatine bone quite uncharacteristic of the insectivores. The auditory chamber, with a simple tympanic ring freely enclosed inside it, also suggests an affinity with the Lemuriformes. The visual apparatus is another piece of this total morphological pattern. There is an expansion of the visual area in the cerebral cortex. The retina includes an area in the central region, without blood vessels, very similar to one found in lemurs. This is apparently the first stage in the evolution of the true macula. (See Chapter 13 for a discussion of this structure.) The brain in general is quite complex, and this complexity approaches that of higher primates. The cerebral cortex in particular has undergone considerable enlargement as a whole, and there is considerable differentiation of various areas in the cortex, as can be demonstrated histologically. There is incipient development of a distinct temporal lobe. Most of these features are not present in the Insectivora, and all contrast with that group. The pyramidal tract (a part of the spinal cord involved in conveying voluntary control of the muscles) of *Tupaia* has been compared with that of the slow loris, *Nycticebus*. The path of the neurons is different in the two animals. The olfactory apparatus of *Tupaia* is reduced absolutely and is small relative to other bony structures and to the rest of the brain. The turbinal bones within the nasal cavity are not well developed, and there is a relatively small olfactory center in the brain. The brain-to-body ratio of *Tupaia minor* is 1:26; that of man is about 1:47.

The dentition is primitive. The dental formula is $\frac{2.1.3.3.}{3.1.3.3.}$. The molars and premolars are relatively simple. The three incisors in the lower jaw are typical of the primitive mammalian dentition. However, the third incisor has been reduced in size, and in one genus, *Urogale*, it has almost disappeared. This reduction in incisor size indicates a dentition well on its way to becoming a primate dentition. The lower incisors are procumbent to a degree, and when the jaw is closed they find a place in a gap that exists between the upper incisors. The lower canine teeth are not part of a toothcomb or tooth scraper. However in one genus, *Anathana*, the canine is quite procumbent and probably related to an early evolutionary stage of the modern toothcomb of lemurs and lorises.

The hands and feet of these animals suggest those of primates. The extremely mobile thumb and big toe may be abducted widely. The animal is able to grasp things relatively well by flexion of the other digits in association with thumb or big toe. This particular character may reflect the beginning of the opposability of hallux and pollex in the higher primates. In one genus, *Ptilocercus*, the hallux is individualized more strongly than in the others. The first metatarsal bone is strongly developed, and it is clear that there is some differentiation of the intrinsic muscles of the foot. The significance of this particular complex in the total morphological pattern cannot be overemphasized. It was this development of prehensile extremities that led to the evolutionary segregation and rise of the Primates. The Tupaiiformes have sharp claws instead of nails on their digits. They are like some other primitive mammals in this respect.

There are very few detailed accounts of the muscles of tree shrews in the zoological and anatomical literature. Those who dissected and described the muscles of *Tupaia* found they resemble those of the Lemuriformes. The anatomy of *Tupaia* is of special interest, for it combines primitive with primate characters. Le Gros Clark once pointed out that comparative anatomy of muscles is of great value for systematic studies of animals, despite a tacit assumption that muscles undergo great modification during evolution. Major changes in way of life are supposed to be accompanied by changes in the muscles. The similarities between muscles of lemurs and tree shrews are often attributed to convergent evolution rather than to phylogenetic affinity. The assumed close correlation between way of life and morphology of muscles is not so close as all that. Burrowing, terrestrial, and arboreal rodents have an almost identical arrangement of muscles. Le Gros Clark examined the muscle anatomy of squirrels and tree shrews. Both are small, quick-moving, arboreal animals. Yet muscles of squirrels are not modified like those of tree shrews. Furthermore in the disposition of the muscles there are no differences between small, long-tailed, completely arboreal tree shrews and larger, short-tailed, primarily terrestrial tree shrews.

In many ways the muscles and the arrangement of the muscles of the tree shrew, *Tupaia*, resemble those of the Prosimii very closely. There are at least seven muscles or morphological features of muscles present in both *Tupaia* and the Lemuriformes and generally absent in the Insectivora. At the same time there are a number of particular morphological features of the muscles in which *Tupaia* is very similar to certain Insectivora. Muscle morphology of tree shrews is similar to that of the insectivores in features which are essentially primitive. Such characters are more relevant to the fact that *Tupaia* is a mammal than to the degree of affinity *Tupaia* may have with the Insectivora.

It is also possible to argue that the Tupaiiformes do not belong with the Primates, and some students have done this, basing their arguments mainly on fossil evidence. The fossil Anagalidae were, until recently, believed to be tupaiids. The terminal phalanges of the foot of a fossil anagalid are flattened and spatula shaped. It is believed that these digits had nails on them, not claws. The presence of nails on an Oligocene tupaiid, such as *Anagale*, added force to the argument that the Tupaiiformes are Primates. However recent authorities believe that Anagalidae have certain features, particularly a tympanic bulla, which do not appear to be characteristic of the anatomy of tupaiids. Some paleontologists believe that comparison of the living Tupaiiformes with ancient extinct Insectivora, rather than with modern Insectivora, will show that they belong in the same group.

Despite such arguments, the total morphological pattern of the Tupaiiformes suggests that they belong with the Primates. They are contemporary descendants of a primitive mammal which was on its way to becoming a full-fledged primate. As we pointed out in Chapter 7, it is very difficult to make an absolute distinction between Prosimii and Insectivora. Such a distinction does not exist in nature, and we must do the best we can in our classification.

The Tupaiiformes are found throughout southeast Asia. The map (Fig. 12.2) shows the present distribution of the five genera and twelve species (Table 12.1). The tree shrews are found in China—Yunan, Kwangsi, Hainan —in Burma, Indochina, Thailand, the Malay Peninsula, Sumatra, Java, Borneo, and on many small islands of the East Indies, all the way to Palawar

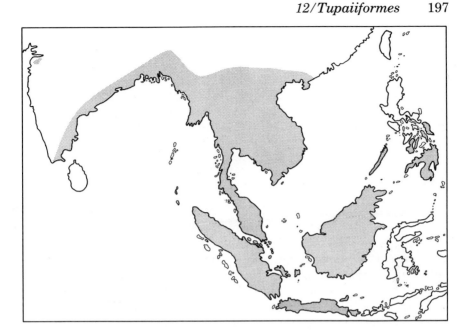

Fig. 12.2. Geographical distribution of Tupaiiformes in southeast Asia. Tupaiiformes are not found continuously throughout the shaded areas. These areas represent the limits of the range in which Tupaiiformes live.

TABLE 12.1
Classification of Tupaiiformes

Tupaiinae   Lyon 1913
  Tupaia   Raffles 1822
    Tupaia glis   Diard 1820
    Tupaia gracilis   Thomas 1893
    Tupaia javanica   Horsfield 1822
    Tupaia minor   Günther 1876
    Tupaia montana   Thomas 1892
    Tupaia nicobarica   Zelebor 1869
    Tupaia tana   Raffles 1821
  Anathana   Lyon 1913
    Anathana ellioti   Waterhouse 1849
  Dendrogale   Gray 1848
    Dendrogale melanura   Thomas 1892
    Dendrogale murina   Schlegel and Müller 1843
  Urogale   Mearns 1905
    Urogale everetti   Thomas 1892
Ptilocercinae   Lyon 1913
  Ptilocercus   Gray 1848
    Ptilocercus lowii   Gray 1848

in the Philippines. They are also found in Sikkim, in India all around the Bay of Bengal, and on the Nicobar islands. *Dendrogale* is found in Cambodia, Annam, Cochin China, and Indochina, *Urogale* in the Philippines.

They live in forest and bush country. Some varieties, such as *Tupaia minor*, live high in the trees, where they emulate the squirrels in habit. Others live in low bushes and cutover land. *Tupaia glis* is believed to live on or near the ground and generally will climb only into low bushes.

Behavior

Some behavioral characteristics have been noted in captive *Tupaia*. The behavior of tree shrews has not been extensively described from studies of free-ranging or wild populations. The animals are usually diurnal in habit. *Ptilocercus* may be nocturnal, since it has a retina composed exclusively of rods, i.e., a scotopic or night vision retina. Tree shrews groom their fur, particularly their tails, with their lower incisors—a trait they share with lemurs and lorises, even though the tree shrews have no toothcomb.

Tupaiiformes are certainly not completely insectivorous; they are probably omnivorous. Our experience with captive tree shrews shows they will eat live baby mice, mealworms, canned dog food, and a variety of fruits and vegetables, as well as a synthetic diet. Recently Vandenbergh found that *Tupaia* kept in a large enclosure killed lizards, on one occasion killed and ate a mouse, but never caught a rat. They also foraged on the ground in litter and, presumably, found and ate insects from the litter.

Tupaia glis, as we said, is believed to live on or near the ground. It runs along the ground and occasionally climbs low bushes into slender, tiny branches. In order to grasp such fine twigs, an extraordinarily variable grip is necessary. The hand of *Tupaia*, when grasping, shows no preferred orientation. *Tupaia* is a very active animal. In captivity it moves in short dashes in various directions, making as many as seventy dashes per minute. Our observations of these animals indicate that their movements are highly stereotyped. *Tupaia* spends a fair amount of time, as do many other primates in captivity, doing somersaults. It seems to specialize in backward somersaults!

As we noted earlier, our observations indicate that *Tupaia* is basically omnivorous. It will use its procumbent lower incisors to bite into relatively soft fruit. It seldom if ever chews or tears with its front teeth. When feeding, it flexes its clawed digits toward the palm, usually using both hands to hold the food, bracing it between digits and palm. It will generally bite off and chew small pieces of meat, fruit, or an insect with its molar and premolar teeth. No social use of the hands by *Tupaia* has been observed.

Tupaia is an extremely aggressive animal. A number of Vandenbergh's animals were wounded and died in fights. Aggression between males was extreme. When a new alien male was introduced into the enclosure, it was killed within four days by the resident dominant male. *Tupaia* will often rub its chest against a piece of twig or shelf in the cage. The males, at least, secrete a sticky substance from their chest glands. Apparently this rubbing occurs immediately after a fight with other males. This is probably a kind of scent-marking behavior. Occasionally, *Tupaia* were seen grooming each other.

Generalizing from the few observations we have of *Tupaia glis,* tree shrews apparently do not form large social groups like many other diurnal primates. There does not seem to be much fighting between the sexes. The implication is that small family groups form as a kind of social unit. A stable family group is unlikely to be maintained because *Tupaia* matures very early, and a young mature male will break up the peaceful coexistence within the larger group. Vandenbergh found that males born in the captive small group were killed by the dominant resident male as soon as they matured. *Tupaia* females in captivity often kill most of their offspring. This happened in our colony and has been reported by Vandenbergh and by Sprankel. The average gestation period for *Tupaia* is about 40 days. Females appear to have many ovulations each year. Our observations indicate that, in captivity, the females are extremely poor mothers.

Systematics

The systematics of the Tupaiiformes is not in good order. If we look at the list of families, genera, and species from the older literature, there are two subfamilies, six genera, and forty-seven species. This seems a bit excessive considering the modern trend to systematize mammals from a lumper's point of view. Simple differences are no longer considered sufficient to make two groups distinctive taxonomically. The Tupaiiformes are not very well known, and it is not possible to make final judgments about the classification of this group. Many of them live on islands, and almost each island variety has been designated as a distinct species by some naturalist. This is not a good practice. In the past not only were forty-seven species erected, but within one or two of the species as many as twelve subspecies were once defined. Today it seems reasonable to follow the recommendation of Ellerman and Morrison-Scott and reduce the number as indicated here (Table 12.1).

The most important systematic problem is the one mentioned earlier. Are the Tupaiiformes a member of the order Primates, or should they be classified as Insectivora? If we accept the concept of total morphological pattern as Le Gros Clark has defined it, it seems clear that the affinities of the living Tupaiiformes are with the living Primates and not with the living Insectivora.

The problem of speciation and of identifying species, which was discussed in Chapter 4, is well illustrated by the Tupaiiformes. A very large number of different species were named and identified by zoologists who first worked with these animals. Hill published a study of a large collection of Malaysian mammals in 1960. One section of this work was devoted to an attempt to determine, by some exact measure, the boundaries of the various taxa, the subspecies, of *Tupaia glis*. One set of measurements was made of body length, tail length, and skull length, but these linear measurements were not particularly useful in discriminating among the various populations. Another objective measure was the degree of reflectance of the pelage (hair) of these animals. Pale colors reflect more light than dark colors, and the relative degree of reflectance can be determined by means of a photoelectric reflectometer. Hill measured the reflectance of the skins of *Tupaia glis* from all over the Malay Peninsula and from a number of different islands. His measurements were made on the upper part of the back between the shoulders and on the rump. Some of the results of his studies are shown in Fig. 12.3 and 12.4. The color intensity of the pelage of *Tupaia glis* changes as one goes northward on the Malay peninsula from Singapore Island to Burma. There is an obvious difference between the most northerly and most southerly population. A generation ago primatologists and mammalogists would have placed these two populations in different species. But a whole set of populations exists intermediate in color, and these form, more or less, a continuum or

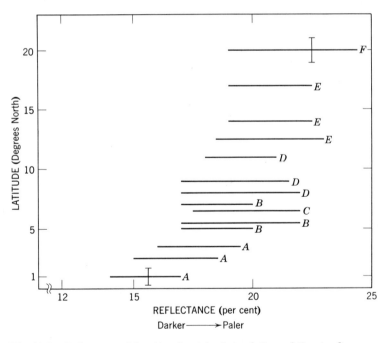

Fig. 12.3.   Reflectance of the skin of mainland populations of *Tupaia glis* correlated with the latitude at which each subspecies lives. The vertical lines indicate the average values for the southernmost and northernmost populations. (A) *Tupaia glis ferruginea*, (B) *T. g. glis*, (C) *T. g. wilkinsoni*, (D) *T. g. clarissa*, (E) *T. g. belangeri*, and (F) *T. g. siccata*. (Adapted from Hill, 1960.)

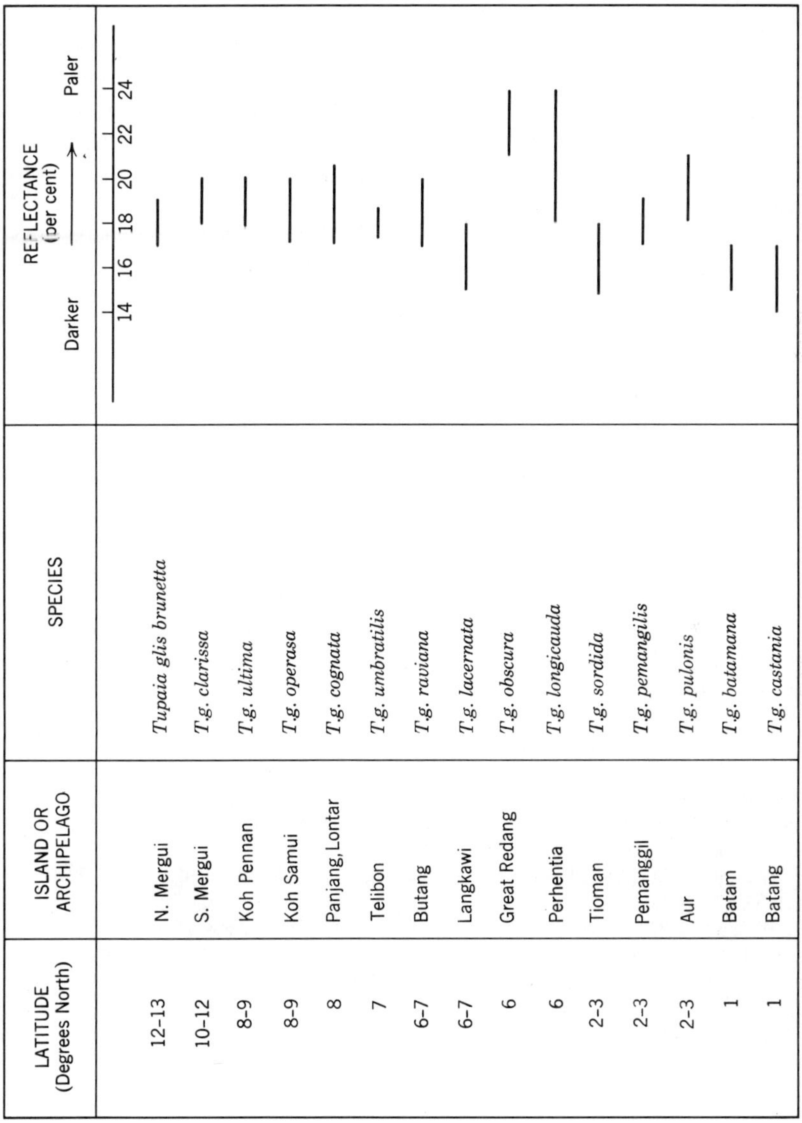

LATITUDE (Degrees North)
ISLAND OR ARCHIPELAGO
SPECIES
REFLECTANCE (per cent)
Darker
Paler
14 16 18 20 22 24
12–13 N. Mergui Tupaia glis brunetta
10–12 S. Mergui T.g. clarissa
8–9 Koh Pennan T.g. ultima
8–9 Koh Samui T.g. operasa
8 Panjang, Lontar T.g. cognata
7 Telibon T.g. umbratilis
6–7 Butang T.g. raviana
6–7 Langkawi T.g. lacernata
6 Great Redang T.g. obscura
6 Perhentia T.g. longicauda
2–3 Tioman T.g. sordida
2–3 Pemanggil T.g. pemangilis
2–3 Aur T.g. pulonis
1 Batam T.g. batamana
1 Batang T.g. castania

Fig. 12.4. Reflectance of the skin of island populations of *Tupaia glis*. (Adapted from Hill, 1960.)

cline of color from north to south (Fig. 12.3). The island populations of *Tupaia glis* vary in color to about the same degree as those of the mainland. Here we have a number of local populations that differ from each other in the expression of this particular character, but the differences overlap throughout the range in which the animal exists (Fig. 12.4). Some would give subspecies designations to each of these populations.

The Tupaiiformes are ambiguously primates, for they have some insectivore characteristics and many other primate characters. We have accepted them as primates for in a sense they forecast the morphology of lemurs and other prosimians. Definition of a separate order for them merely because they are neither clearly Primates nor clearly Insectivora is not a solution to the problem. Since tree shrews possess many traits typical of the Primates, it seems sensible and useful to place them in the same order. The tree shrews of southeast Asia *represent* the kind of mammal that very probably was the kind from which the Primates developed.

Suggested Readings and Sources

Chasen, F. N., A handlist of Malaysian mammals. *Bull. Raffles Museum* No. 15 (1940).

Clark, W. E. Le G., The myology of the tree-shrew (*Tupaia minor*). *Proc. Zool. Soc. London*, 461 (1924).

Clark, W. E. Le G., On the brain of the tree-shrew (*Tupaia minor*). *Proc. Zool. Soc. London*, 1053 (1924).

Clark, W. E. Le G., On the anatomy of the pen-tailed tree-shrew (*Ptilocercus lowii*). *Proc. Zool. Soc. London*, 1179 (1926).

Hill, J. E., The Robinson collection of Malaysian mammals. *Bull. Raffles Museum* No. 29 (1960).

Jane, J. A., C. B. G. Campbell, and D. Yashon, Pyramidal tract: a comparison of two prosimian primates. *Science* 147, 153 (1965).

Kaufmann, J. H., Studies on the behavior of captive tree shrews (*Tupaia glis*). *Folia Primat.* 3, 50 (1965).

Lyon, M. W., Tree shrews: an account of the mammalian family Tupaiidae. *Proc. U. S. Natl. Museum* 45 (1913).

McKenna, M. C., New evidence against tupaioid affinities of the mammalian family Anagalidae. *Am. Museum Novitates* No. 2158 (1963).

Meister, W., and D. D. Davis, Placentation of the pygmy tree-shrew *Tupaia minor*. *Fieldiana, Zool.* 35, 73 (1956).

Meister, W., and D. D. Davis, Placentation of the terrestrial tree-shrew (*Tupaia tana*). *Anat. Record* 132, 541 (1958).

Simpson, G. G., Long-abandoned views. *Science* 147, 1397 (1965).

Sprankel, H., Über Verhaltensweisen und Zucht von *Tupaia glis* (Diard 1820) in Gefangenschaft. *Z. Wiss. Zool.* 165, 186 (1961).

Vandenbergh, J. G., Feeding, activity and social behavior of the tree shrew, *Tupaia glis*, in a large outdoor enclosure. *Folia Primat.* 1, 199 (1963).

Tarsiiformes

13

THE LIVING TARSIIFORMES consist of a single genus, *Tarsius*. Tarsiers are found only on islands of the East Indian Archipelago from the southern Philippines and Celebes on the east to Sumatra on the west. They have been found on the following islands: Sumatra, Banka, Billiton, Karimata Islands, Natuna Islands, Borneo, Celebes, Salayer, Pulo Peleng, Great Sangir, Mindanao, Bohol, Leyte, and Samar. They have also been reported from Savu, an island that lies between Sumba and Timor. The present distribution is shown on the map in Fig. 13.1.

Characteristics

Tarsiers are very small animals with very long tails and very long hind-limbs. They have enormous eyes which give them a most startling appearance (Fig. 13.2). Their dentition has certain generalized features of primitive mammals and is not typical of other prosimians. The lower incisors do not form a toothcomb or tooth scraper. The crowns of the upper molars are tritubercular. The dental formula is $\frac{2.1.3.3.}{1.1.3.3.}$. The foramen magnum is further to the front of the base of the skull than it is in other prosimians. This is related both to the expansion of the brain, especially the visual centers, and to the erect position in which the trunk is held when tarsiers hop.

The tarsus (bones of the arch of the foot) is elongated, and this is supposed to be the basis for the animal's name. The hands are specialized with great pads at the tips of the digits. These special pads allow *Tarsius* to cling to a flat smooth surface and even to walk along a vertically held sheet of glass. The second and third digits of the foot have very prominent, claw-like nails. Presumably these digits are used by *Tarsius* in making its morning toilet. Most prosimians have a single grooming claw (or toilet claw) on each foot whereas *Tarsius* has two of these.

Tarsiers are nocturnal animals. They are often called crepuscular as well, since they are active during the twilight hours too. They are arboreal and seem to prefer to live in what is known as second growth, where the primary rain forest or jungle has been cleared for agriculture or by sporadic fires. On the island of Borneo they are reported to live at altitudes no higher than 300 feet above sea level.

Tarsius spends most of the day sound asleep, clinging to a vertical branch. At twilight *Tarsius* awakens and hops in a rather frog-like fashion from tree to tree searching for food. It is believed that tarsiers are basically insectivorous. They will feed on grasshoppers, beetles, mealworms, and similar invertebrates when in captivity. According to recent reports by Harrisson of the Sarawak Museum in Borneo, tarsiers are particularly fond of very small

205

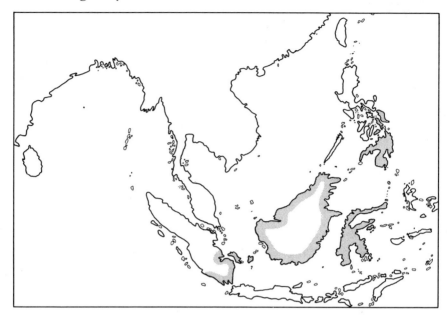

Fig. 13.1.   Geographical distribution of Tarsiiformes on the islands of southeast Asia. *Tarsius* has been reported from many localities within the shaded areas on the map.

lizards and geckos upon which they leap from relatively long distances and devour with gusto. It is not at all certain, yet, that fruits, vegetation, and flowers do not play a part in their diet.

Tarsiers move through the bush and trees with their characteristic frog-like hop. The tail is apparently of some importance in jumping, sitting, and moving. When the animal is clinging to a vertical stem or branch, grasping it tightly with hands and feet, the tail is very firmly pressed against the branch. The tail acts as a kind of strut to enhance the stability of the animal. Tarsiers leap by extension of their powerful lower limbs, and the mechanical action thus resembles that of a frog. Tarsiers also walk on all fours along branches and on flat surfaces. In such semiquadrupedal locomotion the tarsier's tail hangs down loosely. The tail is not prehensile as it is in South American monkeys, but it does have a very specialized area of skin on the ventral surface. Le Gros Clark noted that if the tail of a tarsier that is clinging to a branch is lifted up, the body seems to sag from the branch. When the tail is let loose, it snaps back into the position it had before. This presumably indicates a kind of incipient prehensility. When the animal becomes excited or is about to jump, tail flicks are often observed. When a number of tarsiers sleep together, their tails may intertwine about each other like the tails of lemurs.

The visual sense is probably the dominant sense of tarsiers. However we

Fig. 13.2. *Tarsius*. The animal shown here is a female from Borneo. (Photograph courtesy of the New York Zoological Society.)

would suspect that auditory and tactile impressions are also of great importance. Olfaction is clearly not very important. There is no direct evidence of this based on observations of behavior or on other experiments with live animals; the evidence is wholly morphological. The anatomy of the tarsier's brain and face shows that the olfactory apparatus is very much reduced and the visual apparatus enormously expanded when compared with more primitive mammals, including some other Prosimii. The eyeball of *Tarsius* is enormous relative to other structures in the skull. The retina is typical of nocturnal mammals in that the elements that receive sensory impressions are composed only of rods.

There are two kinds of photoreceptors in the retina—rods and cones. Rods function in what is known as scotopic or twilight vision. They are supposed to be very responsive to objects that move at the periphery of the visual field. The cones, sensitive to a much higher intensity of light, are the basis for photopic or diurnal, daylight, vision. In most diurnal mammals, the cones are more frequent toward the center of the retina and the rods toward the edges. In the retinae of nocturnal mammals, there are only rods.

A very important feature of the retina is the macula lutea. This macula lutea, or yellow spot, has practically no rods at all in higher primates. In the Anthropoidea and in man there is a small depression in the very center of the macula, the fovea, where there is only a single layer of cones. There are no blood vessels in the region of this depression. This single layer of cells and this pit (fovea) permit light to be brought to this single layer without passing through anything else in the retina and without interference of blood vessels. It is the point of greatest visual sharpness in the retina. It is believed to increase the visual acuity and discrimination of the cones of a diurnal mammal.

Polyak made a detailed examination of the retina of a living *Tarsius* and reported a large macula lutea with a very small fovea in it. This was unexpected by most students of the Primates, for the photoreceptors in the eye of *Tarsius* are all rods. In other words, it is a completely nocturnal retina. Polyak's findings become less astonishing when we realize that the nocturnal South American monkey, *Aotus*, which has a rod retina, also has something like the macula lutea with a vestigial fovea. It is difficult to decide whether the ancestral tarsiers were diurnal with the development of a rod retina a secondary adaptation to nocturnal life or whether the macula lutea and the fovea appeared in primate evolution at the prosimian level of phylogeny.

Behavior

It is unlikely that tarsiers live in social groups larger than pairs. In the London Zoo three tarsiers were kept together and, although there was a considerable amount of fighting, probably play fighting, it is reported that they were not "out for blood." There are no reports of hiding places, nests, or sleeping places such as holes in trees. How groups of them might behave in captivity is something we will have to wait for the future to show. At the present time, the available data indicate that tarsiers occur sporadically in pairs, but not in larger groups. This is, of course, based upon reports of aboriginal peoples and those who have been hunting tarsiers in their native bush habitat.

Female tarsiers appear to be good mothers. Infants cling tightly to the abdominal fur and are carried about in that position as the mother leaps from bush to bush. No reports of the mother carrying the infant in her mouth have ever been verified. There is no reliable information available about the period of gestation.

Systematics

The systematics of Tarsiiformes is in an unsatisfactory state, as is that of all the other Primates. There is one genus, *Tarsius*, and the principal problem is whether each island form should be put in a separate species. At the present time, three species, each with a large number of subspecies, have been defined (Table 13.1). This does not seem unreasonable. However, much work is necessary to verify this. The variation among the three species does not seem to be of any greater degree than occurs among various members of the species *Lemur fulvus* or of *Homo sapiens* for that matter.

There are some important problems outstanding in assessing the relationship of the living tarsiers to the fossil tarsioids. As was noted in the section on fossil primates, the distinction between a fossil tarsioid and a fossil lemuroid is difficult to make. *Necrolemur* is the best example of a tarsioid, and it is probable that the lineage from *Necrolemur* to *Tarsius* is more or less unbroken. Another interesting problem is whether the orbit of *Tarsius* (Fig. 13.3) is a development convergent, parallel, or affinitive to higher primates. Although we can become lost in the fine points of systematics here, it is quite clear that the morphology of the orbits should not be taken as evidence of affinity, but as parallelisms, for reasons we will discuss. Some details of the anatomy of the orbit are necessary in order to show why it is now the opinion that the development of the large orbits of *Tarsius* indicates parallelism to rather than affinity with the Anthropoidea.

TABLE 13.1
Classification of Tarsiiformes

Tarsius　Storr 1780	
Tarsius bancanus　Horsfield 1821	
Tarsius spectrum　Pallas 1778	
Tarsius syrichta　Linnaeus 1758	

The most important point about the orbits of tarsiers, and the point which distinguishes the orbits from those of the higher primates, is the way they close. Orbital closure occurs in *Tarsius* from three main centers of bony

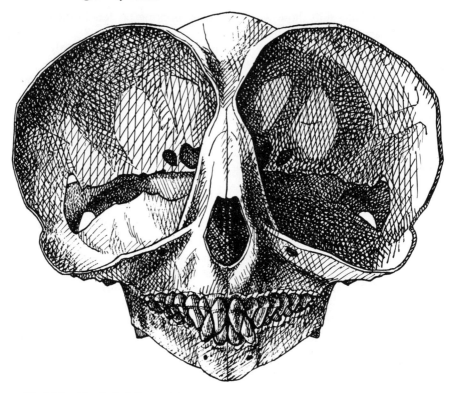

Fig. 13.3.   The skull of *Tarsius*. For a comparison of *Tarsius* and *Necrolemur*, see Fig. 7.6.

growth: (1) The posterior midregion of the orbital plate of the maxilla grows upward in a ventral and lateral direction. (2) The middle part of the post-orbital bar grows in an anterior and posterior direction, lateral to its own center. (3) A flange from the frontal bone starts beneath the postorbital bar on the frontal bone and grows dorsally and laterally. The flanges of bone that surround the orbit (circumorbital flanges) of *Tarsius* are enormous and startling. Studies made by Simons and Russell of crania of *Necrolemur* show that the flanges around the orbit of this fossil are not so pronounced and resemble those of immature tarsiers. In the specimens of *Necrolemur*, at least two of the three growth centers for bony closure of the orbits seem to be developed to some degree as in *Tarsius*. In one specimen with an uncrushed orbital bar, ossification was spreading from the center of the bar in an anterior and posterior direction. The frontal element of the orbits, that is, the dorsal and lateral spreading of a flange of bone from the frontal bone, appears to be reduced in *Necrolemur*. As mentioned earlier, *Necrolemur* cannot be a lineal ancestor of *Tarsius* because of differences in dentition. The molars of *Necrolemur* are more specialized in the anthropoid direction than are those of living *Tarsius;* the molars have developed a hypocone. *Necrolemur* provides evidence, nonetheless, that postorbital closure, characteristic of tarsiers, was developing in primates of the Eocene.

When we compare *Tarsius* to other living primates, it is important to point out that the way postorbital closure occurs is quite different from that seen in ceboids and cercopithecoids. At least the malar and frontal bone components are different. In tarsiers the postorbital closure occurs by an outward and downward growth of a flange of the frontal bone, with a very small expansion of the malar bone. In the higher primates, that is, the ceboids and cercopithecoids, closure of the dorsal and lateral parts of the orbit is produced by the growth of the orbital plate of the malar bone. This difference has suggested that the orbital closure of the tarsiers is a development parallel to that of the Anthropoidea and is not a character of common inheritance.

The Tarsiiformes have been considered primitive living representatives of the prosimian lineage from which the Anthropoidea diverged. The prominent eyes and the bony orbits were believed to be a characteristic which demonstrated a relationship between tarsiers and the anthropoids. However analysis of the structure of the orbit shows that we cannot consider tarsiers as lineal relatives. They are living relicts of the Eocene who managed to avoid extinction.

SUGGESTED READINGS AND SOURCES

Burmeister, H., *Beiträge zur Kenntnis der Gattung Tarsius*. G. Reimer, Berlin (1846).

Clark, W. E. Le G., Notes on the living tarsier (*Tarsius spectrum*). *Proc. Zool. Soc. London*, 217 (1924).

Hill, W. C. O., *Primates, Vol. II. Haplorhini: Tarsioidea*. Wiley, Interscience, New York (1955).

Jones, F. Wood, Some landmarks in the phylogeny of Primates. *Human Biol.* 1, 214 (1929).

Polyak, S., *The Vertebrate Visual System*. Univ. of Chicago Press, Chicago (1957).

Simons, E. L., and D. E. Russell, The cranial anatomy of *Necrolemur*. *Brev. Museum comp. Zool. Harvard* No. 127, 1 (1960).

Lorisiformes

14

LORISIFORMES HAVE BEEN REPORTED from all parts of Africa south of the Sahara, on the island of Zanzibar off the east coast of Africa, and, presumably, on other islands off the African mainland. But they are not found on the Comoro Islands or on Madagascar. They have also been found in parts of Asia, particularly India, Burma, Ceylon, and throughout southeast Asia (Fig. 14.1).

All the Lorisiformes are arboreal and nocturnal. They have developed the tapetum in common with many other nocturnal animals. The tapetum is a layer on the choroid of the eye (the pigmented vascular layer next to the retina) that concentrates dim light, such as the animal is apt to have available in twilight or moonlight. This layer is responsible for the eyes that glow in the dark when one flashes a bright light at a nocturnal animal.

The Lorisiformes may be conveniently divided into two locomotor groups, the slow climbers and creepers and the fast hoppers. The slow climbers and creepers are usually placed in the subfamily Lorisinae and the fast hoppers in the subfamily Galaginae (Table 14.1). The principal differences between

TABLE 14.1
Classification of Lorisiformes

Lorisinae   Flower and Lydekker 1891
   Loris   E. Geoffroy 1796
      Loris tardigradus   Linnaeus 1758
   Nycticebus   E. Geoffroy 1812
      Nycticebus coucang   Boddaert 1785
   Perodicticus   Bennett 1831
      Perodicticus potto   P. L. S. Müller 1766
      Perodicticus calabrensis   J. A. Smith 1860
Galaginae   Mivart 1864
   Galago   E. Geoffroy 1796
      Galago alleni   Waterhouse 1837
      Galago crassicaudatus   E. Geoffroy 1812
      Galago demidovii   Fischer 1808
      Galago elegantulus   Le Conte 1857 ?
      Galago senegalensis   E. Geoffroy 1796

the two groups are structural modifications of the limbs that are consequences of very advanced specializations for arboreal life. The lorises have gone in one direction and the galagos in another. We shall discuss the two subfamilies separately.

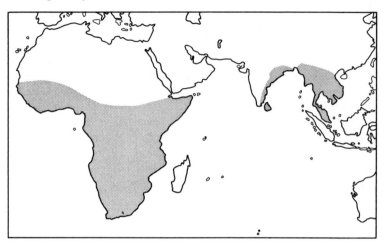

Fig. 14.1. Geographical distribution of Lorisiformes. *Perodicticus* and *Galago* are found only in Africa; *Loris* and *Nycticebus* are found only in southeast Asia.

Lorisinae

The Lorisinae have the wider distribution of the two subfamilies. The genus *Perodicticus* occurs in Africa, and the genera *Loris* and *Nycticebus* in Asia. There are two African lorises, the potto (*Perodicticus potto*) and the angwantibo (*P. calabrensis*). The pottos have very short, stubby tails, reddish brown fur, and look a little like small teddy bears (Fig. 14.2). They have several exposed spines on the cervical vertebrae. Reports that these are used when the animals fight have not been confirmed by reliable observers. The angwantibo, or golden potto, of West Africa is extremely rare. It is smaller than the potto and is not very well known.

There are two Asiatic lorises, the slender loris (*Loris tardigradus*) found in Ceylon and southern India and the slow loris (*Nycticebus coucang*) found throughout southeast Asia. The slender loris looks a bit like a banana on stilts and has no tail (Fig. 14.3). It is wholly arboreal and nocturnal. The slow loris resembles a pink and white potto, although usually larger with a very short, stubby tail.

The dentition of the lorises is typically prosimian (Fig. 14.4). The lower incisors and canines form a procumbent toothcomb. The first premolar closely resembles the usual canine tooth. It is sharp, pointed, and has a single root. The dental formula is $\frac{2.1.3.3.}{2.1.3.3.}$. The digits of hands and feet are quite specialized. The index finger is a mere tubercle with no nail. The second digit of each foot has a large grooming claw rather than a typical primate nail. Close observation shows that this structure is used to scratch the ears, head, and mouth.

Fig. 14.2. *Perodicticus potto.* The potto is the African representative of the lorises.

Lorises walk along branches hand over hand. This progression is accompanied by a side-to-side snake-like wiggle of the back, a mechanical consequence of alternately placing a hand, then a foot, then the other hand, then the other foot, in consecutive order, on a branch. These slow creepers and climbers move with caution up and down branches and tree trunks, extending themselves cautiously, gripping the branch with their prehensile feet, reaching toward a secure platform on another branch. They seldom move very rapidly, although on occasion we have seen them amble hastily along the ground when attempting to escape. The lightning fast movement when

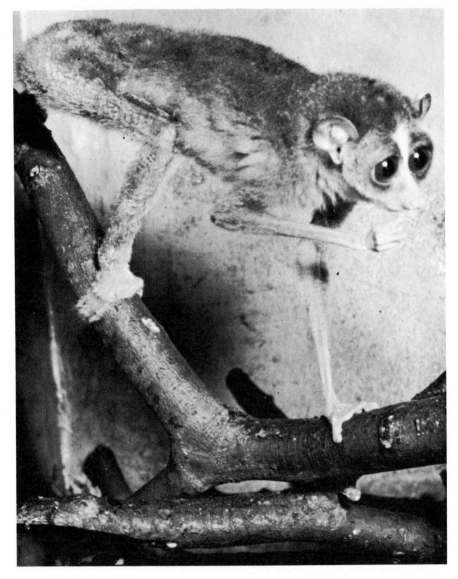

Fig. 14.3. *Loris tardigradus*, the slender loris of Ceylon and India. (Photograph courtesy of A. Jolly.)

they strike at an insect with their forelimb is an astonishing contrast to their slow progression along branches of trees.

One of the specializations of the lorises, especially marked in the potto, is a very powerful grasp. This grasp can be sustained for a very long time. The reduction of the index finger to a mere bump increases the power of the hand. Two persons, pulling with as much strength as can be mustered, are often needed to wrest one potto free of its grasp of a stick or the wire floor of its cage. The thumb and big toe are strong and well developed, and there is a wide angle of divergence between them and the other digits. The maintenance of such a powerful grip for a very long time is due to a special group of blood vessels which form a rete mirabile. These retia are a kind of storage chamber for the blood which passes back and forth from the distal muscles of the limbs. The muscles may remain contracted for long periods of time without the usual fatigue. Respiratory exchange in the contracting fibers of the muscles is more efficient because of these retia mirabilia.

The strength in the limbs and the powerful grasp enable lorises and pottos to suspend themselves by their hindlimbs leaving their hands free to capture or grasp food. They are able to support themselves by one foot, especially if they are suspending themselves upside down. They often walk upside down appearing superficially like sloths. Lorises, like all the Primates, often sit with their trunks erect. They will rise up on their hind limbs when reaching for a branch or twig above their heads.

The diet of the Lorisinae is believed to be insectivorous. Yet they eat fruit, vegetables, flowers, small mammals, birds' eggs, baby birds, and almost anything else offered them, and are essentially omnivorous in captivity.

The social behavior of the Lorisinae has not been closely observed in the field. They are nocturnal and arboreal, a fact that makes field observations difficult. They sleep all day curled up in crooks or holes of trees. They are all quite solitary animals, although they are usually found in congeries rather than at random in their forest habitat. The slender loris is very quarrelsome, and it is not easy to keep pairs in cages. The others are no more sociable, but groups have been kept together successfully.

The females are extremely good mothers, and the young often cling to them for many weeks after birth. They cling tightly as the mother ambulates

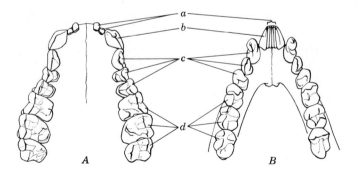

Fig. 14.4. Upper (A) and lower (B) jaws of *Loris tardigradus*; (a) incisors, (b) canines, (c) premolars, and (d) molars. The toothcomb formed by the lower incisors and canines is present and is typical of Lorisiformes and other prosimians.

slowly through the trees. The slender loris often produces two young at birth, but the others usually have single births. There are few available data about the breeding habits of these animals. Females appear to have more than a single ovulatory cycle per year.

Galaginae

The Galaginae (bush babies) are found only in Africa south of the Sahara desert and on certain offshore islands such as Zanzibar. The galagos range from animals the size of large mice to those the size of small rabbits. All have long tails, and those of the larger species, *Galago crassicaudatus* for example, are very bushy and thick. The smaller species such as *G. senegalensis* have rather thin tails with the hair closely packed.

All galagos have prehensile hands and feet with five well-developed digits. The second digit of each foot has a grooming claw, which is probably a specially adapted nail. The dentition is typically prosimian, with a well-developed toothcomb formed by the lower incisors and canines (Fig. 14.4). The first lower premolar is caniniform with a single root. The dental formula is $\frac{2.1.3.3.}{2.1.3.3.}$.

The galagos move through the trees with great facility and grace. Both the very small *G. senegalensis* (and *G. demidovii* and *G. alleni*) and the larger *G. crassicaudatus* can leap from tree to tree with amazing agility for relatively long distances. Their movements are similar to those of a kangaroo, since the powerful hindlimbs propel the animal forward while the forelimbs appear to be tucked against the chest. The forelimbs grasp the branch or tree trunk on which the animal lands. The galagos are also very facile at leaping from vertical support to vertical support, gripping such supports with considerable force. The long bushy tail seems to be strictly a balancing organ, for it is not prehensile. There is no sign of specialized skin on it as there is on the tail of *Tarsius* or on the tails of South American monkeys.

Galagos will move on the ground by hopping kangaroo-fashion. The trunk is held erect, and they hop from place to place on the floor. They will also sit erect while holding in their hands a piece of carrot or eating anything large that requires a number of bites. They are not terrestrial animals, but they obviously can move on the ground from tree to tree if necessary. They will walk across the floor in a kind of quadrupedal saunter. But they seem to prefer hopping. They will jump from chair to bookcase to lamp to desk, seizing upon vertical surfaces or projections much as if they were hopping through trees.

The diet of the Galaginae is quite omnivorous. They have been observed to eat fruit, vegetation, small mammals, geckos, insects, and eggs. In captivity there are few things they will not eat.

The Galaginae have not been intensively studied in their native habitat, for they, too, are completely arboreal and nocturnal. Most of our information comes from studies on captive or tame, hand-reared animals. Galagos do not appear to live in distinct, structured, social groups, yet congeries are usually found in a single place. During the day as many as eight have been observed sleeping together in the high branches of a single acacia tree. They appear to go their separate ways at night, calling and shrieking to maintain contact with each other.

The galagos can be divided into two groups. One consists of small, very active, nervous, quick animals such as *G. demidovii* and *G. senegalensis*. The other includes the larger, relatively slower and more sluggish animals such as *G. crassicaudatus*. The smaller, quick-moving galagos are often found in groups. But whether these are actual social groups or congeries of animals is not yet determined. The females of *G. senegalensis* and *G. demidovii* in captivity breed and raise their young with the males in the same cages. A number of these smaller galagos may be kept caged together with only occasional evidence of fighting and aggressive behavior.

The larger *G. crassicaudatus* appears to be much less social than the smaller species. Pairs have been kept together, but there are occasional fights. During the ovulation of the females, the males become quite agressive. Males usually are taken out of cages of their pregnant mates, for they often attack and kill the infants immediately after birth. The behavior of the females in our colony toward their mates at this period suggests that under natural conditions they would drive the males away in late pregnancy and raise their infants in solitude.

Single births probably are the rule, although a number of multiple births have been recorded—in one case, triplets. Our own experiments in breeding galagos suggest that multiple births are associated with particular lineages or pedigrees and are not distributed at random. Multiple births may also be related to species and subspecies differences. Gestation is about 16 weeks, and there are several ovulations each year.

The newborns cling very tightly to their mothers for at least the first week or two of life. After that, our experience with captive and tame animals suggests that the infants are left behind in tree hollows when the female goes foraging at night. It does not seem reasonable, to us, that a female could jump from tree to tree carrying two or three infants. Even a single one is difficult to manage after the first week or two. Females carry their infants in their mouths by seizing between their jaws a fold of the infant's skin at the lower abdomen (Fig. 14.5).

Systematics

The Lorisiformes are divided into two subfamilies and several genera (Table 14.1). The African Lorisinae were once placed in two genera,

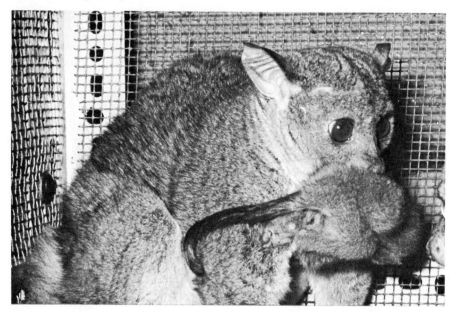

Fig. 14.5. *Galago crassicaudatus*, female with infant. Note that the mother is holding the infant by the skin fold on the infant's abdomen.

Arctocebus, the angwantibo, and *Perodicticus*, the potto. The angwantibo has not been studied extensively, for it is apparently rare. It is probably more reasonable to put the two African lorises in a single genus, *Perodicticus*, and have two species. The Asiatic lorises have been placed in two genera, *Loris* and *Nycticebus*, with a single species in each.

The Galaginae were once split into a number of genera and species. Today it seems most reasonable to put them in a single genus, *Galago*. Many have defended the erection of another genus, *Euoticus*, on the basis of the "needle claws" of these animals (*Galago elegantulus*). They have nails, not claws, but there is a raised keel in the center of each nail which has caused much excitement among those who belong to the splitting school of classifiers. Until more detailed study of living animals is possible, we prefer to place all Galaginae in one genus (Table 14.1).

Suggested readings and sources

Bishop, A., Use of the hand in lower primates. *In* J. Buettner-Janusch (Ed.), *Evolutionary and Genetic Biology of Primates*, Vol. II, p. 133. Academic Press, New York (1964).

Buettner-Janusch, J., The breeding of galagos in captivity and some notes on their behavior. *Folia Primat.* **2**, 93 (1964).

Buettner-Janusch, J., and R. J. Andrew, The use of the incisors by Primates in grooming. *Am. J. Phys. Anthropol.* **20**, 127 (1962).

Butler, H. The reproductive biology of a strepsirhine (*Galago senegalensis senegalensis*). *In*

W. J. L. Felts and R. J. Harrison (Ed.), *International Review of General and Experimental Zoology,* Vol. I, p. 241. Academic Press, New York (1964).

Hill, W. C. O., *Primates, Vol. I. Strepsirhini.* Wiley, Interscience, New York (1953).

Lowther, F. de L., A study of the activities of a pair of *Galago senegalensis moholi* in captivity, including the birth and postnatal development of twins. *Zoologica* 25, 433 (1940).

Petter, J.-J., Nos premiers parents: les Lémuriens. *Sciences* (Paris), No. 23, 8 (1963).

Petter-Rousseaux, A., Recherches sur la biologie de la reproduction des primates inférieurs. *Mammalia,* Ser. A, No. 3794 (1962).

Lemuriformes

15

THE ISLAND OF MADAGASCAR, fourth largest island in the world, is the home of all the Lemuriformes. The distribution of the Lemuriformes on the island is shown in Fig. 15.1. The Lemuriformes are divided into two superfamilies, the Lemuroidea and the Daubentonioidea (Table 15.1). The Lemuroidea are subdivided into two families, the Lemuridae and the Indriidae. The Lemuridae are further divided into two subfamilies, the Lemurinae and the Cheirogaleinae. Daubentonioidea contains a single genus.

The lemurs are so various, so relatively little known, and so fascinating to us that the discussion of them will be more detailed than the discussion of the other Prosimii. In addition, the presence of lemurs on Madagascar presents an intriguing problem in itself. We shall discuss this after we describe some of the members of this group.

Lemurinae

The Lemurinae are perhaps the largest group of the Malagasy lemurs. Three genera have been defined, *Lemur, Hapalemur,* and *Lepilemur.* Of these three genera, *Lepilemur* appears to be completely nocturnal in its habits and the others diurnal, crepuscular, and, for that matter, somewhat

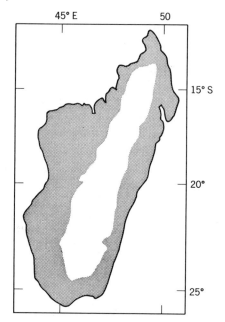

Fig. 15.1. Geographical distribution of Lemuriformes on Madagascar. The Lemuriformes are found only on the island of Madagascar. Some genera, such as *Lemur*, have been reported from almost every place on the island where Lemuriformes are found. Other genera, such as *Indri*, have been reported from restricted areas only.

nocturnal also. Our own field observations of certain members of the genus *Lemur* indicate that these animals lead busy lives, active almost 24 hours per day. It is certain that they have periods of rest, somnolence, or sleep during the hours around midday and presumably around midnight. All of the species in the subfamily Lemurinae are arboreal.

The nocturnal genus, *Lepilemur*, has not been studied extensively. It is a small animal, approximately the size of *Galago crassicaudatus*, with a long tail. It resembles those galagos superficially. Its diet is said to be wholly insectivorous, although planters and others who have kept *Lepilemur* in captivity on Madagascar and in Paris find it eats vegetable matter and synthetic diets as well as insects and newborn mice. It is quite a solitary animal, but it seems to occur in congeries rather than at random in the forests. Woodcutters have found pairs in hollow trees or in holes in living trees. There seems to be no social group larger than the pair.

All the other Lemurinae are completely arboreal and largely diurnal. At one time it was suggested that *Hapalemur* is a nocturnal, or at least crepuscular, lemur. Our own detailed field observations show that it is far more

TABLE 15.1

Classification of Lemuriformes

Lemuroidea   Mivart 1864
  Lemuridae   Gray 1821
    Lemurinae   Mivart 1864
      Lemur   Linnaeus 1758
        Lemur catta   Linnaeus 1758
        Lemur fulvus   E. Geoffroy 1812
        Lemur macaco   Linnaeus 1766
        Lemur mongoz   Linnaeus 1766
        Lemur rubriventer   I. Geoffroy 1850
        Lemur variegatus   Kerr 1792
      Hapalemur   I. Geoffroy 1851
        Hapalemur griseus   Link 1795
        Hapalemur simus   Gray 1870 ?
      Lepilemur   I. Geoffroy 1851
        Lepilemur mustelinus   I. Geoffroy 1851
        Lepilemur ruficaudatus   A. Grandidier 1867
    Cheirogaleinae   Gregory 1915
      Cheirogaleus   E. Geoffroy 1812
        Cheirogaleus major   E. Geoffroy 1812
        Cheirogaleus medius   E. Geoffroy 1812
        Cheirogaleus trichotis   Günther 1875 ?
      Microcebus   I. Geoffroy 1828
        Microcebus coquereli   A. Grandidier 1867 ?
        Microcebus murinus   J. F. Miller 1777
      Phaner   Gray 1870
        Phaner furcifer   Blainville 1839
  Indriidae   Burnett 1828
    Indri   E. Geoffroy and Cuvier 1795
      Indri indri   Gmelin 1788
    Lichanotus   Illiger 1811
      Lichanotus laniger   Gmelin 1788
    Propithecus   Bennett 1832
      Propithecus diadema   Bennett 1832
      Propithecus verreauxi   A. Grandidier 1867
Daubentonioidea   Gill 1872
    Daubentonia   E. Geoffroy 1795
      Daubentonia madagascariensis   Gmelin 1788

diurnal than *Lemur,* the typical genus of the subfamily. *Hapalemur* is extremely active from sunrise to sunset and appears not to move out of its roosting place during the night. Our captive animals have periods of activity at night and rest quietly and sleep for several hours during the day. This is quite unlike their behavior as observed in the Forest of Perinet in Madagascar.

Lemur catta is said by many to be a terrestrial lemur that inhabits rocky promontories, thus occupying an econiche very similar to that of the hama-

Fig. 15.2.   Members of the subfamily Lemurinae. (A) *Lemur catta*, the ring-tailed lemur, (B) *Lemur fulvus collaris*, the whiskered lemur, (C) *Lemur fulvus albifrons*, the white-fronted lemur, (D) *Hapalemur griseus*, the gentle lemur, and (E) *Lemur variegatus*, the ruffed lemur. The four *Lemur fulvus albifrons* (C) are a familial group, adult male and female and infant twin offspring, male and female. Sexual dimorphism is pronounced in this subspecies; the males have prominent white topknots; the faces and heads of the females are completely gray.

dryas baboon of Africa. Our own observations, once again, suggest that *Lemur catta* is largely a diurnal animal that lives in troops of moderate size and is almost wholly arboreal. However, it is the only species of *Lemur* that we have observed crossing open places in the forest and the only one we have seen come out of the trees in the dense forests to eat grass and other terrestrial vegetation. Reliable observers also report that occasionally they have seen a *Lemur catta* race madly across a bush road in southern Madagascar. Most of the terrain and habitat where such observations were made is cutover land, covered with low-growing bushes and with many stands of very tall trees. *Lemur catta* is basically an arboreal animal, but it has adjusted to a considerable amount of terrestrial activity.

All of the other members of the genus *Lemur* seem to be completely

Fig. 15.2 (Continued)

arboreal. They appear to spend most of their time well below the tops of large trees, although toward evening and early morning they may be found in the highest branches of these same trees. There seems to be some stratification of the various Lemuriformes in the forests of Madagascar. For example, the diurnal species of the family Indriidae seem to spend most of

Fig. 15.2 (Continued)

their time high in the crowns of the trees, whereas species of the genus *Lemur* appear to spend most of their time beneath the crowns of the trees.

There is a wide range of variation in size and pelage color and markings within the subfamily Lemurinae. *Lepilemur* is small, generally gray or brown, somewhat like the larger galagos. *Hapalemur* is also a small animal, the size of a very large gray squirrel or a very small cat, brown, with a long bushy tail. It often has a rust-colored triangle of fur on the top of its head. *Lemur variegatus,* the largest member of the genus, is found in two major color forms, black and white or red and black. The black and white form is reminiscent of a small panda. *Lemur variegatus* is about the size of a cocker spaniel and has a long bushy tail and rather large ear tufts. The other members of the genus *Lemur* vary enormously in color although their sizes range from that of a small cat to that of a medium-sized dog. They all have long bushy tails quite obviously used as balancing organs when they leap from tree to tree and branch to branch. Some notion of the wide variation in external appearance of these lemurs may be seen in Fig. 15.2.

All species of the subfamily Lemurinae have prehensile hands and feet with five well-developed digits on each extremity. The second digit of each foot has a grooming claw which resembles that of the Galaginae.

The dentition is typically prosimian (Fig. 14.4) with a very well-developed toothcomb or tooth scraper formed by the lower incisors and canines. The dental formula is $\frac{2.1.3.3.}{2.1.3.3.}$. The first lower premolar is caniniform and in this way resembles that of the galago. *Lepilemur* is an exception, with a dental formula of $\frac{0.1.3.3.}{2.1.3.3.}$. It has two upper incisors in its milk or deciduous dentition but loses them when it matures, and they are not replaced. In all other respects it has a typical lemur dentition with the procumbent lower incisors and canines forming the tooth scraper.

Lemurs move with great facility through the trees. They are among the most spectacular arboreal aerialists known. Anyone who has observed these animals closely in their native habitat sees that the way they leap over extraordinary distances and grasp small branches, as they swing themselves up into a tree at the end of a jump, resembles the precision of the very greatest human trapeze artists. There is little question that binocular vision is well advanced in this group. The evidence is not complete, but, as we noted in the introductory section to these chapters on the living primates, the overlap in the field of vision is sufficient for binocular vision, and the proper neurological structures are in the brain. Whether color vision is present among these animals has often been discussed; it is our opinion that it probably is. None of the evidence that has been presented supports, in any conclusive way, its absence. The wide range of color and the considerable color sexual dimorphism among the many species and varieties of lemurs suggest that color vision is present.

The diet of most lemurs is mainly vegetable matter. They appear to avoid insects and meat. Observation in the field suggests they subsist upon fruit, flowers, bark, and leaves. In the captive colony they eat most synthetic diets and a wide range of vegetable materials. They usually avoid insects, live baby mice, and other animal protein devoured avidly by other prosimians in captivity.

The members of the genera *Lemur* and *Hapalemur* are highly gregarious, social animals. Although not all are wholly diurnal, they are sufficiently active during the day that few would regard their behavior during the daylight hours as atypical. They live in social groups ranging from four to five to as many as sixty. There is evidence that the groups as large as sixty are actually nocturnal nesting or sleeping groups. During periods of diurnal activity, when the animals are actively seeking food and exploring their habitat, they appear to break up into smaller groups. There are few data available about the structure of such bands or troops. The most reliable field observations show that the sex ratio, for example, is almost always 1:1. This is rather different from the distribution found among the troops of monkeys of the Old World in which the sex ratio is seldom 1:1. Among baboons it may even be as high as 10 females:1 male (Chapter 17).

The diurnal lemurs sleep high in the trees, in groups. Two or more will huddle together on a limb or in the crotch of a tree with their tails wrapped around themselves and each other. The tail is longer than the body of the animal. A lemur will rest, sitting up partly erect, with its tail thrown over one shoulder. If two of them huddle together, the tails often wrap up more than one lemur. Although lemurs are active at night, there are periods when they appear to settle down and sleep or rest. In early morning as the sun rises they are usually asleep. Upon awakening they raise their heads, stretch, yawn, and groom themselves and each other.

The process of grooming is thorough and most interesting. The lemurs do not have a very fine grasp and cannot take hold of each other's hair for grooming. They will grasp each other and groom the part grasped with tongue and toothcomb. Solitary lemurs kept in cages by themselves manage to keep most of their pelage in excellent condition, but none ever looks as well groomed as those that have the opportunity to groom each other. The grooming claw of the second digit of the foot is generally used to scratch the mouth, ears, and head. It is not used for careful grooming of the coat as its name implies.

A controversy over the function of the procumbent lower incisors and canines of lemurs and other prosimians has been revived from time to time. Many have insisted that it is not a scraper or comb for the pelage. Recently, the controversy has been laid to rest, we hope. The animals do not use the procumbent lower incisors in any manner similar to that of the rodents. This procumbent tooth structure is used almost exclusively in grooming. Fine hairs from the coats of the lemurs have been found wedged between incisor and canine teeth in this procumbent tooth structure after a bout of grooming.

Grooming is an essential part of the social life of lemurs. It is one of the most common activities in which lemurs mutually indulge. Its importance is probably as much social as it is to remove scurf, parasites, and filth from the coat. The intense grooming of the infant lemur by its mother is indicative of this.

As the sun warms the branches upon which lemurs have been sleeping, and it is quite cold at night in a Madagascar forest, the animals will sit erect, stretch their arms out and expose their abdomens to the warmth of the sun. *Lemur variegatus* is noted for this behavior, to the extent that the native Malagasy consider it a sun-worshipping animal.

Fully awake, the animals leap off through the trees, somewhat in follow-the-leader fashion, seeking food and whatever else keeps a lemur busy and interested during the day. Considerable scent marking is observed as they move from place to place. The males and females will rub their anal scent glands thoroughly on a branch or other object on the daily path. Each animal that follows will do the same. It is possible that the slow, careful somewhat fussy procession of these lemurs through the high trees at night, in the dark, is facilitated by well-marked trails left during the day.

Troops of lemurs appear to be territorial in behavior. They recognize foreign troops of their own species, give warning calls when these foreign troops appear, and either retreat or sit in the trees and make warning or threat gestures. Very few reliable observations exist about interaction of troops of lemurs of the same species. Our few encounters with them in the field suggest that two troops of the same species which meet will retreat and avoid the meeting place for a very long time afterward.

Within a troop there seems to be little aggressive behavior, except during the breeding season. According to available evidence, there is one ovulation and perhaps one estrus per year among diurnal Lemurinae. The males become excited and quite active during this season, particularly in the morning and evening hours, and probably late at night. The females are marked vigorously with the anal scent glands of the male, while, at the same time, the females are approached by other lemurs or humans. This appears to have a proprietary effect. During this time other males are driven off, and the few aggressive traits ever seen in ordinary daily lemur life appear. Females will show considerable aggressiveness and will present their backsides to males. We have noticed that female *Lemur macaco* in captivity become very aggressive and bad tempered during the estrus period. This may have been exacerbated by the captive conditions and isolation, for the three bad-tempered females we observed had no male lemur with them. There are reports that shortly after the breeding season troops of lemurs will have a number of individuals with torn ears and scarred flanks, suggesting that a considerable amount of violence accompanies estrus.

Mating may occur at night, particularly during the monthly period of the full moon. This has often been dismissed as a legend, but recent observations have led some to suggest that general activity is greater on those nights with a full moon. We found in two different Madagascar forests that *Lemur fulvus* was just as active during pitch dark nights as it was on nights when the moon was full. Some believe there are more than one estrus period per year in the genus *Lemur*. No conclusive evidence exists, and all available data suggest there is only one ovulation per year. The ovulation must, if birth records are reliable, occur at the same season every year.

Gestation lasts from 120 to 135 days, according to observed copulations and pregnancies in zoos and in captive colonies. The actual date of conception may not be determined accurately, but there is not likely to be more than a 2 or 3 day error in estimating the gestation period. Mating behavior, at least during the day, occurs only on a very few days in the year.

The usual number of offspring among the Lemurinae is one. In our limited experience we are aware of five sets of twins and one set of triplets, all born to captive animals, although one set of twins must have been conceived before capture. There is no body of data against which to compare these numbers, and we have no way of determining at the moment whether this is an abnormal frequency or we happened to have sibships that have a familial tendency for multiple births.

The young are relatively helpless when born and cling tightly to the fur of the mother's belly. We wonder how helpless they are, for they are able to cling tightly enough for the mother to leap through the trees and keep up with the troop during the day. Our observations of captive lemurs demonstrate that Lemurinae females make extremely good mothers, very solicitous of their young. They drive off all other lemurs who approach or disturb them and their infants. The males of pairs that have been together in cages for a long time are permitted to groom the infant from the day of birth. Other females and males in such a cage are driven off by the mother. At approximately 2 months of age, the infant lemur is occasionally seen away from its mother. But it seldom completely loosens its hold, usually touching the mother with foot or hand. At about 10 weeks, the young lemur moves around by itself, at least this is so in captivity.

There is one exception to this general mode of infant behavior and maternal care among diurnal lemurs. *Lemur variegatus* produces a completely helpless and partially hairless infant. Shortly before birth, the female has been observed to pull hair from her flank and make a small nest. She adds leaves and torn paper, if she is in captivity. The helpless infant is left in this nest while the mother forages for food. The infant *Lemur variegatus* is almost completely hairless as late as 15 days of age. The mother will carry it by a skin fold at the lower part of the abdomen. This is exactly the same place female galagos carry their infants. There is no attempt to carry infants by the scruff of the neck.

Infant lemurs are extremely attached to their mothers. If they are removed from their mothers, they seek some hairy or furry object upon which to cling. We have removed infant lemurs between the ages of 3 and 5 weeks and found that for a considerable period of time they require a security blanket or a surrogate mother. Even as they grow tame and accustomed to humans, they rush back to their cage to lick and touch the surrogate every few minutes. It is not clear whether the phenomenon of imprinting, as known from ducks, is a feature of the learning of infant lemurs. But there are some analogies to imprinting in the way baby lemurs attach themselves to certain humans after they have been removed from their mothers and the way ducklings imprint upon humans.

Cheirogaleinae

The Cheirogaleinae are small, totally nocturnal lemurs of Madagascar. Since their nocturnal habits make it difficult to estimate their numbers, they may be more numerous than Lemurinae. We were surprised at the large number observed in certain forests in southern Madagascar. There are three genera and possibly five species. They range in size from an animal smaller than some mice, *Microcebus*, whose weight ranges from 50 to 90 grams, to an animal the size of a red squirrel, *Cheirogaleus major*. *Microcebus* is the smallest lemur and the smallest living primate. As is the case with the nocturnal Lorisiformes, the nocturnal Cheirogaleinae may be divided into two groups, faster and slower moving species. *Microcebus* and *Phaner* are the faster of the three genera, and *Cheirogaleus* the slower. *Microcebus* lives in low trees and bushes and moves rapidly. It can jump considerable distances and run rapidly along very tiny twigs. The animals are usually found in pairs, and we have some evidence, as is the case with most nocturnal primates, that congeries of them are grouped in the forest.

The genus *Phaner*, the fork-marked lemur, is a little known but extremely interesting nocturnal creature. There is apparently only one species, *Phaner furcifer*, and it is relatively widespread in southern Madagascar. Very few specimens have ever been captured, and almost no living specimens have been studied. *Phaner* is about the size of a small *Galago crassicaudatus*. It has a long feathery tail and resembles galagos in many superficial ways. Its specific name is derived from the forked band of dark color that runs down its back from the top of the head to the tail. We have been able to observe and photograph this animal on a number of occasions in the early evening in forests of southern Madagascar (Fig. 15.3). It does not appear to be able to leap with the force of galagos since its fore- and hindlimbs are approximately equal in size and strength. Nevertheless, it is very agile, rapid, and difficult to follow and watch as it scampers through the trees. It appears to be a solitary animal, but again congeries rather than random individuals are found. It has a very loud, piercing call often confused with that of some nocturnal bird. These calls are used in situations of alarm and, most likely, in order for the animals to keep in contact wtih each other.

Cheirogaleus medius, the smaller of the two species of *Cheirogaleus*, resembles *Microcebus* very much in its habits, behavior, and distribution,

Fig. 15.3.   *Phaner furcifer,* the fork-marked lemur. This picture was taken about 11:00 P.M. in the forest near Sakaraha in southern Madagascar. Illumination was provided by two powerful spotlights mounted on the roof of a Land Rover. (Photographed by P. Boggess and J. A. Smith.)

although it is about twice as big. *Cheirogaleus major* appears to be a much more torpid and slothful animal but is actually quite swift and agile when fully awake at night. *Microcebus* and *Cheirogaleus* do not jump as do the galagos. They run rapidly along the branches and twigs of low trees and bushes. They have extremely effective prehensive power in their hands. The palmar and plantar pads have coarse ridges which probably increase the strength of their prehensile grasp. The hands and feet of the Cheirogaleinae are typical of the Lemuridae with a grooming claw on the second digit of the foot.

The dental formula of the Cheirogaleinae does not vary from the general lemur formula of $\dfrac{2.1.3.3.}{2.1.3.3.}$. The lower incisors and canines form the typical tooth scraper or toothcomb of the Lemuridae.

There is one physiological peculiarity which is characteristic of the genera *Cheirogaleus* and *Microcebus* among the Cheirogaleinae. These animals go through a torpid phase each year. It seems to be a period during which physiological activity practically ceases and the animals exist on the fatty tissue stored in their tails and rumps. The tail, which becomes enormously fat at a certain period of the year, is responsible for their common name, fat-tailed lemur. The term for the period of torpor is aestivation. It does not occur during a true winter and hence cannot, etymologically, be termed hibernation.

Evidence based only on *Cheirogaleus* and *Microcebus* is that Cheirogaleinae are polyestrus during certain seasons of the year. In other words, there appear to be more than one estrus during certain months of the year. These then alternate with periods of sexual inactivity. The best data available indicate that the duration of an estrous cycle is about 30 days in *Cheirogaleus* and 50 days in *Microcebus*. Nevertheless most sexual activity seems to be seasonal, and births appear to occur seasonally, once a year. Twins are extremely common among *Microcebus*. Observations of captive animals suggest that gestation ranges from 59 to 70 days. As far as the reliable available data show, neither of these two genera have infants that are carried on the mother's back or that cling to the fur of the belly. Both *Microcebus* and *Cheirogaleus* females carry the infants in their mouths by means of a skin fold on the lower abdomen or flank. Occasionally infants are seized by a bit of skin fold on the back, but never have they been observed being carried by the scruff of the neck. Females of both genera appear to be extremely good mothers. The infants are left behind in their nests when the mother goes out to forage for food. All of the information available about the general behavior and social behavior of *Microcebus* and *Cheirogaleus* has been summarized here.

Indriidae

The Indriidae, the largest of the lemurs, are found in most of the forested parts of the island. The nocturnal genus, *Lichanotus*, has not been well studied and is not well known. The two diurnal genera, *Indri* and *Propithecus*, are fascinating animals with magnificent aspect and enormous grace in trees. These have been studied relatively intensively (Fig. 15.4).

Lichanotus has a very wooly pelage and looks a bit like an owl. The body is approximately a foot or a foot and a half long with a tail somewhat longer than the body. It is probably completely vegetarian, eating leaves, bark, twigs, and flowers. Little is known of its social behavior, except that it is a rather solitary animal. Females with infants are found together with

Fig. 15.4.   Indriidae. (A) *Indri indri* and (B) *Propithecus verreauxi*, the sifaka.

no male nearby. Infants cling very tightly to their mothers' bellies. As is the case with many other nocturnal primates, *Lichanotus* is found in congeries, and it is difficult to tell whether or not it occurs in groups. During the breeding season, the males chase the females around in the forest until the females are seized and copulation occurs.

The diurnal Indriidae are placed in two genera, *Propithecus* and *Indri*. *Propithecus* is a large spectacular-looking animal with a very long tail often curled tightly under. This tight curl resembles a watch spring, suggesting to some that the tail is prehensile. It is not. It is a balancing organ. At present, there seem to be two species. There is much variation from region to region in pelage colors and markings. This has led to the definition of many subspecies.

Indri is the largest of the Malagasy primates. It is black and white, with a white patch on its head, prominent, black, tufted ears, and a very short stub for a tail. More than 200 years ago, Malagasy natives pointed out this magnificent animal to a French naturalist, saying, "indris, indris" when they saw it. Endriisi, or indriisi to a Frenchman, means in Malagasy "look at that" or "look here." One of the vernacular names of *Indri* is babakoto, which means "man of the forest" freely translated in Malagasy. The almost tailless *Indri*, sitting erect high in a tree, could easily have suggested a man-like creature to the earliest inhabitants of the island.

All of the Indriidae have prehensile hands and feet with well-developed digits. The second digit of each foot has the modified nail-claw typical of

Fig. 15.4 (Continued)

most Prosimii. The forelimbs are considerably shorter than the hindlimbs. This is unquestionably related to the mode of locomotion of the animals. They leap and jump from tree to tree and branch to branch, much like the galagos. Both *Propithecus* and *Indri* have a kind of orthograde posture, for they sit and stand erect easily. When they walk across the ground, which they seldom do, they hop and walk bipedally. This erect posture, which is probably due to the habitual maintenance of the body in the vertical position, has had interesting consequences for the development of the hand of

Indri and *Propithecus*. The Indriidae and *Lepilemur* (also rather orthograde in its locomotion and posture) have the greatest specialization of the hands of all the lemurs. In particular, the hands are quite elongated.

There are some specializations of the viscera, particularly the long colon labyrinth presumably an adaptation for the high proportion of cellulose present in the diet, for the Indriidae feed exclusively on fruit, bark, leaves, buds, and flowers. Bark is an important part of the diet, and we found it essential to provide fresh, green bark and woody stems to our captive *Propithecus*. When we were unable to provide these foods the animals suffered from various intestinal disturbances, and one animal developed paralysis of the colon and died.

The dentition of the Indriidae differs from that of the other Lemuriformes. The formula is $\frac{2.1.2.3.}{1.1.2.3.}$. The lower canines and incisors form the procumbent toothcomb or tooth scraper typical of the lemurs.

The typical social group of the Indriidae is a family group. There are usually two adults, a male and a female, and one or two infants or juveniles. Reliable field observations show that permanent troops larger than six are seldom found. Occasionally there seem to be situations in which larger and perhaps more complex groups are forming, but no reliable information on the persistence of such groups is available. The Indriidae are subject to many myths and legends told by the Malagasy, explorers, and others. Most of what has been written, until recently, about the behavior and habits of these animals can be dismissed as unreliable.

Indri, the stub-tailed member of the Indriidae, has never been seen, by reliable observers, in groups larger than three or four, apparently family groups. The presumption is that when a male or female offspring reaches sexual maturity the parent of the same sex as the juvenile or the juvenile itself is driven off during the breeding season. A kind of prosimian oedipal situation is believed to exist.

Both diurnal genera of the Indriidae are quite territorial in their behavior. They seem to be aware of the boundaries, as it were, of their normal territory, and they avoid other troops of the same species. Indeed, experiments in the field show that the animals become uneasy when they leave their own familiar territory. They make great efforts to remain within it. When two groups of the same species of Indriidae meet, avoidance gestures, threat gestures, and alarm calls are made, and the groups avoid the place of meeting for a very long time afterward. Part of the territorial behavior involves pronounced and dramatic vocalizations.

The Indriidae occupy an econiche that is almost exactly that occupied by the gibbons of southeast Asia. There is considerable parallelism between *Indri* and the gibbons. They both live in similar family groups, and both exhibit relatively intense territorial behavior. They both have remarkably loud calls with some astonishing resemblances. *Indri* sings a loud and penetrating song on the same general occasions on which gibbons produce their "hoots" or songs. The similarity is evident even in analysis of the sound on a sound spectrogram. *Indri* and gibbons have similar gaits when they come to the ground and some resemblances in locomotion. The gibbons, however, have developed brachiation to a fine degree, but the closest *Indri* or *Propithecus* comes to brachiating is to swing by the hands on rare occasions.

As far as it is possible to determine, gestation of *Propithecus verreauxi* is approximately five months. There is an annual breeding season and probably an annual estrus. The newborn *Propithecus* is quite well developed. The natal coat is complete, and the infant is able to grasp the hair of its mother's belly and back firmly. Observations of captive animals show that during the first few days after birth the infant lies across the belly of its mother grasping tightly to her coat. The mother tends to keep her thighs bent sideways and forms a kind of pocket or space with her thighs and body in which the infant lives for the first few days. During this early period, at least in captivity, the mother has been observed to move very carefully and cautiously. The few field observations of mothers with newborn appear to have been made well after this very early period. By the end of the first month the infant is able to clasp and climb on its mother's back. After a month and a half or so, the infant begins to explore the branches close to the spot where its mother chooses to rest. In its native habitat, at this age, it begins to play with other infants and other juveniles and can jump short distances. It utters distress cries if it gets in trouble with other youngsters, and its mother will come and rescue it. The juvenile is carried by its mother for as long as 7 months. By this time it has attained two-thirds or more of its full growth. We have observed female *Propithecus* and *Indri* making enormous leaps, without hesitation and with almost full-grown juveniles clinging to their backs, through the dense forests of Madagascar. Sexual maturity is probably not reached until the animal is 2 or 3 years of age. Observations on this are ambiguous.

Daubentonioidea

The most unusual of the Lemuriformes is *Daubentonia*, the aye-aye. This strange animal has been the subject of much discussion in the past, primarily because it was not certain whether it belonged in the same group with the lemurs. We are sure now that it is a member of the Lemuriformes. It is a nocturnal and arboreal animal, found primarily in low altitude forests on the east and northwest coasts of Madagascar. It is about the size of a cat and has large naked ears and a long bushy tail. It has large eyes typical of nocturnal Prosimii and is said to have extremely acute hearing. There are stories of its being able to detect the sounds of insects boring into the wood of trees.

The hand of the aye-aye is large, long, and slender. The fingers have been somewhat compressed, and the nails are pointed with the exception of the flattened nail on the opposable hallux. The middle finger of the hand is unusually elongated and on it is a long wiry claw. The claw is said to be fully developed at birth. This animal has very specialized behavior. It appears to listen for the noise made by the larvae of wood-boring insects as they move through timber. It then knocks upon the surface of the wood with the elongated middle finger, its movements and the tapping sound resembling the noise of a woodpecker. Once the larvae are located, the middle finger is used to probe and dig at holes in the wood, or these holes are enlarged by the specialized and enormous incisors.

The dental formula is $\frac{1.0.1.3.}{1.0.0.3.}$. The enormous elongation and robustness of the upper and lower incisors is a most unusual feature. The roots of the incisors grow back under the roots of all the other teeth in the mandible and in the maxilla.

It is believed that the female makes a nest in a tree just before she gives birth, but little is known about gestation or the postnatal behavior of the young. Current field studies should provide us with many data about this quite interesting animal.

Living Fossils

One of the most unusual features of the Lemuriformes is their presence, in splendid isolation, on Madagascar. How did the lemurs reach the island? The island of Madagascar has been divided from Africa by the Mozambique Channel since at least Cretaceous times, and this water barrier may have been in existence as early as Jurassic times. The Mozambique Channel is wide and deep, and the lemurs or their ancestors must either have crossed the channel or been present on the island before the channel was formed.

There is evidence that there was a close connection between Africa and Madagascar at one time. None of it can be used to argue convincingly about the specific epoch when such a connection existed. Geologic data, which are scanty, and theories about continental drift or the formation of the Great Rift Valley suggest the connection was broken in Jurassic times. The flora of the island includes many plants which are characteristic of the ancient fossil flora of other parts of the world. To some extent the modern trees and flowers of Madagascar can be said to be the descendants of plants that reached a high development during the Paleocene and Eocene of Africa and other parts of the world. The traveler's palm (*Ravenala madagascariensis*) is a notable example in the Malagasy flora of an indigenous plant probably related to Paleocene progenitors.

Still, how did the lemurs reach the island? Pleistocene and possibly late Pliocene fossil and subfossil lemurs and other mammals have been recovered from a number of places on the island. Most of them are closely related to species and genera that now inhabit Madagascar. There are also such subfossil semiaquatic mammals as the hippopotamus that exist on the mainland of Africa. All of the fossil lemurs are considered subfossils since they are probably of late Pleistocene age and indeed more recent than that. They were undoubtedly hastened to extinction by *Homo sapiens* who arrived on the island between one and two thousand years ago.

The lemurs most likely arrived on the island of Madagascar as migrants, clinging to rafts of trees and vegetation that floated from the mainland of Africa across the Mozambique Channel. This process of seeding flora and fauna around the world by floating masses of vegetation, or floating islands, is a well-known phenomenon and should cause no surprise. What is surprising is the extent to which the radiation of lemurs occurred on Madagascar as the result of such intermittent migrations, if such they were. We may suppose that, once the ancestors of the contemporary lemurs reached Madagascar, they found an ecological situation ideal for their further development. They developed and filled almost every niche in the trees and bushes of the island.

Why were the diurnal lemurs, in particular, not replaced by later evolutionary developments among the Primates, such as the diurnal cercopithecid monkeys? We shall summarize some of the possible answers here. First, once the lemurs were well established, we might expect that it would be difficult for another form to establish itself in competition with them merely by an occasional migration across the Mozambique Channel. Second, conditions may have changed by the time the cercopithecid radiation developed, and rafts of vegetation across the Mozambique Channel may have become an extremely improbable event. Third, the adaptive radiation of the cercopithecid monkeys may well be a much later radiation than is suggested by their morphology. It is generally supposed that the relatively primitive anatomy and general morphology of the cercopithecid monkeys demonstrates an evolutionary stage intermediate between Prosimii and higher primates, such as the Hominoidea. This has never been demonstraed fully from the fossil record and may not be the case. Fourth, the island of Madagascar itself may not be a suitable environment for the monkeys of Africa. There are many strange features about the composition of the fauna of the island. Some have suggested that the Lemuriformes and other indigenous mammals have adapted to the probable absence of important nutritional elements in the soils of Madagascar. Very few imported mammals have established themselves, not even the rabbit.

Whatever the reasons for the lack of other primates, the lemurs did establish themselves. They are a unique experiment in primate evolution. If we are clever enough to take proper advantage of this experiment, we may continue to learn more about the nature of the mammalian order to which we belong and more about evolutionary processes in general. Almost every kind of primate has developed within the lemuriform radiation except terrestrial and bipedal types.

SUGGESTED READINGS AND SOURCES

Andrew, R. J., The displays of the Primates. *In* J. Buettner-Janusch (Ed.), *Evolutionary and Genetic Biology of Primates*, Vol. II, p. 227. Academic Press, New York (1964).

Attenborough, D., *Bridge to the Past*. Harper, New York (1962).

Bishop, A., Use of the hand in lower primates. *In* J. Buettner-Janusch (Ed.), *Evolutionary and Genetic Biology of Primates*, Vol. II, p. 133. Academic Press, New York (1964).

Buettner-Janusch, J., and R. J. Andrew, The use of the incisors by Primates in grooming. *Am. J. Phys. Anthropol.* **20**, 127 (1962).

Grandidier, G., and G. Petit, *Zoologie de Madagascar*. Société d'Editions Géographiques, Maritime et Coloniales, Paris (1932).

Hill, W. C. O., *Primates, Vol. I. Strepsirhini*. Wiley, Interscience, New York (1953).

Montagna, W., The skin of lemurs. *Ann. N. Y. Acad. Sci.* **102**, 190 (1962).

Owen, R., On the aye-aye. *Trans. Zool. Soc. London* **5** (1866).

Paulian, R., *Les Animaux Protégés de Madagascar*. L'Institut de Recherche Scientifique, Tananarive (1955).

Petter, J.-J., Ecological and behavioral studies of Madagascar lemurs in the field. *Ann. N. Y. Acad. Sci.* **102**, 267 (1962).

Petter, J.-J., Recherches sur l'écologie et l'éthologie des Lémuriens malgaches. *Mém. Museum Nat. Hist. Nat. (Paris)*, Ser. A. **27**, No. 1 (1962).

Petter-Rousseaux, A., Reproductive physiology and behavior of the Lemuroidea. *In* J. Buettner-Janusch (Ed.), *Evolutionary and Genetic Biology of Primates*, Vol. II, p. 92. Academic Press, New York (1964).

Ceboidea

16

THE CEBOIDEA ARE THE NEW WORLD or platyrrhine monkeys. It is highly probable that they are part of an evolutionary lineage descended from a group of prosimian primates of the Eocene, separate from the group which gave rise to the Cercopithecoidea, the Old World monkeys, and descended from a group separate from the ancestors of the Lemuriformes, Tarsiiformes, and Lorisiformes. There are a great variety of New World monkeys (Fig. 16.1), all of which have certain major features they share that are distinct from those of primates of the Old World. However there is much variation among species and genera, far too great to be discussed in any great detail here. There is another reason why we do not give them the detailed attention which they deserve. Relatively few reliable data are available about such important matters as the distribution, range, and characteristics of many of the New World primates.

The various New World monkeys are a fascinating evolutionary development. Like the lemurs, they are a unique experiment in primate evolution. In isolation they have radiated widely into the available arboreal niches of the New World. They *parallel* the Old World monkeys. It is unfortunate that the extent of the parallel development is sometimes mistaken as evidence of close biological affinity. Far too many expectations about the nature of similarities between the Old and New World monkeys rest upon an overenthusiastic impression of the meaning of the parallels and similarities. There are certain characteristics, aside from the obvious complete geographical separateness, by which the New World monkeys are distinguished from those of the Old World. The principal criterion used in the past for the distinction was the shape of the nose, which is presumably flat, hence platyrrhine (flat-nosed) as opposed to catarrhine (sharp-nosed) monkeys and apes of the Old World. The dental formula differs from that of most Old World primates: For the Cebidae it is $\frac{2.1.3.3.}{2.1.3.3.}$ and for the Callithricidae $\frac{2.1.3.2.}{2.1.3.2.}$ (Fig. 16.2). A number of the New World forms have also developed a specialization of great interest, the prehensile tail.

The New World primates are said to be of the monkey grade of organization, and the name monkey has stuck to them. (In colloquial usage a monkey is any quadrupedal primate except members of the Prosimii and the Hominoidea.) Obviously this stresses what is a spuriously close relationship between Old and New World primates. The similarities in Old and New World monkeys are, as many primatologists point out, more apparent than real. Nonetheless it is this very reason that makes the Ceboidea potentially such an important group for students of primate evolution. It is important to remember that the mental and neurological developments and other morphological traits of the New World monkeys as a group are not in any sense intermediate between those of the Prosimii and those of the Ceropithecoidea or the Hominoidea.

245

Fig. 16.1   Some representative ceboids. (A) *Lagothrix*, the woolly monkey, (B) *Ateles*, the spider monkey, (C) *Cacajao*, uakaris and (D) *Cebuella*, marmosets. (Photographs courtesy of the New York Zoological Society.)

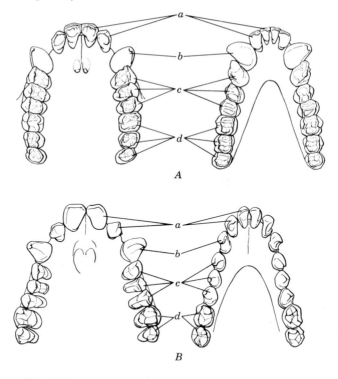

Fig. 16.2. Upper (left) and lower (right) jaws of Ceboidea, (A) *Cebus*, family Cebidae, and (B) *Callithrix*, family Callithricidae. (a) Incisors, (b) canines, (c) premolars, and (d) molars.

Some general observations are made here before we take up each group. The range in size of the American monkeys is great. The largest probably are the woolly monkeys of Panama or the howler monkeys. They are about the size of a collie or a poodle with a long tail. The smallest New World monkey is the pygmy marmoset, *Cebuella*. It is slightly larger than the smallest of the lemurs, *Microcebus murinus*. There is one nocturnal monkey, *Aotus*, the night monkey, but most appear to be diurnal and relatively social. All are arboreal. Many are insectivorous and frugivorous, and most appear to be omnivorous.

The New World monkeys are found throughout the forested areas of South America and in Central America as far north as the state of San Luis Potosi of Mexico (Fig. 16.3). They are heavily concentrated in the Amazon basin. The southernmost range is about 30° south latitude. The heavy forest cover exists and has existed for a very long time throughout southern Central America and South America. Almost all the mammals that developed in the southern continent of the New World are adapted for forest life. No great

Fig. 16.3. Geographical distribution of Ceboidea. The limits of the range of these New World primates are reasonably well established, but the exact areas in which they live have not been adequately delimited.

mammals such as roam the plains, steppes, and veldt of Africa developed in South America.

Two families have been defined in the Ceboidea, the Cebidae and the Callithricidae. There are six subfamilies defined in the Cebidae and five genera in the Callithricidae (Table 16.1). The systematics of the New World monkeys is more confused than that of most primates. Much has already been done by Hershkovitz of the Chicago Natural History Museum to clean up this Augean Stable, and it is to be hoped more will be done soon. In this discussion of the New World primates we shall emphasize only those species and genera most adequately studied. We shall also emphasize only those topics which have been given careful recent examination.

Cebinae

The subfamily Cebinae is divided into two major groups—the capuchin monkeys, genus *Cebus*, and the squirrel monkeys, genus *Saimiri*.

The squirrel monkeys are quite small, approximately the size of squirrels. They are arboreal quadrupeds and appear to live completely in the trees. They have a very long, thick, nonprehensile tail. When the squirrel monkey leaps from branch to branch or tree to tree, its tail serves as a kind of balanc-

TABLE 16.1
Classification of Ceboidea

Cebidae   Swainson 1835
Cebinae   Mivart 1865
   Cebus   Erxleben 1777
   Saimiri   Voigt 1831
Alouattinae   Elliot 1904
   Alouatta   Lacépède 1799
Atelinae   Miller 1924
   Ateles   E. Geoffroy 1806
   Brachyteles   Spix 1823
   Lagothrix   E. Geoffroy 1812
Aotinae   Elliot 1913
   Aotus   Humboldt 1911
   Callicebus   Thomas 1903
Callimiconinae   Thomas 1913
   Callimico   Ribeiro 1911
Pitheciinae   Mivart 1865
   Pithecia   Desmarest 1804
   Cacajao   Lesson 1840
   Chiropotes   Lesson 1840
Callithricidae   Thomas 1903
   Callithrix   Erxleben 1777
   Cebuella   Gray 1866
   Leontideus   Cabrera 1956
   Saguinus   Hoffmannsegg 1807
   Tamarinus   Trouessart 1899

ing rudder. It functions as support when the animal is sitting on a branch, and on occasion it is thrown over the shoulder when the animal is resting.

Saimiri is among the most common of the South American monkeys, and these active and restless squirrel monkeys are often kept as pets. They have been found mostly in the high forests along the river banks of the Amazon valley and in the Guianas. Their range extends from Costa Rica on the north to Bolivia on the south. Their ecological niche is highly specialized, restricted to high forests close to river banks and perhaps to the edges of forests adjacent to open spaces. These segments of the forest are tangled masses of vines and bushes that bear the flowers, fruits, and nuts upon which these animals subsist. They are known to eat insects, and it is said they will descend to the ground to catch them. Reports are that they do not exist in well-defined troops as do many other monkeys. Flocks or crowds of them, as many as five hundred, have been seen gathered together.

The capuchin monkeys, genus *Cebus,* often known as the organ-grinder monkeys, have tough, long, prehensile tails. Their tails do not have areas of specialized hairless skin near the tip. They resemble the storybook, or standard, monkey of Africa more than any other of the New World primates. The superficial resemblance is probably due to their large, round heads, similar to the heads of some cercopithecids of Africa. They are considered to be rather generalized monkeys and represent a kind of generalized development with none of the anatomical, pelage, or social specializations of other New World monkeys. They are arboreal and inhabit dense tropical forests of South America. They come out of the forests to steal fruit and vegetable material from farms. Apparently they are quadrupedal on the ground but sit and stand erect easily.

There are reports that they live in troops containing as many as thirty individuals. Their territory is relatively restricted; they act to drive other troops away. Their diet is quite omnivorous, both in the wild and in captivity. The genus *Cebus* is found in Honduras, continuously through Central America, into Venezuela, the Guianas, and the Amazon basin. They are also found in the west in Colombia, at relatively high altitudes of 5000 feet in Bolivia, in southern Brazil, and as far south as Paraguay and Argentina.

Aotinae

The titis, genus *Callicebus*, are small monkeys with long, woolly, nonprehensile tails and silky pelage. They are usually put in the same subfamily with the night monkey, *Aotus*. They are diurnal and arboreal. They apparently keep to the highest trees in the forest and stay on the finer branches. Many observations indicate that they also occupy those groves or stands of trees in which a dense underbrush or bush flora grows. Their hands and limbs are adapted for grasping and for locomotion on small, thin branches. Their morphology is generally considered to be primitive and not specialized.

It is fortunate that the titis have been given a thorough, systematic review by Hershkovitz. The titis are presently put in a single genus, *Callicebus*. There are probably two species and a relatively large number of local varieties. Hershkovitz showed that populations of each of the two species are sympatric in certain areas. Specimens from these sympatric species appear to diverge less from each other in superficial characteristics than specimens from allopatric populations of the same species. This illustrates an important point about speciation among mammals. Closely related species converge more toward each other when their habitats are similar and diverge more when their habits are dissimilar. Thus sympatric populations or sympatric species gradually grow to look alike. An invader responds to the same environmental pressures that act on its sympatric indigenous relative.

Although the titis are diurnal, their large eyes suggested to early explorers that they were nocturnal. In the nineteenth century Deville remarked that they were very active toward twilight and appeared to spend the day curled up in a ball high in the trees. They have been described as living in small flocks or troops, but very few observations of them in their natural habitat have been made. They appear to be omnivorous in their diet.

The night monkeys, genus *Aotus,* are small animals with a thick and woolly pelage. Their tails are not prehensile. Explorers in the early part of the nineteenth century often commented on their resemblance to lemurs, and the fact that they have eyes extremely well adapted for nocturnal vision may have suggested this similarity. *Aotus* is a completely nocturnal animal and is said to sleep in hollow trees, in the crotch of branches, or in holes in large trees. The night monkeys are found throughout the northern part of South America, in the Amazon basin, in the Orinoco basin in Peru, Bolivia, and Brazil. Explorers who have seen them in their native surroundings report that they are sociable, gregarious animals. Little is known about their social behavior. Scanty observations suggest they move in family bands, an adult male and female with one or two young. There are no reports of larger social groups.

Pitheciinae

This subfamily consists of three genera, *Pithecia, Chiropotes,* and *Cacajao,* all of which have tails that are not prehensile. The tails are always hairy and the long-tailed forms have quite bushy tails. The members of the genus *Pithecia* are completely arboreal. They appear to live in social groups of five to ten or larger. Explorers of the midnineteenth century argued strongly that they were crepuscular or nocturnal in their habits, but this has been denied in recent years. Their diet appears to be omnivorous. They are extremely nervous animals, and no great success has been achieved keeping them in captivity. Apparently their range is quite restricted: they are found in the Amazon basin, as far inland as Ecuador and Peru, in the Guianas, and on the south bank of the Orinoco but not in the forests of southeastern Brazil which impinge on their normal habitat. They also have been reported from Colombia.

The bearded sakis have been placed in the genus *Chiropotes.* The distinguishing features are the bearded face, the absence of a swelling on one part of the nasal bones, and the medium hair-whorl on the tip of the tail instead of on the proximal region. Practically nothing is known about the behavior of the bearded sakis in their natural habitat. They probably live only in the forests and have been observed mainly along river banks. Whether their ecological range is restricted to river banks or whether this observation is made because it is easier for explorers and naturalists to follow rivers in the forests of South America is not certain. They are arboreal quadrupeds and are quadrupedal on the ground. The diet is omnivorous. They eat nuts, vegetable material, and fruit. Insects have been found in the stomachs of bearded sakis shot in the wild.

The uakaris, genus *Cacajao,* are bald-headed monkeys with relatively short tails. The pelage is generally a bright red color, the face and head quite naked, or at least with very few hairs. They seem to be quite timid and have not been extensively studied in captivity. They apparently live in small bands by river banks in the forests. They seem to stay in the highest trees and are said to be very poor jumpers. As far as is known, they are purely vegetarian.

Alouattinae

At present one genus, *Alouatta*, has been defined, with six species. It is unlikely that this large number of species will be retained forever. The howler monkeys have been extensively studied as subjects for behavioral research by Carpenter and his students and by Collias and Southwick. The howler monkeys are large animals with prehensile tails. When they are full grown they may be as large as a medium-sized spaniel or poodle. They have a magnificent beard and long silky hair. Howler monkeys live in the highest branches of the largest trees in the forest. Troop size varies from four to as many as forty individuals. It is difficult to know what the average troop size is in their natural habitat. Excellent studies of these monkeys have been made on Barro Colorado Island in the Panama Canal Zone, and most of the reliable information comes from these studies.

Since these particular animals are in a protected area, their social behavior may be somewhat different from that of free-living members of the species. But it is unlikely that there is a basic and significant difference. The argument is often raised that little of value can be learned of primate behavior from captive or tame animals or animals in protected reserves. The speciousness of this argument is demonstrated by these studies by Carpenter, Collias and Southwick, by the work of Andrew on primate displays, vocalizations, and facial expressions, and by Washburn and DeVore on the social behavior of baboons. Andrew's studies were made with captive animals and those of Washburn and DeVore were made on baboons in protected game parks.

Life in a howler troop must be most unusual, for the females appear to be dominant over the males. The average sex ratio among adults appears to be 7 females:2 males. Howlers gather in the early morning and early evening to concertize. They howl in concert and create a tremendous racket. The vocalizations of howler monkeys attracted the attention of European travelers and explorers as early as the sixteenth century. The sound has been called a howl, a roar, drum-like, a bark. It has been characterized as melancholy, powerful, intense, indescribable, and insufferable. Howler monkeys have been described as brutish, fierce, savage, and fearsome looking. The contours of the face and skull, which lead to these descriptions, are primarily the result of the enlarged mandible and the hyo-laryngeal apparatus, the structures involved in producing the howler's characteristic vocalization. The hyoid bone (the "throttle bone") which supports the tongue and the larynx is very large. The mandible has a very broad and deep vertical ramus, and it is believed this developed to accommodate the huge hyo-laryngeal apparatus. The larynx and upper respiratory tract are modified to provide an amplifying chamber. There are two irregularly shaped bony cavities, formed from the hyoid bone and thyroid cartilages of the neck, one in front of the other. They connect to the throat and are involved in the production of the howling vocalization.

Howler monkeys are seldom observed on the ground and are said not to descend to drink. Water is obtained from their diet or by licking wet leaves. They cross rivers and streams by leaping through trees whose branches span the water. They jump from one slender branch to another. The social organization of troops of howler monkeys is sufficiently integrated and re- fined so that mutual cooperation is the rule and antisocial or aggressive behavior is minimized or suppressed. Troops of howler monkeys appear to have a distinct and clear knowledge of their territorial limits. When two troops meet each other at the boundaries of their territories they vocalize vigorously. They seem to assert their territorial rights, warning other mem- bers of their troop that a rival troop is near and, at the same time, threaten- ing the other troop.

The locomotion of howler monkeys is typically pronograde, quadrupedal locomotion. The longitudinal axis of the body is parallel to the branch upon which the animal is moving. Carpenter has observed that howlers follow each other in single file through the trees in their daily round of activity. The largest or the leading male usually is in front, and the actions of the leader are imitated by the individuals that follow him.

The hands and feet are of some interest since the thumb is not a com- pletely opposable digit as it is in other higher primates. The tail is a pre- hensile, tactile organ, used to drive away flies and to grasp objects. The diet of the howler monkey seems to be restricted to fruits, nuts, leaves, buds, and bark. The few records of howlers in captivity show they are omnivorous despite their restriction to a diet of vegetation in their natural habitat.

Atelinae

The subfamily Atelinae is divided into three genera: *Lagothrix*, the woolly monkeys; *Ateles*, the spider monkeys; and *Brachyteles*, the woolly spider monkeys. *Brachyteles* probably should not be a separate genus. All members of the subfamily have prehensile tails with an area of specialized naked skin at the end of the tail.

The woolly monkeys are placed in two species of the genus *Lagothrix*. A number of clearly marked races, or allopatric populations, exist. They all have very thick woolly fur and long prehensile tails. It is said they have mild dispositions and are easy animals with which to work. The two species have been reported in bands of fifteen to twenty-five, and as many as fifty individuals have been counted together. Old males have been found living as solitary animals.

They are native to northwestern South America and range from sea level to as high as 9000 feet. Their natural diet seems to be fruit and leaves. In captivity they are quite omnivorous and will eat almost anything. Captive and zoo specimens will stand erect and walk bipedally for a considerable amount of time. There are no reliable observations of this having occurred in their natural surroundings.

The woolly spider monkey, genus *Brachyteles*, seems to be intermediate in morphology between the spider monkey and the woolly monkey. It is relatively large and robust with a woolly coat resembling that of *Lagothrix*. It has very long arms and legs, and the pollex (thumb) is absent or very tiny. It shares these two characteristics with *Ateles*, the spider monkey. The woolly spider monkey is confined to the mountainous forests of southeastern Brazil. Its diet seems to be strictly vegetarian.

The spider monkeys proper are placed in the genus *Ateles*. Their general build, very slender with extremely long arms and legs, gives them the spidery appearance recognized in their common name. The pollex is tiny or absent. The hair is rough, thin, and scraggly, unlike the thick woolly coat of the other members of the Atelinae. Both *Ateles* and *Brachyteles* have flat nails on their hands and feet. Several striking specializations have been achieved by spider monkeys. Their arms are longer than their legs. Their hands have no trace of an external thumb, but a metacarpal bone is buried in the tissues of the hand. Occasionally a single phalanx appears as a tiny external tubercle. The spider monkeys parallel, in a remarkable way, the gibbons of southeast Asia. They are excellent brachiators, and they fill, in a sense, the same ecological and morphological niche in the radiation of New World primates that the gibbons do in the adaptive radiation of the Hominoidea. The spider monkeys have a marvelous adaptation which the gibbons lack—the prehensile tail. This tail is capable of a strong grasp and can act as a supporting organ while the hands and feet are busily engaged in other activities. The feet are extremely strong, and both hands and feet are useful in climbing. In general, the spider monkeys appear to be arboreal quadrupeds, although they have often been observed sitting erect. In the erect posture they usually support themselves by a strong grip with their prehensile tail, on a branch or tree trunk. They normally live in the crowns of the trees in the high forests of South America. The floor of their native forests is covered in some regions with flood waters during part of each year. After the floods have receded, the monkeys are seldom seen on the ground. From reports on captive animals, it appears they do not cross open grassland or unforested terrain.

The spider monkeys have developed an erect or orthograde mode of locomotion more than other New World primates. Brachiation is highly developed. Since there is no thumb, the hands become hooks with which to grasp thin or stout branches. Reliable reports tell us spider monkeys are able to jump distances as great as 30 feet. They are almost entirely frugivorous. In captivity they will eat a variety of foods, but they are not as omnivorously inclined as other primates.

The spider monkeys' northern range is Mexico, and they are found as far south as the Amazon in Brazil. They have also been reported in Colombia. They live in bands which vary in number from ten to forty. The best analysis of the behavior, troop structure, and territorial range of spider monkeys is by Carpenter. He notes that in one species twelve or thirteen is the average number in a troop and there may be subgroups of odd adult males attached to such an "average" troop. They are quite territorial and have specific routes they travel in their daily round of activity through the forest. Troops avoid the territory of and contact with other troops. As Carpenter has pointed out, the territory of a nonhuman primate troop must be defined by time as well as by space. The actual physical territory of two troops may overlap. Yet the time of day or year at which the two troops occupy or use the overlapping area is always different.

Callimiconinae

The single genus, *Callimico*, Goeldi's monkey, is one of the smaller of the New World monkeys. It is quadrupedal with an extremely soft, silky, black pelage. It has a long nonprehensile tail and pointed nails on all digits. These monkeys are often called Goeldi's marmoset. The name marmoset is a colloquial term and has no technical significance. Since most of the animals considered to be marmosets are members of the family Callithricidae, it is probably better not to use this name for members of the genus *Callimico*.

Little is actually known about the behavior of these animals in the wild. They should certainly be more extensively studied. Goeldi's monkey has been placed in a subfamily of the Cebidae largely because its dental formula is that of the cebids. Its range is restricted to the upper part of the Amazon basin.

Callithricidae

The true marmosets may be divided into two principal groups that might eventually be given the status of separate subfamilies. The division into two groups is based upon the morphology of the lower canines. One group consists of the genera *Cebuella* and *Callithrix* which have lower canines resembling lower incisor teeth, hence incisiform canines. The other group includes all the rest of the genera of the Callithricidae which have the normal caniniform lower canines. This group includes the genera *Leontideus*, *Saguinus*, and *Tamarinus*. The dental formula of the Callithricidae is $\frac{2.1.3.2.}{2.1.3.2.}$ and distinguishes them from members of the family Cebidae (Fig. 16.2). The Callithricidae include the smallest New World primate, *Cebuella*, the pygmy marmoset.

Systematics

The biology (morphology, ecology, physiology, distribution, etc.) and the systematics of the Ceboidea have not yet been correlated. Until such time as they are, there is little that can be said about them. Almost every population has been given specific rank, and every species group has been given generic designation. Here is an example of the kinds of obstacles poor taxonomic practices can raise against the understanding of the biology of a group of primates. The classification of the Ceboidea presented here (Table 16.1) lists only those taxa at the level of the genus and above.

The New World primates are another evolutionary experiment deserving much closer scrutiny than it has received. As in the case of the lemurs, an adaptive radiation of a primate lineage occurred in isolation. The New World monkeys have advanced beyond their prosimian ancestors in several respects. Their use of the hand is more elaborate and more skillful. Some have developed prehensile tails. All except one species are diurnal. Most are quite sociable and appear to form structured groups somewhat more complex than those of the lemurs. There are species believed to be quite primitive and others quite specialized. A large number of evolutionary questions may eventually be answered by study of the Ceboidea. As with the lemurs, much intellectual energy has been spent on questions of nomenclature and classification. Many of the interesting biological problems that underlie these taxonomic activities have been neglected.

Suggested Readings and Sources

Carpenter, C. R., A field study of the behavior and social relations of howling monkeys. *Comp. Psychol. Monog.* **10**, No. 2 (1934).

Carpenter, C. R., Behavior of red spider monkeys in Panama. *J. Mammal.* **16**, 171 (1935).

Erikson, G. E., Brachiation in New World monkeys and in anthropoid apes. *Symp. Zool. Soc. London* **10**, 135 (1963).

Fooden, J., A revision of the woolly monkeys (Genus *Lagothrix*). *J. Mammal.* **44**, 213 (1963).

Hershkovitz, P., Notes on American monkeys of the genus *Cebus*. *J. Mammal.* **36**, 449 (1955).

Hershkovitz, P., Type localities and nomenclature of some American primates, with remarks on secondary homonyms. *Proc. Biol. Soc. Wash.* **71**, 53 (1958).

Hershkovitz, P., The scientific names of the species of capuchin monkeys (*Cebus* Erxleben). *Proc. Biol. Soc. Wash.* **72**, 1 (1959).

Hershkovitz, P., A systematic and zoogeographic account of the monkeys of the genus *Callicebus* (Cebidae) of the Amazonas and Orinoco River basins. *Mammalia* **27**, No. 1 (1963).

Hill, W. C. O., *Primates, Vol. III. Pithecoidea: Platyrrhini—Hapalidae.* Wiley, Interscience, New York (1957).

Hill, W. C. O., *Primates, Vol. IV. Cebidae, Part A.* Wiley, Interscience, New York (1960).

Hill, W. C. O., *Primates, Vol. V. Cebidae, Part B.* Wiley, Interscience, New York (1962).

Cercopithecoidea

17

THE OLD WORLD MONKEYS constitute the Cercopithecoidea. Their range is shown in Fig. 17.1. They are found from the Cape of Good Hope in southern Africa to the islands of Japan. Their northern limits are the Himalaya mountains. There is, or was until recently, an isolated group in North Africa. None live in Europe now, and in the Near East there are only a few isolated populations on the Arabian peninsula.

The cercopithecoids are placed in one family which is divided into two subfamilies, the Cercopithecinae and the Colobinae (Table 17.1). Other subdivisions have been proposed, but further splitting tends to obscure the phylogeny of this group. The majority of species are adapted for an arboreal way of life. The most numerous, however, are the terrestrial varieties. The Cercopithecinae are omnivorous, the Colobinae vegetarian. The dental formula of the Cercopithecoidea is $\frac{2.1.2.3.}{2.1.2.3.}$, and it is constant throughout this taxon (Fig. 17.2). The Cercopithecoidea present many problems that deserve discussion. We shall restrict ourselves, for reasons of space, to a few important aspects of their evolutionary history, social behavior, and systematics.

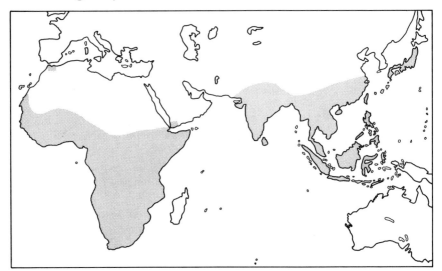

Fig. 17.1.   Geographical distribution of Cercopithecoidea. Both subfamilies, the Cercopithecinae and the Colobinae, are found in Africa and Asia.

TABLE 17.1
Classification of Cercopithecoidea

Cercopithecidae   Gray 1821
   Cercopithecinae   Blanford 1888
     Cercopithecus   Brünnich 1772
       Cercopithecus aethiops   Linnaeus 1758
       Cercopithecus cephus   Linnaeus 1758
       Cercopithecus diana   Linnaeus 1758
       Cercopithecus hamlyni   Pocock 1907
       Cercopithecus l'hoesti   Sclater 1899
       Cercopithecus mitis   Wolf 1822
       Cercopithecus mona   Schreber 1775
       Cercopithecus neglectus   Schlegel 1876
       Cercopithecus nictitans   Linnaeus 1766
       Cercopithecus nigroviridis   Pocock 1907
       Cercopithecus patas   Schreber 1774
       Cercopithecus talapoin   Schreber 1774
     Cercocebus   E. Geoffroy 1812
       Cercocebus albigena   Gray 1850
       Cercocebus aterrimus   Oudemans 1890
       Cercocebus galeritus   Peters 1879
       Cercocebus torquatus   Kerr 1792
     Cynopithecus   I. Geoffroy 1835 ?
       Cynopithecus niger   Desmarest 1820
     Macaca   Lacépède 1799 ?

TABLE 17.1 (*Continued*)

Macaca assamensis M'Clelland 1839 ?
Macaca cyclopis Swinhoe 1862 ?
Macaca fuscata Blyth 1875
Macaca irus Cuvier 1818 ?
Macaca maura Cuvier 1823
Macaca mulatta Zimmerman 1780
Macaca nemestrina Linnaeus 1766 ?
Macaca radiata E. Geoffroy 1812
Macaca silenus Linnaeus 1758
Macaca sinica Linnaeus 1771
Macaca speciosa Cuvier 1825 ?
Macaca sylvana Linnaeus 1758
Papio P. L. S. Müller 1776
Papio cynocephalus Linnaeus 1766
Papio gelada Rüppell 1835
Papio hamadryas Linnaeus 1758
Papio sphinx Linnaeus 1758
Colobinae Elliot 1913
Colobus Illiger 1811
Colobus badius Kerr 1792
Colobus polykomos Zimmerman 1780
Colobus verus Van Beneden 1838
Nasalis E. Geoffroy 1812 ?
Nasalis larvatus Würm 1781
Presbytis Eschscholz 1821
Presbytis aygula Linnaeus 1758
Presbytis carimatae Miller 1906
Presbytis cristatus Raffles 1821
Presbytis entellus Dufresne 1797
Presbytis francoisi Pousargues 1898
Presbytis frontatus Müller 1838
Presbytis johni Fischer 1829
Prestbytis melalophos Raffles 1821
Presbytis obscurus Reid 1837
Presbytis phayrei Blyth 1847
Presbytis pileatus Blyth 1843
Presbytis potenziani Bonaparte 1856
Presbytis rubicunda Müller 1838
Presbytis senex Erxleben 1777
Pygathrix E. Geoffroy 1812
Pygathrix nemaeus Linnaeus 1771
Pygathrix nigripes Milne-Edwards 1871
Rhinopithecus Milne-Edwards 1872 ?
Rhinopithecus avunculus Dollman 1912
Rhinopithecus roxellanae Milne-Edwards 1870
Simias Miller 1903 ?
Simias concolor Miller 1903

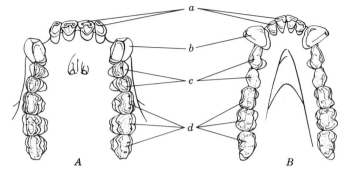

Fig. 17.2.   Upper (*A*) and lower (*B*) jaws of *Colobus,* showing typical cercopithecoid dentition. (*a*) Incisors, (*b*) canines, (*c*) premolars, and (*d*) molars.

Cercopithecinae

The Cercopithecinae (Fig. 17.3) are divisible into three groups—the terrestrial baboons and macaques, *Papio* and *Macaca;* the largely arboreal mangabeys, *Cercocebus;* and the arboreal monkeys or guenons of the genus *Cercopithecus. Cercocebus* and *Cercopithecus* are found only in Africa, whereas the terrestrial members of this subfamily are found in both Asia and Africa.

The members of the genus *Cercopithecus* are arboreal, with the exception of *Cercopithecus* [= *Erythrocebus*] *patas* which seems to be specially adapted for rapid locomotion on the ground. All are diurnal in habit, although some not wholly reliable reports imply that one species, *Cercopithecus hamlyni,* has nocturnal habits. All species that have been well studied live in troops. The number in a troop may vary from six to forty. The size of a troop probably depends upon the availability of food and the size of the forests through which the animals roam. The size of the sleeping group and the diurnal group may not always be the same. There is some suggestion that groups of cercopithecine monkeys break up into smaller groups during the day. This habit is similar to that of certain diurnal lemurs.

Troops of cercopithecine monkeys appear to have two daily periods of active feeding, a morning and a late afternoon or early evening session. Toward noon there is a period of somnolence. These monkeys resemble members of the genus *Lemur* in this daily rhythm of activity. There are occasional reports that they feed at night, particularly in areas much disturbed by trappers, hunters, or farmers. The diet is almost completely frugivorous, but they probably should be classed as omnivorous, because they eat insects avidly and are known to eat birds and birds' eggs and on occasion small mammals. It is probable that such carnivorous behavior is infrequent and occurs only under special conditions.

The size, the structure, and the processes by which troops of *Cercopithecus* are held together have not been well elucidated. By analogy with baboon troop structure, it has been assumed that male dominance is important. The average-sized troop of *Cercopithecus* has been estimated from field studies of only two or three species. These have been relatively well studied in their natural habitats, and reports have been made by reliable observers. The typical social group of these monkeys is supposed to be a family band or harem group. If this is true, it is likely that a dominant male holds a small group of females and immature animals together. Threat displays and some signs of dominance are observable in these monkeys both in the natural habitat and in captivity. Field observations, of course, are very difficult, since these are arboreal monkeys that live in dense bush or forests. They are extremely wary animals, and it is difficult for a human observer to follow them for any length of time. Recent studies have demonstrated that at least some troops of *Cercopithecus aethiops* living in East Africa have a principal dominant male.

Haddow, who spent many years collecting and studying guenons in Uganda, shot a relatively large number of solitary monkeys of one species, *Cercopithecus nictitans* [= *C. ascanius*]. He expected that such solitary monkeys would be aging males or young males that had been driven away from family troops by the dominant animal. But the majority of the solitary monkeys shot by Haddow were adult males; there were very few old males; and there was one adult female. The important point is that the majority of these solitary monkeys, the adult males, were in their physical prime as determined by postmortem examination. This observation has been confirmed by Haddow's examination of the records of the monkeys he shot on other occasions and of those shot by others. There are as yet no reliable reports that would suggest there is sexual competition between adult male *Cercopithecus* in their natural habitat. Indeed no fights were observed at all during a very long period of time while monkeys were being collected by Haddow's staff. Very few adult males of the *Cercopithecus* species group he collected were found to have scars or injuries. A similar collection of adult male baboons would probably bear many more signs of battle.

Troops of *Cercopithecus* exhibit some territorial behavior, although different troops of the same species appear to tolerate each other's presence. Troops of different species have been attacked when they encroached, but a peaceful conclusion to at least one such battle has been observed. Booth reported that a small troop of *Cercopithecus aethiops* moved into trees occupied by a small troop of *C. mitis*. After a tremendous fight, the *C. aethiops* attached themselves to the troop of *C. mitis*, and mutual play among individuals of the two species was observed.

Haddow's extensive observations over many years demonstrate that various species and genera of arboreal monkeys of East Africa are found together during their daily round of activity. Guerezas (*Colobus*), guenons

Fig. 17.3. Cercopithecinae. (A) *Cercopithecus mitis,* the Sykes monkey; (B) *Cercopithecus aethiops,* the vervet or green monkey; (C) *Macaca silenus,* the lion-tailed macaque; and (D) *Papio cynocephalus,* the common baboon of East Africa.

(*Cercopithecus*), and mangabeys (*Cercocebus*) have been observed in friendly association. Mixed bands, made up of animals of two or three of these genera, have been observed.

The female guenons are rather good mothers. The infants are capable of walking independently several days after birth. Haddow found that about the time the umbilical scar heals infants are capable of a reasonable amount of muscular coordination. They appear to be weaned at 2 or 3 months. The milk dentition is believed to be complete at about 6 months. At this time the infants have been observed acting independently of their mothers.

The color and texture of the pelage of newborns in all species of *Cercopithecus* are quite different from the adult animals'. This natal pelage or coat changes, in both color and texture, to a juvenile pelage at about age 8 weeks. This change is apparently coincidental with the eruption of the second molars of the milk dentition. The natal coat produces a strong reaction from the adults of the species. A striking demonstration was reported by Booth when she took an infant *Cercopithecus*, still in its natal coat, into an area in which wild troops of *Cercopithecus aethiops* lived. She was threatened and closely approached by the most dominant male in a troop of the wild monkeys. The general impression is that the natal coat of the infant stimulates adult monkeys to rescue and care for the infant. The juvenile coat is quite similar to that of the adult, with one exception, *Cercopithecus neglectus*, whose juvenile coat is very different.

There is an interesting juvenile trait in males of the species *Cercopithecus aethiops*. The scrotum of juvenile males is blue. The adults have a very brilliant turquoise colored scrotum. It has been suggested by one investigator that these color characteristics of pelage and scrotum function as a signal so that an adult male will not attack and injure a juvenile. Observations suggest that *C. aethiops* and *C. neglectus* exhibit the most aggressive and violent traits of any of the guenons.

Infant guenons of some species have a transient prehensile power in the tail. It is usually lost by the time the monkey has become independent of its mother, and the adult monkey uses the nonprehensile tail as a balancing organ. *Cercopithecus patas* is able to use its tail like the third leg of a tripod. It presents a remarkable appearance propping itself up on its two hindlimbs and tail while it manipulates objects with its hands.

Many species and subspecies of guenons have been defined. The confusion in the taxonomy of the African monkeys parallels that of the New World monkeys, particularly the marmosets. Every color difference was believed by some writer to warrant species rank, and every species group was awarded generic status.The arboreal monkeys of Africa radiated into the trees and speciated as the forest became discontinuous. The countless variations in pelage color, facial markings, and even genital colors are correlates of life in the trees.

Cercopithecus is a widely distributed genus in Africa. Although it is considered an arboreal animal, several species can and do exploit the floor of the forest, sisal plantations, and the scrubland and bush within the open savannas of certain parts of Africa. Most species of the genus *Cercopithecus* appear to have a rather restricted distribution and to be the most conspicuously arboreal in habits. Those species found all over Africa, for example *C. aethiops*, are much less tied to a forest habitat. They are no less arboreal as far as locomotor adaptations go, but they are found on the edge of savanna country, in plantations and other areas settled by man, and in the low bush and scrub country which often impinges upon the grassy plains of Africa. Speciation was probably relatively rapid and extensive in the groups tied to a forest habitat. Isolation of a segment of an allopatric population of *Cercopithecus* was, and is, a likely event. Every major break in the forest cover is a potential barrier separating populations of an arboreal species. Repeated occurrence of breaks between forest stands would have led to recurrent isolation and would have been a powerful factor in the formation of the large number of species that now exist. The myriad varieties and species of the genus *Cercopithecus* may be a consequence of a very recent adaptive radiation into the forests of Africa.

The mangabeys, genus *Cercocebus*, are relatively large, partly terrestrial monkeys that are found throughout the forested parts of Africa. They appear to be most numerous in swampy and riverine forests. It is reported that they are relatively unaggressive, particularly when their behavior is compared to that of macaques and baboons. Mangabeys are found across the forest belt of Africa from the Tana River in Kenya on the east to Gambia on the west. Their northern range is not well delineated, and they are found as far south as the borders of Zambia.

The mangabeys have very long tails, and they have ischial callosities as do the baboons, the macaques, and most species of *Cercopithecus*. These ischial callosities are thick pads with rough surfaces. Their location suggests they might as well be called seat pads. Mangabeys, like baboons, will sit on a rough tree branch with their ischial callosities firmly in contact with the tree.

The mangabeys are a group of monkeys that should receive far more study in the future. They are adapted to an arboreal habitat, yet they have many of the characteristics of a terrestrial primate. Their hands, for example, are very similar to those of baboons rather than to those of wholly arboreal primates. Exploitation of the ecological niche they occupy is not simply a matter of a terrestrially adapted animal taking to the trees. Rather, their habits seem to resemble those of the langurs, if the few reports available are reliable. The mangabeys are arboreal, but they spend most of the day on the ground; they are relatively unaggressive; and they seldom leave the shelter of the trees that are their refuge.

The terrestrial monkeys of Africa and Asia belong to two major groups, the baboons (*Papio*) and the macaques (*Macaca*). The baboons are found throughout Africa south of the Sahara and on the Arabian peninsula. The macaques are found in India, Pakistan, in both continental and island south-

east Asia, and in China and Japan. Baboons and macaques are closely related and should probably be placed in the same genus, since fertile hybrids
are not at all uncommon.

Analysis of the systematics of baboons and macaques suggests that they
are a widespread terrestrial population with two major groups of species—
one African, the other Asian. Specialized populations such as *Papio gelada*
and the black ape (*Cynopithecus niger*) link the two groups. The gelada
baboon is a macaque-like animal in Africa, and the black ape is a baboon-
like animal in Asia.

The morphology of the skulls of these primates forms a continuum, with
the mandrill at one end and the stump-tailed macaque of Asia at the other
(Fig. 17.4). The sketches in the figure are idealized to some extent and
are based upon skulls of males. There is pronounced sexual dimorphism
in the skulls of these animals. Males show the greatest differences from
species to species. The skulls of female mandrills (*Papio sphinx*) and common baboons (*Papio cynocephalus*), for example, are often indistinguishable.

The terrestrial members of the Cercopithecinae, the baboons and the
macaques, have developed a number of specializations that appear to be
related to life on the ground. Similar specializations of the hands and limbs
are also found in *Cercopithecus patas*. The precision grips of terrestrial
baboons and macaques and of *C. patas* resemble the human precision grip.
Life on the ground was probably a major selective force leading to increased
fine control of the hands. Using *C. patas* as an experimental animal, Bishop
showed that the precision and power grips of this animal are more sharply
distinguished from each other than are the precision and power grips of
arboreal cercopithecids. Details of this are discussed in Chapter 20. The
point to be made here is that terrestrial life led to certain specializations
and developments that are significant to an understanding of the evolutionary story of *Homo sapiens*.

Baboons and macaques are sexually dimorphic in a number of traits. The
canine teeth of males are considerably larger than the canines of females.
The canines are often used by males in threat displays and, probably, in
other social situations. They are, of course, important weapons and probably
function as tools. Females of the terrestrial Cercopithecinae have a most
striking development of the sexual skin. The sexual skin turns red, and there
are correlative changes in other morphological features during the estrous
cycle. During estrus the perianal region of the female swells enormously
and turns a brilliant red or reddish pink. A male baboon can sight a female
in estrus from far away. In certain populations and individuals—and indeed
this is quite variable from individual to individual—the skin of the chest
and the face will also turn the same brilliant red color during estrus. Sexual
contact and copulation are not restricted to females in estrus, but it is clear
that such a female stimulates sexual behavior on the part of the male.

Baboons and macaques have a very well-studied and interesting social
life. Their daily activities take place on the ground, even though they sleep

in trees. Thus they are far easier to observe than arboreal primates. There are many features of baboon and macaque troops that suggest the kind of adaptation which may have been part of the behavioral repertory of early hominids living on the grassy plains of Africa. We can develop hypotheses about trends in the early evolution of human social behavior by carefully studying baboons (Chapter 20).

The size of baboon troops varies widely. The total numbers range from less than ten to more than a hundred. The largest troop for which there is an accurate count had one hundred eighty-five members. The sex ratio in baboon troops is not easy to determine. The best data available show that there is considerable variation, the sex ratio (females to males) ranging from 1:2 to 10:1. The sex ratio in most troops probably ranges between 2:1 and 4:1. All of these ratios are based on counts of adult animals, for it is often very difficult to determine the sex of infant and subadult baboons. The age composition of a baboon troop is also difficult to determine. The best estimates indicate that from one-half to two-thirds of the members of most troops are not fully grown.

The structure of baboon troops has been studied extensively by many scholars, particularly Zuckerman, Hall, Washburn, DeVore, and the author of this book. One of the first general conclusions reached in early studies of the social life of baboons, notably the conclusions of Zuckerman, was that the primary bond, the most important social cement, is sexuality. In Chapter 20, we discuss the nature of the sexual cycle and our reasons for still believing that primate sexuality is one of the fundamental biological bases for primate social life.

As more and more studies of baboons in their African habitat are completed, it is clear that baboons are intensely social animals. Relationships of various kinds, not only sexual, exist among individuals of different age-grades within a troop. This sociability and these individual interrelationships maintain the stability of a baboon troop. The entire life of a baboon is organized around such relationships. As Washburn and DeVore point out, life in a troop is the only feasible way of life for a baboon.

Troops maintain their structure, stability, and integrity by a variety of mechanisms. There are several important forces other than those primary biological and social factors just noted that act upon baboon populations. One of the most important is the food supply, which is controlled by general environmental conditions. There is a close correlation, we believe, between certain kinds and facets of baboon troop structure and the ecological situation in which the troops live. The diet of baboons is made up almost entirely of vegetable matter. They should be classed as omnivorous, nevertheless, for they eat insects, birds, and eggs, and are known to kill and eat small animals occasionally. But it is the vegetarian habits of baboons which require that a troop have available a relatively large area in which to feed. The territory through which a baboon troop passes during its daily round of activities must have a sizable food supply. During severe drought or when

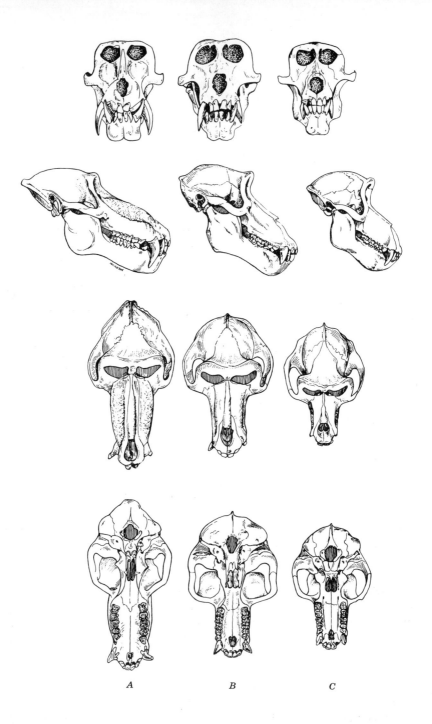

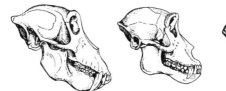

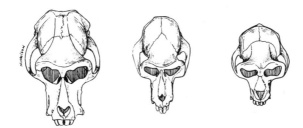

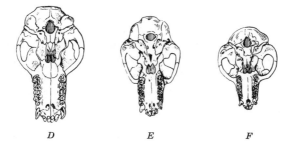

D   E   F

Fig. 17.4. Skulls of Cercopithecinae. (A) *Papio sphinx,* mandrill; (B) *Papio cynocephalus,* baboon from West Africa; (C) *Papio cynocephalus,* baboon from East Africa; (D) *Macaca nemestrina,* pig-tailed macaque; (E) *Cynopithecus niger,* black ape; and (F) *Macaca speciosa,* stump-tailed macaque. (Redrawn after Gregory, 1951.)

other drastic assaults upon the vegetation occur, the behavior of a baboon troop is affected. The troops in a large area modify their behavior toward each other. Another important external force that acts upon baboon troops is the terrain in which they live. The troop structure reported for the hamadryas baboons, which live on the treeless, bushless, high, rocky plains of central Ethiopia, is rather different from that of the baboons of the forest and bush country of East Africa.

Baboon troops are organized around mature adult males who are arranged hierarchically. This hierarchy, usually called the dominance hierarchy, is the stabilizing social unit in a troop. It is often stated, in popular accounts of baboon behavior, that the dominance structure is rigid and that there is a physically powerful dictator baboon at the head of the troop. Such silly notions are unfortunately widespread. The dominance hierarchy is relatively stable, but several males may be codominant with other males, and their exact position in the hierarchy may fluctuate. The tenure of the most dominant male is of uncertain duration. It is not likely that this is a long-term, stable position. Females will often support males in a situation requiring dominance behavior. Status in the hierarchy is a function of physical condition, fighting ability, and codominant relationships with other males. A group of codominant males outranks any particular individual, despite the fact that any one of the codominants may be weaker than a male outside such a group. The primary function of the dominant group seems to be to maintain stability and social peace within the baboon troop and to guard and defend the troop from enemies without. Where there is a clear dominance hierarchy, fighting within a troop is rare. Squabbles among juvenile animals, for example, are usually stopped quickly by the movement of a dominant animal to the vicinity. Occasionally small troops with abnormal age and sex distributions have been observed. Establishment of dominance relationships, unsettled or unclear in such a troop, is the source of severe outbreaks of fighting between members.

Since most baboons live in bush country where the stands of trees are small, the animals are under constant danger of attack. Leopards, pythons, large hawks, and wild canids are among their natural enemies. The troop provides protection from these predators. Solitary baboons are seldom seen, and it is highly unlikely that a lone animal can survive in open bush country. Occasionally a lone animal was observed during the day, but this animal usually rejoined its troop at night. If it did not, it was seldom seen again.

Washburn and DeVore point out that dominance relationships are deducible from the way animals interact with each other. The two situations in which these relations are most obvious are in feeding and in grooming. When food is made available, or tossed, to a group, the most dominant animal will pick it up and eat it. Those subordinate to him will leave it alone and may even avoid looking at it. The arrival of a dominant baboon near some particularly desirable food is enough to cause subordinate animals to move away rapidly. The most dominant animals are groomed most frequently. Whenever a more dominant animal approaches, one who is subordinate moves away. A baboon will leave a less dominant animal to groom one with a higher position in the hierarchy. When the females go into estrus, the relationships within a troop are temporarily disrupted and consort pairs are formed. These pairs apparently move to the edge of a troop, and observations indicate that very few fights occur over females in estrus. Nevertheless occasional fighting at this time does break out in those troops in which the dominance relationships have not been clearly defined. Most observations suggest that a female is never monopolized by a single male. However it is possible that sexual contact leading to fertilization may be restricted to certain males in a troop, although this is not clear from the data available now. Among baboons, other than hamadryas baboons, no harem-like association between a male and a number of females has been observed.

The hamadryas baboons, in the rocky open regions of Ethiopia, live in two kinds of social groups. One, a sleeping group, has no constant membership and may vary from day to day. The other, the one-male group, seems to have a constant membership. It is a kind of harem, with a single adult male leading a group of females and young. The larger sleeping groups appear to have no dominance hierarchies. When several one-male groups have been seen moving together during the day, no dominance hierarchy among the males has been observed.

Newborn baboons are part of another important complex mechanism which appears to maintain social order and stability in a baboon troop. A female with a newborn infant is the object of much interest and concern on the part of other members of the troop. She is frequently groomed by other adult animals, and much attention is paid the infant. Each mother protects her own infant, grooming it often and keeping constant watch over it. When an infant leaves its mother, usually to play with other infants, the mother is constantly aware of it. Males in the troop also show interest in the infants and concern over their safety. Play groups of young and maturing juveniles are formed and are often carefully overseen by dominant adult baboons. But let us not anthropomorphize too much about motherhood among baboons. A female will almost never give food to her infant. Females have been observed to snatch food away from their own infants and will, on occasion, handle their own infants roughly. A baboon female will very often ignore her newly born infant if she is kept in isolation, in a maternity cage for example. In one captive colony all the newborns were lost until the females were allowed to have their babies in the colony. The interest of other females in the newborn apparently triggers an intense, possessive, maternal response.

There are many other relationships within a baboon troop which we need not specify here. The important point is that baboon troop life is highly organized and has evolved to such a state that problems of internal discord, development of aggressive behavior by maturing males, disruptive fights over females, aggressive competition for food, and protection of other members, have been solved.

The ways in which baboon troops form *de novo* are not fully understood yet. The range in troop population size is great, and this is puzzling. It is difficult to understand how a new troop forms and maintains itself over long periods of time with the stability and integrity usually exhibited. Our field observations suggest that under certain ecological conditions large troops of baboons are subdivided internally into smaller groups which are, in a sense, incipient troops. These smaller groups cluster around a few dominant males and females and are structured as subunits of the larger troop as it moves around its territory feeding, copulating, playing, and resting during the day.

Baboons appear to be highly territorial, and they know their territory well. There seems to be little overlap between the area of one baboon troop and that of another. The territoriality of a troop and the aggressive defense of segments of its home range have been much exaggerated by writers and casual observers. Some anthropologists are uncomfortable with the concept of territory in this context. They probably fear that some people will assume baboons have real estate instincts. It is possible that we imply excessive attachment to a particular area, with specified boundaries known to the members of the troop, when we use the expression territory. The territory of a baboon troop is not so well defined as the real estate holdings of an American farmer, but the behavior with respect to the home range is not dissimilar. Under conditions of drought when food supply is limited, territorial behavior is intensified, whereas under conditions of abundant food, the degree of territoriality is reduced.

Baboon troops are also extremely efficient in protecting members against predators and other enemies. Baboons have been observed to move in a kind of symbiotic group with other animals—particularly the smaller herbivorous ungulates of the great open plains of Africa. Baboons are themselves to some extent herbivores, and many aspects of their social life may be understood in part in relation to an herbivorous mode of existence. Washburn observed a group of baboons and herbivores grazing together when a pack of hunting cheetahs approached. The large male baboon in the troop moved to the edge of the mixed group of animals and made a few threatening displays in the direction of the cheetahs. The cheetahs wheeled about and hurried away in search of more amenable prey.

The social life and the troop structure of macaques are not quite as spectacular or well defined as those of baboons. Or at least they are not as dramatically described. The amount of dominance in various primates has been rated on a dominance gradient by Carpenter. Baboons and macaques, and probably langurs, have a sharp dominance gradient, that is, each member of a troop has a well-defined position in a hierarchy. New World monkeys, gibbons, lemurs, and others have a very low gradient. However the structure of troops of macaques is not as easily defined; neither is macaque social behavior or troop behavior as complex and supportive or protective as that reported for baboons. One interesting feature of macaque social life that has no analog among baboons is the formation of separate bands by young and recently mature males. These bands, called outlaw bands (for want of a better term), move around on the outskirts of the major troops of macaques.

Extensive studies of macaque troops in a protected habitat, on Cayo Santiago, an island near Puerto Rico, demonstrate that a dominance hierarchy is extremely well defined and every troop appears to have a single, completely dominant male. Koford, in describing the dominant males, states that they are almost always in superb physical condition and they stride assertively with their tails held high. The peripheral and subordinate males walk in an apprehensive manner, with a slight crouch and their tails held low. Males that are subordinate to the dominant leader of a troop become extremely assertive when they associate with males subordinate to them. On Cayo Santiago where the animals are relatively crowded there are often displays of aggressive, quarrelsome, and competitive behavior over food. From the descriptions of reliable observers it appears that baboons have solved the major problems of monkey social organization better than have the other terrestrial species. In general, however, the behavior of macaques does not sufficiently vary from that of baboons to warrant additional discussion here. There are, of course, differences in details, some of which may be found in the references cited for this chapter.

The black ape, heretofore placed in a separate genus, *"Cynopithecus,"* is an Asiatic monkey found on the Celebes. It resembles baboons in many social and some physical characteristics. The gelada baboon, *Papio gelada,* is a macaque-like animal that lives in remote areas in Ethiopia. Its distribution and habitat suggest it is a remnant population of a species that once had a much wider distribution. Presumably it is being replaced by the common baboon and the hamadryas baboon. This view is probably conditioned by a number of dramatic, and quite implausible, stories about battles between geladas and other baboons. Female gelada baboons have a secondary sex character very similar to that of female macaques. The skin on the chest and forehead turns bright pink or red during estrus. When field and laboratory studies of the black ape and the gelada baboon are completed, the systematic biology of all the terrestrial Cercopithecinae should be less obscure.

The Cercopithecinae include a wide range of monkeys. Although they seem to fall into two groups, the arboreal guenons and the terrestrial baboons and macaques, splitting this subfamily obscures phylogenetic relationships within the Cercopithecoidea. As we said earlier, choice of level in the taxonomic hierarchy, above that of the species, is largely a matter of taste. We have a lumper's taste and prefer to emphasize evolutionary relationships and affinities.

Colobinae

The Colobinae are found in Africa and in Asia. The African guerezas, genus *Colobus,* are essentially forest-dwelling animals, seldom found in bush country. Nevertheless we have seen them sitting in some of the few remaining trees in cleared farming regions of Uganda. In West Africa, the guerezas have been observed in the lower stories of thick, mature forests. The red, the olive, and the black and white colobus are probably distinct species. The red colobus is said to be restricted to the highest parts of the trees, whereas the black and white colobus is found in the lower segments. Because the ranges of various color forms overlap, it has been assumed that they are different species. The danger of this assumption was pointed out earlier.

The hand of the colobus monkey is specialized for brachiation. The thumb is reduced to a node or tubercle, and the four fingers are elongated. The hand is capable of fine control, and the fine precision grip of all Cercopithecoidea is present. The guerezas are often seen in the highest parts of the trees, the upper story of the forest as Napier puts it. Their diet consists wholly of vegetable matter. They presumably live on leaves, as do their Asiatic relatives, *Presbytis, Simias, Nasalis, Pygathrix,* and *Rhinopithecus,* commonly known as the leaf monkeys or the langurs. (*Nasalis* is often called the proboscis monkey.) Relatively little is known about social life and troop structure of *Colobus.* We do know that they are found in troops of moderate size and these troops seem to be structured.

The Asiatic members of the subfamily Colobinae have been much more extensively studied. These leaf monkeys or langurs have been placed in several genera (Table 17.1). *Presbytis* has been studied in India, and its social life has been well described. Langurs live in organized troops whose membership is constant. Any changes are due primarily to births and deaths. As is the case with both Old and New World monkeys whose social life has been extensively studied, the membership of a langur troop varies little during a year. Troops vary in size from five to fifty individuals. The ratio of adult females to males is usually 1:2, but the ratio of total number of females to total number of males is usually close to 1:1. The male langurs mature more slowly than the females. In their daily cycle of activity, langurs stay close to the trees. The protection afforded by trees is undoubtedly a factor behind certain differences between baboon behavior and langur behavior. Many solitary langurs have been observed, whereas it is inconceivable that a solitary baboon will live long. Langur troops interact peaceably with each other, although each troop tends to remain within its own home range. Jay observed that adult male langurs do no more than grind their canine teeth in a kind of aggressive display when they encounter another troop. A loud call is given by langurs at certain times during the day. This call presumably functions to maintain or specify distances between adjacent troops, much as does the song of *Indri* or the hoot of the gibbon.

Among the langurs, there is a poorly defined dominance hierarchy of the relation of females to each other and to males. The hierarchy is ever-changing, for the dominance status of a female changes with her sexual and pregnancy state. Her status usually rises when she is in consort relation with a male, and she appears to withdraw completely from the hierarchy when she gives birth. The male dominance hierarchy is much more clearly defined. It is established and maintained with a minimum of fighting and aggressiveness. Jay reports that there is no correlation between successful copulations by males and position in the dominance hierarchy. During estrus, females aggressively solicit males.

The newborn langur is a focus for the attention and social relations of the adult females in a troop. Female langurs, unlike baboons, "loan" their infants to other females. During its early months, the newborn langur has intimate, protective contact with many adult females. Again unlike baboons, the adult male langurs appear to take no interest in the newborns. Adult females do not hesitate to threaten or chase males which appear to be interfering with an infant.

The differences in the nature of troop organization and social life of various highly organized terrestrial or semiterrestrial monkeys suggest that the general ecological situation may be one of the fundamental determinants of the organization of social life. The intense predator pressure under which baboons live requires the tight organization and aggressive behavior they exhibit. Macaques live under equally stressful circumstances, although rather less predator pressure may be present. Langurs do not have the same cares that the terrestrial monkey, which roams from the trees each day, has. Although langurs do spend most of their time on the ground, they dwell in woods and forests and always have the trees as a retreat.

The problems of living together have been solved in different ways and with varying success. A terrestrial primate, in order to survive predators and competitors, must be aggressive, tough, strong, and quick to react to threats. But many of these reactions make for difficulties in social groups. Macaque troops appear to cast out the animals which do not fit in peaceably, whereas baboons and langurs seem to have overcome these problems in a more efficient and more complex manner.

Suggested readings and sources

Allen, G. M., A checklist of African mammals. *Bull. Museum Comp. Zool. Harvard Coll.*, 83 (1939).

Altmann, S., A field study of the sociobiology of rhesus monkeys, *Macaca mulatta. Ann. N. Y. Acad. Sci.* **102**, 338 (1962).

Bolwig, N., A study of the behaviour of the Chacma baboon, *Papio ursinus. Behaviour* **14**, 136 (1959).

Booth, A. H., Speciation in the Mona monkey. *J. Mammal.* **36**, 434 (1955).

Booth, A. H., Observations on the natural history of the Olive Colobus monkey, *Procolobus verus* (van Beneden). *Proc. Zool. Soc. London* **129**, 421 (1957).

Booth, A. H., The Niger, the Volta, and the Dahomey Gap as geographic barriers. *Evolution* **12**, 48 (1958).

Booth, C., Some observations on the behavior of cercopithecus monkeys. *Ann. N. Y. Acad. Sci.* **102**, 477 (1962).

Buettner-Janusch, J., Biochemical genetics of baboons in relation to population structure. *In* H. Vagtborg (Ed.), *The Baboon in Medical Research,* p. 95. Univ. of Texas Press, Austin (1965).

Buxton, A. P., Observations on the diurnal behavior of the redtail monkey (*Cercopithecus ascanius schmidti*) in a small forest in Uganda. *J. Animal Ecol.* **21**, 25 (1952).

Chance, M. R. A., Social structure of a colony of *Macaca mulatta*. *Brit. J. Animal Behaviour* **4**, 1 (1956).

Gillman, J., and C. Gilbert, The reproductive cycle of the Chacma baboon (*Papio ursinus*) with special reference to the problems of menstrual irregularities as assessed by behaviour of the sex skin. *S. African J. Med. Sci.* **11**, *Biol. Suppl.,* 1 (1946).

Gregory, W. K., *Evolution Emerging,* Vol. II. Macmillan, New York (1951).

Haddow, A. J., Field and laboratory studies on an African monkey, *Cercopithecus ascanius schmidti* Matschie. *Proc. Zool. Soc. London* **122**, 297 (1952).

Haddow, A. J., The Blue Monkey group in Uganda. *Uganda Wild Life and Sport* **1**, 22 (1956).

Hall, K. R. L., Social vigilance behaviour of the Chacma baboon, *Papio ursinus*. *Behaviour* **16**, 261 (1960).

Hall, K. R. L., Numerical data, maintenance activities and locomotion of the wild chacma baboon, *Papio ursinus*. *Proc. Zool. Soc. London* **139**, 181 (1962).

Hall, K. R. L., The sexual, agonistic and derived social behaviour patterns of the wild chacma baboon, *Papio ursinus*. *Proc. Zool. Soc. London* **139**, 283 (1962).

Hall, K. R. L., Variations in the ecology of the chacma baboon, *Papio ursinus*. *Symp. Zool. Soc. London* **10**, 1 (1963).

Imanishi, K., Social organization of subhuman primates in their natural habitat. *Current Anthropol.* **1**, 393 (1960).

Jay, P., Aspects of maternal behavior among langurs. *Ann. N. Y. Acad. Sci.* **102**, 468 (1962).

Koford, C. B., Group relations in an island colony of rhesus monkeys. *In* C. H. Southwick (Ed.), *Primate Social Behavior,* p. 136. Van Nostrand, Princeton (1963).

Kummer, H., Soziales verhalten einer Mantelpavian-gruppe. *Beih. Schweiz. Z. Psychol.* No. 33 (1957).

Kummer, H., and F. Kurt, Social units of a free-living population of hamadryas baboons. *Folia Primat.* **1**, 4 (1963).

Maxim, P. E., and J. Buettner-Janusch, A field study of the Kenya baboon. *Am. J. Phys. Anthropol.* **21**, 165 (1963).

Pocock, R. I., A monographic revision of the monkeys of the genus *Cercopithecus*. *Proc. Zool. Soc. London,* 677 (1907).

Rode, P., *Les primates de l'Afrique.* Larose, Paris (1937).

Schwarz, E. L., The species of the genus *Cercocebus*. *Ann. Mag. Nat. Hist.* **10**, No. 1, 644 (1928).

Southwick, C. H., Patterns of intergroup social behavior in primates, with special reference to rhesus and howling monkeys. *Ann. N. Y. Acad. Sci.* **102**, 436 (1962).

Tappen, N. C., Problems of distribution and adaptation of the African monkeys. *Current Anthropol.* **1**, 91 (1960).

van Wagenen, G., and H. R. Catchpole, Physical growth of the rhesus monkey (*Macaca mulatta*). *Am. J. Phys. Anthropol.* **14**, 245 (1956).

Verheyen, W. N., Contribution à la cranologie comparée des primates. *Ann. Musée Royal Afrique Centrale, Sci. Zool.*, No. 105, Tervuren, Belgium (1962).

Washburn, S. L., The genera of Malaysian langurs. *J. Mammal.* **25**, 289 (1944).

Washburn, S. L., and I. DeVore, The social life of baboons. *Sci. Am.* **204**, No. 6, 62 (1961).

Pongidae

18

THE APES, THE PONGIDAE, are man's closest living relatives. These include the chimpanzees and gorillas of Africa, genus *Pan*, the orangutans of Borneo and Sumatra, genus *Pongo*, and the gibbons and siamangs of southeast Asia, genus *Hylobates*. The name *great ape* refers to chimpanzees and gorillas; *ape* refers to all the Pongidae. The list of species is presented in Table 18.1, and their distributions are shown in Fig. 18.1 and 18.2.

TABLE 18.1
Classification of Pongidae

Ponginae   Allen 1925
  Pongo   Lacépède 1799
    Pongo pygmaeus   Linnaeus 1760
  Pan   Oken 1816
    Pan gorilla   Savage and Wyman 1847
    Pan troglodytes   Blumenbach 1799
Hylobatinae   Gill 1872
  Hylobates   Illiger 1811
    Hylobates lar   Linnaeus 1771
    Hylobates moloch   Audebert 1797 ?
    Hylobates syndactylus   Raffles 1821

The apes are among the most interesting primates from our own anthropocentric point of view. There are many fanciful and foolish stories about them in the literature of science, travel, and exploration. Our conception of their essential nature has not been assisted by the often grotesque, anthropomorphic displays many of them have been put through in some of the great zoological parks in the United States and other countries. They are *not* half men, and their behavior is unique and fascinating for its own sake as well as for what it will tell us about the probable behavior of our own remote ancestors. We need not dress them up and teach them to drink tea, ride bicycles, and "ape" certain actions of men. Fortunately a number of extremely good studies of these animals, in their natural habitat and in the laboratory, have been carried out within the last 15 or 20 years.

Our discussion will begin with the gibbon of southeast Asia. Gibbons are slight, slender, graceful creatures with extremely long arms (Fig. 18.3). They swing through the trees with extraordinary agility. They are the essential true brachiators of the order Primates. Brachiation, by definition, is the spectacular manner in which gibbons swing hand over hand through the trees. Gibbons' hands are elongated, and their digits are slender and long. They are capable of very fine precision grips. They progress across the ground, when they are forced to do so, in bipedal fashion. They are not ungraceful, but are rather inefficient bipedalists.

A very large number of species of gibbons has been defined. It is highly

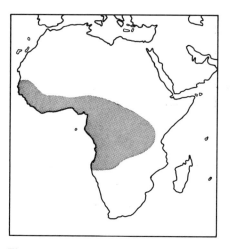

Fig. 18.1.   Geographical distribution of African Pongidae, chimpanzees and gorillas, genus *Pan*. The specific areas in which gorillas have been found are shown in Fig. 18.5.

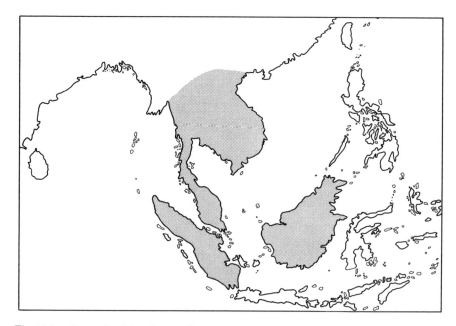

Fig. 18.2.   Geographical distribution of Asiatic Pongidae, *Hylobates* (gibbon) and *Pongo* (orangutan). Gibbons are found only on the mainland and orangutans are found only on the islands of Borneo and Sumatra.

probable that no more than three should remain and quite likely that this number could safely be reduced to two. Gibbons, like so many other groups of primates, have been put into many species on the basis of minor variations in pelage markings and color.

The social life of gibbons was brilliantly described by Carpenter a number of years ago, and some of the salient features of his work follow. The reader should make an effort to look at Carpenter's monograph. Gibbons appear to live in small family groups. An adult male and an adult female with one or two immature offspring make up the social group. Carpenter reports that the largest group he observed was an adult pair with four immature animals. It is believed that when the juvenile male or female offspring in a family group reaches sexual maturity, the parent of the same sex attempts to drive it out of the family group. Here is the primitive oedipus complex in action! New family groups are formed when the newly adult gibbon, driven forth from its family, meets another gibbon of opposite sex, also newly driven forth from its family group. Presumably when the parent animal grows old, a mature offspring is able to drive off its parent.

The phenomenon of territoriality is a fundamental behavioral character of free-ranging gibbons. The ranges of gibbon groups are, as are those of almost all other primates, really overlapping mosaics. They are not definite territories with specific boundaries. The area through which a gibbon family group ranges may shift from time to time, for a number of reasons. The familiarity with a particular area, the food supply, the trails, the trees for sleeping, and the places for resting, playing, and otherwise keeping busy are important stabilizing factors. The extreme territoriality of the gibbons would appear to inhibit extensive migration. The defense of territory is efficient, and it is physical and vocal. Many gibbon vocalizations appear to have developed, to use Carpenter's phrase, as a buffer to pugnacity. Gibbons are rather different from many other primates, for there seems to be equal dominance in both sexes. Since there is almost no secondary sexual dimorphism among gibbons, the lack of obvious male dominance is probably closely related to the fundamental similarity of the two sexes in strength, size, and appearance.

Gibbons fight by biting with their long, sharp canine teeth and by clawing with their fingers. They are not very strong animals, but a bite from a canine tooth of a gibbon is no minor affair. Gibbons are said to be gentle and unaggressive creatures except when they are defending their territory. They vocalize beautifully, hooting in an enchanting and haunting manner as they swing through the trees. The hooting is a kind of territorial signal which helps to space family groups of gibbons in the forest. In this trait they resemble langurs of India, howler monkeys of South America, and *Indri* of Madagascar. As we said earlier, the hoot of the gibbon and the song of *Indri* are extremely similar in many respects. Gibbons have a regular daily cycle of vocal activities. Carpenter observed that the cycle varied when there was competition between family groups for territory or food, or on

windy and stormy days. When there was competition, the frequency of vocalizations increased, and on windy and stormy days, the number of calls decreased.

The orangutan, genus *Pongo*, is found today only on the islands of Borneo and Sumatra. The orangutan is found on the mainland only as fossil or sub-fossil remains. Orangutans also seem to live in small family groups, and recent studies by the Harrissons suggest that they also occupy caves on Bornean hillsides. They appear to be gentle, pensive creatures, completely vegetarian in their diet, ungainly, covered with a scraggly coat of reddish hair. They are popular zoo animals, and it is estimated that over half of the total world population of orangutans is now in captivity.

The orangutan and the gibbon are believed by many to be more closely related to each other than they are to gorillas and chimpanzees. It is clear from the fossil record that gibbons differentiated well before the other pongids reached their modern form. Orangutans and gibbons are similar in some respects, for they both have long arms. The ratio of arm length to leg length is very large in both. The anatomical and other morphological differences of orangutans are unlikely to warrant definition of a separate family or subfamily.

Perhaps the best known of the Primates are the chimpanzee and gorilla of Africa. The distribution of these two animals is indicated in Fig. 18.1. The gorilla (Fig. 18.4) is the largest of living primates and, perhaps, contrary to popular mythology, the shyest and gentlest. Chimpanzees are rather rowdy and boisterous animals. There are many excellent sources for detailed

Fig. 18.3. *Hylobates lar*, the gibbon of southeast Asia, (A) close-up and (B) with limbs out-stretched. These are two different animals.

descriptions of the general morphology, biology, and anatomy of these animals, and we shall restrict ourselves to a discussion of their social behavior and general natural history.

Schaller's account of his field studies of the mountain gorilla of the eastern Congo is one of the most extensive ever written. Our discussion is based largely on his excellent work. Gorillas occur in two forest habitats in Africa, one in West Africa and the other in central Africa (Fig. 18.5). It is possible that these two separate populations were once connected by a population that extended from south of Lake Albert to the present day west African home of the lowland gorilla. The modern lowland and mountain forms may actually be the ends of what was a continuous distribution. Schaller divided the permanent habitat of mountain gorillas into three forest types—a lowland rain forest (moist, evergreen), a mountain rain forest (moist, montane),

A

and a bamboo forest. In the Congo basin he found that 75 per cent of the gorilla range lies in the lowland rain forest.

Gorillas are quadrupedal and terrestrial, but there is no question about their use of the arboreal habitat for sleeping, for they make nests in the trees. As a number of naturalists have observed, whenever there is danger or when gorillas become alarmed, they descend to the floor of the forest and flee. Bipedal locomotion on the ground is quite rare, but the animal spends a great part of the day in a more or less erect or vertical position. The animal squats with its trunk erect as it plays or beats its chest or feeds. The gorillas, like the other great apes, are called brachiators, but they are brachiators anatomically, not in their normal locomotion. They are certainly too large to leap from tree to tree easily, and there are no reliable reports that they do.

The temperament of gorillas is placid. In psychodynamic terms they have an introverted or shut-in personality, as Yerkes observed. Schaller believes that the terms "reserved," "stoic," and "aloof" express gorilla temperament. These are excellent descriptive terms for the observed way in which they behave. Whether they are stoic, reserved, and aloof in an anthropo-

Fig. 18.4.   *Pan gorilla*, the gorilla, (A) male and (B) female.

morphic sense is another matter entirely. Perhaps we may summarize the gorilla as an amiable, placid animal, with the temperament, character, and diet of a large herbivore.

The size of free-living gorilla groups ranges from eight to twenty-four. Generally the sex ratio for adults is about 1 male to 2 females, and it may even be higher. Adult male gorillas have a fascinating hooting display, which resembles advertisement calls of other arboreal primates, such as gibbons, *Indri,* and *Lemur variegatus.*

The daily cycle of gorillas is diurnal. They start feeding shortly after they awaken and feed intensively for 2 or 3 hours. Toward noon they become somnolent and rest for a few hours. In the afternoon they again feed and move about for several hours. Then they make nests and go to sleep. Schaller found, in one area where he observed gorillas, that most of the sleeping nests were located on the ground, but in other forests only 20 to 50 per cent were on the ground. Building a nest is a rapid process, generally taking no more than 5 minutes. The animal pulls and breaks the nearby vegetation, wraps the broken vegetation around its body, then sleeps.

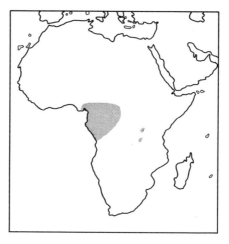

Fig. 18.5.   Areas in Africa in which gorillas have been found. The lowland gorilla ranges through the shaded area in western Africa, and the mountain gorilla has been found in several localities near Lake Tanganyika and Lake Albert.

Schaller never observed gorillas using any kind of tool or engaging in tool-using behavior. The gorilla has some powerful built-in tools. The strong arms and hands and powerful front teeth pull up and tear apart the vegetation.

Each gorilla troop has a large silverbacked male as leader. The silverbacked males are the oldest males in the group; the silvery color is the result of aging. The dominance hierarchy, if such there is within a gorilla troop, is correlated with the size of the animal. All silverbacked males are dominant to all the other animals, and females and blackbacked males are dominant to all the juveniles and to those infants not in continual contact with their mothers. In the juvenile and infant group, too, dominance and size seem to be correlated. When Schaller found more than one silverbacked male in a group, there was a linear hierarchy among these older males. Explicit dominance behavior was seldom observed, and it was less often observed by Schaller among gorillas than among any other closely observed primates. The most common sign of dominance behavior was exhibited by animals seizing the right of way along a trail or the right to a particular sitting or resting place.

Mutual grooming was rarely observed between adults, and Schaller believes it has little or no social significance. He suggests it has the utilitarian significance of keeping the animal clean. Most grooming behavior was observed between females and much younger animals, usually infants. On rare occasions Schaller observed an infant grooming a female and deduced that this activity established a kind of social relationship or social intercourse between the two.

A great deal of active play was observed among infants and between infants and juveniles. But by the time animals are 6 years old, there is very little mutual play. Young animals also indulge in solitary play—running, climbing, sliding, swinging, and jumping around. There seems to be almost no overt sexual behavior in a gorilla troop, although infants are born every year or so to females in the proper age group. During his entire study which lasted over a year, Schaller saw only two copulations. To one who has studied baboons in the field, the gorilla appears to be an extraordinarily restrained animal.

Gorilla females are extremely protective mothers. The infants are help less and completely dependent, for a considerable time, on their mothers for food, protection, transport, and social and psychological satisfactions. Gorillas develop about twice as fast as human infants. By the time the animal is 6 or 7 months of age, it is quite alert and active and can be weaned. They have been observed to suckle as long as a year and a half. Infants 3 months or older are often seen to ride on the backs of females. Before that they are usually carried by the female and cradled close to her chest in one arm. The mother provides a kind of home base for the infant, to which it can return for safety, protection, and comfort. Gorilla females have been observed to carry dead infants for as long as 4 days before discarding them. This behavior has been observed among other primates as well. Sexual physiology of gorillas is known only from observations of captive animals. The estrous or menstrual cycle is estimated to be about 30 days. Four reliable zoo records suggest that the gestation period ranges from 250 to 300 days.

A general statement about gorilla social and individual behavior is difficult to make. As animals that have been the subject of such extraordinary interest on the part of man, gorillas seem rather disappointing, for their repertory of behavioral traits implies they are restrained and placid. They probably have found an almost perfect ecological niche. They are able to lead a quiet, peaceful life. They are so large that it is unlikely that most predators and carnivores would attack them. Any carnivore would have quite a battle on its hands. Despite their strength, it is amazing how unaggressive gorillas are and how seldom their strength is used for anything except tearing up vegetation. Very few carnivores occur in the same habitat as gorillas, with the exception of leopards. It is not known to what extent leopards prey upon gorillas, but Schaller believed they did so rarely. Man is the most persistent predatory enemy of gorillas and will be the principal cause of their extinction.

Chimpanzees are smaller and more active animals than gorillas. They, too, spend much of their time on the ground but also appear to build nests and sleep in trees. We are fortunate that Goodall has recorded an enormous amount of chimpanzee behavior on movie film. Her field studies are slowly being published, and they are providing much fascinating material for anthropologists. Chimpanzees appear to be gregarious and sociable animals. They have an intense curiosity which apparently overcomes any wariness they may have. Gorillas tend not to poke their noses into other primate business. Chimpanzees, like baboons, cannot resist so doing. The diet of chimpanzees is largely vegetable matter, but they will catch and kill arboreal monkeys. One such incident has been recorded on film. They have been seen to drag very small ungulates out of dense bushes and eat them. Whether the chimpanzees find the animals dead or kill them has not been positively determined. Nonetheless the obvious implication is that they kill them.

A very intriguing discovery was made by Goodall during her field studies of chimpanzees in Tanganyika. She demonstrated that chimpanzees engage in a kind of tool-using behavior. It is claimed by some who interpret her observations that chimpanzees make and save these tools and teach infant chimpanzees how to make them. Since these observations have been recorded on film, it is possible for any number of people to examine them and come to a variety of conclusions, as scholars usually do. The fact of the matter is that the chimpanzees do take small pieces of grass or small twigs and prepare them for use. They poke the sticks into termite nests, catch and eat the termites. They do use a kind of sponge made of leaves to take up water from the boles of trees. On Goodall's films, a very young chimpanzee, who apparently got his signals crossed, uses a termite stick instead of a crushed leaf sponge to attempt to get water. Despite this rather impressive demonstration of a kind of tool-using, it is unlikely that we must change our definition of man, the genus *Homo*, and remove the criterion of toolmaking simply because chimpanzees use tools. As we have said so often before, ancestors and descendants and close relatives should be expected to be very similar. No one will argue that the chimpanzees are *not* closely related to man. Retrospectively, as usual, it is not wholly surprising that chimpanzees exhibit such behavior. It is perhaps surprising that other great apes do not. But, as we pointed out, the huge anterior dentition of the gorilla is all the tool that animal needs. The chimpanzee with his extroverted disposition probably manipulates his environment more and stumbled into making tools. There is probably sufficient neurological development for this to have become a teachable bit of behavior. The circumstances under which Goodall's observations were made were good, and the observations do not suggest that toolmaking is an instinctive kind of behavior. But it is not toolmaking to the same degree that man makes tools, and it is not toolmaking to the same degree that the australopithecines made tools, if they did indeed make the stone tools found in the same strata in which they were found. It could be predicted that a long and tiresome discussion of this point would take place in the anthropological literature. We shall ourselves participate in this discussion in Chapter 21.

Chimpanzee groups appear to be relatively small, somewhat smaller than those reported for gorillas. There seems to be a reasonably well-organized hierarchy of dominance in the troop, but the structure of the troop seems considerably looser than that of most other primate troops whose members live such social lives. The dominance hierarchy is not as pronounced and perhaps not as fundamental, explicit, or crude as it is in baboon troops. Goodall discovered that most chimpanzees are relatively nomadic, moving about in groups of three to seven individuals. It is possible that during the day two or three or more such small groups may move together for a while or even for a few days. During times of the year when certain fruits or trees are in season or food of a particularly desirable sort is plentiful, as many as five or six groups may be seen together. When two groups join temporarily and later separate, individuals often have been exchanged between the two groups. This suggests that there may not be a very tight group structure. Chimpanzees, more extroverted than most primates, express their hostilities or aggressive feelings in loud vocalizations and a considerable amount of physical action. Fighting between individuals is rare nonetheless.

Grooming among chimpanzees is much more important than it is among gorillas. It is as important, if not more so, than among baboons. Adult chimpanzees have been observed to groom each other for as long as two hours during a day. They will sit quietly, searching each other for foreign particles that cling to the hair. This behavior facilitates social interaction between individuals, it serves to keep the coat clean, it keeps the animal free of parasites, and it is an important way to prevent disease.

Baboons and chimpanzees have been observed to interact. They appear to tolerate each other, even though they are competitors for approximately the same food plants. On occasion baboons will be chased by chimpanzees, but in general they seem to feed together without much intergroup aggression.

The pongids are man's closest living relatives, but they are not, in all probability, very similar to the hominoid from which man's lineage arose. Each of the apes is a particular adaptive radiation, with anatomical modifications for a particular ecozone. All are basically herbivorous, although it is clear from our discussion that the chimpanzee is something of an exception to this. All are adapted for life in the trees. Even though the African great apes are almost completely terrestrial, their anatomy and locomotor adaptations are arboreal. They very likely developed out of a lineage that was well adapted for arboreal life. They are quadrupedal most of the time on the ground. Like all primates, they are able to hold their trunk erect, and they can move more or less bipedally for a considerable period of time. Nevertheless their long arms and relatively short legs are an indication that they are basically adapted for efficient locomotion through the trees. Man, on the other hand, has a very different adaptive relationship with his environment and fills a quite different ecological niche. The separation between the adaptive system of the great apes and the adaptive system of Hominidae is extreme, and it has become more pronounced with the passage of time. The systematics of the Hominoidea (Chapter 10) reflects this separation.

SUGGESTED READINGS AND SOURCES

Bernstein, I. S., and R. J. Schusterman, The activity of gibbons in a social group. *Folia Primat.* **2**, 161 (1964).

Bingham, H. C., Gorillas in a native habitat. *Carnegie Inst. Wash. Publ.* **426**, 1 (1932).

Carpenter, C. R., A field study in Siam of the behavior and social relations of the gibbon (*Hylobates lar*). *Comp. Psychol. Monog.* **10**, 1 (1940). [*In* C. R. Carpenter, *Naturalistic Behavior of Nonhuman Primates*, p. 145. Penn. State Univ. Press, University Park (1964).]

Coolidge, H. J., Notes on a family of breeding gibbons. *Human Biol.* **5**, 288 (1933).

Du Chaillu, P. B., *Explorations and Adventures in Equatorial Africa.* John Murray, London (1861).

Falk, J. L., The grooming behavior of the chimpanzee as a reinforcer. *J. Expt. Anal. Behavior* **1**, 83 (1958).

Goodall, J. M., Nest building behavior in the free ranging chimpanzee. *Ann. N. Y. Acad. Sci.* **102**, 455 (1962).

Goodall, J. M., Tool-using and aimed throwing in a community of free-living chimpanzees. *Nature* **201**, 1264 (1964).

Gregory, W. K. (Ed.), *The Anatomy of the Gorilla.* Columbia Univ. Press, New York (1950).

Harrisson, B., *Orang-utan.* Collins, London (1962).

Köhler, W., *The Mentality of Apes* (Second edition). Routledge and Kegan Paul, London (1927).

Lang, E. M., *Goma, the Gorilla Baby.* Doubleday, New York (1963).

Miller, C. S., The classification of the gibbons. *J. Mammal.* **14**, 158 (1933).

Nissen, H. W., A field study of the chimpanzee. *Comp. Psychol. Monog.* **8** (1931).

Reynolds, V., An outline of the behaviour and social organisation of forest-living chimpanzees. *Folia Primat.* **1**, 95 (1963).

Reynolds, V., *Budongo. An African Forest and Its Chimpanzees.* Natural History Press, Garden City, New York (1965).

Schaller, G. B., The orang-utan in Sarawak. *Zoologica* **46**, 73 (1961).

Schaller, G. B., *The Mountain Gorilla.* Univ. of Chicago Press, Chicago (1963).

Schaller, G. B., *The Year of the Gorilla.* Univ. of Chicago Press, Chicago (1964).

Schultz, A. H., Observations on the growth, classification and evolutionary specializations of gibbons and siamangs. *Human Biol.* **5**, 212 and 385 (1933).

Schultz, A. H., Some distinguishing characters of the mountain gorilla. *J. Mammal.* **15**, 51 (1934).

Tomilin, M. I., and R. M. Yerkes, Mother-infant relationships in chimpanzees. *J. Comp. Physiol. Psychol.* **20**, 321 (1935).

Yerkes, R. M., Conjugal contrasts among the chimpanzees. *J. Abnormal Social Psychol.* **36**, 175 (1941).

Yerkes, R. M., and A. W. Yerkes, *The Great Apes.* Yale Univ. Press, New Haven (1929).

Variations on a Theme

19

IT IS A FACT that no two individual mammals are identical. Individual differences or variations among members of a population must be understood by all who attempt to assess the degree to which any two animals are related. *Differences* between two similar organisms, for example any two species of primates, does not mean they are *distinct* phylogenetically. It is very difficult to find the point at which differences between two groups of organisms are large enough that they must be considered members of distinct taxa. This will come as no surprise to those who have read the preceding chapters.

Paleoanthropologists are not, or act as if they are not, fully aware of the extent to which living descendants of fossil primates vary. If they were, they would not have created the excessive number of species and genera which now clutter up our accounts of human evolution (Chapter 10). In this chapter we select some traits from some primates and show how they vary within groups of related animals.

A trait may vary in direct relation to several fundamental biological factors. It may vary with age—the dentition of young animals is different from that of adults; because of environmental factors—size and color may depend upon varying amounts of certain nutritional factors available in an animal's diet; or with sex—females may be one color, males another.

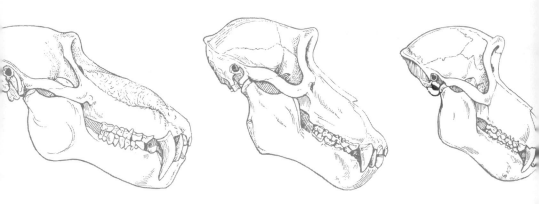

This is called sexual dimorphism. In general females are smaller than males, other things being equal. These are some of the simple causes of variation in which the relationship between the variation in expression of a trait and the agent causing it are known.

More difficult to analyze are cases where the trait in question varies continuously along some scale. Often the variation is continuous between two extreme values on the scale. Tooth size, height, body length, face length, skull depth may be described as such continuous variables. It is often possible to express variations of this sort by an average value and a certain range on each side of the average. It has been found that most members of a population do not deviate extremely from the average value of a trait measured on some linear scale.

Variability such as we have listed is of sufficient importance that it is recognized by a special name—polymorphism (Greek: many forms). The Primates, especially the Hominidae, are an extremely polymorphic group of animals. Failure to recognize this has had important consequences as we shall see and as we have noted already. Polymorphisms are often mistaken for traits of taxonomic significance. When sufficient information is available, the polymorphism becomes apparent and its taxonomic significance vanishes.

As an animal grows after birth, many changes occur which may be characteristic of a particular species, genus, or family. The sequence of events during growth is at times significant. The sequence of eruption of the permanent teeth is one example. The sequence in which the epiphyses are closed during growth is another. The epiphyses are cartilaginous plates at the ends of the long bones. These do not fuse with the rest of the bone until linear growth is completed. Delay in fusion of these plates enables growth of the long bones to take place. The age at which various body systems reach maturity varies, as do the ages of menarche and spermatogenesis and the age at which body growth is completed. Some records available indicate that the ages at which some of these growth processes are completed have changed within recent history (Fig. 19.1). We must always be aware that there is intragroup as well as intergroup variation in such traits and that some of this variation is large. Not only do the sequences of growth events differ among various species, but also the rates of growth differ in different parts of the body. For example in primates length of the hand relative to the total length of the upper limb changes, and the length of the thumb changes relative to the length of the hand.

There is much individual variation of traits more difficult to define, such as the "general aspect" of a skull. This general aspect is often described by measurements such as length of face, distance between orbits, thickness of brow ridges. However such measures often lend a spurious precision to the definition of distinct groups. Figures 10.1, 10.2, and 19.2 illustrate particularly pertinent and fascinating examples of the consequences of not taking into account the individual variation in a single population of a single

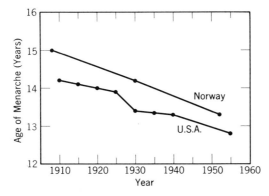

Fig. 19.1. Change in age of menarche in Norway and
the United States since 1900. This is only one example
of variations in growth processes in human populations.
(Data from Tanner, 1962.)

species related to a fossil species. When the variation in similar features of
several fossil australopithecine skulls was first evaluated, three genera were
defined. These genera were *Australopithecus, "Zinjanthropus,"* and *"Paranthropus."* The differences among the three were emphasized and taken
as indicating taxonomic distinctions. The variation in the skulls and faces
of chimpanzees (Fig. 10.1 and 10.2) and gorillas (Fig. 19.2) is notable. The
number of genera and species which might have been described would have
been immense had skulls of chimpanzees and gorillas been treated in the
same manner as those of australopithecines.

A striking example of the variation of expression of a minor morphological
feature of primate mandibles, the mylohyoid groove, was analyzed by
Straus. The mylohyoid groove is a depression on the inner surface of the
ramus of the mandible. In living animals it contains the mylohyoid nerve
and associated blood vessels. Generally the groove runs from the mandibular foramen to the submandibular fossa, beginning beneath a small, thin,
bony projection, the lingula mandibulae (Fig. 19.3). Until Straus completed
his study, it was assumed there was a different arrangement among many
nonhuman primates, that there was a "simian" as opposed to a human type.
This groove was used by Robinson at one time as one of the traits distinguishing *"Paranthropus"* from *"Telanthropus."* Since then Robinson has
lumped the two genera.

Straus analyzed a large number of primate mandibles and found there
were four major types of groove arrangements. Three of these were divisible
into two subtypes each (Fig. 19.3). This impressive analysis of variation is
summarized in Table 19.1. In addition asymmetry was often found among
mandibles of the Anthropoidea. One type of groove was found on one side
of the mandible and another type on the other. The salient conclusion is
that the variation in relationship between groove and foramen is sufficiently

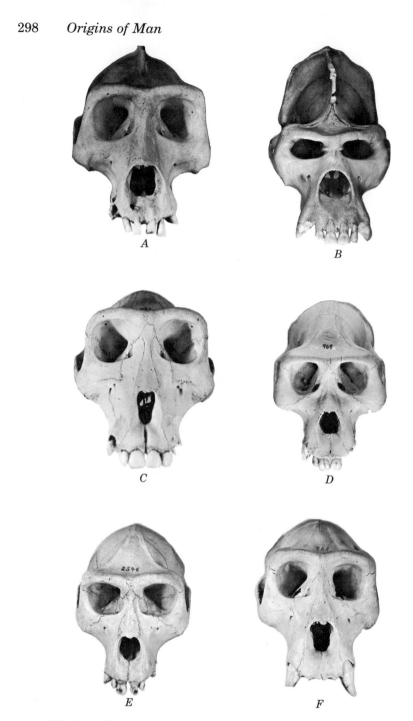

Fig. 19.2.   Variations in size and shape of skulls of gorillas. (B) and (F) are from males, (B) is Gargantua's skull. (A) and (C) are probably male skulls, (D) and (E) probably female.

great that the groove is a character of little significance in making phylogenetic distinctions.

Primate body size varies enormously. The absolute adult size ranges from that of the tiny mouse lemur (*Microcebus murinus*), whose weight ranges from 50 to 90 g, to that of gorillas, whose weight reaches 200,000 g. The fossil lemur, *Megaladapis,* was even larger than the modern gorilla.

Variations directly related to sex include pelage color, size of the canine teeth, total body weight, limb proportions, and a number of other such secondary sexual characteristics. Sexual dimorphism in mammals is, fundamentally, related to the chromosomal sex-determining mechanism. The sex hormones, chemical regulators of many metabolic processes, mediate

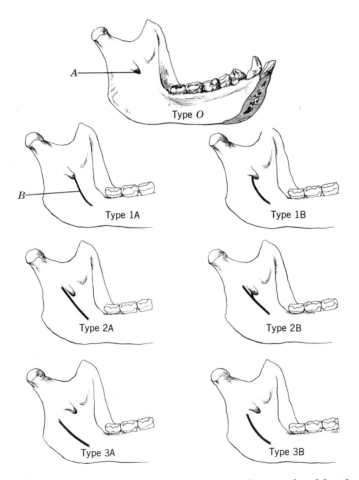

Fig. 19.3. Four types, and their respective subtypes, of mylohyoid groove in primates. The inner surface of the mandible of an adult male orangutan is shown here. (A) Lingula mandibulae and (B) mylohyoid groove. (Redrawn after Straus, 1962.)

TABLE 19.1
Variations in Mylohyoid Groove
(From Straus, 1962.)

Genus	Number of half-mandibles	Percentage of total			
		Type of groove[a]			
		1	2	3	0
Prosimii					
Tupaia	29	27.6	3.4	10.4	58.6
Tarsius	32	6.3	—	3.1	90.6
Lemur	32	6.3	—	—	93.8
Propithecus	22	4.6	13.6	22.7	59.1
Galago	20	25.0	—	—	75.0
Ceboidea					
Cebus	32	21.9	56.2	21.9	—
Saimiri	28	—	3.6	28.6	67.9
Alouatta	44	47.7	25.0	27.3	—
Ateles	32	9.4	56.3	6.2	28.1
Aotus	29	13.8	24.1	55.2	6.9
Cercopithecoidea					
Cercopithecus	30	26.7	40.0	23.3	10.0
Macaca	40	97.5	—	—	2.5
Papio	82	73.2	20.7	6.1	—
Colobus	32	53.1	40.6	6.3	—
Pongidae					
Pongo	142	23.2	38.8	36.6	1.4
Pan	123	8.9	32.5	51.3	7.3
Hylobates	92	45.6	28.3	22.8	3.3
Hominidae					
Homo	291	82.1	17.9	—	—

[a]The distinguishing features of the four types of grooves are illustrated in Fig. 19.3.

many of the sexual dimorphisms we observe. There are hormones specifically male and others specifically female. These influence hair growth, patterns of distribution of body hair, body proportions, and even emotional states. These differences are pronounced in some animal groups, but among the Primates marked sexual dimorphism is not the rule. When it does occur, it is usually as exaggerated characters in the male. The sexual dimorphism of *Homo sapiens*, unlike that of many other primates, includes an exaggeration of some characters in the female as well as exaggeration of characters in the male. This unusual and typically human sexual dimorphism is probably related to the lack of highly noticeable changes in the female at estrus. Fertilization of human females depends upon regular, frequent copulation, since there is no obvious way to signal the desirability of copulation at ovulation as is the case with seals, baboons, and most other mammals.

Exaggerated characters in the females of *Homo sapiens* may provide the signals.

Average body dimensions (weight, height, etc.) are usually different in males and females. Sometimes the ranges around the average values overlap, but the averages are usually different. It is for this reason that body measurements and growth rates are often presented in two groups, one for males and one for females.

One of the more impressive cases of sexual dimorphism among primates is the difference between the skulls of male and female baboons and their close relatives, the Asiatic macaques. Comparisons of female skulls from most of the species of baboons and macaques show no obvious interspecific distinctions. Indeed it is often not possible to distinguish a female mandrill skull from a female baboon skull. Skulls of males of these species differ markedly from female skulls and show obvious interspecific differences. However, the sharp distinction that was once thought to exist between Asiatic and African terrestrial cercopithecines disappears when sufficient skulls from a sufficient number of populations are examined (Fig. 17.4). A long time ago William King Gregory pointed out this gradient of skull forms in baboons and macaques from Africa to Asia. Use of secondary sexual characteristics of the males to define species will result in misleading ideas about phylogenetic relationships, unless the extent of sexual dimorphism is known or can be estimated. Proper assessment of primate fossil discoveries should take account of the probability that sexual dimorphism existed in the remote past, too.

Males of a species are usually heavier than females, but this is not always the case (Table 19.2). In this sample, female spider monkeys and female marmosets were very slightly heavier than males.

For the Neandertals and pithecanthropines, fourteen mandibles have been examined, and the ratio of the breadth of mandibular symphysis to its vertical height has been determined. This symphysis is usually described as being massive and high in the fossils, as might be expected with forms that have massive teeth. Yet Garn has demonstrated that height and massiveness of the mandibular symphysis and the ratio of breadth to height fall within the range of these same measurements on skulls of Americans of European descent. He has also shown that the thickness of the skull, considered at one time to be greater among Neandertal and other hominid fossils, varies considerably among fossils and among living Americans. The range of thickness overlaps in the two groups.

An almost infinite number of examples of variability in the Primates could be cited. Some of them have been mentioned here; others were discussed earlier. Let us re-emphasize that each individual in any population is genetically unique (except for monovular twins), and any measurable trait will vary from individual to individual. When the trait has been measured in a sufficiently large number of individuals, the numerical value assigned the trait for any single individual will lie on what is essentially a continuum. Never-

TABLE 19.2

Relationship of Body Weight of Females and Males

(From Schultz, 1956.)

Genus	Weight of females as percentage of weight of males[a]	Number of specimens examined
Ceboidea		
Cebus	77.1	61
Saimiri	91.9	30
Alouatta	81.0	198
Ateles	103.2	127
Aotes[b]	87.9	11
Leontocebus[b]	104.2	34
Cercopithecoidea		
Macaca	69.0	25
Nasalis	48.5	25
Presbytis	88.7	38
Pongidae		
Pongo	49.2	21
Pan	92.1	34
Hylobates	92.9	80
Hominidae		
Homo	83[c]	

[a]All of the animals were adults. Most specimens of *Pan*, some of *Pongo*, and all of *Macaca* were captive animals; all of the other specimens were animals shot in the wild.

[b]In the classification preferred here (Table 16.1), *Aotus* [=*Aotes*] and *Leontideus* [=*Leontocebus*].

[c]Approximate value.

theless most individual values will be found to cluster fairly close to an average value. Since much of our understanding of human evolution is based upon the morphological analysis of relatively few individual primate fossils, it is important that the ends of the continuum for each particular trait in a closely related living group be known. A single, large, robust skull does not warrant the definition of a new taxon. "*Pithecanthropus robustus*," for example, is merely a large pithecanthropine. When there are a number of fossils available, as there are for australopithecines, pithecanthropines, and Neandertals, variability must always be kept in mind. It is no less important to use the same caution in working on the systematics of living primates. When we assess the taxonomic significance of traits of living primates, the extremes of the range are as important a consideration as are the most common value, the usual color, or the common appearance.

SUGGESTED READINGS AND SOURCES

Frisch, J. E., Dental variability in a population of gibbons (*Hylobates lar*). *In* D. Brothwell (Ed.), *Dental Anthropology*, p. 15. Pergamon, Macmillan, New York (1963).

Frisch, J. E., Sex-differences in the canines of the gibbon (*Hylobates lar*). *Primates* (Kyoto) **4,** 1 (1963).

Garn, S. M., Culture and the direction of human evolution. *Human Biol.* **35,** 221 (1963).

Garn, S. M., and A. B. Lewis, Tooth-size, body-size, and "giant" fossil man. *Am. Anthropol.* **60,** 874 (1958).

Gregory, W. K., *Evolution Emerging*, Vol. I–II. Macmillan, New York (1951).

Koski, K., and S. M. Garn, Tooth eruption sequence in fossil and modern man. *Am. J. Phys. Anthropol.* **15,** 469 (1957).

Pyle, I., and W. Sontag, Variability of ossification in epiphyses and short bones of the extremities. *Am. J. Roentgenol. Radium Therapy Nucl. Med.* **49,** 795 (1943).

Schultz, A. H., Age changes and variability in gibbons. *Am. J. Phys. Anthropol.* **2,** 1 (1944).

Schultz, A. H., Variability in man and other primates. *Am. J. Phys. Anthropol.* **5,** 1 (1947).

Schultz, A. H., Studien über die Wirbelzahlen und die Körperproportionen von Halbaffen. *Vierteljahresschr. Naturforsch. Ges. Zürich* **99,** 39 (1954).

Schultz, A. H., Postembryonic age changes. *Primatologia I,* 887. S. Karger, Basel (1956).

Stewart, T. D., Sequences of epiphyseal fusion, third molar eruption and suture closure in Eskimos and American Indians. *Am. J. Phys. Anthropol.* **19,** 433 (1934).

Straus, W. L., Jr., The mylohyoid groove in primates. *In* H. Hofer, A. H. Schultz, and D. Starck (Ed.), *B bliotheca Primatologica* **1,** 197 (1962).

Tanner, J. M. (Ed.), *Human Growth.* Symp. Soc. Study Human Biol. III, Pergamon, New York (1960).

Tanner, J. M., *Growth at Adolescence* (Second edition). Blackwell, Oxford (1962).

Todd, T. W., *Atlas of Skeletal Maturation.* C. V. Mosby, St. Louis (1937).

Washburn, S. L., The sequence of epiphyseal union in the opossum. *Anat. Record* **95,** 353 (1946).

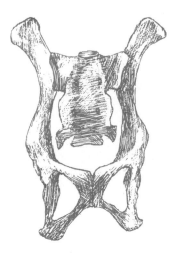

Functional Developments

20

THE SPECIFIC INFORMATION, events, and sequences we have been discussing up to this point now enable us to consider the functional developments that characterize the evolution of the order Primates and, specifically, the evolution of *Homo sapiens*. Locomotion, manual dexterity, visual acuity, social behavior, and intelligence are some of the important functional complexes which developed during primate evolution. Discussion of these functional complexes will integrate much of our information about fossil and living primates.

Locomotion

Locomotor adaptations are important and fundamental. As we have pointed out, the primates took to the trees with grasping hands. They then developed arboreal locomotion in a variety of ways, the most notable specialization being brachiation. One group of primates became terrestrial and adapted in a particular way for quadrupedal locomotion on the ground. Finally the particular taxon to which we belong, the Hominidae, developed erect bipedal walking in the middle Pliocene, if not earlier. Since anthro-

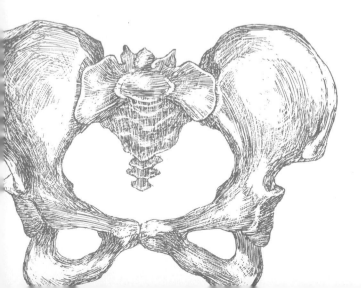

pologists generally agree that the basis for the human adaptive radiation was a locomotor adaptation, the fundamental question we are trying to answer is—Did man evolve from an arboreal lineage that brachiated or from a lineage that was basically terrestrial and quadrupedal?

The similarities between man's anatomy and the anatomy of brachiators are rather detailed. These similarities led some anthropologists, notably Washburn, to suggest that an arboreal brachiating primate was the progenitor of man. We cannot accept or reject this view of man's relationship to brachiators without considering brachiation and its possible connection with the evolution of erect posture and bipedal locomotion. During true brachiation the arms are fully extended over the head. The arms alone propel the body forward and suspend it from a horizontal support. Other terms have been defined—modified brachiation, probrachiation, semibrachiation. This kind of classification with its multiplicity of terminology obscures the functional nature of variations on this mode of locomotion, as Erikson has pointed out. Actually there is a continuum from an essential pronograde quadruped to an orthograde brachiator and, finally, to an orthograde biped.

The brachiating complex, as Washburn defines it, includes the following changes from the body plan of a pronograde quadrupedal animal. There is a reduction in the number of lumbar vertebrae and in the number of deep muscles in the back; the trunk is shortened; the forelimbs are lengthened; and there are a variety of changes in the positions and orientation of the viscera. The brachiating complex includes the development of muscles and joints in the shoulder girdle so that pronation and supination of the forelimb are efficient and skillful. The shoulder is adapted for greater abduction and flexion than is possible in quadrupedal primates. Washburn suggests that these changes are closely associated with the development of a brachiating way of life.

An important feature that is part of the brachiating complex is extreme elongation of the forelimbs. The intermembral index is an efficient way to express this elongation (Table 20.1). The intermembral index is equal to $100 \times$ the ratio of the length of the humerus plus radius to the length of the femur plus tibia. This index in the true brachiators is greater than 100.

The hands and feet must also be considered. The hands of brachiators are elongated and usually do not have a very large or strong thumb. Some species have completely lost the distal phalanx of the thumb. The feet of brachiators are prehensile, and the tarsal bone is relatively longer than it is in terrestrial primates.

An excellent analysis of the various kinds of brachiation has been made by Erikson. Of the apes, gibbons (Fig. 18.3B) consistently use brachiation as their mode of locomotion. The other apes have the shoulder girdle expected of brachiators, but they seldom brachiate. The great apes of Africa, as is well known by now, have many extreme adaptations of trunk and limbs, perhaps more extreme than the gibbons. These adaptations are related to their particular econiches, and it is possible to regard the great apes as animals that

TABLE 20.1

Intermembral Indices for Representatives of the Major Primate Taxa
(From Schultz, 1956.)

Genus	Index[a]	Genus	Index[a]
Tupaiiformes		Ceboidea	
Tupaia	93	*Aotus*	100
Tarsiiformes		*Ateles*	137
Tarsius	86	*Cebus*	106
Lorisiformes		*Lagothrix*	124
Loris	104	Cercopithecoidea	
Perodicticus	113	*Cercopithecus*	106
Galago	89	*Papio*	115
Lemuriformes		*Presbytis*	110
Lemur	96	Pongidae	
Microcebus	99	*Pan gorilla*	138
Propithecus	81	*Pan troglodytes*	136
Daubentonia	111	*Pongo*	172
		Hylobates	165
		Hominidae	
		Homo	88

[a]Intermembral index = $100 \times$ [(length of humerus + length of radius)/(length of femur + length of tibia)].

have evolved to a stage beyond the typical brachiator. They have become slow moving, large bodied, arboreal-terrestrial primates. They do not represent a transitional form between an arboreal brachiator and a terrestrial biped.

There is a group of primates which have evolved a sufficient range of locomotor types to make it possible to study variations in kinds of arboreal locomotion, including brachiation. These are the New World monkeys. Among the Ceboidea, one finds a variety of brachiators, climbers, jumpers, and other kinds of arboreal quadrupeds. Erikson divides them into three locomotor groups. (1) The springers, of which *Aotus* is the typical form. These are small arboreal quadrupeds that jump a great deal. They have long slender trunks and short arms and legs. (2) The climbers, of which the organ-grinder monkey, *Cebus*, is typical. These are medium-sized monkeys, with the legs and arms not as short relative to the trunk as is the case with the springer group. (3) The brachiators, of which the spider monkey, *Ateles*, is typical (Fig. 16.1*B*). These are relatively large but slender monkeys with prehensile tails. They have short, stout, inflexible trunks which are relatively shorter than those of monkeys in other groups. A spider monkey occasionally stands and walks bipedally. It is not likely that it does so in its normal arboreal habitat. Careful examination of a variety of measurements and characteristics of the anatomy of these various locomotor types shows no sharp distinctions between the groups. Nevertheless the averages of such measurements are different, and the differences reflect differences in modes of locomotion. These measurements, more importantly, indicate that brachiation developed independently in the New World.

An arboreal environment undoubtedly provides a number of ecozones in which orthograde quadrupedal locomotion is advantageous. Climbing, hopping, jumping, springing primates do move about in the trees with their trunks erect—up and down the tree trunks, across the branches, from tree to tree, from tree to ground and back again. Arboreal primates will sit on a branch, eat a handful of leaves, groom a fellow, rest, fall asleep, holding their trunks erect, although there is considerable variation in the extent that trunks are held erect. It is not uncommon to find *Cercopithecus aethiops,* basically an arboreal monkey, rising to peer through tall grass when it is gathering food on the ground. Some arboreal primates have a greater tendency to sit, hop, climb, or sleep erect than others. The habit is present in all. But the habit of holding the trunk erect is not confined to the arboreal primates. An erect stance is also commonly observed among terrestrial monkeys, such as baboons.

The adaptation for orthograde posture requires a number of muscle and bone rearrangements and changes from the primitive mammalian form, the pronograde posture. It also requires a relatively large number of reorganizations in nerve pathways, internal organs, and blood vessels. Erect posture is also a kind of preadaptation for the development of and the selection for efficient bipedal locomotion. Indeed an erect trunk unquestionably preceded the developments in the forelimbs characteristic of brachiation and the developments for complete extension of the legs characteristic of bipedalism. The fact that the trunk is erect led to a trend, noticeable in all living non-human primates, for the forelimbs to be of somewhat greater importance in locomotion, particularly in climbing, than the hindlimbs. Straus notes that this makes all primates potential brachiators, a fact which makes it much easier to explain why brachiation, with its extreme concomitant specializations, has parallel developments in several primate groups, such as the New World monkeys and the Asiatic apes. The similarities between man's anatomy and that of brachiators may be secondary to the evolutionary emphasis on an erect trunk, increased size, and markedly different functions for forelimbs and hindlimbs.

It is to the fossil record that we must appeal in our discussion of which of the two alternative origins of erect bipedalism has the greater evidence in its favor. The fossil record of the lineage leading to the most efficient and specialized brachiator, the gibbon, should have some critical bearing upon this question. The fossil gibbon, *Pliopithecus* of the early Miocene, has an intermembral index that is about 95. Even though this is a low index relative to 165 in living gibbons, the cranial and dental anatomy of *Pliopithecus* indicates this is the ancestor of living gibbons (Chapter 8). If shorter forelimbs than hindlimbs were characteristic of ancestors of the gibbons, it is unlikely that other Anthropoidea of the Miocene would have had an intermembral index greater than or equal to 100. Since there is now considerable evidence to support the old view that the gibbon lineage segregated from the other pongids in early Miocene times, it follows that the elongated forelimbs developed independently in the Ponginae and the Hylobatinae.

Straus' analysis of the fossil *Oreopithecus* shows that this is a form in which elongated forelimbs were developing independently of the Hylobatinae and Ponginae. Certainly the elongated forelimbs of the New World primates are another independent development. Our own observations of Malagasy lemurs, particularly *Lemur catta, Lemur fulvus,* and *Propithecus,* suggest that suspension of the body by the forelimbs from branches or cage tops is not an unusual posture. Tendencies and habits of life for which the brachiating complex may be adaptively significant are seen among living prosimians. Simons' view that at least three and as many as six *independent* parallel developments of elongated forelimbs occurred during primate evolution is given support by this evidence.

The fact that adaptations of the limb skeleton consistent with brachiation are deducible from the fossil record and the observation that brachiation is present among widely separated groups of living primates do not demonstrate that a brachiator per se was the direct ancestor of man. The arms, shoulder girdle, and thorax of *Homo sapiens* form an anatomical complex which has many similarities to that of the great apes, the gibbons, and the brachiating monkeys of the New World. But this shoulder girdle-thorax complex of man is not that of a true brachiator, even though there is a resemblance. The evolutionary source of this functional complex of the upper trunk and shoulder girdle of modern man may not have been a brachiating ancestor but a frightened Miocene man-ape who had to stand erect as he peered over the tall African grass to look for predators. The specializations of the living brachiators are such that it would be necessary to propose reverse evolution before a brachiator could become an erect biped. It is not necessary to assume that an orthograde ancestor of man was so specialized that he developed the extreme modifications of the hand, shoulder girdle, and long bones which characterize modern brachiators.

The erect posture of man could have had its origins in a number of situations with which his ancestors were forced to cope. The erect trunk of man has many parallels among his relatives. Straus has pointed out that man's relatives in the seven other major taxa of living primates include species able to sit, stand, sleep, and even walk with the trunk held upright. This pronounced tendency for holding the trunk erect may be viewed as a basic primate characteristic and probably should be included in the list of essential primate characteristics (Chapter 5).

The bipedal gait of man, on the other hand, has certain unique characteristics not so obviously a consequence of the need to adapt to situations that other living primates have had to meet. Man can *stand* erect on his two feet, and he can walk slowly forward (or backward) with legs fully extended.

Many animals are bipedal, but in different ways and on different occasions. Birds are bipedal, so are galagos, so are great apes, and so is man. And furthermore there are rather essential differences among these various animals. We will not learn much about hominid evolution and phylogenetic affinities by comparing birds with primates; any schoolchild knows that a monkey and a robin are different. Comparisons of the terrestrial locomotion of birds and that of primates, merely because both are bipedal, are not apt to produce answers to our questions.

Among the living primates, there are four kinds of bipedalism. These are bipedal hopping, bipedal running, bipedal walking, and bipedal standing. Bipedal running and hopping occur in almost all groups of Primates. We have seen baboons, *Indri, Propithecus, Lemur,* and gibbons running in the bipedal position with a bent-knee gait. Since these animals run bipedally with this gait, the relationship between hindlimbs and pelvis must be like that in the pronograde quadrupedal primates. Bipedal walking is found among the members of the genus *Homo* and the genus *Pan.* The African apes use a bent knee gait when they walk bipedally. True, erect, bipedal standing is possible only for members of the genus *Homo.* Any number of other primates may, on occasion, stand in this manner, but it is a momentary or unusual thing and the bent-knee position is typical. A film has been made of an orangutan walking bipedally, with its legs fully extended and its knees locked. Full extension of the legs is commonly seen in captive orangutans, but the bipedal gait is habitual for one particular orangutan (Fig. 20.1), and it is not the usual method of orangutan locomotion. When man stands, he stands with the legs extended and the knees locked. *Homo sapiens* is the only primate who stands in this way for any length of time.

The vertebral column is an important part of the anatomical complex which determines the mode and the efficiency of locomotion. Very few primate fossil vertebrae have been found, and almost none from the Miocene-Pliocene, the epochs when it is believed the critical changes occurred. Therefore we must examine the vertebral columns of living animals and attempt to reconstruct and deduce the structural changes which led to or made possible new kinds of locomotion and posture.

A quadrupedal animal has a vertebral column shaped like a bow or an arch. The highest point of this arch is at the middle of the column or back. At or near the center of the column is the anticlinal vertebra, that is, a vertebra with a spinous process more or less perpendicular to the column. On the vertebrae between the anticlinal vertebra and the head, the spinous processes slope caudally, toward the anticlinal vertebra. On the vertebrae between the midpoint of the back and the tail, the spines slope cranially, toward the anticlinal vertebra. In quadrupedal animals, the vertebral column flexes at the center of the back. The anticlinal vertebra is the point of greatest flexion, and the action of the spinal column, in pronograde quadrupeds, may be visualized as that of a great bowspring which bends back and forth when the animal gallops along.

The anticlinal vertebra occurs in pronograde primates, but in orthograde primates the anticlinal vertebra is not always clearly identifiable and is no

Fig. 20.1.   Male orangutan (*Pongo pygmaeus*) standing erect. (Photograph courtesy of J. Sorby.)

longer in the middle of the back. The spinous processes of *more* than half the vertebrae point caudally. This change in orientation is related to the rearrangement of the muscles of the back that are involved in holding the trunk erect. Orthograde posture, orthograde quadrupedal locomotion, and orthograde brachiation all required changes in the bony supports of the muscles of the back.

The evolutionary change from the pronograde to the orthograde position of the trunk shifted the position of the vertebral column. The spine moved ventrally. The weight-bearing axis came closer to the trunk's center of gravity. The shift in position of the vertebral column led to changes in the functions of the column, particularly supportive and weight-bearing.

The semierect orthograde posture of the great apes may be taken as the kind of postural preadaptation which led to the habitual erect posture and bipedal locomotion of man. The transformation of the vertebral column of a great ape (or of a baboon, for that matter) into that of man requires the addition of the lumbar curve (Fig. 20.2). The vertebral column is not a rigid

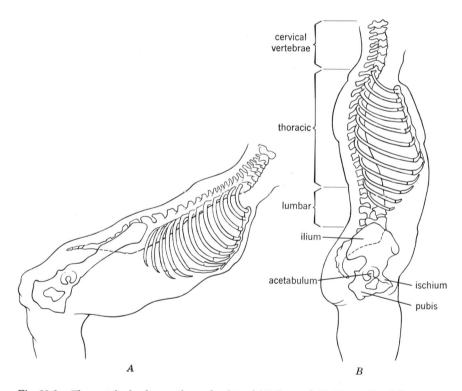

Fig. 20.2.   The vertebral column, ribs, and pelvis of (A) *Pan* and (B) *Homo*. The differences in the overall shape of the pelvis and the relative size of its parts in quadrupedal and erect posture can be seen readily in these drawings. The curvature of the vertebral column is also noticeably different in the two forms. (Redrawn after Schultz, 1961.)

structure but a flexible curved rod. It functions as a flexible strut rather than as a nicely balanced bowspring. The human lumbar curve and other much less pronounced curves in the vertebral column of man are some of the major changes in morphology of the vertebral column related to the assumption of erect posture. The lumbar curve, in particular, is an inefficient response to a change in posture and locomotion. The lumbar curve develops *de novo* in each individual as he matures. Not until a child sits up and walks does it appear. It is not, as far as we can tell, a genetically controlled trait. Curvature in the lumbar region of the spine of some orthograde monkeys, especially those that habitually sit erect, may be seen during dissections of freshly killed animals. But the curve is not manifest in the prepared vertebral skeletons of these monkeys.

The pelvis changed rather more than the vertebral column during the evolution of habitual erect posture and the bipedal gait (Fig. 20.3). The ilium became shortened and bent back on the ischium. The angle between ilium and ischium grew smaller. The broadening of the ilium and the reduction of the angle between ilium and ischium (the bending of ilium back upon the ischium) brings the gluteus maximus behind the joint of the hip, and gluteus maximus becomes a powerful extensor of the leg (Fig. 9.4).

The three bones that make up the pelvis—ischium, ilium, and pubis—did not develop and change at the same rate. Just as the entire organism is an example of mosaic evolution, so is the differential evolutionary development of the pelvis an evolutionary mosaic. Throughout our discussion of the pelvis, emphasis was placed upon the ilium and not the ischium or pubis. The ilium

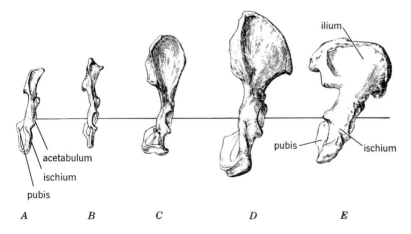

Fig. 20.3. Comparison of the pelves of several living primates. (A) *Tupaia*, (B) *Lemur*, (C) *Hylobates*, (D) *Pan gorilla*, and (E) *Homo*. The most impressive trend is the widening of the ilium. The ischium and the pubis have been less modified. The orientation of the pelvis relative to the horizontal line through the center of the acetabulum is noticeably different in *Homo*.

has changed the most (Fig. 20.3). Washburn is quite correct when he states that man's ischium is almost the same as that of the living Pongidae, whereas the ilium has changed considerably.

As Le Gros Clark, Napier, Zuckerman, and Washburn have each indicated, there are certain measurements on the pelves of living great apes and of the australopithecines which may well help us to understand the mode of locomotion of long-dead australopithecines. For example the linear distances from the anterior superior iliac spine to the acetabulum and from the acetabulum to the ischial tuberosity are very long in the Pongidae (Fig. 20.3). Analysis of the pelvis and the long bones of the australopithecines shows that the line from the anterior superior iliac spine to the acetabulum is short, and the distance from the acetabulum to the ischial tuberosity is long. In modern man both distances are short. What this kind of measurement demonstrates (and it is one of Washburn's great contributions to physical anthropology that he has seen this so clearly and emphasized it so often) is that a structure like the pelvis is not just one bone or one great unit that modifies, changes, and adapts in evolution. It is many different things and has many different functions at the same time. What we might call the final form of the pelvis in modern man is a compromise between various demands that are placed upon it by the organism. It must function so that the trunk may be held erect and stable in bipedal walking. It must provide support to keep the trunk from falling over backward when man is standing upright. It must provide an adequate birth canal through which the large-headed and large-brained fetus may pass at parturition, and it must provide a structure to support a variety of viscera.

Although the postural and locomotor demands placed upon the pelvis have been met, it is possible to argue that they have not been met completely. Man's pelvis is not the totally efficient and comfortable structure an engineer might design for him, had man a choice. Many structural disorders to which man is subject are due to the incomplete adaptation of the back and pelvic girdle to erect bipedalism. The lumbosacral region of the spine is structurally weak, and the mechanical strains will produce many severe ailments, such as herniated intervertebral discs and spondylolisthesis. The adequate but imperfect pelvis is the product of haste. An erect stance and a bipedal gait provided an enormous advantage to the first hominids. The subsequent events (enlargement of the brain, increased manual dexterity) took place quite rapidly. These developments made a completely perfect lumbar region and pelvis unnecessary and compensated for some disadvantages which "low back pain" and an inefficient pelvis may confer on man.

A crucial part of the analysis of the evolution of man's posture and gait is whether the pelvis of the australopithecines is truly a pelvis of the hominid kind or whether it is more like that of the pongids. Unfortunately the australopithecine musculature has vanished. All we can do is express opinions about the pelvic muscles and their probable relationship to a probable mode of locomotion (Chapter 9). At least one characteristic of the pelvic remains of the australopithecines is clearly hominid; the ilium is bent back, shortened, and broadened so that the iliac blade has the characteristics expected of a pelvis which is part of the bipedal complex. It seems clear to us that the australopithecine pelvic bones are hominid and not pongid (Fig. 9.2 and 9.3).

The hand, arm, pelvis, shoulder girdle, and leg are all essential anatomical structures which are part of the locomotor complex. The foot, too, is important. The feet of primates vary greatly. Although all are constructed of essentially the same bones, they differ enormously in relative size and special morphology. Much of this difference can be correlated with modes of locomotion. Man, like *Tupaia*, has a foot that is not prehensile. Man's planti grade foot was derived from the prehensile foot common to the other primates. The modifications are numerous, most of them directly related to the role of the foot in erect posture and bipedal locomotion. The foot of erect bipedal man must be capable of completely supporting the weight of the body, and it must be strong enough to lift the body by its lever action (Fig. 20.4–20.6).

The structural modifications of the prehensile primate foot make the human foot a strong platform. The big toe converges with the others and is not opposable. The metatarsal bone which supports the big toe is parallel to the long axis of the foot. The plantar surface of the big toe rests upon the ground. The other toes are shorter and much less mobile than are the grasping toes of other primates. The big toe is still the longest, and it is the most robust of the toes. The base of the big toe with its metatarsal is one of the principal supports of the foot. Indeed at least half the weight on the foot is supported by the first metatarsal. The tarsal bones are stout, firmly articulated to the arch, and quite strong. They are not flexible and mobile and in no way correspond to their analogues in the wrist. Stability is important. The talus, the ankle bone, is elongated and enlarged.

The human foot may be considered a two-armed lever, as Schultz has described it (Fig. 20.6). The force of contraction of the calf muscles is transmitted through the Achilles tendon attached to the talus. This force lifts the heel and ankle when man walks. Man's foot has been a sturdy platform with a robust nonprehensile big toe since the late Pliocene or lower Pleistocene. Man's foot is essentially similar to the foot of a fossil australopithecine found at Olduvai Gorge, Tanganyika (Fig. 9.5). A convenient way to characterize the foot is to calculate indices that relate the total length of the foot to the length of the trunk. Schultz found that this index varies enormously among Prosimii, whereas it varies much less among Anthropoidea. The length of the prosimian foot ranges from about 20 to 80 per cent of the length of the trunk. Among the Anthropoidea the length ranges from 30 to 60 per cent. The different relative lengths of the feet of various primates are related to the way the lever functions in locomotion.

Schultz dramatically illustrated the theoretical effect of different lever

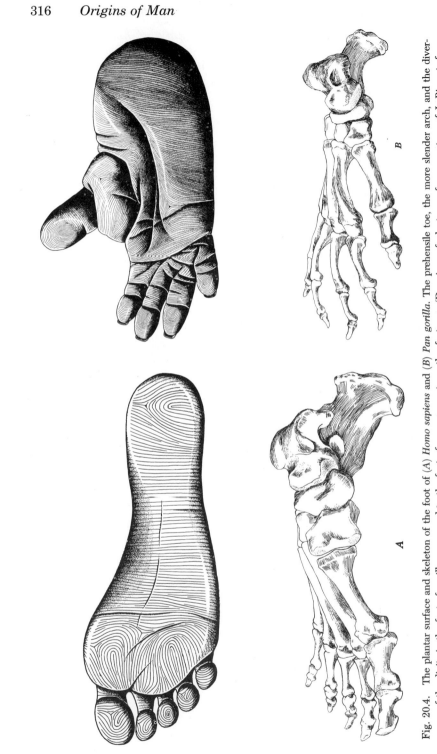

Fig. 20.4.   The plantar surface and skeleton of the foot of (A) *Homo sapiens* and (B) *Pan gorilla.* The prehensile toe, the more slender arch, and the divergence of the digits in the foot of gorilla, compared to the foot of man, are noteworthy features. (Drawings of plantar surfaces courtesy of J. Biegert, from Biegert, 1961.)

lengths by contrasting the feet of the gorilla and the tarsier. The power required to life the body is 3.4 × the load (the body weight) in the gorilla, which has a relative power arm length of 44. The power required to lift the body is 8.3 × the load in the tarsier, for it has a relative power arm length of only 12. In functional terms this means the calf muscles of the tarsier must contract with force sufficient to lift about 1100 g, assuming a weight of 130 g for this animal. The gorilla weighing about 200,000 g would have to contract the calf muscles with force sufficient to lift about 700,000 g. The tarsier must exert more than 8 times its body weight to move its body by lifting the foot, whereas the gorilla need exert only 4 times its body weight.

Elevation of the foot, such as is required in jumping, is produced by powerful movement of the distal end of the load arm. The shorter the power arm relative to the load arm of the lever, the more efficient is this movement. The gorilla, who has a power arm of relative length 44, lifts the foot at the ankle only 2.3 cm when the calf muscles are contracted 1 cm. The tarsier, who has a relative power arm length of 12, lifts its foot at the ankle 8.4 cm with a 1 cm contraction of the calf muscles. The foot of a heavy, largely orthograde primate such as the gorilla is nicely adapted for its primary function, lifting and supporting the heavy body. The foot of the tarsier is also beautifully arranged for its primary function, levering the animal into space as it jumps from tree to tree. The gorilla has a foot that is poorly arranged for jumping, and the tarsier's foot is poorly adapted for erect, plantigrade, bipedal locomotion. Similar analysis of the feet of other primates shows that there is often a close correlation between the body size and locomotion of the animal and the proportionate lengths of the lever arms of the foot. In spite of these correlations, accurate proportions of the lever arms in the foot cannot be predicted from body size and mode of locomotion, and exact mode of locomotion and body size cannot be predicted from meas-

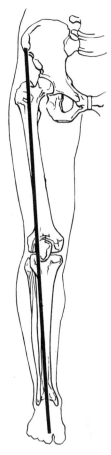

Fig. 20.5. The human leg and foot and the axis of transmission of the body weight, the load, to the foot. This weight-bearing axis is indicated by the heavy line from the pelvis to the foot.

urements of the lever arms. There are a combination of factors related to an animal's mode of locomotion, but the fortunate find of a hominid fossil foot might be helpful in postulating mode of locomotion and posture.

Unfortunately there are very few primate feet to be found in the fossil record. If a foot of *Ramapithecus* were available, it might show us how long the load arm is relative to the power arm of the lever of the foot. The longer the relative length of the load arm, the more likely it would be that *Ramapithecus* was a terrestrial orthograde, bipedal, erect primate. If the

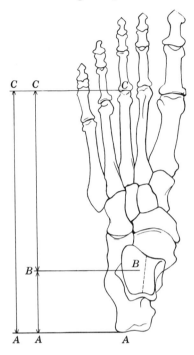

Fig. 20.6.   The lever of the foot, illustrated on a human left foot. (Redrawn after Schultz, 1963.) The distance AC is the lever length; the fulcrum of the lever is at the center of the calcaneus (B); the distance AB is the length of the power arm; the distance BC is the length of the load arm. The power (force) required to lift the load (body weight) $= \dfrac{BC}{AB} \times$ load. The load is transmitted as illustrated in Fig. 20.5.

power arm of this hypothetical foot lever of *Ramapithecus* were as long, relative to the load arm, as it is in the gibbon or some other arboreal jumpers and brachiators, it would be evidence that a jumper and, by implication, a brachiator was directly ancestral to man.

Since the identification of *Ramapithecus* as the earliest known hominid, some of the obvious difficulties in interpreting functional changes in hominid evolution are gone. By placing the origin of the Hominidae in the Miocene or in the very early Pliocene, there is plenty of evolutionary time, almost 14 million years, for a great many developments to have taken place. *Ramapithecus*, the Miocene member of the Hominidae, may have been well on its way toward the body plan of an erect, orthograde, even possibly brachiating primate. There is no direct evidence to support this (see Chapter 8 for the indirect evidence). The erect bipedal form we possess today may have descended from a body structure adapted for brachiation, incipient or full. It is not unreasonable to suppose that a primate well adapted to arboreal existence, with the shoulder girdle and lower back refined for branch sitting and possibly brachiation, would have been forced to come to the ground with the disappearance of the trees in which it was accustomed to live. There was probably sufficient time for it to have adapted to erect bipedal existence on the ground. The discovery at Olduvai Gorge of a hominid foot (Fig. 9.5) adapted for bipedal locomotion, well developed by the end of the Pliocene or in the early Pleistocene, is believed by many to support this general view. But this hominid foot is not the foot of an arboreal jumper or a brachiator. This foot is a very important and fundamental part of the system that is the basis for the adaptive radiation we call erect posture and bipedal locomotion. It has little if any direct resemblance to the foot of a brachiator.

The hand is also part of the locomotor complex. The evidence we find in primate hands suggests that adaptation for brachiation by man's ancestor is not the source of the anatomical similarities between man and pongids. It is clear that the hands of the brachiating apes of the Old World and of the brachiating monkeys of the New World are very different from human hands (Fig. 20.7). In the brachiating hand, digits two through five function together as a hook. Most brachiators, except the gibbon, have very small, poorly developed thumbs. There is a crease all the way across the palm of the hand of a brachiator. This is called a simian crease, and it is the line along which maximum power is exerted in a hooked grasp by the hand of a brachiator. The comparable crease in man's palm extends from the medial edge of the palm to the region between digits two and three. In man digits three through five are used together for exertion of maximum power—chopping wood, shoveling snow, rowing, pole vaulting. The gibbon is the only brachiator with a well-developed thumb. The thumb does not appear to be typical of or highly specialized for brachiation. The gibbon's thumb has features which lead to skillful use of the hand and precise movements of the thumb in conjunction with the other digits.

There are close similarities between the hand of man and the hands of terrestrial primates. The distinction between the precision and the power grip (Fig. 20.8) is very sharp among the terrestrial Cercopithecoidea. The distinction is believed to be the result of adaptation for terrestrial life. Certainly the prehensile patterns of terrestrial Cercopithecoidea resemble those of man more than do the prehensile patterns of the brachiating Asiatic apes. Baboons, for example, are able to control the index finger separately from the thumb and may be able to control each digit separately as do the Hominoidea. This suggests that adaptation for terrestrial life may have been the source of the fine control man exerts over his digits. Although it is not implausible that our hand developed from a brachiating extremity with a gibbon-like thumb, it is evident that a primate adapted for terrestrial existence has a hand from which ours could have stemmed.

The reader will see that we can find evidence to support at least two major theories about the adaptive radiations and the exact structural features which were directly ancestral to *Homo sapiens*. Although we may state in somewhat positive and even dogmatic form the nature of the adaptive radiations which led to modern man, the reader should be aware that there is not very much direct fossil evidence to support either theory, and the evidence that is available, both from paleontology and the comparative study of living primates, can be marshalled to support alternative views. The most reasonable theory in our view as indicated earlier is that a terrestrial, social, diurnal primate was the ancestor of man.

Although we lack sufficient data, we do have considerable evidence with which to postulate the types of adaptations preceding man's erect posture and bipedal gait. We are quite certain that the resemblances between *Homo sapiens* and those animals that are well adapted for brachiation are real

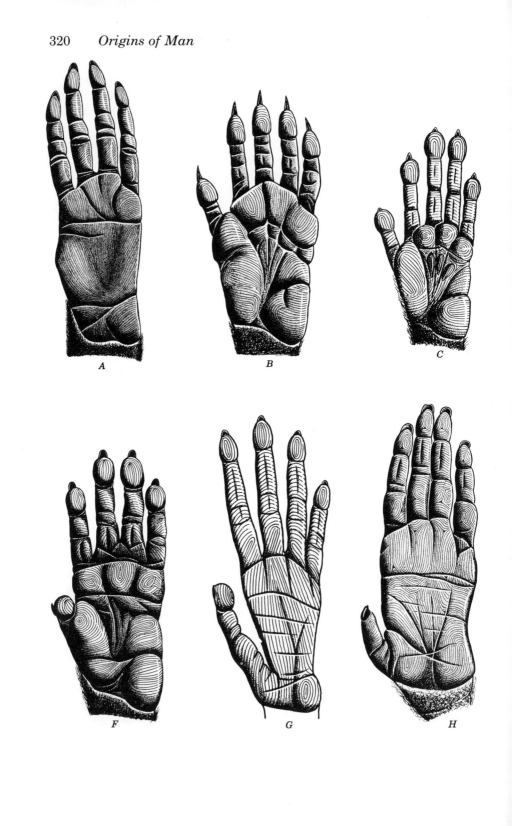

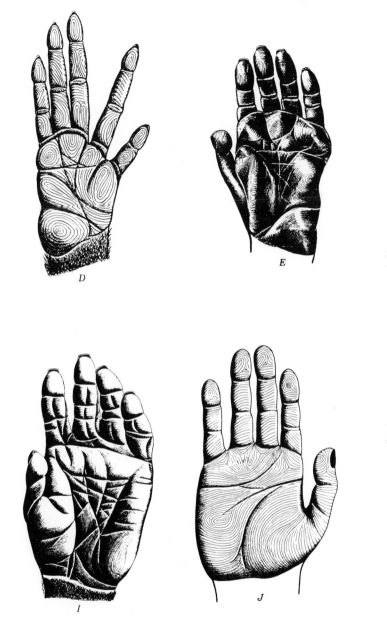

Fig. 20.7. Palmar surfaces of hands of higher primates. (A) *Ateles*, (B) *Callithrix*, (C) *Callicebus*, (D) *Cebus*, (E) *Papio*, (F) *Pygathrix*, (G) *Hylobates*, (H) *Pongo*, (I) *Pan gorilla*, and (J) *Homo*. The right hand is shown for *Homo* and for *Cebus*, all others are left hands. (Drawings courtesy of J. Biegert, from Biegert, 1961.)

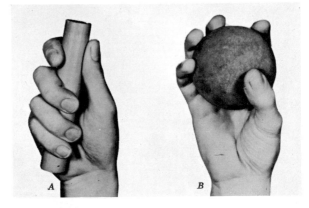

Fig. 20.8.   The two grips of the hand, (A) the power grip and
(B) the precision grip, shown here for the human hand. (From
Napier, 1956.)

resemblances. But it is indeed unlikely that these resemblances mean that
a brachiator came to the ground and developed into an erect biped. We
prefer not to postulate the amount of reverse evolution required to remodel
an erect arboreal brachiator into an erect terrestrial biped. The weight of
evidence suggests that brachiators are so highly specialized that the primate
lineage which led directly to *Homo sapiens* from the common stock of
pongids and hominids was not a group of brachiators. It seems more plausible to us that the postulated ancestral animal developed resemblances to
the brachiators as secondary consequences of the primary adaptive emphasis
on an erect trunk and the precise and skillful use of the forelimbs and hands.
The feet of living brachiators are quite unlike the foot of man. The human
hand, when analyzed functionally, is only superficially like the hand of a
brachiating ape. We believe that the total evidence available suggests that
the Miocene-Pliocene ancestral hominid was a vocal, social, terrestrial
animal.

Manual Dexterity

Man's hand is an impressive, useful organ which distinguishes him from
all other primates. The hand may be used in many nonlocomotor and nonprehensile ways, such as shoving, clubbing, or poking with the fingers. But
it is prehension and the precise movements of the human hand that are
the unique developments of man. The human hand is probably the most
elegant and skillful biological organ that has ever been developed through
natural selection. The hands of man's living relatives give us some notion
of what a variety of possibilities were open to primates.

The hands of living primates vary enormously, for they have been a most adaptive organ throughout primate evolution (Fig. 20.7 and 20.9). Each of the many different kinds of hands is more or less beautifully adapted to the several functions it must perform. The shapes of these hands demonstrate no obvious trends from lower to higher primate, for each is highly specialized. All the primates, except man, must use their forelimbs and hands as major organs of locomotion. Man alone has freed his hands for manipulation through specialization for erect posture and bipedal locomotion.

All primates have prehensile hands. Prehension may be described by defining the positions the hand assumes when it holds an object or the ways in which the hand reaches for an object. The former are called prehensive grips, the latter prehensive patterns. Napier was able to divide the prehensive actions of the hand into two kinds, the power grip and the precision grip (Fig. 20.8). The hand must hold an object securely, and stability in the grip must be achieved. The power grip produces stability when an object is held in a kind of clamp. This clamp is formed by partial flexion of the fingers and the palm with counterpressure applied by the thumb. The power grip is the position of the hand when it exerts maximum pressure on an object it is holding. The precision grip produces stability when an object is pinched between the flexed fingers and the opposing thumb. The precision grip is the position of the hand when it holds an object with maximum accuracy of control.

Prehensive patterns assumed by the hand as it reaches to grasp an object are also important in studying the use of the hand and the evolution of manual dexterity, as shown by Bishop. A prehensive pattern is the changing position of the hand as it reaches to grasp an object. In general the Ceboidea, Cercopithecoidea, and Hominoidea have evolved power patterns that are distinct from their precision patterns. Most of the lower primates, the Prosimii, have variable grips, and these can be described as parts of a single prehensive pattern. A precision pattern can be discerned when an animal uses special movements of the hand or fingers for grasping small objects. The hand takes on a different shape for small objects (a penny, a needle, or a small pebble) than it does when it is used to hold large objects or to hold an animal on a branch. These requirements for different functions, consequences of the highly variable environments of primates, are the background against which we reconstruct the development of power and precision patterns and grips.

Precision evolved, as far as we can tell, through development of voluntary control over each digit of the hand. This means that fine motor control must have increased in primate evolution. Not only have a variety of bone and muscle changes occurred, but also the neuroanatomical basis for voluntary control developed. The centers in the brain concerned with association and control, essentially those in the cerebral cortex, became highly devel-

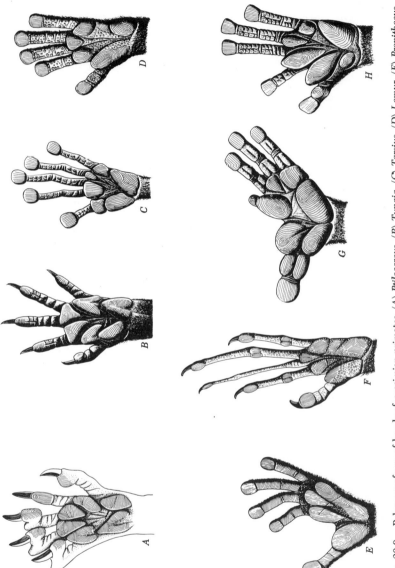

Fig. 20.9.   Palmar surfaces of hands of prosimian primates. (*A*) *Ptilocercus*, (*B*) *Tupaia*, (*C*) *Tarsius*, (*D*) *Lemur*, (*E*) *Propithecus*, (*F*) *Daubentonia*, (*G*) *Perodicticus*, and (*H*) *Galago*. The right hand of *Ptilocercus* is shown; the left hand is shown for all other genera. (Drawings courtesy of J. Biegert, from Biegert, 1961.)

oped as precision patterns and grips developed in the order Primates. The evolution of the hand and brain were concomitant events. One could not evolve without the other. As greater manual dexterity was achieved, additional neurological advances must have occurred.

All prehensile movements of the human hand combine the two basic grips of precision and power. The simplicity of this analysis leads many to conclude that man's hand is indeed primitive and generalized when compared to the more specialized hands of pottos, gibbons, or guerezas. The arrangements of muscles and the relative lengths of the fingers of man are primitive in the sense that they resemble a less drastic specialization away from the primitive mammalian condition than do the hands of most other primates. But it is more a metaphorical generalization than a scientific statement that the hand of man and the marvelously refined movements it makes in writing, watchmaking, surgery, needlework, and thousands of other activities resemble a primitive mammalian condition. The enormously complex cerebral cortex and the human nervous system, too, are basic factors in the development of such skills.

The hands of primates have evolved in a variety of directions. The use of the hand and its particular development has been influenced by the mode of locomotion of the animal. But feeding and social habits have also had a role in establishing particular developments. Classifications of primate hands attempt to relate to function or to phylogenetic sequence. Napier's classification puts hands into three groups: convergent, prehensile with pseudo-opposable thumbs, and prehensile with opposable thumbs. The convergent hand is the typical mammalian appendage, with the digits arranged in a fan when the fingers are extended. When the fingers are flexed against the palm, they converge toward each other. Napier says that if an animal habitually picks up and holds food in one hand, it has a prehensile type of hand. Tree shrews are able to pick up food in one hand, but they rarely do so. The hand of the tree shrew, thus, may be considered convergent despite the fact that the thumb or the first digit is somewhat divergent. All other primates hold food or other objects in one hand and have divergent thumbs, when they have them.

A truly opposable thumb is one that rotates at the carpometacarpal joint so that it opposes digits two through five. If the thumb moves in only one plane at this joint, that is, if it does not rotate, it is pseudo-opposable. In *Tarsius*, the thumb may have great freedom at the metacarpophalangeal joint, but by definition it is pseudo-opposable. All the Prosimii and the Ceboidea have pseudo-opposable thumbs, whereas the Cercopithecoidea and the Hominoidea have opposable thumbs. These major advances in the anatomy of the hands recognized in this classification are the convergence of the fingers, divergence of the thumb, and true opposability of the thumb.

Tupaiiformes have hands characteristic of small agile mammals adapted for moving quickly and easily along small branches and twigs (Fig. 20.9).

The length of the palm relative to the length of the digits is greater in Tupaiiformes than in most other primates. The digits are clawed, more or less parallel, and symmetrically placed. The hand of *Tupaia* resembles the primitive mammalian appendage more than the hand of any other primate does.

The hands of Lemuriformes and Lorisiformes (Fig. 20.9) may be considered together, and special features, such as the almost total absence of the second digit (index finger) in the potto, do not alter the general propositions discussed here. The hands of these prosimians are clearly and sharply distinguished from those of other mammals. They have fingernails and a pseudo-opposable thumb. Careful analysis shows that most can bring the ball of the thumb into contact with the ball of the second and third digits, although this contact is not head on. Even in man's hand, with its completely opposable thumb, the balls of the digits usually meet at an angle and not head on. For this reason thumbs of Lemuriformes and Lorisiformes might well be considered opposable. Despite this opposability, these animals seldom if ever pick up objects by pressing fingertips together. Objects are held in a firm grasp between the digits and the distal pads of the palm.

The locomotor grips of Lemuriformes and Lorisiformes are as highly variable as their many modes of locomotion. Their manual dexterity varies. Galagos, particularly the small species, attain a high level of precision and accuracy in striking at and seizing objects. The Lemuriformes are somewhat less skillful at picking up small objects from flat surfaces. When grasping objects, the prosimians have excellent whole-hand control. The prosimian hand is capable of a multiplicity of grips, but there is little evidence of tactile sensitivity. These factors may contribute to the lack of fine control of the hand.

The pattern of prehension in the Lemuriformes and the Lorisiformes is more complex than in *Tupaia*. All prosimian primates except *Tupaia* open their hands and close their fingers around an object. They do not merely flex their fingers in a single plane. This implies that there is an advance in the nervous system of lemurs, lorises, and tarsiers which permits finer control over the digits than Tupaiiformes have. But the major advance seems to be in the structure of the hand, not in the further differentiation of the nervous system, for there is only a single pattern of prehension in prosimian primates.

The Ceboidea (Fig. 20.7) do not oppose the thumb. Even the rather limited way in which prosimians oppose the thumb to the other digits does not occur. Under certain conditions the thumb is flexed with the other fingers, in the same plane. Nevertheless there are differences in prehensive patterns in this group. For example the Atelinae are brachiators, and in one genus, *Ateles,* the thumb has been lost (Fig. 20.7), whereas in the other, *Lagothrix,* the thumb is elongated and operates in concert with the other fingers. Yet both *Ateles* and *Lagothrix* can pick up very tiny objects, such as pebbles or raisins, by closing two fingers together in a scissors grip. The brachiating Atelinae are the most highly specialized of the Ceboidea. They have also the most developed use of the hands; two precision patterns and grips can be distinguished. Less specialized Ceboidea such as *Cacajao* have developed a precision grip that is distinct from the power grip but have not developed the precision pattern.

The marmosets (Callithricidae) are a special case. The hand of the marmoset has a long palm, and the digits have claw-like appendages rather than nails (Fig. 20.7). The hand is short relative to the length of the arm, and when the animal runs along branches the hand is used in a manner that resembles that of *Tupaia*. Marmosets flex their fingers toward the distal pads of the palm when they grasp raisins, grapes, or other food. To reach something desirable that is relatively inaccessible, they stretch their hand and arm straight forward and scrape it toward themselves with their claws. The prehensive gestures of marmosets resemble those of *Tupaia*, but marmosets have one-hand control, which they use more consistently than do the tree shrews.

Among the Ceboidea that walk on the tops of tree limbs, there is no sharp differentiation between the power and the precision grips, with the possible exception of *Callicebus*. *Callicebus* (Fig. 20.7) has whole-hand control and uses a scissors-like grip to grasp objects. Its thumb is used consistently with the other digits when it walks on very thin branches. These Ceboidea that walk on the tops of branches, *Aotus, Callicebus, Saimiri*, are a relatively primitive, generalized group. They show no specialized advances in the use of the hand beyond prosimians, except that they use their fingers more and appear to have more sensitive fingertips.

All of the Cercopithecoidea except *Colobus* have and use the opposable thumb (Fig. 20.7). There is no distinction possible between a primitive and specialized use of the hand as there is among the Ceboidea. There are two kinds of advanced fine control of the hand among the cercopithecoids. The first kind is the precision grip of the arboreal cercopithecines, the use of the thumb against part of the second digit or the side of the hand. The second is typical of the terrestrial monkeys, baboons for example, that control the index finger separately from the thumb. Indeed baboons may be able to control each digit separately. This ability would place them with the Hominoidea who possess a third kind of fine control of the hand—each digit is controlled separately. Bishop suggests that fine control of the hand, as found among the higher primates, has its source in two activities of prosimians, play and social grooming. These activities might well have led to the variability of pattern and grip which preceded development of fine control. During social grooming the lemurs and lorises take hold of the hair of another animal with both hands and scrape away with the toothcomb. The cercopithecoids part the hair of the animal they are grooming and pick out tiny bits of dirt or parasites with their fingers. Furthermore in order to grasp the hair of the grooming partner, lorises and lemurs must be able to bring their distal phalanges in close proximity to the palmar touchpads.

Control and tactile sensitivity are concentrated on the touchpads. Since grooming is one of the principal social activities of lemurs and lorises as well as of monkeys, it is reasonable to suggest that selection acted upon this facet of manual activity to produce the superior ability found among monkeys.

Play unquestionably involves greater motor variability than does locomotion. Lemurs and lorises are capable of a surprising amount of variability in hand patterns when they play, particularly when they play with objects—pencils, telephone cords, test tubes, pipes, window shades. Selection may have acted upon the hand and brain (intelligence) quite markedly through play. Both intelligence and a tendency to play are required as the animal exploits new situations, new combinations of objects in its environment. During play, prosimians tend to vary patterns of control of the hand. The play situation may have been a preadaptation—if fine control of the hand was advantageous in the exploitation of new environments, any increase in fine control would have provided a functional complex on which selection operated favorably.

The argument just outlined implies that selection pressure upon play as such led to increased intelligence. It can also be argued that problem-solving ability (intelligence) was the primary object of selection and play is a secondary manifestation of the whole selection process. We can only speculate about these matters, for we have only the end products—the living prosimians with which to work.

The distinction between the precision and the power grip is sharp among terrestrial Cercopithecoidea. Many think that terrestrial life has made this distinction. The prehensive patterns of terrestrial Cercopithecoidea are far more like those of man than are the prehensive patterns of the brachiating Asiatic apes. Man and the African great apes have a high degree of fine control of the hands, with varying amounts of independent control of each digit. The power and precision grips and patterns are completely differentiated from each other. Among the great apes the precision grip is usually formed by digit one being placed against the side of the first phalanx of digit two. Digit two is flexed so that a small object is caught under the first interphalangeal joint. This particular grip is a consequence of the short thumb and lengthened digits two to five. The precision grip of man is slightly different, in that the thumb is much more completely opposable to the other digits when they are flexed. The grips of the great apes (gorillas and chimpanzees) are very likely conditioned by the specialization in their hands that goes with brachiation. The gibbon hand is extremely specialized for manipulating objects. It is stronger and has a more complex set of muscles than the hands of the other apes. Since the thumb is not used by the gibbon during locomotion, it was probably developed for fine control of objects. Because of differences between the hands of apes and man (Fig. 20.7), Napier and others argue that the human precision grip and the human hand derive from a ground-living ancestor rather than a tree-living brachiator.

Special Senses

The earliest primates, we believe, took to the trees when mobile digits and a prehensile hand were in the early stages of development. The demands of arboreal life very likely produced powerful and continuous selection for vision as the dominant sense. Stereoscopic vision is of obvious advantage to arboreal primates. The fossil record gives us a relatively good evolutionary sequence of the development of the bony structures—skull, orientation of orbits, bony enclosure of eyeballs—which support the soft tissues of the visual apparatus (Fig. 7.3 and 7.4). But again we must turn to living primates for data relevant to the appearance of stereoscopic and color vision and visual acuity.

Stereoscopic vision requires two modifications of the primitive mammalian eye structure. The first is osteological—forward rotation of the orbits; the second is neuroanatomical—sorting of the optic nerve fibers at the optic chiasma in the brain. The orbits must be rotated toward the front of the skull so that the fields of vision of the two eyes overlap. In the Anthropoidea, all of which have excellent stereoscopic vision, the orbits have rotated so that the optical axes are parallel. Overlap of the fields of vision is almost complete. The second major modification involves sorting of the optical nerve fibers. The fibers cross completely in the brains of lower vertebrates. This is called total decussation of the optic fibers. This means all the nerve impulses from the right eye are carried to the left hemisphere of the brain, and those from the left eye to the right hemisphere. Vertebrates with stereoscopic vision have some optic fibers that do not cross over. This is called partial nondecussation and means that some nerve impulses from corresponding points in the visual fields of each of the two eyes must terminate in the *same* hemisphere of the brain.

Among prosimians, the rotation of the orbits is sufficient so that some overlap of the visual fields occurs. The reduction of the snout and of the olfactory sense is correlated with the increased size of the orbits, their forward rotation, and the expansion of those parts of the brain concerned with vision. But the prosimian snout is still prominent, and this does not permit the rotation of the orbits sufficiently for the optical axes to become parallel. Considerable lateral divergence of the optical axes occurs, but not to the extent of the divergence in lower mammals. In many lower mammals, where divergence is complete, each eye picks up a different image.

The diurnal lemurs have efficient, stereoscopic vision. The optical axes converge more than in tree shrews, and the optical centers in the brain of *Lemur* are larger than those in *Tupaia*. The proportion of nondecussated optical fibers is large, probably as large as in monkeys. The diurnal lemurs, such as *Lemur macaco* and *Propithecus*, are remarkable arboreal aerialists. They must have been under strong selection pressure to develop efficient, functional, stereoscopic vision. When *Lemur macaco* leaps 15 to 30 feet across a gap between two trees, gracefully takes hold of a thin branch near the top of the second tree, and propels itself forward across another 10 feet of space, the conclusion that a high level of visual acuity and stereoscopic vision are helpful to it does not seem unreasonable. A very few days watching *Indri* and *Propithecus* propel themselves around the treetops of the Malagasy forests convinced us that visual acuity, a high level of precision in striking objects (eye-hand coordination), and stereoscopic vision would not be undesirable and undoubtedly are possessed by those species. Visual acuity, good eye-hand coordination, and stereoscopic vision would be an advantage to any creatures that hop and jump in an arboreal habitat. The animal should be able to judge the distance it must leap; it should be able to grab a very small branch; it must be able to coordinate the movements of its extremities with the information it receives through its eyes.

The enlargement of the optical region relative to other segments of the brain is more or less progressive throughout the order (Fig. 20.10). Detailed discussion of the neuroanatomy of primate vision will be found in the suggested readings at the end of the chapter.

Color vision is believed to be present among the Anthropoidea. Unfortunately, experimental determination of the occurrence of color vision in a species is difficult. Most of the experiments do not enable us to distinguish whether intensity or wavelength is the stimulus to which the animal responds. Conclusive results are possible if monochromatic beams of light are used as conditioning stimuli. Gibbons and orangutans were put into an experimental situation which forced them to discriminate between colored and gray patches of paper. They were given food if correct choices were made and so presumably were conditioned to expect a reward if they chose correctly. They did choose colored papers, and this was interpreted to suggest that they are able to discriminate colors. Despite absence of convincing proof of color vision among the Anthropoidea other than man, there is little reason to doubt that some ability to distinguish colors must be present. After all, there is the indirect evidence of the wide variation in color of pelage, including considerable sexual dimorphism, among some species.

The prosimians are usually listed as not possessing either color or stereoscopic vision. The latter is very obviously present in most of the diurnal genera, and the former probably is also. There is no conclusive evidence that color vision is not present.

Primate neurological evolution includes more than reduction of the olfactory and enlargement of the visual centers of the brain. Efficient exploitation of and survival in the arboreal habitat required the development of fine control of limbs and of movement and balance, as well as elaboration of sensory perception of the environment.

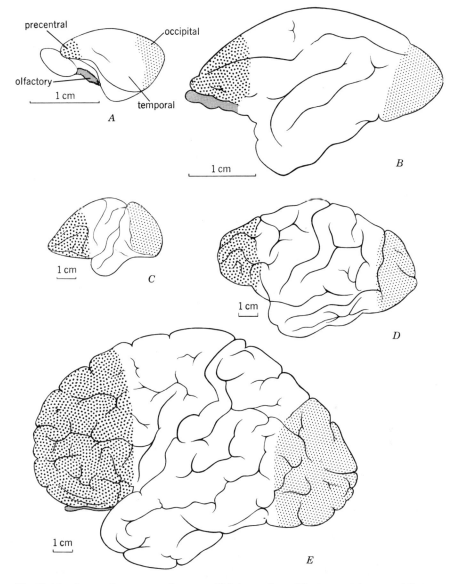

Fig. 20.10.   Brains of primates, side views of left hemisphere. (A) *Tupaia,* (B) *Lemur,* (C) *Cerco-pithecus,* (D) *Pan,* and (E) *Homo.* In all the Primates, the area of the brain concerned with vision (occipital lobe) is relatively large. The precentral cortex, concerned with motor control, is larger in the higher primates. The olfactory lobe, concerned with the sense of smell is relatively larger in the lower primates. The temporal lobe is more complex and larger in the higher primates.

The relative development of the brain and its various functional parts among living primates is illustrated in Fig. 20.10. The occipital lobe of the brain expanded as vision became refined and visual acuity increased. This area is associated with the interpretation of visual stimuli. The precentral cortex increased in complexity and size, for it is the region associated and concerned with motor control of muscles and vocal cords. Part of it is an important association area. The temporal lobe enlarged and became complex. It is the center for sound discrimination, important when vocalizations become more complex and important. The cerebellum enlarged and so did its interconnections with the motor area of the cortex.

The evolution of the primate brain can be described as the extraordinary development of voluntary control as well as the retention of the basic features of the mammalian cortex. The cerebral cortex of mammals has several basic functions which are reflected in its anatomy. There is differentiation of sensory areas which receive impulses from various sense organs, such as the eye, the ear, and the hand. There is a motor area which emits impulses that initiate and control voluntary movements. There are association areas between the sensory and motor areas which interrelate the two areas. The progressive changes in the brain, summarized above, are primarily concerned with one of the major features of primate evolution—the development of the brain as an organ of voluntary control as well as of association.

Social Behavior, Intelligence, and Displays

Most of our ideas about the evolution of social behavior, intelligence, and vocal and facial displays of Primates must come from comparative studies of living animals. The fossil record can provide only minimal evidence about these important complexes. The discussion on locomotion and manual dexterity of the Primates may have suggested that those functional complexes are the bases for all that happened in primate evolution. Because of the availability of fossil materials and the deductions made therefrom, there is something to this point of view. Nonetheless research on and analysis of primate behavior are becoming increasingly important in evolutionary studies.

In the evolutionary context, intelligence is a characteristic which is best considered as the ability of lower primates to solve problems and of the hominid line to make tools. There is no obvious selection pressure on mammals that explains the high level of intelligence reached by primates. Marsupials have a very low intelligence when compared to placental mammals. Thus, when an opossum is compared to a white rat on simple conditioning tests, the white rat turns in a superior performance. The ability to solve problems varies widely among mammals. Some, such as marsupials, have managed to survive since the Eocene with no obvious change in the gross structure of the brain that might be correlated with increasing intelligence, that is, ability to solve problems.

Among the Primates there is a relatively wide range in problem-solving abilities and, by definition, intelligence. Andrew has shown that lemurs, New World monkeys, and Old World monkeys differ markedly in problem-solving abilities. All three groups occupy similar arboreal econiches, have similar diets, and show similar social behaviors. All three are derived from a common ancestry. Yet lemurs perform poorly on choice tests, much more poorly than macaques. The probable reasons for the differences may suggest how selection for intelligence occurred.

Two general reasons have been advanced for the lower problem-solving ability of lemurs. First, lemurs do not use their hands to part the hair or pick tiny bits of dirt from the skin during grooming. An alternative method of grooming developed. Lemurs use the toothcomb (Fig. 2.1B and 14.4) for that part of grooming which the Cercopithecoidea do with their fingers. The lemurs have only one prehensive pattern, and they are relatively inefficient at grasping small objects. The level of manipulative skill reached by lemurs, certainly lower than the level attained by monkeys, may well have been a factor in limiting further development of intelligence. Regardless of the level of intelligence and manipulative ability achieved by lemurs, natural selection has operated to produce animals which are well adapted to their particular habitat.

The second reason is that the lemurs had very few other kinds of mammals with which to interact on Madagascar. Isolated and alone in the lush forests of the island, lemurs did not need to develop extraordinary manipulative skills and intelligence in order to exploit their econiche successfully. If a primate has to develop increased intelligence in order to cope with an econiche it is entering, it will be able to compete in econiches to which its ancestors had no access. Lemurs had no need to develop further than they did until *Homo sapiens* colonized Madagascar. By that time, the gap between these two primate lineages was great, so great that there was no chance for lemurs to compete, and they were set upon the path to inevitable extinction.

The wide range of intelligence present among mammals suggests that increased intelligence must be the result of potent selection pressure. The length of time since divergence from the common ancestor of the living primates seems to be correlated with intelligence. The more primitive lemurs, for example, are less intelligent than baboons, and both of these are less intelligent than man. The relative rapidity with which intelligence evolved in the Primates is another indication that selection must have emphasized this trait. Much of such selection pressure seems to be provided by other species of intelligent primates. The interesting example of the lemurs also suggests that manual dexterity is very important among primates. The implication is that the intelligence of the higher primates and the development of toolmaking abilities by man are the result of competition with other intelligent primates and of possession of a high level of manual skills.

Brain size certainly is related to the evolution of intelligence. It is clear from the fossil record and from the comparative anatomy of living primates that larger brains go with the more intelligent primates. But the critical question, discussed in Chapters 9 and 21, cannot be answered by examining the gross size of the brain. At what brain size did hominid intelligence become great enough to create culture? Until we know a great deal more than we do about the detailed function of brain structure in all the Primates, we can only guess and appeal for our answer to the archeological record of toolmaking.

The evolution of social behavior was touched upon by implication in Chapters 17 and 18, and little more will be added here. One of the most important things to emphasize is the sharp break between social behavior and social groups of the nonhuman primates and of man. The difference is most likely due to the development of symbolic vocal communication in the hominid line. Other aspects of primate behavior, unquestionably, have been and continue to be of value in evolutionary studies. However we shall probably learn much more about the evolution of human behavior by comparative study of primate vocal communication than by the study of almost any other component of primate psychogenic behavior.

Now we shall turn our attention to vocalizations and related facial displays of primates. We define a display as a pattern of effector activity which serves to convey information. The passage of this information is advantageous to the individual of the species, to the social group to which he belongs, to others of the same species, and to other species. The study of vertebrate displays originated with Darwin's *The Expression of the Emotions in Man and Animals*. The study of primate vocal and facial displays has been much neglected until recently. Primate vocalization is one aspect of primate behavior that is apt to lead to a better understanding of the evolution of human vocal communication and the invention of culture by man. Facial expressions are closely related to primate vocalizations, and we shall discuss their evolution in conjunction with vocalization.

Two questions have usually been asked about vertebrate displays and about primate displays in particular. What is the function of these displays, and why did they evolve? What is the cause of the display in the individual? We shall not discuss causation but shall emphasize the evolution of the function of displays. Selection pressure which works on animal displays is to a large extent determined by the degree to which the animal is social. There will be much stronger selection pressure on the vocal and facial displays of highly social primates than on those of the more solitary, nocturnal species.

Among the important characteristics of the Primates which are related to the evolution of vocal and facial displays are the following: increased visual acuity, particularly stereoscopic vision and color vision; increased facial mobility with correlative decreased mobility of the ears; increased manual dexterity and manipulative ability; increased problem-solving ability; and the development of permanent social groups or societies. These characteristics, among the salient features of primate evolution, have evolved to a different degree, relative to each other, in different groups of primates.

Primates continue to use certain calls of their nonprimate ancestors, for some calls are present throughout the order in one form or another. Twitters (high, short calls) were probably used by the common ancestors, for twitters can be found among *Tupaia*, Ceboidea, and in certain Lemuriformes, Cercopithecoidea, and Hominoidea. They are also found in many Insectivora. Noisy expiration as a threat call is so widespread among primates, and for that matter other mammals, that it is believed to be primitive. Modifications of the twitter that sound like clicks are also considered primitive. The techniques for studying primate calls are described in some of the references at the end of the chapter.

The functions of calls and facial expressions in the social vertebrates have been studied intensively in such groups as birds. The generalizations reached by these studies apply as well to calls and facial expressions of the Primates. There are close-contact calls which individuals give while a social group is moving together. There are distant-contact calls when an individual is isolated or separated from members of his social group. Mobbing calls are given when a predator or stranger is seen from a safe distance. These mobbing calls are easy to localize. Intense-warning calls produced when the animals are frightened are difficult for the hearers to localize. Threat calls and precopulatory calls are also present. With Primates close-contact calls are among the most variable. They function to greet fellow members of the troop and to make it clear that no attack or flight is planned. It is likely that the calls produced by an animal when it is fleeing from another are derived from or similar to contact calls and function to show that the animal will be submissive or will continue to flee.

The evolution of warning calls may well be related to two things that happened among the Primates. Visual acuity developed to a high degree, and the length of time of association between mother and infant and among juveniles was lengthened. The territorial behavior of primates together with contiguous territory and contiguous troops may have also played a role. In other words, related animals that recognize each other would benefit from warning calls, and these warning calls would have a selective advantage. Primates developed an arboreal way of life, and selection probably favored the mobbing and the intense-warning calls. The former is very easy for other members of the species to localize. The intense-warning call functions when an animal is frightened, and it is much more difficult to localize. This call probably serves as a signal when more dangerous predators appear.

Distinctive vocalizations are produced by nonsocial animals, primates included, in certain situations, such as mother-infant interactions and male-female interactions during estrus. The calls used in these situations may have acquired significance in social situations and may be used in more general contexts. As permanent societies of diurnal primates developed, it is quite likely that strong selection for giving these calls to fellow members of the troop occurred. One reason for believing this is that the calls which are elicited in friendly greeting, by one social fellow to another after a very brief separation, resemble those given by the infant when it is searching for the mother's nipple. Among the Cercopithecoidea, greeting calls have reached a very elaborate development, especially in baboons. Baboons' greetings sound like human grunts, and Andrew believes these have been adapted to convey information about mouth and tongue positions. Baboons appear to form much more coherent social groups than do any other terrestrial monkeys, such as macaques, for immature males do not have to leave a baboon troop in order to relieve social tension. Selection pressure on baboons may well have operated to produce much more informative social displays, including facial and vocal displays, than is the case among the macaques. The information conveyed is probably related to the social status of the immature male in a baboon society. Baboon society might be expected to splinter and be rather unstable, given the presumed tensions that would arise between the adult dominant males and maturing males. Yet baboon society, which is selectively very advantageous for baboons, is not unstable. It is likely that the more elaborate and informative displays of young baboons were quickly selected for in order that the society might preserve itself.

Another interesting evolutionary development among the vocalizations of primates is the wailing, territorial song of *Indri* and gibbon (Chapters 15 and 18). Both animals seem to live in very small groups, probably family groups, high in the forests. This wailing song may well have evolved because it conveys information over great distances about the location of groups in a dense forest.

A very striking feature revealed by examining facial and vocal displays of primates is the amount of parallel evolution that has occurred. Almost every feature of vocalization and facial expression has developed in parallel among the major primate stocks. The grin and lip rounding which have evolved independently as facial displays among the Ceboidea, Cercopithecoidea, and Hominoidea are most remarkable cases of parallel evolution. The formation of prolonged calls from series of twitters has occurred in the Lemuriformes, the Ceboidea, and the Cercopithecoidea. The pitch of vocalizations has changed. The pitch of calls of both apes and some species of baboons is lower than in prosimian primates, probably as a result of greater size of the former.

Primate vocal and facial displays are two of the few sources of information about the origin of human symbolic vocal communication—language. There are many theories about the origin of language, but some of them are not well substantiated or directly relevant here. Man's imitation of sounds he hears in his environment, including sounds heard from other animals, must be one of the roots of origin. It is peculiar that there should have been any argument about this point, for it seems obvious that a vocal language could arise only as a modification of existing vocal systems. It is reasonable to postulate that language arises in an animal in which auditory control over vocalization is developed sufficiently so the animal is able to imitate sounds. This is not at all related to the question of whether onomatopoeia was involved. Onomatopoeia may well have been involved in the development of words, but this is a question that has to do with later events in the evolution of vocal communication, events that occurred after vocal communication was well established.

What is the selection pressure which produced vocal imitation good enough for some kind of transmission of symbolic vocalizations from one generation to another in *Homo sapiens?* We must look for those features of vocalization over which man has considerably more control than do his fellow primates.

Control of vocalization in man is, in a large part, control of changes in the resonance properties of the buccal cavity (the mouth). It is humanoid grunts which take advantage of all of the resonance properties of the upper respiratory tract and the buccal cavity. Contractions of orbicularis oris would have been selected for in order that such variability and range of sound in the buccal cavity would be available. Baboons produce humanoid grunts. Andrew suggests that the reasons for this lie in certain similarities of the social groups of baboons and early man. Any change which makes the transfer of information by vocal or facial expression more explicit and less ambiguous also acts to enhance the advantages of living in a social group. If the direct ancestors of man were terrestrial plains-living creatures, it is reasonable to assume that they were subjected to similar or indeed the same selection pressures which act upon the baboon. The requirement that the social group be a refuge and a fortress in a plains' habitat emphasizes selection for a coherent and well-organized society. The grunts of *Homo sapiens* may well have evolved in a manner similar to that in which they evolved in baboons, in order to convey information simultaneously with facial and vocal expressions. The grunts of baboons and apes are probably due to what may be the beginning of changes in the respiratory tract, changes that developed to a far greater extent in man. The edges of the vocal cords are somewhat blunter among baboons and apes than among lower primates. The drop in pitch of vocalization in these groups is related to this change in the vocal cords. The vocal cords of man are very blunt, but the principal difference in the vocal apparatus of man is the relationship of the position of the larynx to the rest of the upper respiratory tract. Man's larynx is lower in the throat and farther away from the soft palate than it is in other primates, including baboons and apes. It is no longer in contact with the soft palate. The position of the human larynx is in part the result of the erect posture of man. The position of the larynx changed as the foramen magnum moved forward on the base of the skull and as the size of the mandible became smaller. The descent of the larynx creates a long, uninterrupted resonating cavity. This tubular cavity makes possible the low-pitched speech of man. The faculty of the human ear for pitch discrimination probably evolved in close connection with the evolution of the vocal apparatus. This faculty is most acute at discriminating those pitches within the range of normal human conversation.

The appearance of grunts that are like humanoid grunts is not enough to lead to the development of language. Symbolic meanings, crude as they may be, would not be attached to particular sounds if one animal could not imitate certain features of the sounds made by others. The development of vocal mimicking must have been an important feature in the origin of language. Mimicking may have developed from the ability to answer a call with an imitation of the call itself. Among lemurs, *Lemur catta* is able to match or mimic sounds only in a crude way. A wail is elicited not only by the wail of another *Lemur catta* but by many high-pitched, prolonged sounds. The bark of *Lemur catta* is elicited by barks of other *Lemur catta* and by many other loud sounds as well. Baboons match grunt for grunt. The gibbon is believed to mimic more elaborately and extensively, for it answers four of its own calls with the same calls. This mimicking or matching could be extended by selection to the ability to learn to monitor the sound produced by one's own larynx so that the sound could be made to conform to new variations of calls already learned. Among baboons or baboon-like primates, this could have been facilitated by the fact that certain facial gestures elicited themselves. Tongue protrusion might be elicited in response to tongue protrusion and this would, incidentally, improve the mimicking or matching of associated grunts.

Further development of mimicking or matching may well have depended upon the ability to manipulate the behavior of others with vocalizations. Chimpanzees give particular calls that direct the attention of others to an object. If influencing the behavior of another individual led to the attachment of symbolic meanings to sounds, the pressure of natural selection would have become great and led to the perfecting of imitation and vocal control as we see it in *Homo sapiens*.

The evolution of facial displays, closely related to the evolution of vocal displays, is partly understood by changes that occur in the relevant musculature. The muscles of the lips have changed during the evolution of primates, even though the basic plan is more or less the same. In man and chimpanzee the zygomaticus pulls the corners of the mouth back and upward, and the platysma pulls the corners back and downward. The orbicularis oris is a sphincter which closes and protrudes the lips and converts the buccal cavity into a tube.

The separate components of facial expressions of primates and probably many other mammals very likely have evolved from protective responses given to strong stimuli. The most complete facial response of this sort to a strong stimulus can be seen in man after a very unpleasant taste is experienced. The zygomaticus contracts first, sometimes in concert with the levator muscles; platysma contraction will then occur as retraction of the corners of the lips becomes intense and tongue movements appear. The upper lip is raised by the contraction of levator labii superioris. These particular movements occur in nonhuman primate displays largely in association with intense vocalization. Intense vocalizations occur in certain situations of great stress. As primate evolution went on, the elicitation of these facial displays was facilitated until they occurred without vocalization. At least this is the way the evidence from living primates is read.

Among the Lemuriformes and Lorisiformes the initial primate condition for facial displays is probably seen. The zygomaticus of lemurs produces barely perceptible grins, while the zygomaticus and platysma together produce intense grins. In *Lemur fulvus* the intense grin, given generally with shrieks, is elicited by attacks of a superior animal. *Hapalemur* and *Lemur catta* grin in association with calls of somewhat lower intensity. The Lorisiformes grin only with very intense clicks when there is a vigorous attempt made by the animal to reach a social fellow. Among the Ceboidea the marmosets grin only during intense twitters or squeaks. In the more evolved Ceboidea, such as *Cebus,* the grins may appear without associated vocalization. Friendly greetings among Ceboidea seem to elicit almost all the protective facial responses. When the stimulus is of high intensity, vocalizations always appear. More elaborate facial expressions are also seen, including the lowering of the eyebrows in a frown. The Lorisiformes, Lemuriformes, and Ceboidea click their teeth while they grin in response to unpleasant taste stimuli. Such clicking is also elicited when a strange object or very distant moving objects are perceived by the animals. It is well known that humans also click teeth after tasting something unpleasant.

Among the Cercopithecoidea grins may be evoked as silent grins, but they are usually accompanied by some kind of vocalization. Grins often occur in baboons and other Cercopithecoidea when they are frightened by a superior animal. The baboons' grin has reached sufficient autonomy that it is given as a friendly greeting. The grin has been facilitated even more in chimpanzees and in *Homo sapiens* in whom the sight of a desired object may elicit a silent smile.

Suggested readings and sources

Andrew, R. J., Evolution of intelligence and vocal mimicking. *Science* **137,** 585 (1962).

Andrew, R. J., Evolution of facial expression. *Science* **142,** 1034 (1963).

Andrew, R. J., The origin and evolution of the calls and facial expressions of the Primates. *Behaviour* **20,** 1 (1963).

Andrew, R. J., The displays of the Primates. *In* J. Buettner-Janusch (Ed.), *Evolutionary and Genetic Biology of Primates,* Vol. II, p. 227. Academic Press, New York (1964).

Ashton, E. H., M. J. R. Healy, and S. Lipton, The descriptive use of discriminant functions in physical anthropology. *Proc. Roy. Soc. (London) Ser. B* **146,** 552 (1957).

Avis, V., Brachiation: the crucial issue for man's ancestry. *Southwestern J. Anthropol.* **18,** 119 (1962).

Biegert, J., Volarhaut der Hände und Füsse. *Primatologia* **II/1,** Lieferung 3. S. Karger, Basel (1961).

Bishop, A., Control of the hand in lower primates. *Ann. N. Y. Acad. Sci.* **102**, 316 (1962).

Bishop, A., Use of the hand in lower primates. *In* J. Buettner-Janusch (Ed.), *Evolutionary and Genetic Biology of Primates*, Vol. II, p. 133. Academic Press, New York (1964).

Broom, R., and J. T. Robinson, Notes on the pelves of the fossil ape-man. *Am. J. Phys. Anthropol.* **8**, 489 (1950).

Broom, R., J. T. Robinson, and G. W. H. Schepers, Sterkfontein ape-man Plesianthropus. *Tranvaal Museum Mem.* (Pretoria) 4 (1950).

Buettner-Janusch, J., and R. J. Andrew, The use of the incisors by Primates in grooming. *Am. J. Phys. Anthropol.* **20**, 127 (1962).

Clark, W. E. Le G., *The Antecedents of Man* (Second edition). Edinburgh Univ. Press, Edinburgh (1962).

Darwin, C., *The Expression of Emotions in Man and Animals.* Appleton, New York (1899).

Day, M. H., and J. R. Napier, Hominid fossils from Bed I, Olduvai Gorge, Tanganyika. *Nature* **201**, 967 (1964).

Dobzhansky, T., Individuality, gene recombination and non-repeatability of evolution. *Australian J. Sci.* **23**, 71 (1960).

DuBrul, E. L., *Evolution of the Speech Apparatus.* Charles C Thomas, Springfield, Ill. (1958).

Erikson, G. E., Brachiation in New World monkeys and in anthropoid apes. *Symp. Zool. Soc. London* **10**, 135 (1963).

Falk, J. L., The grooming behavior of the chimpanzee as a reinforcer. *J. Exp. Anal. Behavior* **1**, 83 (1958).

Hewes, G. W., Food transport and the origin of hominid bipedalism. *Am. Anthropol.* **63**, 687 (1961).

Hooton, E. A., *Up from the Ape* (Revised edition). Macmillan, New York (1947).

Michael, R. P., and J. Herbert, Menstrual cycle influences on grooming behavior and sexual activity in the rhesus monkey, *Science* **140**, 500 (1963).

Napier, J. R., The prehensile movements of the human hand. *J. Bone and Joint Surg.* **38B**, 902 (1956).

Napier, J. R., Studies of the hands of living primates. *Proc. Zool. Soc. London* **134**, 647 (1960).

Napier, J. R., Brachiation and brachiators. *Symp. Zool. Soc. London* **10**, 183 (1963).

Negus, V. E., *The Comparative Anatomy and Physiology of the Larynx.* Heinemann, London (1949).

Polyak, S., *The Vertebrate Visual System.* Univ. of Chicago Press, Chicago (1957).

Schultz, A. H., Postembryonic age changes. *Primatologia* **I**, 887. S. Karger, Basel (1956).

Schultz, A. H., Vertebral column and thorax. *Primatologia* **IV**, Lieferung 5. S. Karger, Basel (1961).

Schultz, A. H., Relations between the lengths of the main parts of the foot skeleton in primates. *Folia Primat.* **1**, 150 (1963).

Simons, E. L., Fossil evidence relating to the early evolution of primate behavior. *Ann. N. Y. Acad. Sci.* **102**, 282 (1962).

Straus, W. L., Jr., Fossil evidence of the evolution of erect bipedal posture. *Clin. Orthopaed.* No. 25, 9 (1962).

Thieme, F. P., Lumbar breakdown caused by erect posture in man. *Anthropological Papers, Museum of Anthropology, Univ. of Mich.,* 4 (1950).

Tigges, J., On color vision in gibbon and orang-utan. *Folia Primat.* **1**, 188 (1963).

Washburn, S. L., The analysis of primate evolution with particular reference to the origin of man. *Cold Spring Harbor Symposia Quant. Biol.* **15**, 67 (1950).

Washburn, S. L., The new physical anthropology. *Trans. N. Y. Acad. Sci.* **13**, 298 (1951).

Washburn, S. L., Behavior and human evolution. *In* S. L. Washburn (Ed.), *Classification and Human Evolution*, p. 190. Aldine, Chicago (1963).

Zuckerman, S., Correlation of change in the evolution of higher primates. *In* J. Huxley, A. C. Hardy, and E. B. Ford (Ed.), *Evolution as a Process*, p. 300. George Allen and Unwin, London (1954).

Part Two

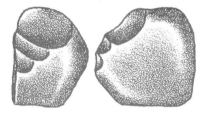

Man's Capacity
for Culture

21

We have described many of the major steps the primate stock took during its evolutionary history from a primitive mammal to contemporary *Homo sapiens*. We have reviewed the fossil and living relatives of man and have discussed each of the major taxa of the living primates except one, the Hominidae. We described briefly some of the important and interesting features of the other seven taxa of living primates in an attempt to suggest to the reader what the progenitors of man are. We tried to present the vast range of characteristics of which man is but a part. Our point of view and our emphasis always has been that man is an animal, a member of the order Primates. The second part of this book will be devoted, in large part, to the eighth major taxon, the Hominidae, and to its single species contemporary *Homo sapiens*.

Culture

The first subject to which we turn our attention is the transition from an anthropoid to a human way of life. We have already discussed the evolution of language as an essential functional part of this transition. We now must consider the origin and differentiation of culture. The evolutionary diverg-

ence of *Homo sapiens* introduced this new factor to the evolution of life on this planet. Man is unique in the possession of culture, and we must evaluate the evidence that bears upon its origin. In the scheme of evolution, culture is a new kind of biological adaptation with a nongenetic mode of inheritance. The mode of inheritance of culture depends upon symbolic contact and transmission rather than upon the fusion of gametes. To some extent its evolution has supplemented the organic evolution of man.

Culture is a biological event; it is a product of the evolutionary process. It is a trait which only one genus of the order Primates developed, and the reasons it developed must be sought in the evolutionary history of that genus—*Homo*. We prefer to use the concept of culture which Tylor, Morgan, Kroeber, and White developed. It is a part of the central organizing principle of anthropology, evolutionism. The classic definition of culture was formulated by Tylor in 1871:

Culture is that complex whole which includes knowledge, belief, art, morals, law, custom, and any other capabilities and habits acquired by man as a member of society.

A more contemporary version of this view of culture has been proposed by White and has its roots in the theories of cultural evolution expounded by Spencer, Tylor, Morgan, and other founders of modern anthropology. White's definition of culture is:

. . . an extrasomatic, temporal continuum of things and events dependent upon symboling. Specifically and concretely, culture consists of tools, implements, utensils, clothing, ornaments, customs, institutions, beliefs, rituals, games, works of art, language, etc.

Once culture developed, it differentiated, grew, and changed much as if it were independent of organic man; indeed the expression "superorganic" was applied to culture by Kroeber. There is no question that culture is man-made, but individuals have hardly any control over it. Culture is one of the most impressive adaptations achieved by any evolving organism. Every human individual is born into a culture of some kind or other. This culture determines the language he will speak, the kinds of clothes he will wear, the rules for choosing a mate, the rituals he will participate in, the musical scale he will consider normal, the standards of interpersonal behavior he must achieve.

Culture is a word much used by anthropologists. It is a word which stands for a concept of fundamental importance in the science of anthropology. There are many ways in which culture, or a particular culture, is learned by individuals. There are many ways in which culture is transmitted from generation to generation. Culture is closely related to behavior, for human behavior is conditioned by and is a function of culture. It is quite possible to develop a sound and useful concept of culture by considering it the sum of all behavior that is uniquely human as opposed to the parts of man's behavior that are instinctual and physiological. We prefer to view culture in its most general form, as a universal biological adaptation of the genus *Homo*. A nonorganic, a superorganic, adaptation was developed by a particular evolving lineage. Culture added a significant factor to this lineage's evolutionary potential, and the subsequent development of this lineage was to a great extent due to the elaboration and evolution of culture.

Culture is based upon an ability, a trait, which appeared during the course of primate evolution, the ability to symbol. A symbol is a thing with physical form that is given meaning by those who use it. The physical form can be a wavelength of light, a frequency of sound, a nod of the head. The physical sensation is transmitted to the brain. It is in this organized neurological tissue that a meaning is attached to the physical sensation. Symbols are given meanings arbitrarily. There is no inherent connection between the physical form of a symbol and its meaning. White's famous example of holy water demonstrates most simply and clearly what a symbol is. Holy water is made up of two things: a material that exists in nature, H_2O, water, and a meaning assigned to it by man. No chimpanzee can distinguish between holy water and drinking water, but any man can be taught the difference. Chimpanzees and other animals can be conditioned to respond to physical objects in such a way that it appears they, too, are giving arbitrary meaning to the objects. The fact that the same physical object may function as a symbol in one context and not in another should not lead to unnecessary confusion. Chimpanzees or dogs can be conditioned to respond to a signal, and the response does not need to have an intrinsic relationship to the physical properties of the signal. An animal can be conditioned to respond to sound of a particular pitch or light of a particular wave length. The cry "halt," a loud sound, or a red light may be established as the cue for a specific dog to stop (or roll over). Man *learns* to respond to many stimuli, particularly vocal stimuli, in much the same manner. He is conditioned to associate a particular vocal stimulus with a particular behavior or response. But man is not a passive participant. Man, unlike a dog or a chimpanzee, can actively determine what meaning the vocal stimulus will have. This is the great difference between chimpanzees and men; man can arbitrarily impose signification upon vocalizations, chimpanzees cannot. A man can teach another man that a red light means "stop"; a man can teach this to a chimpanzee; a chimpanzee cannot teach this to a man, and one chimpanzee cannot teach this to another.

An important problem in evolutionary biology is the determination of what it is that makes it possible for man to associate arbitrary meaning with physical objects. The obvious structure to examine is the brain, the center for association and voluntary control. Unfortunately there are no living animals that are in any way transitional between the primates that do not and the one that does use symbols. Neurological evolution obviously proceeded in the primate line to the point where a primate, which we would unquestionably recognize as a member of the genus *Homo*, was able to symbol. This ability made possible the accumulation and transmission of information from one generation to the next. Culture and symboling, closely connected in their origins, the former dependent upon the latter, may be seen as the consequence of certain characteristics that *Homo sapiens* developed, characteristics not present in other primates.

The distinction between man and other primates lies in the use of tools. Some nonhuman primates make tools even if they do so in a most rudimentary way, and they use tools on many occasions. But there is a fundamental difference between man and the other primates in the use of tools and their manufacture. The use of tools by *Homo sapiens* is a cumulative process, progressive from one generation to another. Among the apes toolusing, toolmaking, does not perceptibly change or progress from one generation to another. The human species, by virtue of its symbolic faculty, can store up information about tools and pass it on to the next generation.

We pointed out that primate evolution may be viewed, in one way, as the development and refinement of erect posture, bipedal locomotion, and manual dexterity. The faculty to symbol and the development of culture are closely associated with man's erect posture, bipedal locomotion, opposable thumb, and gregariousness or sociability. Although animals other than primates are gregarious (sea lions, cattle, and elephants are examples), the social behavior of primates and primate social groups are more complex and more highly structured than those of other mammals. The organ of association and control, the cerebral cortex, developed rapidly as the visual sense grew more complex and refined. At the same time, a fine control over the limbs and digits developed to a greater extent as did voluntary control of the muscles. Man comes from a stock preadapted for symboling and culture.

These brief words on culture are far from all that can be said on this subject. Some of the references listed at the end of this chapter will introduce the reader to the enormous literature about what culture is and about the validity and usefulness of the culture concept in anthropology. We agree with Simpson that many polemics and resulting confusions would be eliminated if, at least for some purposes, culture were viewed as a biological phenomenon. Indeed it seems obvious that it is. For an understanding of human evolution, the cultural evolutionists' view of culture as an extrasomatic trait, which developed out of certain biological characteristics unique to man, is less confusing and more useful than any of the other views of culture. And it is quite valid. Culture, since its invention and differentiation, has had profound effects upon human evolution.

We must ask a number of questions. How do we determine when culture first differentiated; in other words, how do we determine when man first became a toolmaker? What critical changes in the primate nervous system accompanied this change in behavior? Neither of these questions can be answered conclusively, but they can be discussed, and we can make some speculations about the origin of tools, the development of the primate nervous system, and the advent of culture.

Tools and Intelligence

It is unlikely that the earliest appearance of the ability to make tools will be precisely correlated with a specific fossil. By the time symboling had developed and culture had differentiated sufficiently for recognizable tools to appear in the archeological record, the associated primate fossil was far past that stage in his lineage at which the earliest tools were made. Some of the evidence that these abilities are part of man's general primate heritage comes from the work by Goodall, who showed that chimpanzees use objects as tools and appear to pass on some rudimentary information about their use from one animal to another. This does not invalidate our contention that man makes and uses tools in a distinctly different way than chimpanzees.

Most of the evolution of human behavior is based upon the ability to symbol, which in turn is a product of neuroanatomical evolution. During primate evolution the cerebral cortex increased in size (Chapter 20). The major trend, which we have repeatedly emphasized, was the development and elaboration of the special senses and the expansion of those centers of the brain, particularly in the cortex, concerned with conscious control over complex behavior and voluntary control over the muscles. Therefore it is believed that areas of association (areas receiving and sorting complex sensory impressions) and areas of control over voluntary actions have increased in size.

We suggested in Chapter 9 that cranial volume is a trait not especially useful in distinguishing various taxa of the Hominidae after the middle Pliocene. Our point was that by the time of development of the toolmaking capacity, which is the consequence of the capacity to use symbols, the size of the brain was not the critical factor. Once there is evidence that the making of tools to a particular plan has developed, we must assume that symbolic communication and education were part of the repertory of traits possessed by the primate whose fossil bones are associated with the tools. If a small-brained animal, for example an australopithecine, with a brain volume of about 600 cc is found in association with tools, we must assume that the brain volume was large enough for symbolic communication. We also assume that archeologists can tell us whether the fossil primate made the tools. Stone tools are prima facie evidence that there was sufficient neurological material for culture. Incidentally the earliest tools were not necessarily made of stone. Indeed it is unlikely that they were, but stone tools have the greatest likelihood of being preserved. The absence of tools from a site in which early hominids are recovered does not mean the hominid did not use wood or bone tools or that tools were not made.

But this does not answer all the questions we need to ask about evolution of the brain and its consequent effect upon the evolution of symboling and culture. What is there in the fossil record of primates that we can use to make deductions about the increasing capacity for association, control, and, eventually, symboling? All we have are fossil crania. The volume of an endo-cranial cast is such a gross measure that it is very difficult for us to deduce specific neurological functions from it. In recent years this problem has intrigued a number of competent students, and progress has been made in developing a way of answering this question.

Jerison has proposed a theory which can be used to distinguish major groups of primates. His theory is based on estimates of body weight and brain volume. From the estimates he calculates the total number of neurons in the cerebral cortex. The number of neurons associated with the body weight and the number of neurons associated with the adaptive capacity of the animal are then calculated. The calculation that is important here is the one that gives an estimate of the adaptive capacity of the animal (Table 21.1). For simplicity let us call this the number of adaptive neurons. It is not possible to make very refined distinctions among the various primates by using methods based upon this theory, and distinctions between any two closely related species or genera are not possible at present. But the eight major taxa may be distinguished from each other with this method.

TABLE 21.1

Estimates of Adaptive Cortical Neurons in Mammals of Different Brain and Body Sizes

Adaptive neurons are those cortical neurons associated with the adaptive capacity of the brain. (From Jerison, 1963.)

Animal	Brain weight	Body weight	Number of adaptive neurons
Macaca	100 g	10,000 g	1.2 billion
Papio	200	20,000	2.1
Pan troglodytes	400	45,000	3.4
Pan gorilla	600	250,000	3.6
Australopithecus	500	20,000	4.4
Homo erectus	900	50,000	6.4
Homo sapiens	1300	60,000	8.5
Elephant	6000	7,000,000	18.0
Porpoise	1750	150,000	10.0

The quantities in Table 21.1 imply that the number of adaptive neurons has increased markedly in hominid evolution. Even the australopithecines, with only 4.4 billion adaptive neurons (on the average), were toolmakers. Perhaps the development of more elaborate or at least more distinctive stone tools depended upon the increase in adaptive neurons from 4.4 to 6 or 7 billion. Yet we must remember that astonishingly rapid advances and changes in culture took place after the 8.5 billion neuron level was achieved. Perhaps there was a minimum number of these neurons associated with adaptive capacity necessary for great diversity and elaboration of cultural artifacts. There is a relationship between brain size and body size—as the body size increases, brain size generally increases. The size of the brain relative to the size of the body is a different matter. Very small mammals tend to have a larger relative brain size than might be expected. This is probably because of the need for a certain minimum number of neurons for muscular coordination and other important functions. We noted earlier that the living primates have a brain-to-body ratio that is larger than that of other mammals. The "typical primate brain" is about twice the size of the "typical mammalian brain" for any given body size (Table 21.2).

TABLE 21.2
Ratio of Brain Weight to Body Weight in Adults
(From Crile and Quiring, 1940.)

Animal	Body weight	Brain:body ratio
Primates		
Man[a]	61.5 kg	1:47
Chimpanzee	56.7	1:129
Chimpanzee	44.0	1:135
Night monkey	9.2	1:84
Night monkey	8.9	1:75
Sykes monkey	4.9	1:81
Vervet	4.0	1:65
Capuchin monkey	3.1	1:43
Galago	0.2	1:40
Other Mammals		
Deer	65.1	1:310
Bushbuck	35.4	1:253
Beaver	5.8	1:197
Norway rat	0.2	1:122

[a]Average for 35 individuals.

We suggested that there are different levels of capacity for social behavior and for problem solving among various groups of nonhuman primates. But the exact neuroanatomical features associated with such differing levels are, as we stated, not well known. Gross brain size appears to be associated with differing capacities, and this is the only neurological trait that can be determined, even if crudely, for some fossil primates. Jerison's methods of analysis show that gross brain size may be analyzed in such a way that more adaptive neurons are predicted among more capable primates. *Homo sapiens* has the largest number of adaptive neurons.

It so happens that both the porpoise and the elephant have at least as many adaptive neurons as man. How do we explain this? Although there is a rich mythology about the incredibly advanced capacities of elephants for memory and porpoises for speech, there is little evidence to support these stories. Neither the elephant nor the porpoise has the symbolic faculty of man. Yet within the econiche each occupies, each is remarkably successful and well-adapted. There is no need to postulate complex elephant societies with special burial customs or porpoise vocal language to account for what they do with all their neurons. The size of the brain and the number of adaptive neurons cannot be considered outside of the adaptive zone in which the brain must function. As research on elephants and porpoises continues, we have faith that the relationship between the demands of the environment and the neurological adaptations of these mammals will be better understood.

The increase in size of the brain of *Homo* has often been considered the result of increasing manipulative skill (greater effectiveness in making tools) or increasing intelligence. But it does not seem reasonable that the apparent average increase in size of brain from the pithecanthropines to modern man can be accounted for on the grounds that manual dexterity became so great. Since we cannot administer intelligence tests to a pithecanthropine, and since we do not have a very precise notion of what intelligence is in contemporary *sapiens*, it is not profitable to pursue the evolution of brain size as a concomitant of the evolution of intelligence. Evolution of the larger brain of *Homo* may be related to the evolution of culture. Once the neurological capacity to symbol and to make culture evolved, the differentiation and rapid development of culture itself very likely put severe demands upon the brain. The cultural part of the environment of **man** grew more and more elaborate, and the need grew to sort the messages (symbols) coming into the brain. It became necessary to separate those messages that were important from all the sensory information coming through the ears and the eyes. This probably required elaboration of the cerebral cortex, a larger set of association neurons and interconnections between them. As Garn put it,

. . . it may be that our vaunted intelligence is merely an indirect product of the kind of brain that can discern meaningful signals in a complex social context generating a heavy static of informational or, rather, misinformational noise.

Hypotheses about the course of neurological evolution in the hominids are without substance unless material evidence of the products of the evolving nervous system are available. Fortunately there are many tools in the archeological record, products of the evolving hominid nervous system and good evidence for the hominid capacity for culture. Once tools are found in the archeological record and fossil hominids are found associated with these tools, it is clear that the capacity for culture, culture itself, is already developed. The degree to which there is a correlation between neurological evolution and the manufacture of tools is difficult to determine with any kind of certainty today. It appears, from the archeological and fossil records (Fig. 3.1), that there were very long periods when there was little or no change in the tool kit of man. During all of the lower Pleistocene, the same kinds of rather crude tools, stone choppers, were made. In later periods very rapid, abrupt, changes in the catalog of tools occurred. The rate at which tools developed, changed, improved, and became complex increased remarkably during the upper Pleistocene and early Recent epochs. The rapidity of change and the rapid rate of the evolution of culture is an important feature of human evolution. Certain other significant advances, such as the development of agriculture, led to what seems to be an explosive development of cultural inventions.

The archeological record suggests that there was a very long period when tools were extremely simple, merely pebble tools. The tool kit of man is quite simple until the latter part of the lower or the early part of the middle Pleistocene. At this time tools of several different kinds appear. There seemed to be a rapid increase in cultural developments. But there is no evidence at all that there was a large discrete increase in cranial capacity and, hence, in adaptive neurons of the cerebral cortex. The changes in the brain appear to have been gradual. The rather abrupt change in the tool kit of man, if it was abrupt, is not associated with an abrupt change in the physical dimensions of the brain.

It is true that the archeological record may be biased. It very likely represents only a fragment of the material artifacts used by the earliest members of the genus *Homo*. Nevertheless there are some interesting features of the archeological record that can be related to the evolution of the brain. The stone tool kit of man, until the latter part of the lower Pleistocene, consisted of artifacts just a bit above the level of rocks or sticks picked up by a primate for the immediate task at hand. The first stones seen by the primate were seized and used to pound, dig, and cut. The stone and bone tools were made to a plan and imply that the primate who made and used them was a symboling animal. But they do not imply that very detailed blueprints were used in their manufacture. Dart believes that the earliest tools were bones, long bones and mandibles of common antelopes. He proposes that the long bones were used by early man, or by a prehuman ancestor, for clubbing and killing other animals and that the mandibles functioned as knives.

There is another factor which may have influenced the kinds of tools made. Man was restricted to the tropics during the entire early period of his development. The earliest fossil man in Europe is *Homo sapiens;* there is as yet no evidence from Europe of earlier forms such as the australopithecines (Chapter 9). Fossil men are not found in Europe until the brain reached what is essentially modern size. The interpretations by archeologists stress the observation that more elaborate and complex tools are found in Europe than in Africa and Asia. Africa is almost devoid of flint and other raw materials from which were made the complex and diverse stone tools found in Europe in the middle and upper Pleistocene. It is reasonable to suggest that the elaborate flint tools, which occur about the time that hominid fossils appear with brains of modern size, partly depended upon the availability of raw materials. Quartz and quartzite, common in Africa, are, as Rouse points out, excellent materials from which to make pounding tools—choppers—but poor materials for sharp cutting tools—blades. Good, sharp blades can be

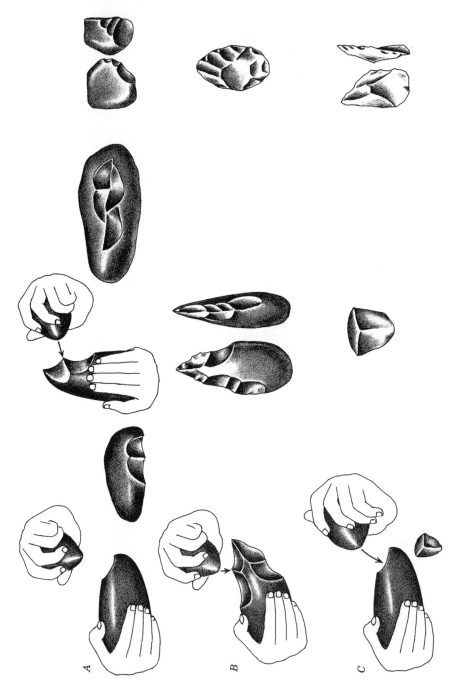

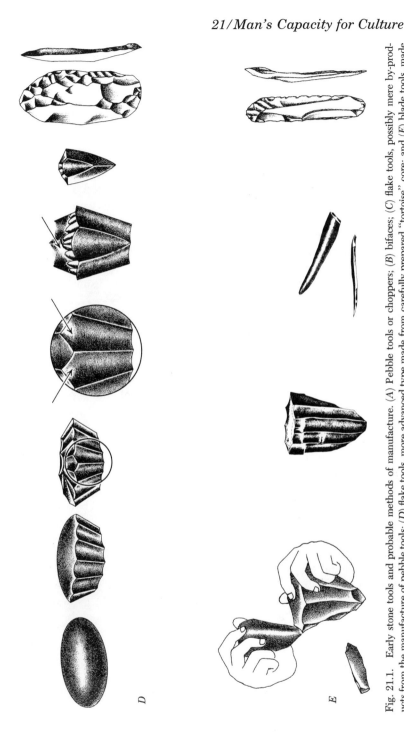

D

E

Fig. 21.1.   Early stone tools and probable methods of manufacture. (*A*) Pebble tools or choppers; (*B*) bifaces; (*C*) flake tools, possibly mere by-products from the manufacture of pebble tools; (*D*) flake tools, more advanced type made from carefully prepared "tortoise" core; and (*E*) blade tools, made from carefully prepared "prismatic" core. The arrows indicate points at which blows are struck. The tools figured on the right are generalized examples of each type.

made of obsidian (volcanic glass), and obsidian is found in Africa. But obsidian tools are not found in sites of the lower Pleistocene. Obsidian is brittle but not much more difficult to work than flint. However the process of working obsidian is more complex than the process of working quartzite to obtain pounding, chopping tools. It is not only the brain and what goes on in the brain that is important, but also what materials are available for it to think about.

The earliest stone tools are called pebble tools. The shape of the original pebble is still present in the tool. Pebble tools are, essentially, stones that have been worked a little at one end, with a few chips knocked off. Such a slightly sharpened stone is a most useful tool, and a new one can be made quickly as the old one wears out. The pebble tool tradition persisted throughout the lower Pleistocene. Pebble tools, often called choppers, are also found throughout the middle and upper Pleistocene in parts of Asia. The effectiveness of such tools was recently demonstrated by Leakey when he skinned and butchered a small antelope in less than 20 minutes by using a pebble tool found at Olduvai.

The next tools in the kit, the flakes and bifaces, may be viewed as byproducts of pebble tools and logical extensions conceived by toolmakers. If the chips knocked off the pebble tool are used and worked, the flake tool results. At its highest development in Europe, it culminated in elegant blade tools. If the pebble itself continues to be worked, the biface or handaxe results (Fig. 21.1). Rouse has suggested that the records of tool types in the Pleistocene can be interpreted as a function of increasing elaboration in conceptual thought. Biface and flake tools are not found until the time when fossil men with relatively large brains are found—brains of almost modern size. It is clear that to make biface or flake tools man must have gone through a more complex intellectual process than the process necessary for making pebble tools. The manufacture of flake and biface tools is an abstract and highly intellectual activity. There are several conceptual stages between the finished flake or biface and the original pebble from which it is made. In comparison the manufacture of the simpler pebble and bone tools found in earlier horizons of earlier strata in the tropics is little more than the utilization of objects at hand.

Man's capacity for culture depended upon the evolutionary development of a brain of sufficient size and complexity to enable him to symbol. The interpretation of fossil and archeological records suggests that brain and tools developed together. Yet there is no clear evidence that the rather rapid increase in diversity and complexity of tools was a function of a brain of a certain critical size. Neither is there evidence that tools had a direct influence on the evolution of a large brain. The availability of raw materials for the hands to use and for the brain to imagine using was also a critical factor. The development of culture also depended upon culture itself. The filling of the tool kit depended on the cultural traditions, on raw materials, and on the brain.

Sex and Society

An important aspect of the evolution of primate societies is the nature of sexual behavior among the various primates, particularly the apes and man. In many species of mammals, sexual behavior is closely tied to cyclical changes in the physiology and anatomy of the female and the male. There is considerable variation in the mammalian reproductive or estrous cycle. Estrous cycle is the term used to describe the reproductive cycle of sexually mature females of many mammalian species. An estrus occurs only during the breeding season, if the species has a breeding season, and it is usually of short duration. During the estrous cycle there is a period of sexual receptiveness or "heat" in the female. During this period of heat (estrus) the female will copulate with the male. The female will copulate at no other time in species with a full estrous cycle. Ovulation coincides with estrus as do other physiological and anatomical changes. Among the higher primates—monkeys, apes, and man—a menstrual cycle occurs. The menstrual cycle is considered a modified estrous cycle; it does not coincide with a breeding season, and there is no definite estrus, i.e., "heat." However among the monkeys some aspects of a typical mammalian estrus exist. For example female baboons develop a large, brilliant red swelling near the base of the tail. This appears at the time of ovulation. This may be a kind of signal to the males that copulation should occur. There is little evidence that a physiologically determined breeding season exists among monkeys, but there does appear to be a periodicity in the fertility of some species. There does not appear to be any cyclical periodicity in the sexual receptiveness of female macaques or baboons. Mounting occurs, and copulation occurs (or appears to occur) throughout the time between observed menstruations in the females. There may be intensification of sexual behavior at the time of estrus, but among the terrestrial monkeys of the Old World this is not the only time when copulation takes place. There is considerable evidence that the copulatory behavior of monkeys is more independent of sexual physiology than that of lower mammals and of prosimian primates.

There is an important question which must be introduced and answered here. How did the cyclical differences in external genitalia and the periodic sexual receptiveness of the females come to be deemphasized? One of the critical factors probably was the development of social life. Solitary primates, such as *Microcebus*, unquestionably are well served by very marked cyclical changes in sexuality. The reproduction of the species is ensured by automatic physiological and anatomical changes which serve to bring males and females together. Continuous contact between males and females, as among baboons, may have reduced the importance of cyclical sexual signals in maintaining the species. But some substitute for cyclical signals must have developed. The marked sexual dimorphism of *Homo sapiens* very likely was substituted for a marked sexual cycle. Sexual dimorphism in man involves characters of both males and females. The most obvious male characters involve the pattern of distribution of the hair—presence of facial hair, baldness at the temples, extension of pubic hair to the navel. The more notable female characteristics are the broadening of the hips by deposits of fat and the development of breasts. The nipples are not placed on obvious permanent prominences in other primates. There are no very noticeable external physical changes in human females during the 25- to 30-day menstrual cycle. If fertilization is to occur, something must stimulate and signal frequent, regular copulation. Characters which are obviously female and sufficiently attractive to males have evolved so that copulation will be regular and frequent. These female traits would have affected the copulatory behavior of the species and reduced the importance of any cyclical differences in the external genitalia or the sexual receptiveness of females. Sexual stimuli, sexual attractiveness, sexual behavior would gradually come to be like that of modern man—dependent upon factors that have no cyclical variation.

Copulation is elicited in many lower primates and other mammals by the cyclical changes in the gonadal hormones. Copulatory behavior can sometimes be prevented in many of these animals by appropriate hormonal injections or by gonadectomy. In chimpanzees and in man, gonadectomy of adults will not prevent copulation. Their sexual behavior is under much greater control of the cerebral cortex. It seems reasonable to argue that a selective advantage would be gained by increased cortical control over sexual behavior and other behavioral activities such as play and sleep. The neurological advances related to this increased cortical control over behavior might have preceded the capacity for culture, the capacity to symbol.

The cortical, hence to a large extent voluntary, control of sexual receptivity of females may not have been essential in forming the family or social units of the first hominids, but voluntary control of sex is relevant to the function of these social units. The family unit in nonhuman primates is, functionally, concerned with sex and reproduction. The human family's dominant function was and is subsistence, and the change in control of sexual behavior may have made this possible. There is a fundamental distinction between nonhuman primate social groupings and human societies. The society of nonhuman primates is wholly dependent upon anatomy and physiology; the society of man is largely governed by culture. Differences among human societies are not the concomitant expressions of variations in the biology of the organism. They are largely, if not entirely, independent of them. Variations in nonhuman primate societies on the contrary are the concomitants of variations in physiology and anatomy.

The principal determinants of primate sociality are probably, as Zuckerman said in 1932, ". . . search for food, search for mates, avoidance of enemies." We base our view on the reasonable assumption that primate social groups are part of the adaptive responses primates developed in order to survive. The capacity to copulate and mate throughout most if not all of the reproductive cycle and throughout all seasons of the year led to the formation of the various kinds of heterosexual groups found among monkeys and apes. Prosimii which have a definite estrus do not have the same kinds of social groups as monkeys and apes. Mammals with short, single mating seasons each year have more transitory heterosexual groups. A new level of social integration developed among the primates as the sexual physiology and receptiveness of females changed. The process of group integration through sex behavior establishes and reinforces heterosexual social bonds.

Defense against predators and avoidance of enemies is certainly a much less significant determinant of primate social order than sex. The search for food is less significant as a dominant force determining sociality in nonhuman primates than it is in man. Among men, even sex has been subordinated to subsistence, to economics, as a determinant. Sahlins points out that primate sexual behavior is used to reinforce economic bonds in human societies.

The reproductive consequences of primate sexual behavior, the helpless offspring which the primate female must carry with her, may have had an important effect upon emerging human social structure. Females with young are handicapped in societies that have a subsistence based on hunting. Carrying the young about at all times would be no great hindrance, as far as we can see, if insects, fruit, flowers, or foliage are gathered. But to hunt even small animals with a helpless clinging infant would be most difficult. A female with young who has a mate that brings her back part of the kill would no longer be at a disadvantage. Permanent pair formation would be an advantage. Among wolves permanent pair formation occurs, and both male and female provide food for the young. This would be a marked advantage to any primate which used hunting as a principal means of subsistence.

Baboons, chimpanzees, and monkeys are relatively successful primates that do not need to hunt other mammals. But let us consider a Miocene hominid like *Ramapithecus*. This creature is a medium-sized anthropoid, with rather smaller front teeth and cleverer hands than most others. It was faced with a massive change in ecozone. It seems likely that this hominid was able to respond by developing a posture and a mode of locomotion that enabled it to see over tall savanna grass and to follow herds of ungulates for long distances over the African, and probably Asian, veldts. The ecological conditions, most likely, were among the primary factors in conferring a selective advantage on an erect biped.

But what was the selection pressure that was the major factor in all this development? Clearly a new adaptive zone was open to the hominids. What was this adaptive zone? It included a change in ecology, from thick forest cover to open bush and savanna, and a change in primate social structure, from a food-gathering group to an organized hunting group. The development of cooperative hunting groups very likely provided the selection pressure, the selective advantage, which emphasized erect bipedalism, skillful use of the hands, symboling, and humanoid social structure. Formation of hunting groups probably emphasized toolmaking, an ability which depends on the unique faculty of symboling. It is our opinion that the recent and often most impressive demonstrations of the use of tools by chimpanzees and baboons do not alter, in any way, our earlier views that this is not the same as human tool-using. Among chimpanzees and baboons use of tools is clearly subject to the restriction, "out of sight is out of mind." Similarly the often dramatic illustrations of meat-eating by chimpanzees and baboons do not suggest in any way an organized hunting party. The chimpanzees, now famous for the capture and eating of a colobus monkey recorded in Goodall's superb films, are gathering food. This activity is not one whit more organized or complex than the banana collecting done by the same animals.

Under the pressure of selection, erect bipeds, with small anterior dentition and skillful hands, still could not create cooperative hunting groups without modifications of the existing social structure. It seems clear that organized hunting as a characteristic of a species requires social differentiation of the roles of those who care for the infants (females) and those who hunt (males). Cooperation among males must develop, and family groups must be permanent and well integrated. While these changes in social structure are developing, certain faculties of the brain are also changing. An ability or mentality must develop that allows for discussion of objects and animals not present, for example, the tools needed for hunting at some future time. The lineage of primates in which all of these capacities were presumably developing would be under strong selection pressure to continue to develop and refine such traits, in an environment rapidly changing from forest to open bush and plains.

George Bernard Shaw once wrote that religion, politics, and sex were the only possible subjects for intelligent conversation. The purely *hominid* way of life became the *human* way of life when our ancestors developed the capacity to make intelligent conversation that met the standards of GBS. Out of sight, out of mind is one way to characterize the quality of prehuman life. Symboling and social life made it possible for the emerging human hominid to plan, to manipulate objects imaginatively, and to convey the results of such imaginative manipulations to his fellows. (Sometimes this ability is called the time-binding ability, an unnecessarily obfuscating phrase.) Let us consider an early human hominid hunting group as it rests after the day's activities. The males are planning the next day's hunt. They ask for the support of supernatural entities; they decide which animals to hunt and how to divide the carcasses among the hunters and their families; and they conclude their planning session by talking about the females who are not present. Of course we dare not speculate about what the females were doing. After this human condition was achieved, nothing could stop the evolutionary success and progress of this fortunate primate, except himself.

SUGGESTED READINGS AND SOURCES

Andrew, R. J., The displays of the Primates. *In* J. Buettner-Janusch (Ed.), *Evolutionary and Genetic Biology of Primates*, Vol. II, p. 227. Academic Press, New York (1964).

Conaway, C. H., and D. S. Sade, The seasonal spermatogenic cycle in free ranging rhesus monkeys. *Folia Primat.* **3,** 1 (1965).

Crile, G., and D. P. Quiring, A record of the body weight and certain organ and gland weights of 3690 animals. *Ohio J. Sci.* **40,** 219 (1940).

Dart, R. A., The predatory implemental technique of Australopithecus. *Am. J. Phys. Anthropol.* **7,** 1 (1949).

Dart, R. A., Myth of the bone-accumulating hyena. *Am. Anthropol.* **58,** 40 (1956).

DeVore, I., and S. L. Washburn, Baboon ecology and human evolution. *In* F. C. Howell and F. Bourlière (Ed.), *African Ecology and Human Evolution*, p. 335. Aldine, Chicago (1963).

Elder, J. H., and R. M. Yerkes, The sexual cycle of the chimpanzee. *Anat. Record* **67,** 119 (1936).

Etkin, W., Social behavioral factors in the emergence of man. *Human Biol.* **35,** 299 (1963).

Ford, C. S., and F. A. Beach, *Patterns of Sexual Behavior.* Harper, New York (1951).

Garn, S. M., Culture and the direction of human evolution. *Human Biol.* **35,** 221 (1963).

Gillman, J., and C. Gilbert, The reproductive cycle of the Chacma baboon (*Papio ursinus*) with special reference to the problems of menstrual irregularities as assessed by the behaviour of the sex skin. *S. African J. Med. Sci. Biol. Suppl.* **11,** 1 (1946).

Hewes, G. W., Food transport and the origin of hominid bipedalism. *Am. Anthropol.* **63,** 687 (1961).

Jerison, H. J., Interpreting the evolution of the brain. *Human Biol.* **35,** 263 (1963).

Kroeber, A. L., The superorganic. *Am. Anthropol.* **19,** 163 (1917).

La Barre, W., *The Human Animal.* Univ. of Chicago Press, Chicago (1954).

Maxim, P. E., and J. Buettner-Janusch, A field study of the Kenya baboon. *Am. J. Phys. Anthropol.* **21,** 165 (1963).

Morgan, L. H., *Ancient Society.* Holt, New York (1877).

Sahlins, M. D., The social life of monkeys, apes, and primitive man. *Human Biol.* **31,** 54 (1959).

Sahlins, M. D., and E. R. Service (Ed.), *Evolution and Culture.* Univ. of Michigan Press, Ann Arbor (1960).

Simpson, G. G., *This View of Life.* Harcourt, Brace and World, New York (1964).

Spencer, H., *The Principles of Sociology.* Appleton, New York (1899).

Spuhler, J. N., Somatic paths to culture. *Human Biol.* **31,** 1 (1959).

Tylor, E. B., *Primitive Culture.* John Murray, London (1871).

Washburn, S. L., Speculations on the interrelations of the history of tools and biological evolution. *Human Biol.* **31,** 21 (1959).

Washburn, S. L., and I. DeVore, The social life of baboons. *Sci. Am.* **204,** No. 6, 62 (1961).

White, L. A., *The Science of Culture.* Farrar, Straus, New York (1949).

White, L. A., *The Evolution of Culture.* McGraw-Hill, New York (1959).

Zuckerman, S., *The Social Life of Monkeys and Apes.* Kegan Paul, Trench and Trubner, London (1932).

Zuckerman, S., *Functional Affinities of Man, Monkeys, and Apes.* Harcourt, Brace, New York (1933).

The Human Integument

22

BETWEEN MAN'S INNER SELF and the world lies a complex protective garment, the integument, the skin. This binds up in one organized package the skeleton, viscera, nerves, and glands. The integument is a wall, a reflecting surface, and a very complex thermoregulatory device. In this chapter we shall present some information about the human integument, its color, its appendages, and its functions.

Differences in the integument of man's fellow primates are used to divide various species into local populations or races. In the case of *Homo sapiens* also races are local populations. Some argue that races do not exist, yet we need only look at people to see the obvious differences between them. The *word* race has been given to subgroups of *Homo sapiens* and is a word with far more connotations than denotations.

Race is an analytic category. We seek to analyze the composition of the species *Homo sapiens* by applying this concept to it. Race is also a biological event, but it is very difficult to make the analytical category and the biological event coincide. Populations differ from each other, and these differences are called racial differences. As long as we stick to an abstract discussion such as this we have no problem. But when we seek to classify real populations of *Homo sapiens* into racial categories, we find we cannot do it in a way that satisfies more than a single person, ourselves. This is not due

to the highly subjective character of the race concept but to the absence of clearly discernible, definable boundaries around many actual subgroups of *Homo sapiens* in nature. There are no difficulties in determining the separateness of Australian aboriginals and Eskimos. They are two populations delineated by quite unambiguous boundaries. But it is not so simple to define the biological boundaries around local populations in Europe or to distinguish the Pygmies of the dense forests of the African Congo from the taller Africans who live on the edges of these forests. Pygmies speak the language of tribes to which they are attached economically and socially, and it is difficult not to believe that they are merely short Africans (Chapter 24). The species presents a much less complicated problem in definition, but this does not solve our problem either. Species are genetically bounded, closed systems. Races are subgroups within a species, but they are genetically unbounded, open systems. We shall return to this subject again.

Skin Color

Certain characteristics of man's integument have been used as criteria for classifying populations and individuals into racial categories. We shall begin with one of the more easily noticed characteristics of skin, its color.

Skin color is one of the more obvious human polymorphisms. Differences in skin color have been known for a long time, but it is technically difficult to specify them. And, to complicate the problems, the genetic bases for skin, eye, and hair color are extremely complex and do not allow for simple analysis.

Although there are several primary pigments which are involved in the pigmentation of the integument, the hair, and the eyes—melanin, pheomelanin, hemoglobin, carotene, etc.—the observed color of skin is largely a function of the amount and concentration of melanin (the dark pigment) and the distribution of blood vessels. The skin color of individuals who have relatively little melanin depends upon the amount of hemoglobin in the superficial blood vessels and the extent to which this hemoglobin is oxygenated. In dark-skinned individuals, the concentration of melanin is the important factor for it masks the effect of the blood vessels.

Melanin occurs in specialized cells of the skin, the melanocytes. These melanocytes produce melanin, small bits or granules of pigment, which is secreted into the cells of the outer layer of skin, probably through specialized processes on the cells. It is worth emphasizing that the number of melanocytes per unit volume of skin is essentially the same in dark-skinned and light-skinned people. The major differences in skin pigmentation are apparently not related to variation in number of pigment cells. It is true that the number of melanocytes per unit volume will vary somewhat in different parts of the body, but the average number is not significantly different in light-skinned and dark-skinned people. Dark-skinned people, especially Africans, have many more and larger pigment granules in their melanocytes. There are only a few, rather small brown granules in the melanocytes of light-skinned people, such as Europeans. No qualitative differences have been demonstrated in the pigments of the skin of human subjects from various populations or races.

Methods are available for determining the amount of melanin present in human skin, but it is not easy to apply them to the massive population surveys necessary for specifying the exact distribution of pigmentation among human populations. However some comparisons among local groups have been made using these various methods. The best-supported generalization is that dark skins reflect less and absorb more solar radiation than light skins. But even the best-controlled and most technically advanced studies must take account of the immediate effect of the environment. If an individual is repeatedly exposed to sunlight, "tanning" occurs. This increases the amount of melanin in the skin, and it is possible that, in some comparative studies of properties of skin, tanning will tend to minimize variation. Older methods which attempted to assign human subjects to various grades, such as black, white, and yellow or light, medium, and dark, were highly subjective. Not only were errors in judgment possible, but there were no standards against which to compare the results.

The inheritance of skin color has not been studied in a satisfactory way. Barnicot has pointed out that studies of offspring of matings between Africans and Europeans merely demonstrate the lack of a clear, simple, genetic segregation. Undoubtedly there are several genes involved, but there are many opinions about the most probable number of genes. There is evidence that skin color in some families of lightly pigmented individuals segregates according to Mendel's law. The pigmentation does not, however, appear to segregate as a simple genetic trait. The distribution of skin color, measured by reflectometry (see Chapter 12), in a population that is genetically relatively homogeneous is a continuous distribution, much like that of stature, weight, or other quantitative characteristics. It is interesting that hybrids between dark- and light-skinned people, where they have been carefully studied, appear to give a reflectance value of skin pigmentation that is approximately halfway between the averages for the light- and dark-skinned parents. There is a folk belief that a light-skinned person, no matter how light skinned he may be, may produce a black-skinned offspring if he has an African ancestor. This is incorrect. The general opinion of geneticists and anthropologists who have worked on the problem is that pigmentation is due to a number of genes which have an additive effect. Therefore hybrids, if hybridization continued in future generations with lighter and lighter skinned people would have the effect of decreasing the number of genes for dark pigmentation, although we cannot, with our present knowledge, predict with certainty the skin color of hybrids.

Skin Color and Solar Radiation

The geographical locations of human populations with dark skins suggested to many that there is a correlation between skin color and such environmental variables as solar radiation, temperature, and humidity. If it were possible to estimate accurately the amount of solar or radiant energy a man receives, it would be easier to discuss the relationship between skin and sunlight. We are aware that the load of solar radiation varies enormously. The radiant energy which reaches the ground and an individual standing on it is different at different times of the year, at different times of the day, and with different atmospheric conditions. Solar radiation may be very high in humid, tropical rain forests and in the Arctic. At present we do not know the variations in amount of solar radiation that strikes any part of the world during the year. The effective load of solar radiation on a man is determined by direct ultraviolet, visible, and infrared radiation from the sun and by the amount of energy diffused and reflected from such things as clouds, cement, asphalt, and trees.

As we said earlier, dark-skinned people reflect considerably less light and absorb more than light-skinned. The difference in the amount of reflected energy between light- and dark-skinned people is greatest in the region of the visible spectrum. This is not the whole story, for the amount of radiation effectively absorbed depends upon the amount of skin surface open to the radiation. However, pigmented skin apparently does give some protection against ultraviolet radiation. Only part of the total ultraviolet radiation reaches the ground, and it is, presumably, responsible for certain lesions of the skin. Until recently it was believed by some who studied the mechanism of sunburning that ultraviolet radiation is absorbed by the stratum corneum, the horny layer of the skin. These workers believed that the thickness of this horny layer, not the pigment lying deeper in the skin, was important in protecting against damage caused by ultraviolet radiation. Work done in Nigeria demonstrated that the skin taken from dark-skinned Africans transmitted less ultraviolet radiation than skin taken from light-skinned Europeans. There was no difference in thickness of the skin, but the pigmentation of the skin of the Africans was greater than that of the Europeans. Barnicot has argued that melanin protects dark-skinned people even though ultraviolet radiation may not reach the basal epidermis in which the melanin is concentrated. The fact that there is considerably more melanin in dark- than in light-skinned people makes it probable that pigment gives significant protection from ultraviolet radiation in dark-skinned people.

If it were possible it would be interesting to show that there is a difference in the incidence of malignant melanomas when light- and dark-skinned populations are compared. Melanomas are pigmented tumors of the skin which are often thought to be started by inflammatory reactions due to sunlight.

There have been a number of studies of the relationship between the incidence of skin cancer, melanomas in particular, and skin pigmentation. The results, as far as we know, are not conclusive. Some of the difficulty is due to an inability to obtain clear-cut data from the obvious place—tropical Africa. If it turns out that Africans, or other dark-skinned people, have fewer melanomas or are protected against the development of melanomas, greater weight would be given the evidence that melanin pigment does indeed protect against the deleterious effects of sunlight.

But there are other things involved. Almost half of the energy of the sunlight that reaches the surface of the earth lies in the visible wavelengths. It is this part of solar radiation that is absorbed in greater amounts by dark-skinned people. An African, in contrast to a light-skinned European, unquestionably must adjust to a much greater heat load from the same amount of solar radiation. It is, of course, very difficult to calculate what this difference is since it depends upon a large number of variable factors. However the results from controlled experiments indicate that an African in desert conditions absorbs more heat per hour through his skin than a European. Very careful experiments with Americans of European and of African descent were made in the desert near Yuma, Arizona. It was found that the Americans of African descent produced more sweat and had higher rectal temperatures than did the Americans of European descent. This suggests that darker skin color resulted in the experiencing of a higher heat load.

In order to explain distribution of skin color in relation to environment, both the reflecting ability of the skin and its absorbing properties have often been cited. However it is not even as simple a matter as that. The use of clothing or shelters as well as environment will modify the amount of solar radiation. The geographic distribution of dark- and light-skinned members of the species is not directly correlated with obvious environmental features such as heat, humidity, and solar radiation which might be expected to have the greatest effect upon the organisms.

Hulse, for one, has suggested that cultural-economic factors led to the present distribution of skin color among members of the species *Homo sapiens*. The argument that culture and economics are the principal factors responsible for the present-day distribution of human skin color can be summarized best by a little story. Let us imagine that a physical anthropologist from Mars made a survey of the color of the epidermis of the species *Homo sapiens* on the planet Earth. Our Martian anthropologist conducted his study several thousand years before agriculture was developed. *Homo sapiens* consisted of many small populations of practically naked nomadic hunters and gatherers. It is likely that an enormous variety of different kinds of skin pigmentation was found in all parts of the planet. Our Martian probably found a reasonably close fit between the degree of pigmentation and solar radiation, humidity, and temperature of the environment in which these relatively small populations of hunters and gatherers lived.

We are fairly certain that agriculture appeared in two or three areas among populations of light- and dark-skinned peoples. Once an agricultural way of life was achieved and permanent habitation developed, those particular populations that developed or acquired agriculture first reproduced at an accelerating rate when compared to those who were still hunters and gatherers. Had another Martian anthropologist made a second survey of skin pigmentation a few thousand years after the agricultural revolution, he may well have documented what we presume happened. Permanent habitations and the production of a surplus of food through agriculture developed among those people with certain skin pigmentations (light or dark), and they outbred everybody else on the planet. Once a village way of life and food production developed, it is unlikely that the advantages or disadvantages of a high or low degree of pigmentation were critical in the survival of any particular population.

There is more than a hypothetical possibility that the present distribution of skin and hair color is the result of technological development. Hulse has described the effect that technology has had upon the expansion of populations of western Europe and, as a consequence, upon the increase in numbers of people with pale skin and blond hair. The size of the populations of the British Isles, Scandinavia, and the Netherlands was below 10 million at the end of the medieval period. All three parts of this western European population expanded, but the rapid growth in numbers of the population originating on Great Britain and Ireland was especially great and is well documented. In 1600 the population of Britain was probably below 3 million. A combination of technological, geographical, and political factors produced an enormous expansion. The Industrial Revolution, the development of the overseas colonial trade, and the enormous emigration of Britons to the colonies led to a fiftyfold increase in the British stock, so that about 150 million descendants now exist. The world population in general has only increased about sixfold in that time. The expansion of the stock that resided in the Netherlands and Scandinavia has also been great. Altogether these expansions during the last 300 or 400 years produced the large number of pale-skinned and blond-haired people that exist today. The part of Europe with the highest proportion of pale-skinned blondes had no more than 3 per cent of the world's population at the end of Middle Ages. Today 12 per cent of the world's population is descended from this small group. A technologically supported population explosion occurred in Asia and Africa also, although the documentation of its details are less precise than that for Europe. The expansion of the populations was not a function of skin color; the present-day distribution of skin color is a function of technological advance! Cultural advances meant that these populations were beginning to mold the planetary environment sufficiently so that many of the powerful effects of selection upon *Homo sapiens* were being lessened. Natural selection has never stopped acting upon *Homo sapiens*, but some of the gross effects of environmental selection have been minimized.

Hair Color

The color of the hair of an individual is also due to the granular pigment, melanin, which lies in certain cells in the shaft of the hair. The pigment granules are formed by the melanocytes that are found in the outer layer of the hair bulb from which the hair grows. Reflectance spectrophotometry, again, is the simplest and most accurate objective measure of hair color. There is a continuous series of hair colors in *Homo sapiens*, ranging from dark to light, depending upon the amount of melanin in the hair. The hair of albinos, presumably without pigment, reflects the greatest amount of light, while the very dark hair of Africans and Melanesians absorbs the greatest amount of light. The exception to this continuous series of colors is red hair. Pigmentation of red hair has never been satisfactorily explained. Melanin and an iron containing red pigment, trichosiderin, are found in red hair. A relatively high concentration of iron is a distinguishing characteristic of red hair. But it is unlikely that trichosiderin makes red hair red, for after it has been extracted the red remains! The color may be due to the amount of melanin, the oxidation-reduction state of melanin, or the combination of melanin with some other compound such as an amino acid. The red color is probably due to some as yet unrecognized property of melanin or of a derivative of melanin.

The distribution of hair color throughout the world is actually quite uniform, for most hair color is dark. It is in Europe, particularly in the northwestern part, that people with blond and red hair are most frequently found. Nonetheless it is true that there is much variation in the actual amount of pigment present in the hair of dark-haired people, but it is difficult to detect by the eye and even by refined optical methods. The darker the hair, the less easy it is to distinguish any two particular samples by optical methods.

There are some interesting variations in hair color from region to region within certain countries and possibly throughout certain continents. Most of the studies of this variation have been made within a particular political unit, and therefore generalizations about continents are not easy to make at the moment. For example there seems to be a gradient for blond hair in Italy. That is, more and more blond-haired people are found as one travels from southern Italy to northern Italy. It is believed that this parallels the distribution of many other physical and morphological characters. Variation in the distribution of hair color also occurs in other parts of the world. There seem to be more red-haired people in certain parts of England than in other parts of the British Isles. It has also been observed that very dark-haired people, such as the Australian aborigines, appear to have reddish or blond hair when young, and there are many stories of blond Indians in South America. Most of these stories, when traced by competent physical anthropologists or biologists, turn out to be the result of a misapprehension of what hair color is. Studies of the presumed blonds or redheads among Australian aboriginals show that this hair contains both melanin and trichosiderin. The difference between a red- and a dark-haired Australian aboriginal is probably a quantitative one.

There is also the well-known change in hair color of many individuals from birth to maturity. A towheaded European child often grows up to have dark brown hair. Among the populations of Asia and Africa the hair is usually heavily pigmented with melanin from birth.

The genetics of hair color has not been studied adequately, with the exception of studies on the inheritance of red hair. The variations in color, which are probably due to the variation of the amount of melanin at various times of life and possibly to varying environmental conditions, have complicated attempts to understand just how hair color is inherited. There are, unquestionably, a number of genes involved which probably affect the rate of production of melanin and the amount of darkening that results. Hair color is an interesting characteristic, but it is impossible, at the present time, to relate it to any environmental conditions. When the genetics has been worked out, hair color may prove to be another important trait in understanding the relationships among populations and the variations within them.

Eye Color

The particular color of the eye is determined by the amount of melanin in the iris and by the response that the pigment gives to the light incident on the iris. Dark eyes have a large number of heavily pigmented cells in the top tissue layer. Light eyes have much less pigment in this layer, and some of the light is reflected from what pigment remains and from the colorless, but not transparent, layer of tissue on top of the iris. Human eye colors probably form an almost continuous series that ranges from brown to green to gray to blue. This seriation is related to a progressive decrease in the amount of pigment in the top layer of the iris.

There are many difficulties in recording eye color accurately and therefore in analyzing it. The iris is not pigmented uniformly. The anterior layer of the iris is thicker in some places than it is in others, and, for that matter, the melanin is not uniformly distributed. Thus all kinds of variation in eye color will occur even though it is likely that the same genetic basis for pigmentation is present.

There is a common assumption that dark eye color is controlled by genes that are dominant to genes for light shades. This assumption is the result of the fact that it was necessary, when analyzing eye colors, to arbitrarily put them in definite categories. These color categories are highly artificial and do not correspond well with the actual phenotypes. There are many other difficulties. It is not known whether or not the same set of genes controls pigmentation of the hair and the eyes. Since in some areas of the world people with light-colored eyes have very dark hair and some light-haired people have dark eyes, the association of the two traits is not demonstrated.

The adaptive significance of pigmentation of the iris is not obvious. The pigment in the iris reduces the amount of light entering the pupil of the eye and may protect the retina against possible damage from ultraviolet radiation. But there is no clear evidence that the variation in pigmentation which occurs in living human populations has any consequence for visual function. It is clear that extreme deficiencies of pigmentation, such as are found in albinos, may be and often are associated with defective sight and visual functions. It is also clear that there is strong photophobia in albinos. Albinism is an extreme, rare defect, and there is no good evidence that milder cases of photophobia are generally associated with decreased amounts of pigment in the iris.

Climatic Adaptation

The normal body temperature of *Homo sapiens* is maintained within extremely narrow limits around 37°C. There are diurnal variations, but they are usually slight. There are some very slight differences in the average temperature of human inhabitants of various climatic zones. Europeans that live in tropical climates sometimes have been shown to have a small but consistent elevation in the average body temperature when compared to their own relatives living in temperate Europe.

Variations in production of heat by the organism are the result of changing rates of metabolism. *Homo sapiens* is usually not in a state of rest, the condition in which a basal metabolic rate is determined. Muscular work, the effects of the metabolism of food, and many other factors affect production of heat. In order to maintain a stable body temperature, heat is lost to the environment by conduction and convection. The mechanism is simple. The blood is cooled as it flows through the skin. This, of course, is not possible if the temperature of the air is higher than that of the skin. Heat is also regulated by infrared radiation to the environment. The rate of heat loss by radiation depends upon the temperature of the skin and the mean temperature of the surroundings. A variety of studies indicates that man's thermal (or heat) equilibrium occurs in a very narrow range of environmental temperature, between 27° and 30°C. Above this temperature range the accumulated heat within the body is not easily lost to the environment by the usual mechanisms of conduction and convection. The secretion of sweat by the sweat glands also plays a role. Sweat production is stimulated as the temperature rises, and sweat is produced in larger and larger quantities. There is a limit to this. If the heat load is too high for sweating and other cooling mechanisms to operate properly, the equilibrium state cannot be reached. The body temperature will rise to dangerous levels and the circulatory system may collapse. It is possible for the body to secrete sweat at a great rate, as high as 2 liters per hour, but this rate cannot be maintained for very long because of the excessive loss of water and salts from the body. Dehydration and collapse follow when there is an excessive strain on the mechanism for heat regulation. On the basis of this it would be suspected, a priori, that populations of *Homo sapiens* which have always lived in tropical areas would have a larger number of sweat glands and a more efficient temperature regulating mechanism. The older literature in physical anthropology does indeed assert that Europeans have fewer sweat glands per unit area of skin than do dark-skinned people who live in the tropics. In recent years it has been demonstrated that there is no difference in the average number of sweat glands in the skin of living Europeans and Africans. Although there is a great deal of variation within each group, there is no significant difference in the average number of sweat glands. Furthermore there is no evidence that there is a different distribution of the density of sweat glands of Europeans and Africans. Research into climatic adaptive mechanisms of man is still in a preliminary stage. So far there is no good evidence of a significant difference in morphological features related to the heat regulating mechanisms of Europeans and Africans.

Another extreme climate in which *Homo sapiens* lives, and has lived for a long time, is in the cold Arctic region. The cold regions, like the tropical, provide a relatively wide range of thermal conditions. They are not consistently cold. The intensity of cold varies during the year, and there is no constancy of climatic conditions. When an individual is cold, he must increase his rate of metabolism in order to maintain body temperature. When the environmental temperature falls below 27°C, the metabolic rate must begin to increase if the individual is to remain in thermal equilibrium with the environment without a fall of body temperature. At the same time water loss is important. Despite the fact that the secretion of sweat is minimal until the environmental temperature reaches 30°C, at lower temperatures water still diffuses through the skin and lungs and evaporates. This will account for almost one-fourth of the heat loss of a man under standard conditions.

Heat loss is combated principally by insulation (clothes, hair) and by increased metabolic rate, that is, heat production within the organism. The shivering reflex, involuntary muscular action, will increase the heat of metabolism but not by much more than two or three times the heat of normal resting metabolism. Voluntary exercise is the principal method for increasing heat production. It is well known that Eskimos run for long distances for long periods behind their sled dogs at just the right speed to increase their metabolism and heat production without exhausting themselves.

However the most efficient method for minimizing heat loss in cold climates is to insulate the skin. Man is unfortunate for he is not covered with a coat of fur. Arctic species of mammals have extraordinarily efficient fur coats. Man can and does take advantage of this by using the skins of these animals for clothing. Insulation of human skin itself is increased by reflex constriction of the veins of the skin and by accumulation of subcutaneous fat, just below the skin. Many populations undoubtedly do not have sufficient caloric intake to allow for the deposition of significant amounts of subcutaneous fat. Cultural adaptations are unquestionably the most important ones in man's adaptation to the cold. Laughlin's report of a behavior pattern of Eskimos substantiates this. If two Eskimos are hunting or fishing and one falls into the water, the two exchange some of their clothing. The result is two half-wet, mildly uncomfortable Eskimos instead of a single, dry, warm survivor.

There are some physiological mechanisms present in members of certain human populations that seem to be adaptive to life in the cold. Eskimos, for example, have been observed to use their hands with skill and effectiveness at very low outside temperatures. Careful studies demonstrated that there is a vasomotor reaction which warms the hand and is sufficient to account for these superior manipulative performances. The arms and hands of Eskimos were immersed in water at about 5°C, and it was found that the blood flow in the hands and forearms of the Eskimos and of Europeans used as controls fell at first. This is a common expected reaction. But in a few moments the flow of blood increased to much higher levels among the Eskimos than among the Europeans. The European controls were able to tolerate immersion of the hand in cold water only for about an hour. The Eskimos, on the other hand, complained of slight pain and were able to endure this for a considerable period of time. Two of the subjects went to sleep. The blood pressure response to the artificial cold was a familiar one, at first. Among the Europeans there was a sudden rise of blood pressure followed by a fall. But among the Eskimos there was a consistent rise in blood pressure throughout the experiment.

Some interesting differences in the vasomotor responses of African, dark-skinned, subjects was discovered when they were compared with Eskimos and Europeans. When the arms of subjects of African ancestry were placed in water at close to 0°C, the temperature of their fingers fell to lower levels than in the European controls, and there was little or no vasodilation which flushes the hand with warm blood. In other experiments Americans of African descent and Americans of European descent were exposed to cold environments, approximately −12°C. The fall in the rectal temperature and the mean skin temperature was similar in the two groups. In these experiments the temperature of the fingers of the American subjects of African descent fell more slowly but to lower levels. Other studies demonstrated that the shivering reflex started later in Americans of African descent who were exposed nude to 17°C than it did in Eskimos and Americans of European descent. The most interesting fact about these studies is that the subjects were all born and raised in the United States. Some were of African and some of European descent, but their climatic environment from birth was essentially the same. Thus there does seem to be a genetic factor involved in certain physiological responses to heat and cold. It is clear that a great many other experiments are needed before we will understand these physiological mechanisms, but it is evident that there are some differences related to climate.

The importance pigmentation has assumed in anthropological studies is largely a historical and sociological accident. We have briefly reviewed some of what is known about pigmentation, especially skin pigmentation. Some possible correlation with climate and solar radiation was discussed, and we pointed out that there are many problems still unsolved. If we look at human skin as it is, an important functional organ of the body, the superficial nature of much classical work on "racial" variation in skin color becomes apparent. Pigmentation is only one of many aspects of the human integument.

Man's skin, when compared to other primates', has morphological peculiarities that suggest it has undergone enormous adaptive changes. Man does not have a pelage effective in thermal regulation. His pelage functions ornamentally. The skin of man, although quite hairy, is covered with *rudimentary* hairs except on the scalp and face.

The naked skin of man is unlikely to have been an accident. Montagna believes the distribution of hair on the human body suggests that man is striving to achieve total nakedness. The absence of hair is balanced by many fine adjustments which not only replace the functions of the hairy coat of other primates but also probably improve upon them. The vascular system and blood supply of the skin is far greater than is required for its metabolism. This helps the skin serve as a thermoregulating device. The sweat glands of man, morphologically and physiologically, are extremely sensitive to stimulation by changes in temperature. Sweat glands function on the friction surfaces of the skin, such as the palms, to keep these surfaces moist. This improves tactile sensitivity and enhances the grip. As Montagna puts it:

> Far from being a generalized assemblance of all the known cutaneous appendages, it [human skin] has modified the structure and function of each of these in such a way that each has become idiosyncratic.

But whether there are significant differences in these important structures of the skin and their correlative functions among various human populations is not known.

Anthropologists and others have been overly concerned with the color of human skin, which they have measured by extremely crude means. As research on the functions of the skin develops, it is clear that human skin is a very complex structure. Its primary functions probably are thermoregulation and protection against the energy of solar radiation. We do not know whether the gross differences in skin and hair reported among human populations are significant functionally. That they were at some time in the past, we do not doubt. But the evidence available is not consistent with any theories about race per se, about population differences, or about the relation between population differences and climate.

SUGGESTED READINGS AND SOURCES

Adams, T., and B. G. Covino, Racial variations to a standardized cold stress. *J. Appl. Physiol.* **12,** 9 (1958).

Adolph, E. F., and associates, *Physiology of Man in the Desert.* Interscience, New York (1947).

Baker, P. T., The biological adaptation of man to hot deserts. *Am. Naturalist* **92**, 337 (1958).

Barnicot, N. A., Reflectometry of the skin in southern Nigerians and in some mulattoes. *Human Biol.* **30**, 150 (1958).

Barnicot, N. A., Climatic factors in the evolution of human populations. *Cold Spring Harbor Symposia Quant. Biol.* **24**, 115 (1959).

Blum, H. F., Physiological effects of sunlight on man. *Physiol. Rev.* **25**, 483 (1945).

Brown, G. M., and J. Page, The effect of chronic exposure to cold on temperature and blood flow of the hand. *J. Appl. Physiol.* **5**, 221 (1952).

Davenport, C. B., Heredity of skin color in Negro-white crosses. *Carnegie Inst. Wash. Publ.* **188** (1913).

Fitzpatrick, T. B., P. Brunet, and A. Kukita, The nature of hair pigment. *In* W. Montagna and R. A. Ellis (Ed.), *The Biology of Hair Growth,* p. 255. Academic Press, New York (1958).

Hardy, J. D., and A. M. Stoll, Measurement of radiant heat load on man in summer and winter Alaskan climates. *J. Appl. Physiol.* **7**, 200 (1954).

Hulse, F. S., Technological advance and major racial stocks. *Human Biol.* **27**, 184 (1955).

Laddell, W. S., Disorders due to heat. *Trans. Roy. Soc. Trop. Med. Hyg.* **51**, 189 (1957).

Lerner, A. B., Hormones and skin color. *Sci. Am.* **205**, No. 1, 99 (1961).

Macdonald, E. J., Malignant melanoma among Negroes and Latin Americans in Texas. *In* M. Gordon (Ed.), *Pigment Cell Biology,* p. 171. Academic Press, New York (1959).

Montagna, W., *The Structure and Function of Skin.* Academic Press, New York (1956).

Montagna, W., Phylogenetic significance of the skin of man. *Arch. Dermatol.* **88**, 1 (1963).

Montagna, W., and R. A. Ellis, Histology and cytochemistry of human skin. XXI. The nerves around the axillary apocrine glands. *Am. J. Phys. Anthropol.* **18**, 69 (1960).

Newburgh, L. H. (Ed.), *Physiology of Heat Regulation and the Science of Clothing.* Saunders, Philadelphia (1949).

Rennie, D. W., and T. Adams, Comparative thermoregulatory responses of Negroes and White persons to acute cold stress. *J. Appl. Physiol.* **11**, 201 (1957).

Roberts, D. F., Basal metabolism, race and climate. *J. Roy. Anthropol. Inst.* **82**, 169 (1952).

Roberts, D. F., Body weight, race and climate. *Am. J. Phys. Anthropol.* **11**, 533 (1953).

Rothman, S., *Physiology and Biochemistry of the Skin.* Univ. of Chicago Press, Chicago (1954).

Stern, C., Model estimates of the frequency of white and near-white segregants in the American Negro. *Acta Stat. Gen. Med.* **4**, 281 (1953).

Thomson, A., and D. Buxton, Man's nasal index in relation to certain climatic conditions. *J. Roy. Anthropol. Inst.* **53**, 92 (1923).

Weiner, J. S., Nose shape and climate. *Am. J. Phys. Anthropol.* **12**, 1 (1954).

Wilber, C. G., Physiological regulations and the origin of human types. *Human Biol.* **29**, 329 (1957).

Wyndham, C. H., and J. F. Morrison, Heat regulation of MaSarwa (Bushmen). *Nature* **178**, 869 (1956).

Wyndham, C. H., and J. F. Morrison, Adjustment to cold of Bushmen in the Kalahari desert. *J. Appl. Physiol.* **13**, 219 (1958).

Human Genetics

23

Genetics is the study of inherited characteristics. We have referred to genetics on many occasions in earlier chapters. Now we shall define the technical terms used in discussing genetic problems. Today we consider evolution as a change in the genetic composition, the gene frequencies, of a population. In order to understand and explain evolution in genetic terms we must attempt to explain certain aspects of the formal theories which describe how traits are inherited, how genes are transmitted from generation to generation, and how gene frequencies fluctuate within populations. We shall emphasize abstract models and theoretical situations, for there is no single, real human population so well studied genetically that we may see the operation of all the mechanisms of heredity or all the factors which affect gene frequencies in it.

In this chapter the discussion is limited to formal Mendelian genetics. We shall present, in effect, the story of Mendel's peas. In a later chapter the wave of the future—molecular genetics with deoxyribonucleic acid (DNA) as the genetic material—will be discussed. The modern molecular point of view is useful in elucidating many problems in physical anthropology. At the same time it is well nigh impossible for a student of physical anthropology to understand human evolution without some grasp of classical, formal, statistical genetics, of which Mendel was the progenitor.

Formal Genetics

Mendel, an Austrian monk who studied the inheritance of certain traits in peas about 100 years ago, was the first to demonstrate that inheritance is not a blending of the characters of the parents. His data, presented in simplified and summary form in Fig. 23.1 and 23.2, were best interpreted as demonstrating that traits are inherited as discrete units. Mendel's work with peas showed that traits are transmitted as discrete particles which do not contaminate or blend with each other (Fig. 23.1) and that traits will express themselves in succeeding generations. This part of Mendel's work was later formulated as the rule of segregation of traits. The discrete units that Mendel proposed were eventually named genes. Mendel further showed that two traits, simultaneously considered, will sort and recombine independently of each other. The critical evidence for this rule of independent assortment or recombination was statistical (Fig. 23.2).

Mendelian genetics is based on these two "laws"—the law of segregation and the law of independent assortment or recombination. It is also based on the theory of chromosomal inheritance. This theory provided the structural basis for Mendel's laws and is still the foundation of formal genetics.

The chromosomes are small but prominent elongated structures which are present in each cell of an organism. Chromosomes (Greek: colored bodies) are vividly and deeply colored when cells are stained for examination under the microscope. When a cell divides during normal cell reproduction, each chromosome divides longitudinally to produce two chromatids. The chromosomes have reproduced (replicated) themselves. The chromatids line up along a plane that cuts through the center of the cell. The chromatids separate from each other at this plane, which becomes the plane of division. They are pulled or moved to opposite ends of the cell. The cell then divides into two new or daughter cells, each of which has an exact duplicate of the chromosomes of the parent cell. Mitosis is the name given this process of cell division illustrated in Fig. 23.3. Such regularity in the division of the chromosomes was observed long before the discovery of modern genetics. Eventually when the regularity with which inherited traits were passed from generation to generation was noticed, the theory that inherited variations were carried by some kind of controlled mechanism on the chromosomes was proposed. This theory has since been verified experimentally by demonstrating that specific changes in chromosome structure are associated with specific inherited abnormalities.

There is another kind of cell division known as reduction division or meiosis. Meiosis occurs only in the germinal cells of bisexual organisms. The germinal cells are the spermatozoa or sperm of males and the ova or eggs of females. Reduction division is the process by which the number of chromosomes normally possessed by the somatic cells of the organism is reduced by one-half. The number of chromosomes in the somatic cells is called the diploid number $(2N)$ and the number in the germinal cells the haploid num-

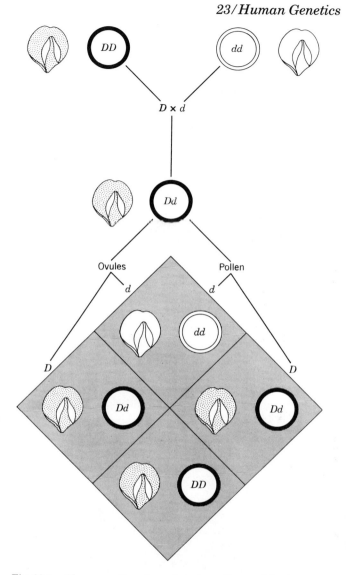

Fig. 23.1. The segregation of genetic traits as illustrated by the cross
between two varieties of sweet peas, those with purple flowers and
those with white flowers. *D* represents the allele for purple color, *d*
represents the allele for white color. *D* is dominant over *d*. (Redrawn
after Dobzhansky, 1955.)

ber (*N*). *Homo sapiens* has 46 chromosomes; 44 are called autosomes, and 2
are called sex-determining chromosomes. Chromosomes occur in pairs, man
has 22 pairs of autosomes and 1 pair of sex chromosomes. In man, $2N = 46$,
the diploid number, and $N = 23$, the haploid number. The sex chromosomes

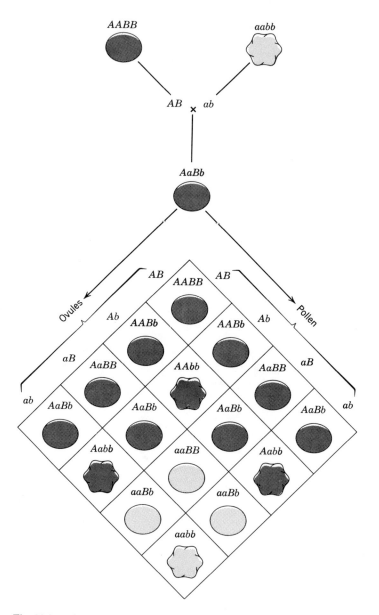

Fig. 23.2.   The independent assortment and recombination of genetic traits as illustrated by the cross between two varieties of sweet peas, those with yellow and smooth seeds and those with green and wrinkled seeds. The alleles for yellow and green are represented by *A* and *a*, the alleles for smooth and wrinkled by *B* and *b*, respectively. (Redrawn after Dobzhansky, 1955.)

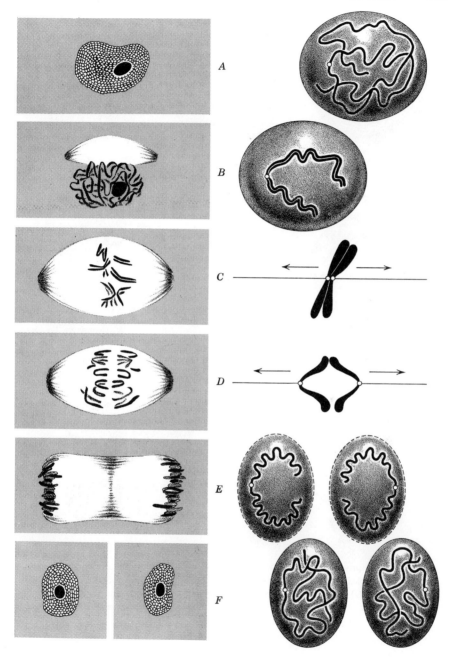

Fig. 23.3. The process of mitosis. The figures on the left show the stages in mitotic division in a cell in the root tip of the common onion. The drawings on the right are diagrams of the behavior of a single chromosome during mitosis. (A) Interphase, (B) prophase, (C) metaphase, (D) anaphase, (E) telophase, and (F) daughter cells. (Redrawn after Dobzhansky, 1955.)

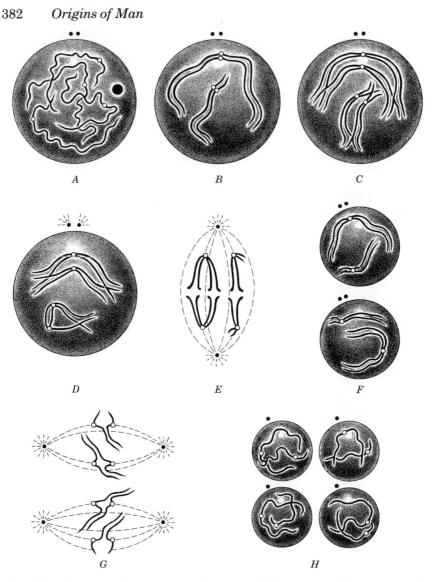

Fig. 23.4. The process of meiosis shown diagrammatically for two pairs of chromosomes. (A) Interphase, (B) pachytene, (C) diplotene, (D) diakinesis, (E) anaphase of first meiotic division, (F) interphase, (G) anaphase of second meiotic division, and (H) the four haploid daughter cells. It is during the transition from pachytene to diplotene phase that crossing over occurs. (Redrawn after Sinnott, Dunn, and Dobzhansky, 1958.)

are named, by convention, the X-chromosome and the Y-chromosome. Normal human males have 1 X-chromosome and 1 Y-chromosome; normal females have 2 X-chromosomes.

During spermatogenesis or oogenesis, that is, when new spermatozoa or ova are produced, meiosis or reduction division occurs (Fig. 23.4). Chromosomes with similar genes or alleles on them are known as homologues, and during one phase of meiosis the chromosomes of a pair of homologues become tightly paired or synapsed. Exchange of genetic material—crossing over—between the members of an homologous pair of chromosomes occurs during this phase of pairing or synapsis. Crossing over results in an exception to strict Mendelian segregation. If genes are closely linked on a chromosome and cross over as a unit, the result will not be the same as that of genes sorting independently. After the synaptic phase, the chromosomes become separated or disjoined. The cell divides after disjunction, and each daughter cell receives one member of each pair of chromosomes. Each sperm or ovum receives exactly one-half the total number of chromosomes. A mammalian female will have an X-chromosome in each ovum, for the normal mammalian female has 2 X-chromosomes. Since the normal human male has 1 X- and 1 Y-chromosome in each diploid cell, reduction division produces two kinds of sperm. One-half will have the X-chromosome and one-half will have the Y-chromosome.

Since two types of sperm cells are produced during meiosis, one bearing an X-chromosome, the other bearing a Y-chromosome, it would be reasonable to suppose that the two types could be separated. It would also be reasonable to say that prediction of sex could be made and control could be exercised over the sex of offspring by means of artificial insemination. There have been several attempts to do this in mammals other than primates, but no unequivocally successful methods for separating the two types of sperm have been reported.

A genetic locus is a place on the chromosome occupied by a gene. This gene may occur in several alternative forms known as allelomorphs or alleles. Since chromosomes occur in pairs, an allele will be present at the same locus on each chromosome of an homologous pair. This is the origin of the homozygous and heterozygous conditions. A homozygous individual has identical forms, alleles, of the gene at a particular genetic locus on each chromosome of an homologous pair. A heterozygous individual is a person who has alternative members of the gene pair, alternative alleles, at this locus on the two homologous chromosomes.

A genotype is the actual genetic constitution of an organism. A phenotype is the expression of the genotype in the organism. Since it is not yet possible to specify completely the genotype of an individual, we usually describe genotypes in terms of alleles at a single locus. Thus in the ABO blood group system an individual who is phenotype AB is also genotype *AB*. The genotype of an A individual may be *AA* or *AO*, and, similarly, the genotype of a B individual may be *BB* or *BO*. The genotypes *AA* and *AO* are expressed as the phenotype A; genotypes *BB* and *BO* are expressed as the phenotype B. Individuals of blood group O have the genotype *OO*.

A gene *D* is called dominant to its allele *d* when it is not possible to distinguish the expression of the trait in the heterozygous individual *Dd* from its expression in the homozygous individual *DD*. Actually this does not mean that there are no differences between the two individuals but rather that we are not able to detect these differences. Dominance, to a degree, refers to the limited extent of our knowledge at the present time rather than to any inherent property of the gene.

A gene d is said to be recessive to its allele D when the heterozygote Dd exhibits no discernible manifestation of the gene. Only the homozygote dd is distinguishable from the heterozygote Dd and the dominant homozygote DD.

A gene is called autosomal if it is located on one of the autosomes, one of the somatic chromosomes. A gene is considered sex-linked, under certain special conditions, if it is carried on one of the two sex chromosomes, the X- or the Y-chromosome. It is assumed that each X- and Y-chromosome consists of two segments. One segment is homologous between the two chromosomes; that is, the genetic material of the homologous segments sometimes may be exchanged by the X- and Y-chromosomes during cell division. If there is indeed an homologous segment, genes located on that segment would be considered partially sex-linked for they would be transmitted sometimes on the X-chromosome and sometimes on the Y-chromosome. The nonhomologous segment of the X-chromosome or the Y-chromosome is that part of the chromosome on which the completely sex-linked traits are to be found. Completely Y-linked genes are called holandric. Holandric inheritance is extremely rare, and some of the best known cases are now considered doubtful. Genes on the nonhomologous segment of the X-chromosome are the most common type of sex-linked genes. The term sex-linked is usually used to refer to X-linked traits.

The formal models for determining the mode of inheritance of an autosomal trait, either dominant or recessive, are presented in Table 23.1. These models are sufficiently general so that they describe the inheritance of autosomal genes in any bisexually reproducing organism.

Man is a somewhat refractory subject, however, for genetic studies. Two major difficulties exist. First, the span of life in man is long, thus a rapid succession of generations is not available for genetic studies. Second, human matings occur for various sociocultural, psychological, and economic reasons. They do not occur to provide controlled breeding experiments so desirable in genetic studies. We cannot study the inheritance of hemophilia, a condition in which blood does not clot easily, or the inheritance of blue eyes by breeding a number of hemophiliacs or blue-eyed persons.

In human genetics we are forced to act after the fact, and we depend upon what are essentially retrospective and statistical methods. When we find an individual with a particular inherited trait, we try to construct a pedigree for this individual. The individual with the trait in question who first comes to our attention is called the propositus. The pedigree is a schematic means of showing the occurrence or nonoccurrence of the trait in the propositus and in his relatives. A famous pedigree for hemophilia is shown in Fig. 23.5.

To determine if an inherited trait is due to a dominant, a recessive, or a sex-linked gene we start with the propositus. We make one simplifying

TABLE 23.1

Autosomal Inheritance

The alleles are D and d. D is dominant to d.

Possible genotypes		Matings	Offspring	
			Genotypes[a]	Phenotypes
♂	♀	no. ♂ × ♀		
DD	DD	1 DD × DD	DD	D
Dd	Dd	2 DD × Dd	1DD:1Dd	D
dd	dd	3 DD × dd	Dd	D
		4 Dd × DD	1DD:1Dd	D
		5 Dd × Dd	1DD:2Dd:1dd	3D:1d
		6 Dd × dd	1Dd :1dd	1D:1d
		7 dd × DD	Dd	D
		8 dd × Dd	1Dd :1dd	1D:1d
		9 dd × dd	dd	d

[a]Sample calculations for determining genotypic ratios:

Mating	no. 2	no. 3	no. 5
	D D	D D	D d
	D | DD | DD | d | Dd | Dd |	d | Dd | Dd | d | Dd | Dd |	D | DD | Dd | d | Dd | dd |
Ratio of genotypes	1DD:1Dd	all Dd	1DD:2Dd:1dd

assumption; we assume the trait is rare. This means there will be few, if any, homozygous individuals *DD* or *dd* in the population. We then ask what criteria we can establish to decide if a trait is due to a dominant, recessive, or sex-linked gene. If the gene is an autosomal, rare dominant, a person with the trait will be the product of the following mating: $Dd \times dd$ (Table 23.1, matings no. 6 and no. 8). If this is so, we would expect to find at least three characteristic features in the pedigree and in the population from which the propositus came. First, one of his parents must show the trait, as must one of his grandparents. The trait will be found in each generation in his direct lineage for as far back as the lineage can be traced. A dominant gene does not skip generations. Second, on the average, one-half of the children and as many sons as daughters of the mating $Dd \times dd$ will show the trait. Third, there will be approximately an equal number of males and females with the trait in the population as a whole.

A recessive, autosomal gene will manifest itself only in the homozygote *dd*. If the trait is rare such an individual will be the product of the mating $Dd \times Dd$ (Table 23.1, mating no. 5). The parents of such an individual and even more distantly related ancestors will not show the trait. There will be as many males as females with the trait in the population. Approximately one-fourth of the offspring of such a mating will have the trait. Finally, individuals who are homozygous for a rare, autosomal recessive are often the products of a mating between relatives, a consanguineous mating.

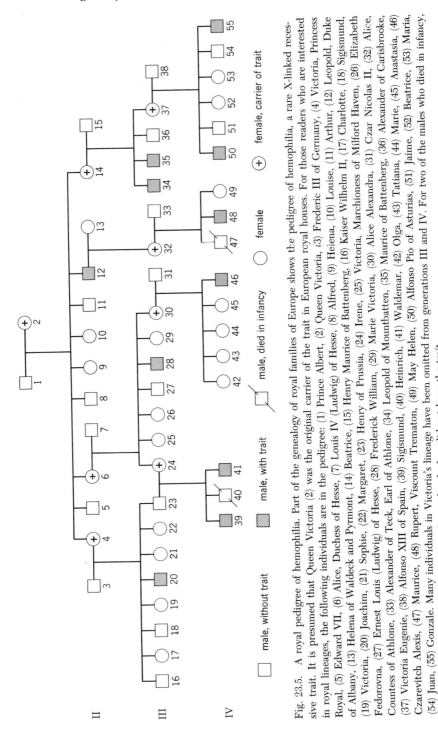

Fig. 23.5.   A royal pedigree of hemophilia. Part of the genealogy of royal families of Europe shows the pedigree of hemophilia, a rare X-linked recessive trait. It is presumed that Queen Victoria (2) was the original carrier of the trait in European royal houses. For those readers who are interested in royal lineages, the following individuals are in the pedigree: (1) Prince Albert, (2) Queen Victoria, (3) Frederic III of Germany, (4) Victoria, Princess Royal, (5) Edward VII, (6) Alice, Duchess of Hesse, (7) Louis IV (Ludwig) of Hesse, (8) Alfred, (9) Helena, (10) Louise, (11) Arthur, (12) Leopold, Duke of Albany, (13) Helena of Waldeck and Pyrmont, (14) Beatrice, (15) Henry Maurice of Battenberg, (16) Kaiser Wilhelm II, (17) Charlotte, (18) Sigismund, (19) Victoria, (20) Joachim, (21) Sophie, (22) Margaret, (23) Henry of Prussia, (24) Irene, (25) Victoria, Marchioness of Milford Haven, (26) Elizabeth Fedorovna, (27) Ernest Louis (Ludwig) of Hesse, (28) Frederick William, (29) Marie Victoria, (30) Alice Alexandra, (31) Czar Nicolas II, (32) Alice, Countess of Athlone, (33) Alexander of Teck, Earl of Athlone, (34) Leopold of Mountbatten, (35) Maurice of Battenberg, (36) Alexander of Carisbrooke, (37) Victoria Eugenie, (38) Alfonso XIII of Spain, (39) Sigismund, (40) Heinrich, (41) Waldemar, (42) Olga, (43) Tatiana, (44) Marie, (45) Anastasia, (46) Czarevitch Alexis, (47) Maurice, (48) Rupert, Viscount Trematon, (49) May Helen, (50) Alfonso Pio of Asturias, (51) Jaime, (52) Beatrice, (53) Maria, (54) Juan, (55) Gonzale. Many individuals in Victoria's lineage have been omitted from generations III and IV. For two of the males who died in infancy, (40) and (47), there is no evidence that they either had or did not have the trait.

A sex-linked trait may be dominant or recessive, but the expression of the gene is modified by the fact that only one locus of the allele will appear in males. Thus a recessive trait will always express itself in males. Males with an X-linked gene are often referred to as hemizygous for that gene.

A dominant X-linked trait, again assumed to be rare, may be found in either males or females. A male with such a trait will have the genotype $X^D Y^o$ and will be the product of the mating $X^d Y^o \times X^D X^d$. If a female with genotype $X^D X^d$ has the trait, she may be the product of either the mating $X^D Y^o \times X^d X^d$ or the mating $X^d Y^o \times X^D X^d$ (Table 23.2). As is the usual case with dominant traits, a sex-linked dominant will occur in every generation in a lineage. There will be twice as many females as males in the population with the trait. The males with the trait, genotype $X^D Y^o$, must have a mother with the trait. These males, if married to $X^d X^d$ females, will have daughters who all have the trait. None of their sons will have it. The females with the trait, if married to $X^d Y^o$ males, will pass the trait on to one-half of their sons and one-half of their daughters.

TABLE 23.2
Sex-Linked Inheritance
X-LINKAGE

Possible genotypes $\male$	$\female$	no.	Matings $\male \times \female$	Offspring Genotypic ratios
$X^D Y^o$	$X^D X^D$	1	$X^D Y^o \times X^D X^D$	$X^D Y^o : X^D X^D$
$X^d Y^o$	$X^D X^d$	2	$X^D Y^o \times X^D X^d$	$X^D Y^o : X^d Y^o : X^D X^D : X^D X^d$
	$X^d X^d$	3	$X^D Y^o \times X^d X^d$	$X^d Y^o : X^D X^d$
		4	$X^d Y^o \times X^D X^D$	$X^D Y^o : X^D X^d$
		5	$X^d Y^o \times X^D X^d$	$X^D Y^o : X^d Y^o : X^D X^d : X^d X^d$
		6	$X^d Y^o \times X^d X^d$	$X^d Y^o : X^d X^d$

		HOLANDRIC INHERITANCE		
$X^o Y^H$	$X^o X^o$	7	$X^o Y^H \times X^o X^o$	$X^o Y^H : X^o X^o$ All males affected, no females with trait

A recessive, sex-linked trait, if rare, will manifest itself in males of genotypes $X^d Y^o$. Such males will be products of the mating $X^D Y^o \times X^D X^d$. Males with the trait will have children without the trait unless they marry heterozygous females $X^D X^d$. In that case one-half of the sons and one-half of the daughters will have the trait. A female who has the trait must be a homozygote $(X^d X^d)$, and all her sons will have the trait. If the trait is rare, there will be many more males than females with it in the population. As is the case with rare autosomal recessive traits, many of the matings that produce children who manifest an X-linked recessive will be between relatives. The pedigree for hemophilia (Fig. 23.5) is one of the best demonstrations of an X-linked recessive trait.

We have assumed, in our discussion, that we are dealing with genes infrequent in a population. This assumption makes it possible to define simple relatively clear-cut criteria for establishing the mode of inheritance of a trait. If a trait is common the determination of the mode of inheritance, from human data, is more difficult. A common autosomal recessive will often be found in the homozygous state *dd*, and matings of the type *Dd* × *dd* and even *dd* × *dd* may be encountered frequently. The frequency with which the trait is expressed among the offspring of these two matings will be about the same, in a human population, as if the trait were inherited as a simple dominant. If an X-linked recessive trait is common, many females heterozygous for it will occur, and many more men with the trait will have sons that manifest it, for the matings X^dY^o × X^DX^d will be fairly common.

Many traits are controlled by alleles that are neither dominant nor recessive to each other. If two alleles *D* and *d* are neither dominant nor recessive, each allele manifests itself in the heterozygote *Dd* (Table 23.3). Some of these traits controlled by nondominant alleles are of the greatest interest to physical anthropologists. A good example of nondominant autosomal inheritance is provided by studies of haptoglobin types in human populations. Haptoglobins are serum proteins of the blood, and in Chapter 31 we shall present some data on these types in several human populations.

TABLE 23.3
Nondominant Autosomal Inheritance

Possible genotypes		Matings	Offspring	
♂	♀	no. ♂ × ♀	Genotypes	Phenotypes
DD	*DD*	1 *DD* × *DD*	*DD*	D
Dd	*Dd*	2 *DD* × *Dd*	1*DD*:1*Dd*	1D:1Dd
dd	*dd*	3 *DD* × *dd*	*Dd*	Dd
		4 *Dd* × *DD*	1*DD*:1*Dd*	1D:1Dd
		5 *Dd* × *Dd*	1*DD*:2*Dd*:1*dd*	1D:2Dd:1d
		6 *Dd* × *dd*	1*Dd*:1*dd*	1Dd:1d
		7 *dd* × *DD*	*Dd*	Dd
		8 *dd* × *Dd*	1*Dd*:1*dd*	1Dd:1d
		9 *dd* × *dd*	*dd*	d

Population Genetics

It is difficult to define a particular population strictly, for the actual boundaries around a specific human population are not always easy to find. A human population is usually found in a particular place, and it is a coherent entity largely because of geographical boundaries. Populations are also defined by economic, social, and psychological boundaries. Regardless of how they are circumscribed, the significance that populations have for evolutionary genetics lies in the web of genetic relationships within and between them—allele frequencies, consanguinity, mating patterns, gene flow, natural selection.

The genetic approach uses the concept of the Mendelian population which Dobzhansky has defined as "a reproductive community of sexual and cross fertilizing individuals which share in a common gene pool." Any Mendelian population may consist of many subunits which are themselves smaller Mendelian populations. The largest Mendelian population known is a species. Species are distinct from each other because they do not share the same gene pool. They are closed genetic systems. Human races are essentially different Mendelian populations of the species *Homo sapiens.* They are recognized as different from each other, because the gene pools of various races differ from each other to some extent. The differences are usually manifested phenotypically. But these populations are not closed to the exchange of genes with each other. In other words, human races are Mendelian populations with potential and actual gene flow between their respective gene pools.

An important point about Mendelian populations is that they are breeding isolates, although they are isolated to varying degrees. A breeding isolate is a population in which the members find most of their mates within the group. A breeding isolate is usually a small population and is isolated for many reasons. Although we generally use the term breeding isolates for very small populations, populations of any size may be isolates as long as the degree of isolation is specified. Geographical barriers are the most obvious factors leading to breeding isolation. Religious, social, and psychological factors are just as important as are geographical barriers.

Populations that are breeding isolates because of geographical barriers include the Eskimos of Greenland; the inhabitants of islands such as Pitcairn in the Pacific, Tristan da Cunha in the south Atlantic, and the Andamans in the Bay of Bengal; the mountain villagers of Switzerland; and the Havasupai Indians of the Grand Canyon in Arizona. Breeding isolates occur in religious groups such as the Hutterites and the Old Order Dunkers of the United States, the Doukhoboors of Canada, the Samaritans of Palestine, and the Cochin Jews of India. Social, linguistic, and psychological factors result in a considerable amount of breeding isolation among segments of the population of New York City or any other large, urban agglomeration. Communities that speak Spanish, Yiddish, Italian, Polish, German, etc. maintain themselves, and mates are chosen from within these communities much more frequently than from without. One need only visit Detroit or Milwaukee, for example, to observe the social, fraternal, and political clubs that are organized around the national origins of the great-grandparents of the members.

It is obvious that genetic differences have developed among human populations. The expression genetically different does not mean genetically exclusive or genetically closed. It means that there are differences in frequencies of various alleles in the populations considered. In the particular province of physical anthropology we can determine frequencies of various genes within a population. We can measure differences in frequencies among various populations, and we can try to discover and describe processes by which these frequencies are changed. Evolution may be viewed as the change in gene frequencies in populations.

The determination of gene frequencies depends upon counting and statistical manipulation of the numbers. It is usually not possible to count each person who manifests a trait in a population. Therefore samples are taken, and methods have been developed to estimate the frequency of the trait in the population from the frequency in the sample. Estimation of gene frequencies in populations is based upon Mendel's two principles and an equilibrium formula which describes population gene frequencies under certain conditions.

At one time inheritance was believed to occur by the blending of parental characteristics. Darwin held this view. If inheritance were not particulate, a population that reproduces sexually would tend to lose its genetically controlled variability at a rapid rate. Yet if evolution is to occur, the species must possess a relatively high degree of genetic variability. Darwin was aware of this contradiction but did not know of Mendel's work which supported the view that inheritance is particulate; that is, the genotype of a child is not a compromise of the traits of its parents but a reassortment of the genes each parent possesses, some of which are transmitted to the child and some not.

If inheritance were the blending of the traits of the parents, variability would be eliminated quickly. We can show this by producing hybrids—by crossing certain kinds of flowering plants, those with red and those with white flowers. When we cross red- with white-flowered plants, the individual plants of the first offspring generation will have pink flowers. If inheritance were not particulate, the crossing of these pink-flowered plants with themselves would produce nothing but pink flowers; crossing them with red-flowered plants would produce dark pink or light red flowers, and crossing them with white-flowered plants would produce pale pink or almost white flowers. But this does not happen. The pink × pink cross produces the following ratio among the offspring—1 plant with red flowers : 2 plants with pink flowers : 1 plant with white flowers. The other two crosses, pink × red and pink × white, produce ratios which demonstrate that the color of the flowers is inherited as if controlled by two factors, a red factor and a white factor. Red flowers are produced when red is in a double dose; white flowers when white is in a double dose; pink flowers when a single dose of each factor is present.

In order to produce and maintain variability in a population, a large amount of variation must be introduced to the population in each generation. Today we should say that if inheritance is the blending of parental characters, a high frequency of recurrent mutations would be required to maintain variability.

Even after the particulate nature of inheritance was widely recognized, the problem of analyzing change in the genetic composition of a population remained. Essentially the problem is to determine the frequency of the various genotypes, with respect to any one set of alleles, in a population that has a mixture of the alleles. Another question often asked was, if a gene is dominant, will it not eventually be the only allele at that locus in a population? This question was answered with a no by an English mathematician, Hardy, and a German physician, Weinberg. They produced what we now call the Hardy-Weinberg law, the root from which modern population genetics grew.

Let us consider a sexually reproducing population with the following properties. (1) Mating is at random; any individual has an equal chance of mating with any other individual in the population. (2) All matings have the same fertility. (3) Natural selection is not acting upon the alleles we are studying. (4) The alleles are stable; they are not mutating. (5) There is no overlap between generations. (6) The population is infinitely large. These six conditions must be assumed if the population is in Hardy-Weinberg equilibrium. Obviously no population meets all six of these conditions, but in order to work with any human population we assume that these conditions hold.

Consider a pair of alleles D and d in this population. In any single generation there will be three genotypes, DD, Dd, and dd. The Hardy-Weinberg formula states that if p is the frequency of allele D and q is the frequency of allele d, where

$$p + q = 1$$

or

$$1 - p = q,$$

the genotype frequencies are given by

$$p^2 + 2\,pq + q^2 = 1,$$

where the frequency of genotype $DD = p^2$, $Dd = 2\,pq$, and $dd = q^2$. This expression of the Hardy-Weinberg law, which gives us the genotype frequencies, then may be used to calculate allele frequencies. Given the necessary information about the frequency of genotypes in a population, we can solve the equation for p or q and obtain a value for the frequency of an allele in a particular population.

The derivation of the Hardy-Weinberg formula may be illustrated very simply by using allele frequencies. If p is the frequency of D, and $q = 1 - p$ is the frequency of d, we can view equilibrium as the random combination of the sex cells, the gametes, and construct a simple multiplication table.

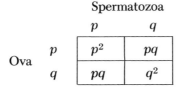

		Spermatozoa	
		p	q
Ova	p	p^2	pq
	q	pq	q^2

Collecting terms, we find that one generation of random mating will produce the following proportion of genotypes:

$$p^2\,(DD) : 2\,pq\,(Dd) : q^2\,(dd).$$

This proportion will be maintained in a population unless something disturbs one or more of the basic conditions of a Hardy-Weinberg population.

A population in which the distribution of genotypes remains stable from generation to generation in this manner is called a panmictic (noun: panmixis), equilibrium, or random mating population. It is obvious that no population is an ideally random mating group. The fact that populations deviate from the model described makes it possible for us to estimate the effects of various evolutionary processes on the gene frequencies or genotypes within a population. A population evolves if the equilibrium is upset. The principal factors that disturb equilibrium are nonrandom mating, mutation, sampling error or genetic drift, migration, and natural selection. It is these factors that we shall now discuss.

Suggested readings and sources

Boyer, S. H., IV (Ed.), *Papers on Human Genetics.* Prentice-Hall, Englewood Cliffs, New Jersey (1963).

Dobzhansky, T., *Genetics and the Origin of Species* (Third edition). Columbia Univ. Press, New York (1951).

Dobzhansky, T., *Evolution, Genetics, and Man.* Wiley, New York (1955).

Fisher, R. A., *The Genetical Theory of Natural Selection* (Second edition). Dover, New York (1958).

Goldschmidt, E. (Ed.), *The Genetics of Migrant and Isolate Populations.* Williams and Wilkins, Baltimore (1963).

Haldane, J. B. S., *The Causes of Evolution.* Harper, New York (1931).

Hardy, G. H., Mendelian proportions in a mixed population. *Science* **28**, 49 (1908).

Kempthorne, O., *An Introduction to Genetic Statistics.* Wiley, New York (1957).

Li, C. C., *Population Genetics.* Univ. of Chicago Press, Chicago (1955).

Mendel, G., Experiments in plant-hybridization [translation by W. Bateson, from original in *Verhandl. Naturforsch. in Brünn, Abhlandl.* **4** (1865)]. *In* J. A. Peters (Ed.), *Classic Papers in Genetics.* Prentice-Hall, Englewood Cliffs, New Jersey (1959).

Neel, J. V., and W. J. Schull, *Human Heredity.* Univ. of Chicago Press, Chicago (1954).

Sinnott, E. W., L. C. Dunn, and T. Dobzhansky, *Principles of Genetics* (Fifth edition). McGraw-Hill, New York (1958).

Sorsby, A. (Ed.), *Clinical Genetics.* Butterworth, London (1953).

Stern, C., *Principles of Human Genetics* (Second edition). Freeman, San Francisco (1960).

Sutton, H. E., *An Introduction to Human Genetics.* Holt, Rinehart and Winston, New York (1965).

Weinberg, W., Über den Nachweis der Vererbung beim Menschen. *Jahresh. Verhandl. Vat. Naturforsch. Würt. Stuttgart* **64**, 368 (1908).

Evolution and Genetic Equilibrium

24

EVOLUTIONARY PROCESSES may operate to maintain equilibrium, but unless allele frequencies change and new alleles are created a population remains static. Populations in genetic equilibrium are not evolving. A static balance might seem desirable, for it would enable the population to cope with the normal amount of mutation and fluctuations of the environment without the need to undergo any change. But such inertia would probably lead to extinction. It is the mechanisms which shake a population out of evolutionary stasis and lead to further evolution that we will now discuss.

Nonrandom Mating

The two most common deviations from random mating within human populations are inbreeding and positive assortative mating or homogamy. Positive assortative mating is simply the old observation that "like marries like." For example a number of studies show that, on the whole, tall people tend to marry tall people, college-educated people tend to marry college-educated people, and most people select mates who were born or live within a relatively short distance. These statements seem incredibly obvious, yet such "common sense" views had to be verified by what we call assortative mating studies. One very strange discovery of these studies was that people with red hair tend to practice negative assortative mating. In other words gentlemen with red hair also prefer blonds. Actually the number of matings between people with red hair is much smaller than predicted for random mating.

Inbreeding occurs when there is a trend within a population for its individuals to mate with relatives. Inbreeding is best analyzed as genetic consanguinity. Consanguinity is not common among human populations, but it does occur, and it occurs more often in some populations than in others. There are rules in many societies that regulate the relationship permitted between two individuals who wish to marry. It is uncommon in modern, urban societies for breeding to be permitted between people who are more closely related than niece and uncle, nephew and aunt, or first cousins. An example of consanguineous marriage is illustrated in Fig. 23.5 by the marriage between Henry of Prussia (23) and Irene (24). In many societies, including those in certain parts of the United States, even some of these forms of marriage are prohibited.

Individuals with rare recessive autosomal or sex-linked traits are often the offspring of consanguineous matings. (Many rare diseases controlled by recessive alleles are discovered only because of consanguinity.) A consanguineous mating is between two individuals who have one or more common ancestors (Fig. 24.1). Such matings are more apt than others to produce individuals who inherit two identical alleles. Alleles are said to be identical, in this sense, if at some time in the past they originated through the replication of the same gene. Again the reader is referred to the pedigree for hemophilia in Fig. 23.5.

Consanguinity has an effect on the frequency of genes and genotypes in a population. The major effect is to reduce the number of heterozygotes and increase the homozygotes. Whether we know it or not, each of us has many a consanguineous marriage in his pedigree. A person who has no inbred ancestors has eight great-grandparents; a person whose parents are first cousins has only six great-grandparents. Thus inbreeding acts to reduce the variation in a population.

Consanguinity in individual pedigrees may not be undesirable. An inbred human lineage that produced many outstanding individuals is the Darwin. Charles Darwin and his wife Emma Wedgwood were first cousins, and the entire Darwin-Wedgwood lineage was highly inbred. We do not know whether the social, professional, artistic, and scientific eminence achieved by many individual members of this lineage is genetically controlled. But the Darwin pedigree shows that there are inbred lineages which do produce persons without severe defects. Consanguinity as a habit in a population may have relatively serious consequences. There are certain Swiss mountain villages in which consanguineous matings are frequent and in which there are elevated frequencies of hereditary defects or diseases controlled by single autosomal recessive alleles.

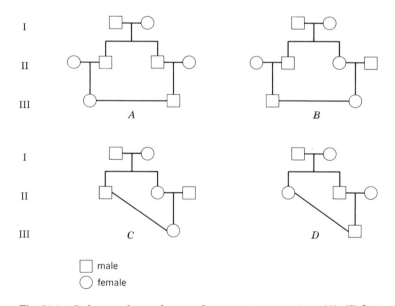

Fig. 24.1.   Pedigrees of several types of consanguineous matings. (A), (B) first cousin matings; (C) niece-uncle mating; and (D) nephew-aunt mating.

The importance of consanguinity in human populations is easily understood if we make a calculation of the number of ancestors any of us would have if no inbreeding occurred. The number of ancestors of an individual is 2^n; n is equal to the number of generations or ancestral steps in which we are interested. If there have been no consanguineous matings within a lineage, an individual who traces his pedigree 1000 years into the past could expect to find records of 2^{40} or 1,000,000,000,000 direct lineal ancestors. This assumes four generations per 100 years, a reasonable estimate. Although we have no sound estimates of the world population during the tenth century, it is unlikely to have been as large as a million million. Consanguineous matings reduce the number of ancestors an individual has, and it is quite safe to say that a considerable amount of inbreeding has occurred in human history. Simple calculations such as we just presented indicate that there is a rather large common ancestry for living human populations.

Consanguineous matings occur in all human societies. Their frequency in certain representative groups is listed in Table 24.1. The frequency is determined as the number of marriages between related individuals per total number of marriages. Consanguineous matings are more probable in small, isolated, rural populations than they are in large, industrial, urban groups. Yet marriages between relatives do occur in the great urban centers of the world.

The rate of consanguinity has been decreasing in Europe and America over the past century for several reasons. Breeding isolates have been breaking up as a consequence of several major features of modern society—the great European wars, the mass migrations of peoples across political boundaries, the development of large-scale efficient systems of transportation and communication, and the movement of people away from farms into cities. These vast changes since the eighteenth century have expanded actual and potential human contacts. Another consideration is the fact that there has been a reduction in absolute size of families, leading to a reduction in the number of close relatives an individual may choose as a mate.

As breeding isolates break up and as consanguineous matings decrease in frequency, changes occur in the gene pool of *Homo sapiens*. Homozygosity decreases, and, seemingly paradoxically, the number of deleterious recessive genes in the human gene pool increases temporarily, as heterozygosity increases. Recessive genes that have a pronounced deleterious effect are more quickly eliminated from populations in which homozygosity is highly probable than from those in which it is improbable. As the smaller isolates break up, the equilibrium which may have been reached is upset. Eventually new genetic equilibria develop.

As average family size decreases, the number of sibs (brothers and sisters) becomes smaller, and the number of cousins which any one individual has becomes considerably smaller. If b is the average sibship size (average number of children per family), the propositus (any one individual) will have $b - 1$ sibs. If b' is the average sibship size of the parents of the propositus,

TABLE 24.1
Estimates of the Frequency of Consanguineous Marriages

Population	Period	Total number of marriages	Frequency of first cousin marriages	Frequency of all con-sanguineous marriages
United States				
Baltimore, Md.	1935–1950	8,000	0.0005	
Watauga County, N.C.	1830–1849	299	0.0408	
Watauga County, N.C.	1870–1889	446	0.0301	
Watauga County, N.C.	1910–1929	439	0.0091	
Watauga County, N.C.	1930–1950	226	0	
Utah	1910–1929	26,325	0.0007	
Utah	1930–1950	15,306	0.0004	
Rochester, Minn.	1935–1950	1,969	0.0005	
Austria				
Vienna Catholics	1929–1930	31,823	0.0053	0.0099
France				
Whole country	1876–1880	1,410,889	0.0103	
Whole country	1914–1919	1,350,683	0.0097	
Loir-et-Cher	1919–1925	19,431	0.014	0.028
Loir-et-Cher	1944–1953	19,861	0.003	0.007
Germany				
Bavarian Parish	1848–1872	5,283	0.0049	
Bavarian Parish	1898–1922	5,193	0.0067	
Bavaria	1926–1933	474,268	0.0020	
Switzerland				
Alpine community	1880–1933	77	0.039	0.195
Alpine community	1885–1932	139	0.007	0.099
Obermatt	1890–1932	52	0.115	0.538

he will have $2(b' - 1)$ aunts and uncles. For the examples used here we assume that $b = b'$, for we also assume that average sibship size does not change significantly in one generation. Our individual will have $b - 1$ sibs and $2(b - 1)$ aunts and uncles; one-half of these aunts and uncles will be on his mother's side, the other half on his father's. Each of these aunts and uncles will have b children. Thus the propositus will have $2b(b - 1)$ first cousins. And he will have $b(b - 1)$ first cousins of the opposite sex, assuming, of course, that there are equal numbers of children of each sex.

Now let us look at two examples of different average sibship size. First, the average size is 6. The propositus will have $2(6)(6 - 1) = 60$ first cousins, and 30 of these he could choose as mates. In the second example, the average sibship size is 2. The propositus will have $2(2)(2 - 1) = 4$ first cousins, 2 of whom are suitable as mates. Thus we see that as average family size decreases the possibility of finding a mate among one's first cousins decreases markedly. In other words the size of the breeding isolate becomes smaller.

It is possible to estimate the average size of genetic isolates if we know

TABLE 24.1 (*Continued*)
Estimates of the Frequency of Consanguineous Marriages

Population	Period	Total number of marriages	Frequency of first cousin marriages	Frequency of all con- sanguineous marriages
England				
Hospital patients	1880–1925	49,315	0.0061	
Hospital patients	1925–1939	10,236	0.0040	
Ireland				
County Londonderry	1954	717	0.0139	
County Tyrone	1954	3,000	0.0020	
Sweden				
Parish of Pajola	1890–1946	843	0.0095	
Parish of Muonionalusta	1890–1946	191	0.0680	
Denmark				
Whole country (sample)	1900–1920	498	0.012	
Japan				
Nagasaki	1949–1950	16,681	0.0503	0.0829
Hiroshima	1948–1949	10,547	0.0371	0.0649
Nishi Nagashima	?–1948	1,433	0.143	0.223
Suigen	?–1948	776	0.124	0.195
Kikuchi	?–1948	888	0.094	0.159
Fiji Islands	1850–1895	448	0.297	
India				
Bombay, caste of Parsees	1950	512	0.129	
Bombay, caste of Marathas	1950	137	0.117	

the frequency of consanguineous marriages. The formula, which was proposed by Dahlberg, is

$$n = \frac{2b(b-1)}{c},$$

where n is the size of the isolate, b is the average sibship size as defined above, and c is the frequency of consanguineous matings. In this case we consider only first cousin marriages, and we assume panmixis. In parts of western Europe the custom of registering all births, deaths, and marriages in the parish churches makes it possible to determine with considerable accuracy the frequency of consanguineous unions. For most North American urban populations, the best we can do is to estimate c. The available marriage records are probably inadequate for accurate determination of degree of relationship between the marriage partners. However in some modern urban societies, where consanguinity has been studied intensively, the frequency of first cousin matings is no greater than 0.0005, or 0.05 per cent (Table 24.2). If the average sibship size is 2, the size of the breeding isolate (n) is

$$\frac{2(2)(2-1)}{0.0005} = 8000;$$

if the sibship size is 3,

$$n = \frac{2(3)(3-1)}{0.0005} = 24000.$$

In both of these examples, the size of the breeding isolate is large.

Theoretical values for the size of genetic isolates are listed in Table 24.2 for some selected populations. It is evident from these estimates that the size of the breeding isolate has been increasing in Western societies over the past 100 years; that the size is smaller in rural than in urban areas; and that sociocultural barriers are important in maintaining breeding isolation.

It would be most interesting to determine the actual number of consanguineous matings and the size of the breeding isolate in many primitive societies. There are not very many data applicable to this problem, and our discussion must remain on a theoretical and speculative plane. The kinship terminologies, marriage rules, marriage preferences, and social structures of many aboriginal peoples suggest that deviation from random mating should occur. Deviations that are due to social rules should continue from generation to generation. One expected consequence would be an increase in consanguineous matings and a relatively high degree of inbreeding. It is well known that many societies prescribe and proscribe marriages among relatives in various ways. Furthermore most societies reckon kinship according to social rules and not according to genetic lines of descent. In the United States we tend to reckon kinship relatively closely in accord with biological descent. Many societies have mating rules that, ideally, require a person to marry his (or her) cousin. If such a rule is obeyed by most members of a society, the genetic makeup of the population will be affected to a considerable extent. In many societies cross cousins are prescribed as mates and parallel cousins are proscribed. We would expect that consanguineous matings would reach a high level in such a society. The degree of inbreeding should be extremely high. We do not know yet whether most of the marriages in a society that has rules which prescribe cousin marriages are actually marriages between cousins, defined genetically. Our discussion of the way we can estimate the size of a genetic isolate suggested that, if populations are small or if the size of sibships is small, there simply will not be enough genetic cousins to go around. This rather simple-minded approach was confirmed by Kunstadter and his colleagues who did a computer analysis of the theoretical effect of demographic variables upon the frequency of cross cousin marriages in model populations. The frequency of cousin marriages in populations with such a preference was found to be insignificantly different from the frequency in those populations which did

TABLE 24.2
Theoretical Sizes of Breeding Isolates in Selected Populations

Population	Period	Frequency of first cousin marriages	Size of isolate[a]	
			A	B
United States				
Baltimore, Md.	1935–1950	0.0005	8,000	24,000
Watauga County, N.C.	1830–1849	0.0408	98	294
Watauga County, N.C.	1870–1889	0.0301	133	399
Watauga County, N.C.	1910–1929	0.0091	440	1,319
Austria				
Vienna Catholics	1929–1930	0.0053	755	2,264
France				
Whole country	1876–1880	0.0103	388	1,165
Whole country	1914–1919	0.0097	412	1,237
Loir-et-Cher	1919–1925	0.014	286	857
Loir-et-Cher	1944–1953	0.003	1,333	4,000
Germany				
Bavarian Parish	1898–1922	0.0067	597	1,791
England				
Hospital patients	1880–1925	0.0061	656	1,967
Hospital patients	1925–1939	0.0040	1,000	3,000
Ireland				
County Londonderry	1954	0.0139	288	863
County Tyrone	1954	0.0020	2,000	6,000
Sweden				
Parish of Pajola	1890–1946	0.0095	421	1,263
Parish of Muonionalusta	1890–1946	0.0680	59	176
Brazil				
Sao Paulo urban	1930–1950	0.0058	690	2,069
Sao Paulo suburban	1928–1946	0.0178	225	674
Japan				
Nagasaki	1949–1950	0.0503	80	238
India				
Bombay, caste of Parsees	1950	0.129	31	93
Bombay, caste of Marathas	1950	0.117	34	102

[a]The figures in column A are calculated for a theoretical sibship size of two; those in column B for a size of three. The formula used is given in the text.

not have a preference. The most significant variables appeared to be the size of the population and the average number of children per family. The *demographic* conditions, in other words, determined the possibility of cousin marriages and their frequencies. Under the demographic conditions existing in most populations, even if the ideal of cousin mating were followed rigidly, it is impossible for more than a minority of the matings to take place between genetic cousins. It seems reasonable to suggest that specific social rules about preferential marriages do not alter appreciably the genetic composition of a population.

There are certain social rules which may have developed because close inbreeding often has extremely deleterious genetic effects in small populations. All societies have rules which forbid matings between relatives of various kinds. These rules are called the incest taboo. The universality of social rules against the mating of relatives, of the incest taboo, has puzzled scientists for generations. Many explanations for the *origin* of the taboo have been given. Of the many explanations advanced, two seem reasonable to us. (1) The prohibition of mating with close relatives removes sexual competition from within the family group of an animal that matures slowly and has a cultural way of life. (2) The incest taboo prevents the genetic consequences of a high degree of inbreeding. But once the taboo originated, its *functions* may have become wider. Probably the most important function was the development and extension of bonds of cooperation among families and larger social groups. Marrying out of the immediate social group leads to the establishment of social and economic bonds between the groups from which the mated pair come. Thus the incest taboo has both social and genetic consequences.

This excursus into sociocultural hypotheses about the origins of kinship and marriage rules has a genetic point. If we had the requisite genealogical data from a number of aboriginal societies, we might be able to determine whether the ideal rules of choosing mates have a significant effect upon the genetic compositions of the populations.

Mutation

New traits, new genes, are introduced into human populations by mutation. A mutation is a change in the basic genetic material that produces a new, inherited, phenotypic trait. In contemporary jargon it is a change in the deoxyribonucleic acid (DNA), a change in sequence of nucleotides or a deletion or insertion of nucleotides in the DNA molecule. In Chapter 25 we discuss DNA as the genetic material.

Mutations can be caused experimentally in fruit flies, mice, and other laboratory animals by a variety of physical agents. These agents presumably act upon the genetic material. They include radiation, certain chemicals, and heat. Rise in external temperature will induce an increased mutation rate in fruit flies. Chemicals that have produced mutations in laboratory animals include mustard gas, phenol, formalin, bile salts, copper sulfate, 1,2,5,6—dibenzanthracene, and urethane. The radiation which appears to have the greatest mutagenic property has wavelengths shorter than 10^{-6} cm. These include roentgen or X-rays, cosmic rays, and gamma rays. Ultraviolet radiation of slightly longer wavelength, between 10^{-5} and 10^{-4} cm, is also a potent mutagen. Human mutation rates are unlikely to be affected by heat, for man maintains a constant body temperature in the face of considerable fluctuations in environmental temperature. Many chemical mutagens will kill a man before they have been in contact with the proper tissues long enough to cause mutations. Of the known mutagens radiation is the only one apt to have a major effect upon man.

A mutation is usually detected as a variation or a change of a trait already present in the population. A mutation of gene d to gene d' may in turn back mutate to d, for mutation is a reversible process. Gene d', the mutated form of the original allele, may again mutate to gene d''. It is usually assumed that mutations are random phenomena, at least with respect to the adaptive requirements of the organism. A mutation does not occur simply because it would be appropriate for a particular trait to occur in a population of organisms when the environment changes. Mutations are random, but there are no infinite possibilities among them. It is highly unlikely that a mutant gene will affect a trait other than the character or system that the gene in its original state controlled or affected. A gene that determines hemoglobin, for example, will not in all probability mutate to affect brow ridges or eye color or blood groups.

Most mutations are deleterious. Organisms are finely and efficiently adapted to their environment. Any change in the genetic structure is apt to upset such a balance and cause difficulties for the organism. If these difficulties are great, it is likely that the changed trait will be eliminated. Mutated genes that become part of the gene pool of the population, part of its genetic heritage, do so only through the action of natural selection. Mutations with which we are concerned occur in the germinal cells, the ova or spermatozoa. We shall not discuss mutations in the somatic cells.

The rates at which mutations occur in man may be determined. Determination of mutation rates is quite complex, and the methods are almost always indirect. Mutation rates are generally expressed as a frequency of detected mutations per genetic locus per generation. Mutation can be determined or measured only if the change in the gene produces a distinct change in the phenotype of the character it controls. In order to determine mutation rates we must be able to analyze the trait as controlled through the inheritance of a single gene. Furthermore the mutated trait must be frequent enough in the population so that we can find it and collect enough data about it to calculate the probable rate at which the gene controlling it mutates. The trait must, however, be sufficiently rare so that new cases can be distinguished from those inherited in family lines in which the trait already occurs. In human genetics most mutations and mutation rates that have been determined are for abnormalities or for genetically controlled disorders.

The work on human gene mutation rates is based upon a sample that is biased. A few mutations of some of the many thousands of genes that exist in man have been studied. Almost all of what we say about mutation rates applies most especially to deleterious conditions which require medical care. These have received the greatest attention. Nonethless the methods are generally applicable. Autosomal dominant genes apparently mutate rarely or their mutants are selected against. Mutant autosomal dominant genes express themselves immediately in the phenotype. A dominant mutant gene will usually be quickly eliminated, since we assume any mutation is apt to upset the dynamic equilibrium an organism has achieved with the environment. A recessive mutant gene, on the other hand, may stay "hidden" in the population. It will be eliminated only when it occurs in a homozygote. In a sense it hides phenotypically unexpressed until such time as expression does not lead to elimination.

A mutation rate for a dominant gene may be calculated directly or indirectly. By the direct method we simply count the number of individuals who have the trait and who are the only persons with the trait in their families. Since the trait is due to a dominant gene, such persons presumably are mutants. The actual mutation rate of an autosomal dominant gene is one-half the total number of cases of the trait found among newborn infants per total number of infants born during a specified period of time. The gene will be a mutant only if the person in which it occurs is unique in his family. The factor of one-half is used because the mutant will almost certainly be a heterozygote DD' or $D'd$. The indirect method for calculating mutation rates involves some assumptions. The population is assumed to be in Hardy-Weinberg equilibrium. The fertility of persons with the trait is estimated as some fraction of the fertility of those without it. This fraction is usually based on the fertility of the brothers and sisters (sibs) who do not have the trait. Then we calculate how often mutation would give rise to new occurrences of the trait in order to maintain a stable number of genotypes in a population. The formula is $R = (1 - F)q$. R is the mutation rate per gene per generation; F is the relative fertility of heterozygous persons with the trait; $(1 - F)$ is the selective disadvantage of having the trait; q is the frequency of the gene. In this case q is one-half the ratio of individuals with to all individuals without the trait born within a given period of time. All individuals with the trait are counted regardless of whether the parents have the trait.

A recessive gene that mutates cannot be detected easily. Only when two persons heterozygous for the trait mate and produce an offspring homozygous at that locus will there be evidence that the mutation occurred. On the assumption that a population is in equilibrium, the formula for calculating the mutation rate for a recessive gene is $R = (1 - F)q^2$. R is the mutation rate; F is the relative fertility of homozygous individuals with the trait; $(1 - F)$ is the selective disadvantage of such individuals; q^2 is the frequency of the trait.

The formula has to be applied with caution, and the two major reservations we have in applying it are as interesting for physical anthropology as the determination of mutation rates itself. First, it is not safe to assume that the trait is completely recessive. We now know that many genes once considered recessive actually have some effect in the heterozygous state. If such a gene affords even a very slight selective advantage to heterozygotes, this advantage will maintain the gene in the population. Even a slight advantage will compensate for the loss of mutant recessive genes which occurs in homozygotes; in the homozygous state, the mutant genes have a deleterious or even lethal effect. If the gene happens to affect the heterozygote adversely, the mutation rate required to maintain the gene frequency will be much increased over the rate estimated by this method.

Second, the equilibrium population which satisfies Hardy-Weinberg conditions seldom, if ever, exists. Human populations are not random mating populations. Inbreeding occurs, and the more inbreeding the more often individuals homozygous for a recessive trait will appear. In other words when inbreeding occurs the likelihood increases that a mutant recessive gene will appear in the homozygous state. As a result the gene will be eliminated at a greater rate than under equilibrium conditions. A population is in equilibrium only if the rate at which those alleles being eliminated by selection is exactly balanced by the rate they are being created by mutation. Thus estimates of mutation rates of recessive genes in human populations will almost always be too low.

Mutation rates for recessive genes are estimated from enumeration of homozygotes. As we have seen the number of homozygotes does not necessarily correlate in a simple manner with mutation rates. Patterns of inbreeding, selective advantages of heterozygosity, genes that are not completely recessive, etc., make the determination of mutation rates a difficult task. These same factors are sources of error and qualification in estimating the rates. For physical anthropologists many of the variations in patterns of inbreeding provide fascinating studies in themselves. Variations in gene frequencies or maintenance of deleterious genes can sometimes be explained by the selective advantages in heterozygotes. The movements of populations and the breakup of isolates in modern times means that inbreeding is decreasing. As a result there will be an increase in the number of deleterious or other mutated recessive genes in the world population. There will be a relative excess of heterozygotes for many of these alleles until a new equilibrium state is achieved. All of these problems, which we must handle if we are to determine reasonably good estimates of gene mutation rates, are extremely interesting in physical anthropology for their own sake.

Mutation is the process by which new genes enter a population. Mutations are not the source, however, for different allele frequencies in different populations. Mutation rates are too low to cause significant changes in allele frequencies in a human population. Selection is the agent which does that. A number of mutation rates for human genes have been calculated in recent years (Table 24.3). The rates range from 4×10^{-6} to 1×10^{-3} mutations per gene per generation. Since all these rates were calculated for deleterious traits, the question is, how closely do these rates represent human gene mutation rates generally? We do not know really. Such mutation rates are basically a function of three factors—the number of alleles at a locus, the total frequency of mutations at that locus, and the manifestation of the mutation or our ability to detect it. A mutation need not give rise to a clear-cut, readily determined difference from the normal condition and probably seldom does. Mutations may merely modify the expression of other genes, or they may be very hard to detect. It is probable that the most important mutations are almost undetectable individually. Collectively and cumulatively these may have important effects. The material for evolution is unlikely to exist among the somewhat spectacular abnormalities whose mutation rates are relatively easy to calculate. It is the "little, hidden" mutations that are the material for selection.

Sampling Error (Genetic Drift)

Sampling error or genetic drift is sometimes called the Sewall Wright effect after the famous population geneticist who provided a precise description of the process. It is a purely statistical matter and a property of the size of the population. Sampling error is unlikely to affect the frequencies of genes in large populations, but it was and is effective among very small ones. When drift operates the resulting shift in allele frequencies is not due to any adaptive value of the genes involved. Its effects are manifest only when certain conditions exist. The requirement that the population be small is fundamental. If the population is small, sampling error will be important if (1) the frequency of the allele in question is low, (2) there is recurrent migration out of the population, (3) there is a considerable fertility differential. For that matter a high incidence of fatal accidents in a very small population might have an appreciable effect upon the incidence of an allele that has a low frequency.

Although sampling error is a rather abstruse and mathematical concept, it can be explained simply with an example. Let us consider a small population of 100 individuals, 95 of its members are blood group O and 5 are blood group A. The probability that various accidents will eliminate a significant proportion of the 5 who are blood group A is high. *Any* events that prevent these individuals from mating—accident, illness, no proper mates—will reduce the number of A genes in the population. These accidents are not related to the action of the gene. These accidents are genetic drift.

The effect of drift in small populations is a function of the number of individuals who breed and the variability of family size. Family size affects the frequency of genes in a small population in the following way. If a few of the marriages produce a large number of offspring and the rest only a few offspring, there will be less genetic heterogeneity from generation to

TABLE 24.3
Estimated Mutation Rates for Deleterious Human Traits

Trait	Mutation rate per generation	
	Direct calculation	Indirect calculation
AUTOSOMAL DOMINANTS		
Neurofibromatosis (tumors of nerve tissue)	$1.3\text{--}2.5 \times 10^{-4}$	$0.8\text{--}1.0 \times 10^{-4}$
Muscular dystrophy		8.0×10^{-5}
Chondrodystrophy (dwarfism)	4.9×10^{-5} 4.2×10^{-5} 1.0×10^{-5}	4.3×10^{-5}
Pelger's nuclear anomaly (abnormal nuclei in white blood cells)	2.7×10^{-5}	
Retinoblastoma (tumor of eye)	2.3×10^{-5} 1.4×10^{-5} 4.3×10^{-6}	
Epiloia (tumor of brain)	$4\text{--}8 \times 10^{-6}$	
Huntington's chorea (involuntary uncontrollable movements)	5.4×10^{-6}	
Aniridia (absence of iris of eye)	4×10^{-6}	
Microphthalmos (abnormally small eyes in absence of mental defect)	5×10^{-6}	
Marfan's syndrome (spider limbs)	5×10^{-6}	
Waardenberg's syndrome (developmental anomalies)	3.7×10^{-6}	
AUTOSOMAL RECESSIVES		
Cystic fibrosis of pancreas		$0.7\text{--}1.0 \times 10^{-3}$
Microcephaly (abnormally small head)		3.0×10^{-5}
Albinism		2.8×10^{-5}
Congenital total color blindness		2.8×10^{-5}
Amyotonia congenita (muscle disorder)		2.0×10^{-5}
Infantile amaurotic idiocy		1.1×10^{-5}
Ichthyosis congenita (scaly skin)		1.1×10^{-5}
Epidermolysis bullosa dystrophica lethalis (lethal skin disease)		5.0×10^{-5}
X-LINKED RECESSIVES		
Childhood progressive muscular dystrophy	1×10^{-4}	1×10^{-4} $4.5\text{--}6.5 \times 10^{-5}$
Hemophilia (defective blood clotting mechanism)		3.2×10^{-5}

generation than if all the various marriages made an approximately equal contribution to succeeding generations.

The various factors affecting the magnitude of genetic drift can be combined and expressed as a quantity which can be determined by study of a group—effective population size. The effective population is a perfect breeding group, that is, one with an equal number of each sex, no reproductive inequalities among individuals, no inbreeding, and no periodic reduction in total size. The effective population size is a number smaller than the population, and, although it is an abstract, analytical category, it is a measure of the reproductive potential of a population.

The actual number of parents and the variation in number of offspring (the gametic contribution to subsequent generations) are the most significant features of a population from an evolutionary point of view. These two characteristics of a population, if known, enable us to calculate the effective size of the breeding population. The formula we use to determine the effective size of the population is

$$N_e = \frac{4N - 2}{s_k^2 + 2},$$

where N_e is the effective population size, N is the number of parents, the breeding individuals of one generation, and s_k^2 is the variance in number of offspring per family. Variance is a measure of the variation around the mean or average number of offspring per mating.

As indicated by the formula, effective population size is a function of the number of pairs that has produced offspring. When the variance (s_k^2) is greater than two, the effective size of the population (N_e) will be less than the number of parents (N). Large variances occur in human groups when there are large differences in fertility of pairs and in viability or survival of offspring.

What about those societies that permit polygynous matings? In such societies where males are permitted to and do have more than one mate, there will be two different effective population sizes—one for males and one for females. The variance (s_k^2) in family size for males will obviously be different from the variance for females. However most societies that permit or encourage polygyny actually do not have anything like total polygyny. Fortunately we can usually ignore this complicating factor.

Genetic drift is calculated by the following formula,

$$\sigma_{\Delta q}^2 = \frac{q(1 - q)}{2N_e},$$

where N_e is the effective size of the population and q is the frequency of the allele for which drift $(\sigma_{\Delta q}^2)$ is being calculated. We are able to calculate the effective population size for a number of communities and, using these values, to determine the magnitude of genetic drift for any given allele frequency. We have made these calculations for arbitrary frequencies of

$q = 0.01$ and $q = 0.5$ for several populations (Table 24.4). These data suggest that in human populations, even as small as those which have an effective population size of 100, sampling error plays almost no role in shifting allele frequencies.

TABLE 24.4
Theoretical Magnitude of Genetic Drift for Two Different Gene Frequencies

Population	Total size	Effective size	Drift (σ^2) $q = 0.5$	$q = 0.01$
North America (Arizona)				
Ramah Navajo	614	129	0.00097	0.00004
Havasupai	177	39	0.00320	0.00013
Caribbean				
Providencia Island	2,140	517	0.00024	0.0000096
Central America				
Paracho	4,593	967	0.00013	0.000005
South America (Brazil)				
Camayura	110	23	0.00543	0.00022
Europe				
Swiss alpine village	250	96	0.00013	0.000005
Africa (Sudan)				
Dinka village	375	109	0.00115	0.000045

Genetic drift and effective population size will be significant quantities only if the total population is small. The theory that sampling error affects frequencies of alleles is a logical consequence of Mendelian mechanisms of inheritance. This does not necessarily mean that there exist conditions in human societies whereby the effects of sampling error do shift or affect frequencies of alleles.

It is unlikely that genetic drift is a particularly important factor in controlling the frequencies of genes in populations of *Homo sapiens* now. Neither has it been a major factor for a long time. But there are certain cases where it may have operated. For example among inhabitants of Sicilian villages, Swiss alpine communities, Icelandic farming settlements, and religious groups such as the Hutterites and Old Order Dunkers in the United States, sampling error may have had an effect upon allele frequencies. In the distant past too, drift may have been significant in small nomadic hunting bands and among small agricultural groups living in isolated areas.

Sampling error or genetic drift is a recurrent phenomenon. If it is to have any appreciable effect upon allele frequencies, it must act for a long period of time. The continuing action of genetic drift should not be confused with the founder principle, as Dobzhansky calls it. Populations are often established by small groups who wander away from their homeland and found new communities, hence new populations, in a strange country.

Any such migratory group is a sample of the original population, and the allele frequencies in such a sample may be quite different from the frequencies found in the parent population. Many generations later a physical anthropologist may study the frequencies of the blood group genes, for example. If, as is sometimes the case, they are very different from frequencies found in related populations, genetic drift may be cited as a possible mechanism that led to such differences. Yet it is not at all unlikely that the founder principle was at work. An example often used is the absence of the gene for blood group B, of the ABO system, among American Indian populations. It is assumed that these populations were founded by small groups of nomadic hunters and gatherers who came across the Bering Straits toward the end of the Pleistocene and founded what became a rather large population of New World Indians. This original small group, or sequence of small groups, stemmed from a population which probably had relatively low frequencies of blood group B. The founding groups may have had no B alleles at all. It is also highly likely that natural selection, operating on the ABO phenotypes, was a potent factor in eliminating the B allele from American Indian populations. The point is that genetic drift unless it is well documented must not be called upon to explain population gene frequencies, for other equally likely and perhaps more plausible explanations can usually be found.

Migration or Hybridization

Migration or hybridization is a process by which genes from one gene pool or population of a species may be brought into another population. Hybridization merely changes the frequency of genes already present within local populations of the species; it does not produce new genes. Nevertheless it can have profound effects upon the genetic composition of a population. Migration can upset or reverse the effect of genetic drift.

Migration is expressed as the amount of admixture of genes from two parent populations in a descendant hybrid population. The calculation is a simple one. If we know the frequency of an allele in the three populations concerned, the one *into which* gene migration is believed to occur, the *ancestral* population, and the one *from which* the migrating genes come, we can estimate the admixture,

$$m = \left| \frac{q_1 - q_2}{q_3 - q_2} \right|.$$

The allele frequencies in the three populations are q_1, q_2, and q_3. The vertical lines on the right side of the equation indicate that we take the absolute value of the result, that is, m is always a positive number. The amount of admixture m has been calculated for many populations (Tables 24.5–24.7). In the population of Americans of African descent, for example,

TABLE 24.5

Estimates of Gene Migration (m) from Americans of European Descent to Americans of African Descent

It is evident from these estimates that the values for m depend on both the allele and the frequency of the allele in each population. (After Roberts, 1955.)

Allele	System	Ancestral population	Gene frequency (q_2)	American-African population	Gene frequency (q_1)	American-European population	Gene frequency (q_3)	m
R_0	Rh blood groups	Ewe I	0.547	Baltimore	0.446	New York I	0.028	0.195
		Ewe II	0.480	Baltimore	0.446	New York I	0.028	0.075
		Ewe II	0.480	Washington	0.449	New York I	0.028	0.069
		S. W. Nigeria	0.602	Baltimore	0.446	New York I	0.028	0.272
		N. Nigeria	0.539	Baltimore	0.440	New York I	0.028	0.182
		N. Nigeria	0.539	Baltimore	0.446	New York II	0.031	0.183
R_1	Rh blood groups	Ewe I	0.077	Baltimore	0.145	New York I	0.420	0.198
		Ewe II	0.086	Baltimore	0.145	New York I	0.420	0.177
		Ewe II	0.086	Washington	0.142	New York I	0.420	0.168
		S. W. Nigeria	0.058	Baltimore	0.145	New York I	0.420	0.240
		N. Nigeria	0.099	Baltimore	0.145	New York I	0.420	0.143
		N. Nigeria	0.099	Baltimore	0.145	New York II	0.434	0.137
S	MNSs blood groups	S. W. Nigeria	0.124	New York I	0.160	New York III	0.337	0.169
		N. Nigeria	0.139	New York I	0.160	New York III	0.337	0.106
Jk^b	Kidd blood groups	W. Africa I	0.217	New York II	0.269	Boston	0.477	0.200
T	Ability to taste phenylthio-carbamide	W. Africa II	0.813	Ohio and southern	0.697	Ohio	0.455	0.324

q_1 is the frequency of the allele in the American population of African descent, q_2 is the allele frequency in certain West African populations, and q_3 is the frequency of the same allele in Americans of European descent.

We make three assumptions about the populations involved when we calculate admixture. First, the migration of genes is entirely from one population into the other. Second, assortative mating with respect to the allele considered does not occur. No selection or rejection of mates is made because of the allele in question. Third, those zygotes produced by a mixture between the two parental populations are fertile. If they were not, no gene migration could occur. No matter how much hybridization went on, if the hybrids did not reproduce, there would be no flow of genes from one population into the other.

So far there are no insurmountable difficulties. The problems arise when we try to determine which are the ancestral or founder populations. The magnitude of the calculated admixture depends upon the frequency of the particular allele in each of three populations. If a mistake is made in choosing a population to represent the ancestral group, a considerable error in the calculated amount of mixture is possible.

It is not always easy to determine whether migration or selection is the agent that led to the observed allele frequencies in a population. We know from historical and sociological studies that interbreeding has occurred between Americans of European descent and Americans of African descent. For obvious reasons it is difficult to determine the magnitude of such interbreeding directly. We are able to make an estimate, however, by determining the amount of gene flow that occurred between the group consisting of Americans whose ancestors came from Europe and the group consisting of Americans of African descent. We know that gene frequencies vary slightly throughout the United States and the best estimates of gene migration can be made from comparisons of groups living in the same area. This knowledge should also enable us to discover whether the magnitude of interbreeding is greater in one part of the country than in another. Several alleles should be used in making our calculations, for the magnitude and direction of the gene flow should be the same for all alleles. If it is not we must examine alternative explanations for the differences.

Various estimates have been made of the amount of mixture between Americans of European descent and Americans of African descent (Table 24.5). It is usually assumed that all the gene flow is from the European group into the African ancestral group to produce the new allele frequencies characteristic of the population we know as Americans of African descent. It is not certain which populations in West Africa should be sampled for allele frequencies that are representative of the frequencies present in the peoples who were enslaved and brought to the New World. Slaves were brought from many places in Africa. It is believed that most of them came from West Africa. Hence it is not unreasonable to use allele frequencies of living West African populations to represent the ancestral component from which are derived the African elements in the American population. This leads to another assumption: The frequencies of the alleles we use in these calculations are today approximately what they were in the seventeenth, eighteenth, and nineteenth centuries, the period when slaves were brought from Africa. We must be cautioned today against blithely accepting these assumptions. Exactly where in Africa the ancestors of Americans of African descent lived is uncertain. The historical records are variable in accuracy and reliability. The compositions of the human populations that today inhabit most of Africa are probably not the same as they were during the seventeenth, eighteenth, and nineteenth centuries. The gene frequencies we find in modern African populations may well be unlike the gene frequencies of populations eight or ten generations ago. Movement of peoples has occurred inside Africa since the seventeenth century, and natural selection may have had some effect upon gene frequencies. Nevertheless reasonably good estimates have been made of the probable amount of genetic mixture that exists in contemporary Americans of partial African ancestry (Tables 24.5 and 24.6).

TABLE 24.6

Estimates of Gene Flow among Three American Populations

These estimates are based on eight blood group systems. The values are the amount of admixture (*m*) for gene flow for a total of six generations. (After Roberts and Hiorns, 1962.)

| Gene flow | | *m* |
From	To	
African	European	0.054
European	African	0.306
African	Indian	0.048
Indian	African	0
Indian	European	0.042
European	Indian	0.300

One valuable contribution which physical anthropology makes to a discussion of race is its discriminating analysis of the degree to which various racial groups have mixed. An example of the difficulties we may encounter in estimating the magnitude of gene flow between human populations, as well as the important results we may obtain, follows. Gene flow from the European stock in America into the African stock in America has been going on for as long as ten generations. But the magnitude of the gene flow varies from area to area, and detailed studies of local groups are among the best approaches to specific answers.

The frequencies for a number of allelic systems in groups now living in Georgia have been determined, and the gene flow has been estimated. The estimates of the gene migration for a number of the alleles in the population of Evans and Bullock counties, Georgia, fall into two groups (Table 24.7). In one set of alleles the estimated gene migration ranges from 9 to 22 per cent. In the other set the estimate ranges from 34 to 70 per cent. The values obtained for the amount of migration of the first group of alleles falls within estimates made using other systems and other American and African populations. The very large *m* estimated for the second group of alleles is due partly to adaptive differences and partly to gene migration. There is good reason to believe that adaptive values for some of these alleles are different in the North American environment and in the African environment. Another reason which makes the second set less useful for migration studies is the fact that the same African ancestral samples were not compared for each allele.

The amount of mixture from American Indians into Americans of African

TABLE 24.7

Estimates of Gene Migration (m) from Americans of European Descent to Americans of African Descent in Rural Communities of the United States

The American-Africans and the American-Europeans lived in the same area, Evans and Bullock counties, Georgia. The ancestral populations were West African. (From Workman, Blumberg, and Cooper, 1963.)

Allele	System	m
A	ABO blood groups	0.107
B	ABO blood groups	0.218
R_0	Rh blood groups	0.113
R_1	Rh blood groups	0.095
r	Rh blood groups	0.129
S	MNSs blood groups	0.143
Jk^a	Kidd blood groups	0.167
P	P blood groups	0.094
Fy^a	Duffy blood groups	0.109
T	Ability to taste phenylthiocarbamide	0.466
Hp^1	Serum haptoglobin	0.42 –0.70
Hb^s	Hemoglobin	0.46 –0.69
G6PD	Glucose-6-phosphate dehydrogenase	0.34 –0.44
Tf^{D_1}	Serum transferrin	0.50

descent was once considered large and significant. The grounds upon which this was assumed were nongenetic. College students of African ancestry have repeatedly reported that they had an American Indian great-grandparent, grandparent, etc. These reports were never verifiable directly, and recent genetic studies suggest they are not correct. The probable admixture among the American Indian, European, and African components of the greater American population is presented in Table 24.6. These figures are based upon eight allelic systems and are the best estimates we have today. These studies demonstrate that no noticeable flow of genes has occurred from American Indian groups into the Americans of African descent.

Two human populations which are brought into contact with each other seldom form, at first, a Hardy-Weinberg or panmictic population. There are many cultural barriers which prevent or delay random mating. Different languages, physical characteristics, social customs, economic positions, and psychological orientations will act to maintain the separateness of two groups and to prevent mating between them. Even two human groups living together in the same locality may remain genetically isolated from each other for a considerable period of time. It is fortunate from our point of view that sociocultural factors often maintain genetic barriers between groups. Such isolation, in the absence of potent selection, maintains the gene frequencies of the two original groups. The hypothesis that a particular group of people has migrated from some other place to the country in which they now live may often be tested by genetic methods. The formula used to estimate the magnitude of intermixture between Americans of European and of

African descent is applicable if we are relatively certain that gene flow between two groups has occurred. The problem of determining whether a group is indigenous to the area in which it lives requires that we compare the gene frequencies of at least three populations—the special group we are interested in, the population among whom they live, and the population from which they are purported to have come. Several examples will make this clear.

The origin of the Gypsies of Europe was once a matter of controversy. The Gypsies claim they are Asiatic Indian in ancestry, and analysis of their language supports this view. Most Gypsies lived in Hungary at one time, and it was believed that they maintained themselves as a breeding isolate. Social customs, the nomadic way of life, economic activities, and the attitudes of the greater Hungarian population all served to maintain a high degree of isolation between Gypsies and the rest of the peoples of Hungary. Examination of the frequencies of the four phenotypes of the ABO blood group system showed a remarkable difference between the Gypsies and the Hungarians and a remarkable resemblance between the Gypsies and the Asiatic Indians. The blood group frequencies (Table 24.8) are consistent with the hypothesis that the Gypsies are a group of Indian origin who mixed to some extent with the peoples amongst whom they now live. This is about the most that can be said on the matter, and strong as this evidence is there are some possible qualifications. Most of this work was done shortly after the 1914–1918 European war, before the twenty or more other blood group systems were discovered. The data are, thus, from a single allelic system. The populations may not have been compared as carefully as we might wish. The Indians used as the ancestral group were soldiers in the British Army stationed in Salonika, Greece. The frequencies of ABO phenotypes in this sample are close to those found among various Indian populations tested since then. Despite such possible objections, the genetic data strongly support the hypothesis which was originally based upon the observations that Gypsies did not look like Hungarians and that they spoke a different language.

The origin of the Pygmies who live on the Andaman Islands and some Oceanic islands such as Papua has led to much speculative writing about presumed migration into the Pacific islands from Africa. If the Oceanic Pygmies

TABLE 24.8
A Comparison of the Frequencies of ABO Blood Group Phenotypes in Gypsies, Indians, and Hungarians
(From Boyd, 1963.)

| | Frequency | | | Differences between | |
| | | | | Gypsies and | Gypsies and |
Phenotype	Gypsies	Indians	Hungarians	Indians	Hungarians
O	0.343	0.313	0.311	0.030	0.032
A	0.210	0.190	0.380	0.020	0.170
B	0.390	0.412	0.187	0.022	0.203
AB	0.057	0.085	0.122	0.028	0.065

are descendants of an African population that brought skin color, hair types, lip form, and nose shape from Africa, they surely would have brought some of the same blood group gene frequencies also. Table 24.9 shows that this is not the case. There never was any good reason to reject the hypothesis that the skin, hair, and facial morphology of the Pygmies of Asia are due to the action of natural selection. Those who wrote that the Pygmies of Oceania and the Pygmies of southeast Asia were migrants from Africa apparently assumed that traits appear once and then spread through migration. Thus they assume that if two human groups look alike (however superficial this may be) they must be fairly closely related. And they assume that local environments play little if any part in determining the physical type of a group. These assumptions are not in accord with modern evolutionary and genetic theory, and the evidence from the blood group gene frequencies of the Papuan and Andamese Pygmies shows this. Furthermore some doubt is cast on the uniqueness of the Pygmies by these studies. The blood group gene frequencies suggest that Papuan Pygmies are simply short Papuans and African Pygmies are short Africans.

TABLE 24.9
*A Comparison of the Frequencies of Some Blood Group Genes in
Three Populations of Pygmies*
(From Boyd, 1963.)

Allele	System	Gene frequency		
		Belgian Congo	Papua	Andaman Islands
A	ABO	0.198	0.075	0.54
B	ABO	0.249	0.139	0.08
O	ABO	0.553	0.786	0.38
R_0	Rh	0.630	0.030	0
R_1	Rh	0.074	0.850	0.92
R_2	Rh	0.194	0.119	0
r	Rh	0.101	0	0
M	MNSs	0.468	0.102	0.61
N	MNSs	0.523	0.898	0.39

Physical anthropologists once thought that morphological traits, such as skin color, hair type, and facial form, were of great significance in answering questions about human migration. But such traits have not yet been analyzed genetically and their value is clearly not as great as was once supposed. This view of morphology is strengthened by the report that Bantu speakers who took up residence deep in the Congo rain forests show a shift in general morphology toward that of the Pygmies, yet there is no major shift in blood group gene frequencies.

Migration or gene flow is an important process in changing the frequencies of alleles in populations. We have seen that it is often difficult to demonstrate migration of genes and in some cases the genetic method makes it possible to reject a migration hypothesis.

Natural Selection

Natural selection is the name of all processes whereby some genotypes develop an advantage over others in fertility and in viability. Natural selection is the most reasonable explanation for the distribution of frequencies of most genes observed in human populations. Selection must act on gene combinations already present among individuals of a population. Despite the complexity of the action of selection on the whole organism, on the phenotypic expression of the genotype, we can measure selection best as a change in frequencies of particular alleles.

The remarkable genetic diversity of man and the large number of combinations of alleles upon which selection may act is illustrated by the following examples. If there are 3 alleles at a genetic locus, there are 6 possible genotypes, for only 2 alleles may be present in any individual. The formula for determining the number of possible genotypes N when multiple alleles occur at a single genetic locus is

$$N = \frac{n(n + 1)}{2},$$

where n is the number of alleles. Thus if we find a locus with 8 alleles, the number of genotypes we could find in a population is

$$N = \frac{8(8 + 1)}{2} = 36.$$

When we consider several loci at once, the number of possible genotypes becomes enormous. If there are L loci, each of which may involve N possible genotypes, the number of possible gene combinations or genotypes is N^L, since the law of independent assortment (Chapter 23) holds.

There are a number of known hereditary disorders of the blood which manifest themselves as anemias or as defects in the blood clotting mechanism. We will consider 6 of them: thalassemia, sickle-cell anemia, hereditary spherocytosis, ovalocytosis, hemophilia C, and parahemophilia. Each of these 6 disorders is probably due to a single gene, thus 2 alleles may occur at a locus, and the number of possible genotypes is

$$N = \frac{2(2 + 1)}{2} = 3.$$

Since there are 6 loci (or allelic systems) involved with 3 genotypes possible for each, the number of possible different combinations of genotypes one may encounter in a population is $3^6 = 729$. Such a large number of combinations of genotypes is an indication of the variability possible in a human population.

If we now consider the large number of inherited human blood groups, we can indicate how enormous this potential variability is. At least 17 blood groups are detectable today, most of which are controlled by single pairs of genes. Despite the fact that several are known to involve 4 or 8 alleles, in this example we will assume that each is due to a pair of alleles. This means that there are 3 genotypes possible for each blood group system and $3^{17} = 129,140,163$ possible genotypic combinations. If we consider only 6 inherited blood disorders and 17 inherited blood groups, the total number of genotypic combinations possible is the product of the possible genotypic combinations of the 2 systems. Thus for the 6 blood disorders and the 17 blood groups, the number of possible genotypic combinations $= 3^6 \times 3^{17} = 3^{23} = 94,143,178,827$, or more than 94 billion combinations.

Since there are many loci with more than 2 or 3 alleles and man must have many more loci than 23, the number of possible genotypes, for all practical purposes, is infinite. There is almost no end to the possible combinations of genes in a human population. Large populations will have an enormous genetic storehouse of potential variation upon which selection may act.

A species, a population, may contain many genotypes that lose their survival value when the environment changes. But if the gene pool has a large potential for producing different genotypes, the population will be able to undergo genotypic recombination and may come up with new adaptive genotypes. The population reacts to adapt to new and changing situations. Adaptation is a response of a population, not of an individual, to the environment.

Selection's main effect on a population is to change the frequency of the alleles controlling the trait studied. Selection may act at any stage of the life cycle. The way in which selection operates may be difficult to determine and may vary enormously in different cases. Subtle and complex actual genetic situations can be described by relatively simple numerical expressions. We shall describe some of these in considerable detail when we discuss human hemoglobins and blood groups.

The differences in frequencies of alleles among various populations are often explained by three partially conflicting theories, namely, race, migration (gene flow), or natural selection. It is often believed that the variations in frequencies are the result of fundamental racial differences that exist between two populations. This is a kind of theory that explains nothing. It is remarkably similar in logical content to the statement that opium puts one to sleep because of its dormitive virtue.

Next is the migration theory. We indicated earlier that the present frequencies of certain genes in populations of Americans of African descent are probably best explained as the result of the migration of genes from

American populations of European descent into the original African gene pool. But there are migration hypotheses that hide rather than solve problems. Certain blood group phenotypes change in frequency as one examines populations along some geographical route. One of the most notable cases is the change in frequency of the gene for blood group B, of the ABO blood group system, as one moves westward from Asiatic to European populations. More persons with blood group B are found in Asia than in Europe, and more B is found in eastern Europe than in western Europe. One common explanation has been that the *B* allele in Europe is due to mixture with Asiatic populations. Recurrent invasions of nomadic herders and warriors from the steppes of Central Asia are supposed to have brought the *B* allele to Europe. There are several shaky assumptions behind the migration story, some of them fairly well hidden. It is assumed that the *B* allele originated as a mutation in Asia and was carried into the European gene pool, which had no or few *B* alleles. It is also assumed that the *B* allele and the other alleles of the ABO system are unaffected by natural selection. However today there is as much reason to accept relatively different adaptive values for the *B* allele in Europe and Asia as there is to state explicitly that the *B* allele is a result of migration from Asia. Since the ABO blood groups are part of man's primate heritage, it is improbable that the *B* allele is a mutation which occurred only in Asia and spread from there. If B blood groups are the result of a mutant gene in Asiatic populations, the relatively high frequency of the allele in such groups must be the result of very potent selection in its favor.

The migration hypothesis, if not rigorously scrutinized, directs our attention ever elsewhere in search of the causes or origins of a trait or gene. We are often asked to look to Asia or the Near East or Central Africa as the center from which a gene spread. Yet when we ask why it spread from such centers we are told that it is a racial characteristic of the peoples of that region. Indeed if some of the more extreme migrationist ideas are pursued to their logical conclusions, we are driven to the view that the traits came from Mars with an interplanetary invading force. Migration as an explanation for most gene frequencies in modern human groups is a *deus ex machina* kind of explanation. The role of selection in maintaining ABO blood group gene frequencies is still not fully appreciated because of the migrationist point of view. It has been far too easy to use the human blood groups as the traits on which to base our interpretations of the actual relationships between two or more human populations. They *are* useful in such studies, as we have shown. Migrations have occurred, and gene flow between populations is measurable. Migration theories, when first proposed, were tenable considering the amount of information available, but today we recognize that migration is probably not the primary process by which allele frequencies are shifted.

The third kind of theory explains that gene frequency differences are the result of the action of selection. The genotypes have a different relative fitness in the two populations, and we must look for physiological properties which give the one gene an advantage over its alleles. Genotypes express their advantage by an increased fertility of matings and increased viability of offspring relative to other genotypes.

The proportion of offspring a genotype contributes to the next generation is called its Darwinian fitness, selective value, or adaptive value. Selection operates on a particular allele if the differences in fitness of genotypes are the result of the presence or absence of the allele. Darwinian fitness, adaptive value, and estimates of fitness are more or less the same thing. They are a measure of the contribution one genotype makes to the next generation's gene pool.

If, in a gene system with 2 genotypes, 1 of the 2 produces on the average 100 offspring whereas the other produces 90, the adaptive value or the estimate of fitness of the first is 1.00, that of the second is 0.90. One also may say that the second genotype is opposed by a selection coefficient of $s = 0.10$. In other words relative fitness RF can be symbolized by $RF = 1 - s$, where s is the selection coefficient against 1 of the 2 genotypes. Estimates of fitness must be made independently for each set of alleles studied. The entire organism, the phenotypic expression of its genetic constitution or genome, is what selection acts upon. But we cannot yet handle efficiently more than one set of alleles at a time.

Methods for calculating the coefficient of selection differ slightly with the mode of inheritance and the kinds of data we can collect. The most interesting and probably most useful case for physical anthropology today is the method for estimating the selection coefficient when selection favors the heterozygote. When natural selection operates on a gene that is partially or completely dominant, it tends toward the elimination of one of the alleles. The gene frequency, if mutation does not occur, will eventually be 1 or 0. When selection is in favor of the heterozygote, rather than either of the homozygotes, the frequency in the population will tend to reach equilibrium at some intermediate value. Both alleles will be maintained in the population, even if there is no further mutation. We may determine the selection coefficient from the following formula

$$\Delta q = \frac{pq\,(s_1 p - s_2 q)}{1 - s_1 p^2 - s_2 q^2}.$$

The coefficients of selection against alleles with frequencies p and q are, respectively, s_1 and s_2, and $p + q = 1$. The quantity Δq is the change in gene frequency after one generation, given the operation of selection of magnitudes s_1 and s_2. The equilibrium condition, in which there is no change in frequencies, obtains when $\Delta q = 0$. Setting the quantity

$$\frac{pq\,(s_1 p - s_2 q)}{1 - s_1 p^2 - s_2 q^2} = 0$$

and solving the equation for s_1, we have $s_1 = s_2 q/p$. The gene frequencies, if equilibrium obtains, are

$$\frac{p}{q} = \frac{s_2}{s_1} \quad \text{or} \quad q = \frac{s_1}{s_1 + s_2}.$$

Therefore it is possible to calculate gene frequencies or selection coefficients given the relationship between them as defined by the equation. Because we usually are able to estimate selection coefficients only crudely, it is wise to determine gene frequencies as accurately as possible.

There are populations in Liberia that have a relatively high frequency of the allele for hemoglobin S. We represent the allele for hemoglobin S by Hb_β^S and the allele for normal hemoglobin A by Hb_β^A. The allele for hemoglobin S produces severe anemia and death in the homozygous condition, Hb_β^S/Hb_β^S. However the heterozygotes Hb_β^A/Hb_β^S, in areas in which malaria is prevalent, appear to be at an advantage relative to the normal homozygotes Hb_β^A/Hb_β^A, despite the fact that the heterozygote has a minor abnormality, the sickling of red cells when there is a reduced amount of oxygen in the blood stream. These alleles provide a good example of the calculation of relative fitness.

The fitness of Hb_β^S/Hb_β^S homozygotes, relative to that of the Hb_β^A/Hb_β^S heterozygotes, is estimated to be no more than 0.25, $RF = 0.25$. This estimate of fitness is based upon vital statistics—age at death of various genotypes, their general viability, and their relative fertility. The selection coefficient against the Hb_β^S/Hb_β^S homozygotes, then, is $s_2 = 1 - RF = 0.75$. In order to determine the relative fitness of the Hb_β^A/Hb_β^S heterozygote, we calculate next the selection coefficient against the normal homozygote Hb_β^A/Hb_β^A. We know s_2, the selection coefficient against the Hb_β^S/Hb_β^S homozygote; $s_2 = 0.75$. The frequency of Hb_β^A is known; $p = 0.794$. The frequency of Hb_β^S is known; $q = 0.206$. We assume that the population is in equilibrium with respect to Hb_β^S; in other words $\Delta q = 0$. We have all the data necessary to calculate s_1, the coefficient of selection against the Hb_β^A/Hb_β^A homozygote.

$$s_1 = \frac{s_2 q}{p} = \frac{0.75\,(0.206)}{0.794} = 0.196.$$

Relative fitness of the normal homozygote is $RF = 1 - s_1 = 0.804$. For every 100 offspring produced by Hb_β^A/Hb_β^S heterozygotes, 80 (theoretically 80.4) will be produced by normal Hb_β^A/Hb_β^A homozygotes. Finally the fitness of the heterozygote relative to the normal homozygote can be calculated.

$$RF = \frac{1}{1 - s_1} = \frac{1}{1 - 0.196} = \frac{1}{0.804} = 1.24.$$

This means that the Hb_β^A/Hb_β^S heterozygotes have an advantage, relative to the homozygotes, of approximately 24 per cent.

This example is a rather dramatic illustration of an apparently deleterious gene being maintained in a population by favorable selection on the heterozygotes. We shall return to some of the problems of hemoglobin S and malaria in Chapter 29. The point here is that not only can we demonstrate that selection takes place, but also we can arrive at a quantitative estimate of its effect on specific alleles in specific populations.

When selection favors the heterozygote over the two homozygotes, as it does in the case of $Hb_\beta A/Hb_\beta S$ individuals with the sickle cell trait, balanced polymorphism results. Genetic polymorphisms occur in a population when two alleles at a locus are present with frequencies too great to be accounted for by recurrent mutations. Genetic polymorphisms are termed balanced if selection favors the heterozygotes. When selection favors the heterozygotes, a stable equilibrium may be achieved and substantial frequencies of both alleles may be maintained in one environment.

Genetic polymorphisms are termed transient or fortuitous if the forces which maintain the alternative alleles at a locus act more or less independently of each other. For example in the ABO blood group system it is possible that selection acts differently on the A and B phenotypes and on the O phenotype.

The situation of balanced polymorphism of the alleles for hemoglobins A and S is an example of heterosis or hybrid vigor. On the face of it this may sound silly, for $Hb_\beta S$ is deleterious. Yet in the heterozygous state it produces a phenotype that, in certain environments, is more fit than either homozygote. The phenomenon of heterosis or hybrid vigor is commonly observed among strains of domestic plants and animals. Hybrid corn is a famous example of the importance of heterozygosity in establishing more vigorous hybrids. It may seem paradoxical that natural selection by acting on heterozygotes maintains unfit and indeed lethal genotypes in a population. But this is not such a paradox after all. Heterozygotes, in cases of balanced polymorphism, are the fittest genotypes. According to Mendel's law of independent assortment, a sexually reproducing population composed entirely of heterozygotes will produce an offspring generation consisting of 50 per cent heterozygotes (Dd) and 50 per cent homozygotes (DD plus dd). Despite the maintenance of relatively unfit genotypes, natural selection attempts to produce a population in which the average fitness is the greatest possible. It is clear that in some environments this is attainable only at the expense of maintaining some relatively unfit genotypes. This leads to the further conclusion that it is the Mendelian population, not the individual, which is the unit upon which selection acts and through which evolution occurs.

We can construct theoretical models that show how selection of various intensities will affect gene frequencies in a population over a number of generations. We must assume panmixis and simple modes of inheritance. One of the most interesting cases is severe selection against a recessive gene with total elimination of the recessive homozygotes. If q is the frequency of the recessive gene, it can be shown that, with complete elimination of the recessive homozygotes, the amount of change in q per generation is

$$\Delta q = \frac{-q^2}{1 + q}.$$

If q is very large initially, the rate at which it decreases will be rapid. From an evolutionary point of view, such selection will diminish gene frequencies quickly (Table 24.10). Quickly, from this point of view, may mean many generations. This is a most important point when considering human populations, for it demonstrates the difficulty of eliminating a deleterious recessive gene. Many of the detectable human deleterious recessive genes have very low frequencies. For example phenylketonuric idiocy, a mental defect due to the inherited inability to metabolize phenylalanine, has been estimated as occurring in approximately 1 in 25,000 individuals (estimated gene frequency $q = 0.006$). The incidence of glycogen storage disease, which also causes mental insufficiency, is approximately 1 in 280,000 individuals ($q = 0.002$).

If the initial frequency of a deleterious recessive gene for one of the mental deficiencies is $q = 0.02$, which is indeed a high frequency, the proportion of the recessive, presumably mentally defective, genotypes in the population will be $q^2 = 0.0004$. If every such individual is sterilized, the amount that the frequency of q is decreased in one generation is

$$\Delta q = \frac{-0.0004}{1 + 0.02} = -0.00039,$$

and the frequency of gene q after one generation of total sterilization of the recessive homozygotes is $q' = 0.02 - 0.00039 = 0.01961$. It can be demonstrated that it will take $1/q$ generations to reduce the initial gene frequency by a factor of one-half. In this case where $q = 0.02$ it will take $1/0.02 = 50$ generations, or about 1500 years, to reduce the frequency to $q = 0.01$. If the initial frequency is lower, many more generations will be needed (Table 24.10).

It is often argued that persons with severe, incapacitating hereditary defects, such as mental deficiency, should be subjected to potent natural selection, that is, sterilization. (Although some would call this artificial selection

TABLE 24.10
Rates at Which Gene Frequencies Change

Hypothetical initial gene frequency	Change in frequency in first generation	Number of generations in which frequency will be reduced by one-half
(q)	(Δq)	($1/q$)
0.5	-0.167	2
0.1	-0.009	10
0.05	-0.0024	20
0.02	-0.00039	50
0.01	-0.000099	100
0.005	-0.000025	200
0.002	-0.000004	500
0.001	-0.000001	1000

rather than natural, the effect, the end result, is the same.) Yet we see that even such complete selection against those who manifest the trait, the homozygotes, will have no very appreciable effect upon the frequency of the allele in the population. Since individuals with severe inherited physical and mental handicaps seldom produce children, despite much folklore that says they do, active measures to prevent such homozygotes from breeding are not apt to produce the desired results.

A much more effective way to control or reduce the frequency of a deleterious recessive gene would be to prevent the mating of heterozygotes, the carriers of the defect. This is easier said than done, because identification of the individuals heterozygous for a recessive trait (the carriers) is impossible in theory and very difficult in practice. Table 24.11 lists the frequencies of heterozygotes in a population with various gene frequencies given. It is easy to see that a relatively rare recessive gene may be more rapidly eliminated from a population if the carriers do not breed. Furthermore the number of heterozygotes in a population per recessive homozygote increases enormously as the frequency of the gene decreases. It would be very difficult to exercise much control over such a trait in a population. These figures indicate, in addition, that recessive genes, recessive mutations, may hide effectively in populations. Although the recessive homozygotes may be eliminated, the gene frequency is not materially reduced. It is always possible that environmental change will occur so that the homozygotes will eventually have a selective advantage and will not be eliminated.

The fact that most of our examples are deleterious traits is merely the

TABLE 24.11

Ratios of Heterozygotes to Homozygotes for
Different Gene Frequencies

These calculations are for a panmictic population where the Hardy-Weinberg formula holds; that is, $p^2 + 2pq + q^2 = 1$. The deleterious recessive gene is d, the dominant D.

Gene frequency		Genotype frequency		Ratio of
Deleterious gene	Its allele	Recessive homozygote	Heterozygote	heterozygotes to homozygotes
(q)	$(p = 1-q)$	$(dd = q^2)$	$(Dd = 2pq)$	$(Dd:dd = 2pq:q^2)$
0.50	0.50	0.25	0.50	2:1
0.25	0.75	0.0625	0.375	6:1
0.10	0.90	0.01	0.180	18:1
0.05	0.95	0.0025	0.095	38:1
0.025	0.975	0.000625	0.04875	78:1
0.015	0.985	0.000225	0.02955	131:1
0.005	0.995	0.000025	0.00995	398:1
0.0025	0.9975	0.00000625	0.0049875	798:1
0.001	0.999	0.000001	0.001998	1998:1

result of our ignorance and the compelling need to care for those individuals with an unfortunate genotype. The majority of the recessive homozygotes we can detect and, hence, study are those which are first noticed in medical, psychiatric, or genetic clinics. Those individuals who have inherited mental or physical conditions which demand expert attention are most often the subjects of study. Eventually the effect of selection upon favorable traits, upon those not so completely deleterious, or upon those with less dramatic and clear-cut modes of action will have to be considered in detail. There are a large number of human polymorphisms which involve mutant genes and appear to be nondeleterious. It has not been easy to study the effects of natural selection on these traits, but we have been able to investigate them, and we shall discuss some of them in later chapters.

In this chapter we have tried to show that the logical extensions of Mendelian genetics allow us to view evolution as changes in gene frequencies. New alleles are introduced by mutations; gene frequencies are altered by natural selection. Other factors make possible new genotypic combinations upon which selection may exert its effect. Increase in the genetic variability of a human group, increased heterozygosity, results from migration and breakup of breeding isolates, for example. Sampling error and nonrandom mating lead to increased homozygosity, increased homogeneity, and less variability in the human gene pool. Evolution occurs when new genetic traits appear in a population by means of mutation and when the survival and increase or the elimination of alleles results from natural selection.

Suggested Readings and Sources

Aberle, D. F., U. Bronfenbrenner, E. Hess, D. R. Miller, D. M. Schneider, and J. N. Spuhler, The incest taboo and the mating patterns of animals. *Am. Anthropol.* **65**, 253 (1963).

Boyd, W. C., Four achievements of the genetical method in physical anthropology. *Am. Anthropol.* **65**, 243 (1963).

Brenk, H., Über den Grad der Inzucht in einem innerschweizerischen Gebirgsdorf. *Arch. Julius Klaus-Stift. Vererbungsforsch. Sozialanthropol. Rassenhyg.* **6**, 1 (1931).

Buettner-Janusch, J., J. R. Bove, and N. Young, Genetic traits and problems of biological parenthood in two Peruvian Indian tribes. *Am. J. Phys. Anthropol.* **22**, 149 (1964).

Croquevieille, N. J., Etude démographique de quelques villages Likouala. *Population* **8**, 491 (1953).

Crow, J. F., Mutation in man. *In* A. G. Steinberg (Ed.), *Progress in Medical Genetics*, Vol. I, p. 1. Grune and Stratton, New York (1961).

Dahlberg, G., *Mathematical Methods for Population Genetics.* S. Karger, Basel (1947).

Darwin, C., *On the Origin of Species by Means of Natural Selection.* John Murray, London (1859).

Eaton, J. W., and A. J. Mayer, The social biology of very high fertility among the Hutterites. *Human Biol.* **25**, 206 (1956).

Egenter, A., Über den Grad der Inzucht in einer Schwyzer Berggemeinde und die damit zusammenhängende Häufung rezessiver Erbschäden. *Arch. Julius Klaus-Stift. Vererbungsforsch. Sozialanthropol. Rassenhyg.* **9**, 365 (1934).

Ehrenberg, L. V., G. Ehrenstein, and A. Hedgran, Gonad temperature and spontaneous mutation-rate in man. *Nature* **180**, 1433 (1957).

Erickson, R., D. Clark, W. Creger, and K. Romney, The comparison of genetic and anthropological interpretations of breeding patterns and breeding isolates. *Abstracts,* Sixty-second Annual Meeting, American Anthropological Association, p. 16 (1963).

Falconer, D. S., *Introduction to Quantitative Genetics.* Ronald Press, New York (1960).

Fisher, R. A., *The Genetical Theory of Natural Selection* (Second edition). Dover, New York (1958).

Fraser, G. R., Studies in isolates. *J. Génét. Hum.* **13**, 32 (1964).

Freire-Maia, N., Inbreeding in Brazil. *Am. J. Human Genet.* **9**, 284 (1957).

Gessain, R., Démographie et généalogie de différents types d'isolats. *J. Génét. Hum.* **13**, 76 (1964).

Glass, B., Genetic changes in human populations, especially those due to gene flow and genetic drift. *Advan. Genet.* **6**, 95 (1954).

Glass, B., On the unlikelihood of significant admixture of genes from the North American Indians in the present composition of the Negroes of the United States. *Am. J. Human Genet.* **7**, 368 (1955).

Glass, B., and C. C. Li, The dynamics of racial intermixture—an analysis based on the American Negro. *Am. J. Human Genet.* **5**, 1 (1954).

Glass, B., M. S. Sacks, E. Jahn, and C. Hess, Genetic drift in a religious isolate: an analysis of the causes of variation in blood group and other gene frequencies in a small population. *Am. Naturalist* **86**, 145 (1952).

Grob, W., Aszendenzforschungen und Mortalitätsstatistik aus einer st. gallischen Berggemeinde. *Arch. Julius Klaus-Stift. Vererbungsforsch. Sozialanthropol. Rassenhyg.* **9**, 237 (1934).

Haberlandt, W. F., Considerations on the isolate problem based on research in population genetics. *J. Génét. Hum.* **13**, 47 (1964).

Haldane, J. B. S., Natural selection in man. *In* A. G. Steinberg (Ed.), *Progress in Medical Genetics,* Vol. I, p. 27. Grune and Stratton, New York (1961).

Hiernaux, J., Analyse de la variation des caractères physiques humaines en une région de l'Afrique centrale: Ruanda, Urundi, et Kivu. *Ann. Musée Roy. Congo,* Sci. de l'Homme **3**, 131, Tervuren, Belgium (1956).

Hiernaux, J., Comments. *Current Anthropol.* **3**, 29 (1962).

Kunstadter, P., R. Buhler, F. F. Stephan, and C. F. Westoff, Demographic variability and preferential marriage patterns. *Am. J. Phys. Anthropol.* **21**, 511 (1963).

Lasker, G., Mixture and genetic drift in ongoing human evolution. *Am. Anthropol.* **54**, 433 (1952).

Lasker, G., Human evolution in contemporary communities. *Southwestern J. Anthropol.* **10**, 353 (1954).

Lenz, F., Die Bedeutung der Statisch Ermittelten Belastung mit Blutverwandtschaft der Eltern. *Münch. Med. Wochschr.* **66**, 1340 (1919).

Malaurie, J., L. Tabah, and J. Sutter, L'isolat esquimau de Thule (Groenland). *Population* **7**, 675 (1952).

Morton, N. E., Genetics of interracial crosses in Hawaii. *Eugen. Quart.* **9**, 23 (1962).

Muller, H. J., Artificial transmutation of the gene. *Science* **66**, 84 (1927).

Neel, J. V., M. Kodani, R. Brewer, and R. C. Anderson, The incidence of consanguineous matings in Japan. *Am. J. Human Genet.* **1**, 156 (1949).

Neel, J. V., and W. J. Schull, *The Effect of Exposure to the Atomic Bombs on Pregnancy Termination in Hiroshima and Nagasaki. Natl. Acad. Sci.—Natl. Res. Council, Publ.* **461** (1956).

Orel, H., Die Verwandtenehen in der Erzdiozese Wien. *Arch. Rassen. Gesell.* **26**, 249 (1932).

Pollitzer, W. S., The Negroes of Charleston (S. C.), a study of hemoglobin types, serology and morphology. *Am. J. Phys. Anthropol.* **16**, 241 (1958).

Roberts, D. F., The dynamics of racial intermixture in the American Negro: some anthropological considerations. *Am. J. Human Genet.* **7**, 361 (1955).

Roberts, D. F., A demographic study of a Dinka village. *Human Biol.* **28**, 323 (1956).

Roberts, D. F., and G. A. Harrison (Ed.), *Natural Selection in Human Populations.* Pergamon, New York (1959).

Roberts, D. F., and R. W. Hiorns, The dynamics of racial intermixture. *Am. J. Human Genet.* **14**, 261 (1962).

Saldanha, P. H., Gene flow from white into Negro populations in Brazil. *Am. J. Human Genet.* **9**, 299 (1957).

Sanghvi, L. D., and V. R. Khanolkar, Data relating to seven genetical characters in six endogamous groups in Bombay. *Ann. Eugenics* **15**, 52 (1949).

Schull, W. J., The effect of Christianity on consanguinity in Nagasaki. *Am. Anthropol.* **55**, 74 (1953).

Schull, W. J., Empirical risks in consanguineous marriages: sex ratio, malformation, and viability. *Am. J. Human Genet.* **10**, 294 (1958).

Schull, W. J., T. Yanase, and H. Nemoto, Kuroshima: the impact of religion on an island's genetic heritage. *Human Biol.* **34**, 271 (1962).

Slater, M. K., Ecological factors in the origin of incest. *Am. Anthropol.* **61**, 1042 (1959).

Spuhler, J. N., and C. Kluckhohn, Inbreeding coefficients of the Ramah Navajo population. *Human Biol.* **25**, 30 (1953).

Strauss, B. S., Chemical mutagens and the genetic code. *In* A. G. Steinberg and A. G. Bearn (Ed.), *Progress in Medical Genetics,* Vol. III, p. 1. Grune and Stratton, New York (1964).

Sugiyama, S., and W. J. Schull, Consanguineous marriages in feudal Japan. *Monumenta Nipponica* **15**, 126 (1960).

Sutter, J., and L. Tabah, Fréquence et répartition des mariages consanguins en France. *Population* **3**, 607 (1948).

Thieme, F. P., Problems and methods of population surveys. *Cold Spring Harbor Symp. Quant. Biol.* **15**, 25 (1950).

White, L. A., The definition and prohibition of incest. *In* L. A. White, *The Science of Culture,* p. 303. Farrar, Strauss, New York (1949).

Wilson, P. J., and J. Buettner-Janusch, Demography and evolution on Providencia Island, Colombia. *Am. Anthropol.* **63**, 940 (1961).

Woolf, C. M., F. E. Stephens, D. D. Mulaik, and R. E. Gilbert, An investigation of the frequency of consanguineous marriages among the Mormons and their relatives in the United States. *Am. J. Human Genet.* **8**, 236 (1956).

Workman, P. L., B. S. Blumberg, and A. J. Cooper, Selection, gene migration and polymorphic stability in a U.S. white and Negro population. *Am. J. Human Genet.* **15**, 425 (1963).

Wright, S., Size of population and breeding structure in relation to evolution. *Science* **87**, 430 (1938).

Wulz, G., Ein Beitrag zur Statistik der Verwandtenehen. *Arch. Rassen. Gesell.* **17**, 82 (1925).

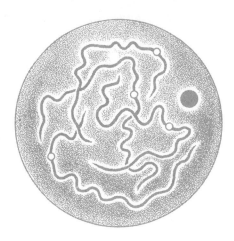

The Chemical and
Structural Bases
of Heredity

25

PHYSICAL ANTHROPOLOGISTS in the past concentrated their efforts for the most part on bones and muscles. These were obviously important, because the fossil record of primate evolution consists of bones. It was necessary that the bones and muscles of the living primates be studied in light of the fossils. Today we attempt to read the evolutionary history of the Primates in tissues and fluids that can never be part of the fossil record. The story is told, in part, by subtle relations and reactions between molecules. We can examine the chemistry and morphology of the chromosomes—the structural units of heredity in the nuclei of the cells. We can describe, on a molecular level, some of the polymorphisms of living populations, and we can reconstruct some of the stages in the evolution of the hereditary material of the Primates.

Our discussion of chromosomes so far has emphasized the "beads on a string" concept—the chromosomes are elongated structures, and the genes are considered as discrete units, arranged like beads on a string. Under the microscope a chromosome looks like a small rod or tiny sausage made up of thin, wiry, thread-like material tightly coiled in a helical formation. The units of heredity which we have been calling the genes are arranged along this coil. The threads of the chromosomes have a definite chemical structure, and during the last decade some of the most exciting work in biology, physics, and chemistry has resulted in our being able to describe chemically the chromosomal material and the morphology of the chromosomes.

We must now ask of what chemical entities the rods and beads are composed and how these structures act to transmit information, inherited traits, from one generation to the next. What are genes and chromosomes, and how do they determine the observed events of inheritance?

Deoxyribonucleic Acid (DNA), the Genetic Material

The chromosomes are primarily nucleic acid; in most organisms they are deoxyribonucleic acid (DNA). This chemical substance is composed of four different nucleotides. Each nucleotide contains a molecule of a sugar, specifically, deoxyribose; a phosphate group, phosphoric acid; and an organic base, which is a compound composed of carbon, hydrogen, oxygen, and nitrogen, a purine or a pyrimidine. The sugar and phosphate groups are identical in each nucleotide—only the four bases differ. These bases are adenine, guanine, cytosine, and thymine, symbolized by A, G, C, and T respectively. The corresponding nucleotides are named deoxyadenylic acid, deoxyguanylic acid, deoxycytidylic acid, and thymidylic acid. Thousands and thousands of these nucleotides linked together form one strand of a DNA molecule. Each nucleotide is linked in the same way to the next nucleotide in the chain through the phosphate group. The DNA molecule (Fig. 25.1) is composed of two complementary strands coiled together in a helix. This structure is known as the Watson-Crick model of DNA. When we say that the two strands of the DNA molecule are complementary, we mean that the sequence of nucleotides in one strand determines the sequence of nucleotides in the other. The nucleotides are paired, A is always paired with T and G with C. Chemical analysis of DNA from many different sources always yields nearly equal amounts of A and T and of G and C. The three-dimensional conformation of the bases makes this pairing possible.

The uniformity in DNA taken from various living organisms is remarkable evidence of the unity of all living things. There are exceptions, of course. The DNA from certain bacteria, viruses, and bacteriophages contains sugars or bases other than deoxyribose or A, G, C, and T. But these are usually simple chemical derivatives of deoxyribose or A, G, C, or T. All living organisms, animal or plant above the level of virus, contain DNA in the nucleus of the cell. The amount of each base may be different in different organisms, but the same four nucleotides are present, and the physical conformation of the DNA molecule is the same.

What is the evidence that DNA is the genetic material, the molecule that passes inherited traits from one generation to the next? We may begin by asking what properties must be present in the chemical material that transmits inheritance. First, this material must be extremely stable. It must be stable in the face of assault from physical and metabolic agents, yet

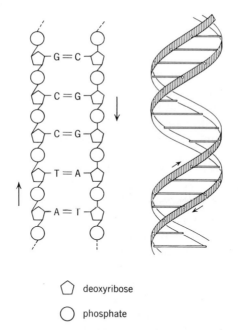

deoxyribose

phosphate

Fig. 25.1.   Schematic diagrams of portions of
a DNA molecule. The four bases adenine, guanine, cytosine, and thymine are represented by
A, G, C, and T respectively.

occasional changes must be able to occur—mutations. Second, the genetic
material must be able to replicate itself with great accuracy. All of the data
available about the inheritance of traits in mammals, and in other forms of
life for that matter, suggest that genes are replicated (or replicate themselves)
with extreme precision and accuracy. Third, the genetic material must be
able to direct the production or synthesis of those molecules, such as proteins, which carry out activities within the cell and which produce what
we call the phenotypic expression of the gene. Fourth, it is necessary that
different parts of the genetic material be able to perform different functions.
There are many thousands of inherited traits in an organism, and the genes
which control these traits are distinguishable from each other. The genetic
material must differ from one individual to another as well as from one
species to another.

DNA meets all of these requirements, and we shall summarize some of
the experimental evidence which leads to the conclusion that DNA is the
genetic material. DNA is a large molecule, highly resistant to attack by
such agents as heat and strong acids. DNA in solution absorbs ultraviolet
radiation most strongly in the region of the spectrum around wavelength

260 mμ. It was observed that mutations could be induced in organisms by irradiation with ultraviolet light. The induction of mutations is most efficient when the wavelength is 260 mμ. This is of course not critical evidence, but it indicates that mutations in the genetic material could probably be induced by the absorption of energy at this wavelength.

Careful measurement of the exact amount of DNA in single cells shows that this amount is the same in every cell within a single species. There is an exception to this—the amount of DNA in the gametes or germinal cells is one-half the amount in the somatic cells. Since the germinal cells have one-half the number of chromosomes found in the somatic cells, this is further, good, indirect evidence that DNA is the genetic material.

The best and direct evidence that DNA is the genetic material comes from some famous experiments with pneumococci, small bacteria responsible for serious lung disease (pneumonia) in mammals. *Normal* pneumococcus bacteria were killed with heat and injected into mice. At the same time live *mutant* pneumococcus bacteria were injected into the mice. After a reasonable period of time living bacteria were recovered from the mice, *normal* pneumococci. A long series of experiments, both *in vivo* (in living animals) and *in vitro* (literally: in glass; outside of living animals) demonstrated that something from the dead bacteria was transforming the living strain so that it appeared to have the traits which characterized the dead strain. The transforming principle was isolated from the heat-killed normal pneumococci, and various physical and chemical tests demonstrated that it was DNA. It behaved like a molecule made up of a large number of similar units, that is, a polymer, and DNA is a highly polymerized molecule. The only components present in the transforming principle were the bases, sugars, and phosphate of DNA. The activity of the transforming principle was destroyed by only one agent—an enzyme called deoxyribonuclease which specifically attacks and breaks down DNA.

The replication of DNA occurs in a way which also provides indirect evidence that DNA is the genetic material. If both strands of the DNA in cells of rapidly growing organisms are marked or labeled with easily detectable radioactive isotopes, it can be shown experimentally that, after the first generation, only one strand of the DNA contains the radioactive marker. The results of these experiments are best described in a diagram, Fig. 25.2. We shall not give the details of the original experiments of Taylor and Levinthal and their co-workers; neither shall we discuss the various ingenious proposals made to describe the manner in which the two strands unwind and rewind. We can accept with little or no doubt the conclusion that, during replication, each strand of DNA remains intact.

These experiments have been confirmed over and over again. There is general consensus that DNA is the genetic material. The postulated structure of DNA does account for the nature of genes and does explain their action.

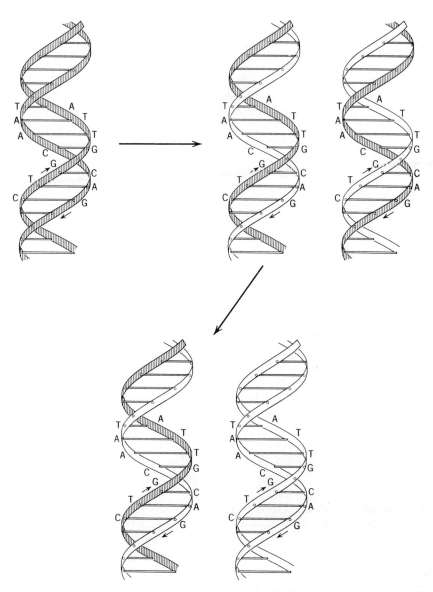

Fig. 25.2. The replication of a portion of a DNA molecule illustrated diagrammatically. A, adenine; G, guanine; C, cytosine, and T, thymine. The shaded strands represent radioactively labeled DNA. The white strands represent nonlabeled DNA. During replication the two strands unwind and two new complementary strands are formed. One of these is complementary to one of the parent strands, the other complementary to the other parent strand.

Transmission of Information by DNA

One fundamental question asked by students of genetics is, how can there be so many genes with so many different kinds of expressions if the hereditary material is the relatively simple DNA? In other words how is it possible for the genes to have their properties specified by a molecule that has only four bases—A, G, C, and T? If we use an analogy, this question can be answered readily. We assume that the four bases A, G, C, and T are the letters of a genetic alphabet, and various combinations and sequences of these letters produce an almost infinite number of words, a large variety of genetic messages. By continuing this analogy, as Dobzhansky has done, evolution of life can be viewed as the action of natural selection on the messages formed by the genetic alphabet. The genetic alphabet remains the same whether viruses, amoebae, corn plants, fruit flies, monkeys, or men are the organisms upon which natural selection is acting.

The genes, then, are segments of the DNA molecule—a word or words made from the four letters of the alphabet. One gene differs from another when the sequence of bases in one segment of DNA differs from the sequence in another. Just as messages are sent over a telegraph wire by use of a code, so genetic messages are transmitted by a code. Instead of sequences of dots and dashes, the genetic code is composed of sequences of nucleotides (Table 25.1).

The synthesis of proteins is one of the primary functions of genes. Most of our knowledge of the chemical basis for the mechanisms of heredity comes from study of proteins, products of the action of genes. Some of the most important constituents of any living cell are its proteins, large molecules composed of various numbers and combinations of about twenty different subunits, twenty amino acids.

Many proteins are enzymes, essential molecules that speed up the rate at which other molecules are built up or broken down. For example the enzyme tyrosinase acts in the synthesis of melanin, the dark pigment of skin and hair. Trypsin, a pancreatic enzyme, helps in the digestion of the proteins which we eat. Other proteins function as hormones, as regulators of many body processes. Insulin which is found in the pancreas regulates the amount of sugar in various tissues. Still other proteins are respiratory pigments such as hemoglobin which carries oxygen to the tissues of the body. These are examples of proteins in mammalian systems. There are analogous proteins in the cells of all living organisms. This brief list should suffice to indicate how important and ubiquitous proteins are.

All proteins are composed of amino acids, organic compounds containing an amide group (NH_2) and an acidic group (COOH) plus a side chain, all attached to a central carbon atom. These amino acids are linked in a chain by peptide bonds (Fig. 25.3). The twenty amino acids commonly found in proteins of living organisms are listed in Table 25.2. The sequence of amino acids in any single protein is unique. The order in which the amino acids

TABLE 25.1
Postulated Genetic Code

Amino acid	Postulated triplets of nucleotides in RNA[a]			
Alanine	(CCG)[b]	(UCG)		
Arginine	(CGC)	(AGA)	(CGA)	(UCG)
Asparagine	(AAC)	(AUA)		
Aspartic acid	(GUA)	(ACG)		
Cysteine	UGU	(UGC)		
Glutamic acid	(ACG)	(GUA)		
Glutamine	(ACA)	(AGA)	(AGU)	(UCG)
Glycine	(UGG)	(AGG)	(GGC)	
Histidine	(ACC)	(UAC)		
Isoleucine	(UAU)	(UAA)	(UAC)	
Leucine	UUG	(UUC)	(CUC)	
Lysine	AAA			
Methionine	(UGA)			
Phenylalanine	UUU	UUC		
Proline	CCC	(CCU)		
Serine	(UCU)	(UCC)	(UCG)	
Threonine	(ACA)	(ACC)	(ACU)	
Tryptophan	(GGU)			
Tyrosine	(UAU)	(UAC)		
Valine	GUU	(GUC)		

[a]Data taken from Nirenberg and Ochoa.

[b]The orders of sequences in parentheses have not been established. The abbreviations are A, adenine; G, guanine; C, cytosine; U, uracil.

are linked together is the same each time the same protein is synthesized by the cell.

The first step in translating the genetic information into the phenotype that is the living organism is the synthesis of proteins. DNA does not synthesize proteins directly. An intermediate substance carries the information to the active synthesizing centers of the cells. The translator and activator of the message is another nucleic acid, ribonucleic acid (RNA). The composition of RNA is somewhat similar to that of DNA. The sugar molecule in its backbone is ribose instead of deoxyribose. The phosphate is the same. Three of the four bases are the same, adenine, guanine, and cytosine. The fourth base is uracil rather than thymine. The RNA alphabet is A, G, C, U instead of A, G, C, T as in DNA. The nucleotides in RNA are adenylic acid, guanylic acid, cytidylic acid, and uridylic acid. RNA appears to be a single strand rather than a double strand.

The actual synthesis of proteins takes place in or on little particles found in the cytoplasm (the material outside the cell nucleus). These particles are called ribosomes and are rich in RNA. Two kinds of RNA are required for

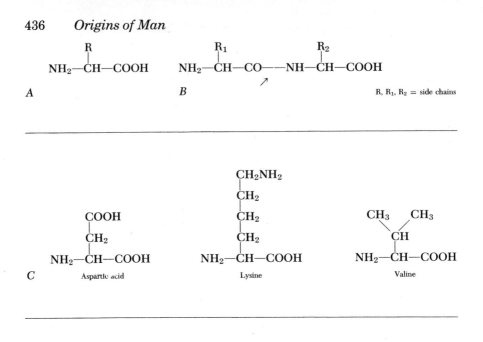

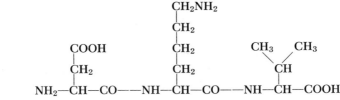

Fig. 25.3.   Amino acids, peptide bonds, and peptides. (*A*) General formula for an amino acid. (*B*) Two amino acids joined by a peptide bond. The peptide bond is indicated by the arrow. A molecule of water (H_2O) is lost when the amino acids are joined in peptide linkage. (*C*) Three common amino acids, one of which is acidic (aspartic acid, with two COOH groups), one basic (lysine, with two NH_2 groups), and one neutral (valine, with an aliphatic side chain). (*D*) Aspartic acid, lysine, and valine bound to form a small peptide, aspartyllysylvaline. Aspartyl is called the amino terminal, NH_2-terminal, or N-terminal amino acid residue, and valine is called the carboxyl terminal, COOH-terminal, or C-terminal residue.

protein synthesis, messenger RNA and transfer RNA. The hereditary information is passed to the site of protein synthesis from the DNA in the nucleus of the cell. A strand of messenger RNA is made on which the particular sequence of letters, organic bases, has been imposed by a section of the DNA. Such a strand, carrying instructions from the DNA, leaves the nucleus and enters the cytoplasm. These instructions are the sequences in which amino acids are to be attached to one another to produce a complete and functioning protein. The transfer RNA, which is believed to be a relatively small molecule, is necessary for activation of the amino acids. Each amino

TABLE 25.2
Amino Acids Commonly Found in Proteins

Amino acid	Abbreviation
Alanine	Ala
Arginine	Arg
Asparagine	Asn
Aspartic acid	Asp
Cysteine	Cys
Glutamic acid	Glu
Glutamine	Gln
Glycine	Gly
Histidine	His
Isoleucine	Ileu
Leucine	Leu
Lysine	Lys
Methionine	Met
Phenylalanine	Phe
Proline	Pro
Serine	Ser
Threonine	Thr
Tryptophan	Try
Tyrosine	Tyr
Valine	Val

acid becomes attached to the end of a transfer RNA molecule. This step produces an array of activated amino acids but no sequence. The messenger RNA becomes associated with the activated but unordered amino acids. The amino acids are put into specific sequence according to the sequence of bases which the messenger RNA is bringing to the ribosomes from the DNA. The amino acids are linked into a protein in the sequence determined ultimately by the DNA. The transfer RNA is dropped off, and a complete, functioning protein molecule is produced. The whole process is illustrated diagrammatically in Fig. 25.4.

We have been speaking of the instructions from DNA which are required for the synthesis of proteins. But how is the information coded? The sequence of nucleotides in DNA and RNA and the twenty amino acids from which proteins are built must be related in some manner. Mathematical postulates about these relationships were made by a number of investigators. The basis for the code that is generally accepted now was established by Crick and his co-workers. Three nucleotides or letters in sequence, a codon, signify one amino acid. This code is nonoverlapping, that is, the words in the genetic message are read in linear fashion from DNA and any one nucleotide belongs to only one three-letter word in the message. The code is called degenerate because a single amino acid may be specified by more than one sequence of three nucleotides. It is believed that such redundancy

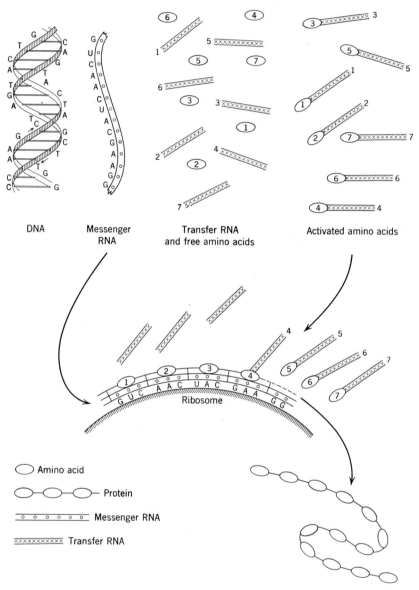

DNA

Messenger
RNA

Transfer RNA
and free amino acids

Activated amino acids

Ribosome

◯ Amino acid

◯—◯—◯— Protein

Messenger RNA

Transfer RNA

Complete functional protein

Fig. 25.4.   Simplified scheme of the manner in which DNA is responsible for the sequence of amino acids in a protein. A, adenine; G, guanine; C, cytosine; T, thymine; and U, uracil. Messenger RNA is complementary to one of the chains of DNA.

reduces the possibility of error in translating the coded message from DNA to protein.

The three-letter code words, the triplets or codons, in RNA have been experimentally determined for a number of amino acids (Table 25.1). The exact sequence of nucleotides, or letters in the code, has not been determined for most of these, but the sequence is believed to be significant. The code for isoleucine, for example, is 1A and 2U's. It is not known whether the sequence is AUU, UUA, or UAU. The codons in messenger RNA correspond to the sequence of bases in DNA and directly determine the sequence of amino acids in a protein. There is a one-to-one correspondence between each amino acid and a nucleotide triplet. A change in a single codon will result in the incorporation of a different amino acid into a protein. A difference of one amino acid in a protein may have an effect sufficient to produce a genetic polymorphism. In man the best example of such a polymorphism, one which has a profound effect on the organism, is that of hemoglobin A and hemoglobin S (Chapter 29).

What Is a Gene?

The code brings us to a fundamental question in biology and genetics—What is a gene? A number of definitions, operational and abstract, have been proposed in the past. Today it is not considered reasonable to think of a single entity as the unit of heredity. Three units, units of genetic action, have been recognized and named: the cistron, the recon, and the muton. The cistron is the smallest unit which functions as a gene; the recon is the smallest segment of the chromosome (DNA) that recombines during crossing over; the muton is the smallest unit that mutates.

In order to understand these definitions we must discuss crossing over and recombination. When describing meiosis we noted that at a certain stage, specifically between pachytene and diplotene phase (Fig. 23.4), the chromosomes of an homologous pair become tightly intertwined or synapsed (Chapter 23). During this stage material from one chromosome strand is exchanged with material from another. The smallest unit, amount, or piece of chromosome which exchanges or crosses over and recombines is the recon. Two units that recombine after crossing over of the chromosomes must occupy different places on the chromosome and hence be nonallelic. Two units that do not recombine must occupy the same site on homologous chromosomes and be allelic (Fig. 25.5). The determination of whether genes are in one recon or not depends upon the availability of a system in which recombinants can be detected. If the frequency of recombination is low or if the organism studied has many chromosomes, it is very difficult to prove recombination unless a very large number of progeny or families are examined. This is the principal reason that almost all the evidence for this

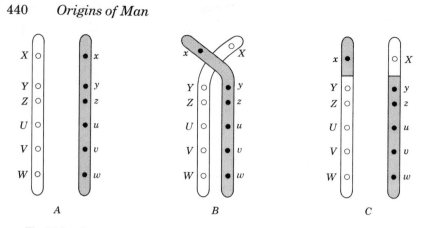

Fig. 25.5.   Crossing over and recombination shown schematically for a pair of homologous chromosomes. (*A*) Homologous chromosomes and genetic loci on these chromosomes; (*B*) crossing over; (*C*) recombined chromosomes. *X* and *x*, *Y* and *y*, *Z* and *z*, etc., are the alleles. *X* will never recombine with *x*, *Y* never with *y*, etc. If two loci are very close together on 1 chromosome, they will not recombine independently. They will cross over as a unit.

mechanism comes from studies on rapidly reproducing systems such as bacteria and fruit flies.

The cistron has been defined as a functional unit on the basis of experiments with bacterial viruses and by borrowing some terminology from organic chemistry. A cistron is a genetic unit of function, and it is divisible into recombination units, the recons. A simple genetic test, usually possible only with rapidly reproducing organisms, will tell us if two recons belong to one functional unit or not. Two loci or two recons are in the *cis* position when they are on the same chromosome, and they are in the *trans* position when they are on different chromosomes of an homologous pair (Fig. 25.6). The terms *cis* (Greek: on this side) and *trans* (Greek: across) have been taken from organic chemistry. In bacterial viruses for example, where normal and various mutant strains can be compared, a cistron is defined by a *cis-trans* test. If the loci in the mutants produce a functioning phenotype when they are in the *cis* position and if they fail to produce a functioning phenotype in the *trans* position (in a double heterozygote, a mutant heterozygous at two loci), the two loci are in one cistron (Fig. 25.6).

The muton is the mutational unit, the smallest element in which an alteration leads to a change in the phenotype, in the gene product. Our discussion of the genetic code suggests that a single change in the code word may change the amino acid that is coded. A mutation may be considered as a change in a single nucleotide pair in the DNA molecule. A muton thus may be no larger than a single pair of nucleotides in DNA.

Today we cannot speak of the gene as a single entity; rather we must

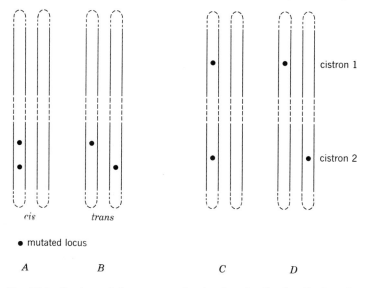

Fig. 25.6.   Portions of chromosomes showing two functional units, two cis-trons. In (A) the two loci are in the *cis* position and both cistrons are functional. In (B) the two are in the *trans* arrangement; the phenotype is not functional, but cistron 1 is functional. The *cis-trans* test defines a cistron. In (C) and (D) both cistrons are functional. (Adapted from Anfinsen, 1959.)

think of three kinds of genetic units—cistrons, mutons, and recons. The functional unit, the cistron, actually fits well the older concept of the gene. The genetic element which produces, for example, normal hemoglobin A in man is probably a cistron. This cistron has within it many mutons, muta-tional sites. A change in any one of these may lead to a change in the amino acid sequence in the protein end product. In the case of human hemoglobin, a change in 1 amino acid out of a total of 146 leads to a hemoglobin with new properties—hemoglobin S (Chapter 29).

The concepts of recons, cistrons, and mutons imply that genes are prob-ably segments of DNA. Unfortunately this approach to genetic analysis has few immediate applications in human genetics. We cannot tell from human data how many, if any, recons there are in a cistron. Yet there is no reason to suppose that the genetic material of man and the other primates behaves differently from that of other organisms. Furthermore these concepts may be helpful in unraveling the apparently complex modes of inheritance in at least one case in man—the Rh blood group system—which will be discussed in the next chapter.

The basic unit of heredity, commonly called a gene, must be viewed today as a segment of a DNA molecule. This segment contains an array of nucleotides in a fixed sequence along the axis of the DNA molecule. The experimental work in this area of biology is producing new results and new interpretations at a rapid rate. Whether we shall hold these same views tomorrow we cannot tell.

Chromosome Morphology

When we view the chromosomes as structures with significant morphological characteristics, we can use them in the investigation of the phylogeny of organisms. The karyotype—the number and types of chromosomes in a cell—is characteristic for a species. The number of chromosomes alone does not specify the karyotype. The number and the sizes and shapes of the chromosomes make up the karyotype of an individual or of a species. The karyotype is closely associated with the genetic makeup, the genome, of an organism. When there is a change in number or form of the chromosomes, that is, in the karyotype, there is a rearrangement in the genetic material. Such rearrangements will generally result in changes in the organism. Rearrangements, changes and breaks in the chromosomes, are often lethal, and those that are not lethal are usually deleterious. It is reasonable that a deviation from the average karyotype of a species which is well adapted to its environment will have strong selection pressure against it. A deviant karyotype will be eliminated quickly from the population in which it appears. We believe that karyotypes are quite stable, and a successful change in the karyotype of a species is rare. A single karyotype is characteristic, in general, of each species, but sometimes unfortunately different species and indeed different genera have the same, or at least indistinguishable, karyotypes.

Analysis of the karyotypes of man and other primates was until recently very difficult because of the technical problems involved, but these problems have been solved. Somatic cells of *Homo sapiens* can be maintained in tissue culture, and chromosome preparations can be made. The methods which have been developed make it feasible to study the chromosome complements of a large number of individuals. These methods are equally applicable to studies of chromosomes of nonhuman primates. Some of the references listed at the end of the chapter describe lucidly the methods used for determining chromosome numbers.

Man has 46 chromosomes (Fig. 25.7). For a considerable period of time (1923–1956), it was believed that man had 48 chromosomes, that is, 24 pairs, 23 pairs of autosomes and 1 pair of sex-determining chromosomes. Despite the occasional reports which suggested that 48 was not the correct number, it was not until 1956 that the correct number 46, that is, 23 pairs, 22 pairs of autosomes and 1 pair of sex-determining chromosomes, was demonstrated.

The chromosomes which we can analyze morphologically are of two kinds, somatic mitotic and germinal meiotic. It would be to our great

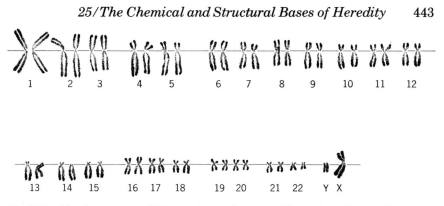

Fig. 25.7.   The chromosomes of *Homo sapiens*, ♂ karyotype. (Drawn from photograph, courtesy of H. P. Klinger.)

advantage to study the meiotic chromosomes, for there are only 23 of them. But they are very tiny, and it is difficult to determine their morphology. Somatic chromosomes from tissue cultures of primate cells are now used primarily. Since the mitotic chromosomes of somatic cells have the diploid number of chromosomes $(2N)$, it is probable that analysis is twice as difficult as analyzing meiotic chromosomes which contain only one-half as many chromosomes, the haploid number (N). Fortunately the mitotic chromosomes are relatively large. We usually find a number of preparations in which the chromosomes will be paired. This is an aid to morphological analysis of the chromosomes.

Each chromosome (Fig. 25.8) consists of two chromatids attached to each other at a constriction, the centromere. The centromere will divide at the beginning of anaphase, and this division separates the two chromatids into daughter chromosomes (Chapter 23). If there is another constriction in the chromosome, the region distal to this so-called secondary constriction is called a satellite. If satellites are present, they are a constant feature of the karyotype of an individual and perhaps of a species.

Each chromosome is characterized by its length and by the position of the centromere (Fig. 25.8). Chromosomes are metacentric (M) if the two arms on each side of the centromere are almost equal in length. If the centromere is at one end or very close to one end of a chromosome, the chromosome is called acrocentric (A). Human chromosomes do not have terminal centromeres, but chromosomes of other species do. Telocentric chromosomes (T) are those with terminal centromeres. Chromosomes with the centromere between the median and terminal positions are called sub-terminal (S). It is often difficult to distinguish S and M or A and S chromosomes. Some cytogeneticists use a classification in which two types are distinguished: acrocentric and metacentric.

To analyze a karyotype, cells in metaphase are found and photographed with a camera attached to the optical system of a light microscope. These

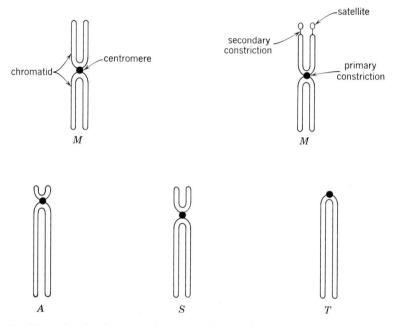

Fig. 25.8.   The classification of chromosomes (at metaphase) according to the position of the centromere—*M*, metacentric; *A*, acrocentric; *S*, subterminal; and *T*, telocentric.

cells are the ones in which the chromosomes are most clearly distinguishable and the chromatids visible. The photographs are enlarged, and the chromosomes are cut and pasted on a piece of paper, in order. The order of placing the chromosomes has been agreed upon by cytogeneticists. The longest chromosome is placed at the beginning of the sequence and the shortest at the end (Fig. 25.7). If two pairs of chromosomes are the same length, the more metacentric pair is placed first, the more acrocentric second. The 46 human chromosomes have been classified into seven major morphological groups (Table 25.3 and Fig. 25.7). The classification is known as the Denver system, for the system of nomenclature was agreed upon in Denver, Colorado, in 1960, by the leading students of human chromosomes.

Many abnormal human karyotypes have been described. They are almost always found in individuals with severe inborn physical and mental abnormalities. These findings suggest that deviations from the normal karyotype of the species are extremely unlikely to be maintained. It is possible that advantageous alterations in karyotypes occur, alterations too small to be discovered by present methods. We do not routinely examine the karyotypes of normal individuals unless we wish to have standards against which to compare karyotypes from abnormal persons. Since the discovery that some major clinical disorders have concomitant chromosome abnormalities, a

TABLE 25.3

Classification of Human Somatic and Sex Chromosomes—the Denver System

Group	Chromosome number[a]	Ratio of arm lengths (long:short)	Morphological classification[b]
A	1	1.1	M
	2	1.5– 1.6	M
	3	1.2	M
B	4	2.6– 2.9	S
	5	2.4– 3.2	S
C	X	1.6– 2.8	M or S
	6	1.6– 1.8	M
	7	1.3– 1.9	M
	8	1.5– 2.4	S
	9	1.8– 2.4	S
	10	1.9– 2.6	S
	11	1.5– 2.8	S
	12	1.7– 3.1	S
D	13	4.8– 9.7	A
	14	4.3– 9.5	A
	15	3.8–11.9	A
E	16	1.4– 1.8	M
	17	1.8– 3.1	S
	18	2.4– 4.2	S
F	19	1.2– 1.9	M
	20	1.2– 1.3	M
G	21	2.3– 6.8	A
	22	2.0– 6.0	A
	Y	2.9–∞ (5.9)[c]	A

[a]Chromosomes are numbered according to size: 1 is the longest and 22 is the shortest. For the somatic chromosomes, the numbers refer to homologous pairs of chromosomes in mitotic figures or karyotypes.

[b]M, metacentric; A, acrocentric; S, subterminal.

[c]Value of 5.9 was determined from mitotic figures of six normal individuals.

large number of individuals with congenital malformations and diseases have had their chromosomes examined. Many of these individuals proved to have unusual karyotypes—Abnormal chromosome morphology and extra chromosomes are common discoveries.

Despite the apparent deleterious effects karyotype alterations produce, many advantageous chromosomal rearrangements must have occurred during the course of primate evolution. Karyotypes vary widely among mammals and among the Primates. These differences must have been selected for during the evolutionary history of our order.

Mechanics of Chromosomal Evolution

A number of rules of chromosome mechanics have been deduced from the study of many thousands of organisms and cells. These rules are descriptive generalizations which have been found to hold in many groups of organisms. The change in karyotypes observed within a group of related species may often be used to deduce the course of evolution in the group, for such changes follow these rules with considerable consistency. The number of chromosomes may change; the shape or form may be altered; or both number and form may change together. Almost all changes in number and form are due to breakage of one or more of the chromosomes. We know from work with nonhuman organisms commonly used in genetics laboratories, *Drosophila* for example, that breakage of chromosomes occurs spontaneously in measurable but low frequency. Evidence from clinical human genetics supports the suggestion that breakage occurs spontaneously in man.

Many rearrangements that have some significance in evolution of the chromosomes involve two simultaneous events, two simultaneous breaks in the same chromosome of the same cell. This seldom happens, yet the probability that it will happen in a population over many generations is sufficiently high to explain the observed differences among the karyotypes of the Primates. The broken chromosomes must also repair themselves in order for a new morphological configuration to develop. The evolutionary process affects the chromosomes as well as other features of the organism. Closely related species may be expected to have more similar karyotypes than distantly related. We know that many karyotypes occur among the Primates (Fig. 25.9–25.14). When the quantity of chromosomal material of the various primates is determined, by measuring either the total length of the chromosomes or the quantity of DNA in the cells, almost no differences are found. The amount of DNA in the cells is not significantly different among the various primates which have been examined. This suggests that differences in chromosome numbers may be due simply to the redistribution of the genetic material. The manner in which this redistribution may occur is of considerable interest to us, for it appears that karyotype evolution occurs only after chromosomes have broken and rearranged themselves.

The mechanics of chromosome rearrangement, after breakage and fragmentation, can be classified as follows: simple deletion or insertion, symmetrical or asymmetrical translocation, centric or tandem fusion, ring formation, duplication, and pericentric or paracentric inversion. Some of these mechanical processes are illustrated in Fig. 25.15.

Deletion of a portion of a chromosome must be a rare event, for if the deleted segment contained alleles vital to the survival of the organism the animal would not live to perpetuate the new chromosome type. Insertion is the reverse of deletion. Translocation, exchange of material between chromosomes, is more likely to produce a surviving configuration than is

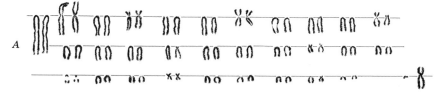

Fig. 25.9.   Chromosomes of Prosimii. Karyotypes of (*A*) *Tupaia glis*, ♂, 2N = 62; (*B*) *Tarsius bancanus*, ♀, 2N = 80; and (*C*) *Nycticebus coucang*, ♂, 2N = 50. (Drawn from photographs, courtesy of H. P. Klinger.)

the deletion of a chromosome or of a segment of a chromosome. When two nonhomologous chromosomes break and exchange material, reciprocal translocation occurs. Reciprocal translocations are asymmetrical when the pieces broken from the two chromosomes fuse so that one of the new chromosomes does not include a centromere. One chromosome with two centromeres (dicentric) and one fragmentary chromosome will appear. This is not believed to be a stable arrangement, and dicentric chromosomes are not known to be typical of the karyotype of any species. Dicentric chromosomes do not segregate normally at anaphase. Symmetrical translocation between two chromosomes produces two new ones that will segregate normally at anaphase. If there is an advantage conferred by the new arrangement, natural selection will lead to its becoming common to the species.

A chromosome that breaks more than once may lead to a variety of new chromosomal forms depending upon the way the broken pieces fuse. A ring chromosome will result if the ends of the chromosome fuse with each other. Although ring chromosomes have been found, they have not been found in karyotypes of normal individuals. If the broken ends of chromosomes fuse with the fragments that do not contain a centromere, inversion results. Pericentric inversion occurs if the centromere is included in the inversion. It is possible, for example, to convert an acrocentric chromosome

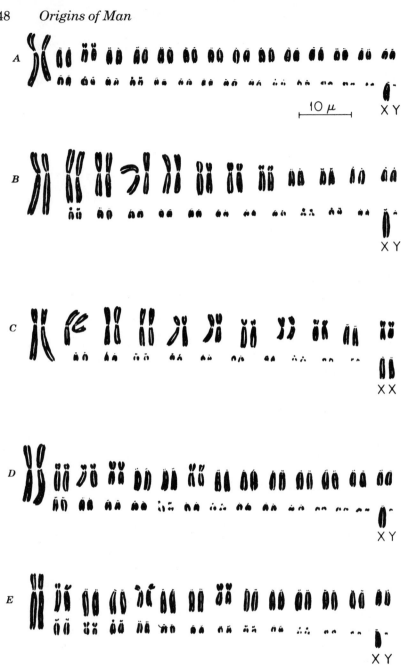

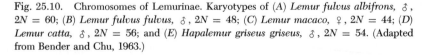

Fig. 25.10.   Chromosomes of Lemurinae. Karyotypes of (A) *Lemur fulvus albifrons*, ♂, 2N = 60; (B) *Lemur fulvus fulvus*, ♂, 2N = 48; (C) *Lemur macaco*, ♀, 2N = 44; (D) *Lemur catta*, ♂, 2N = 56; and (E) *Hapalemur griseus griseus*, ♂, 2N = 54. (Adapted from Bender and Chu, 1963.)

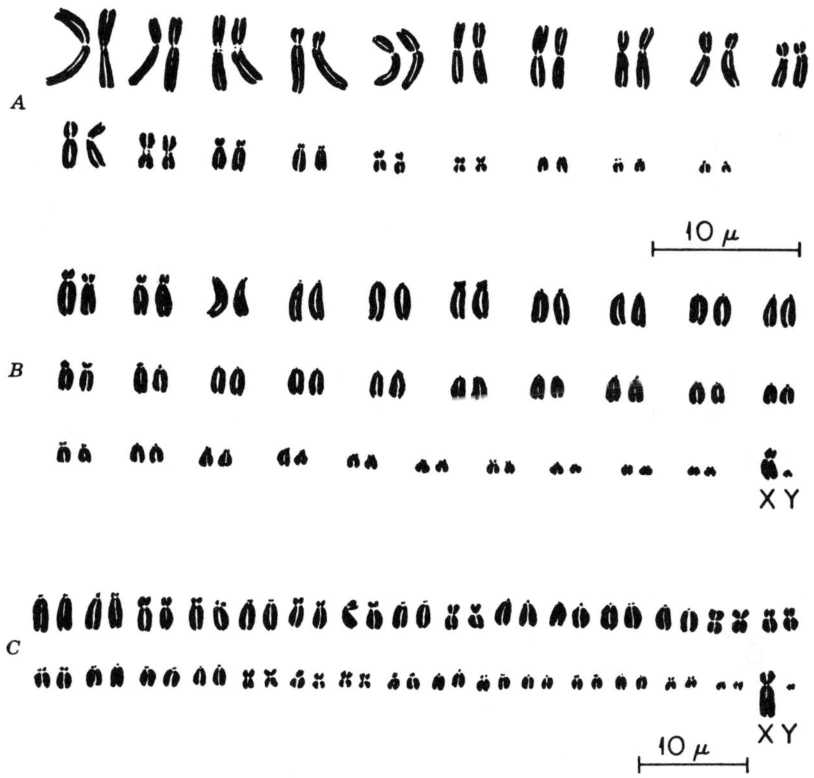

Fig. 25.11. Chromosomes of African Lorisiformes. Karyotypes of (A) *Galago senegalensis*, ♀, 2N = 38; (B) *Galago crassicaudatus*, ♂, 2N = 62; and (C) *Perodicticus potto*, ♂, 2N = 62. (Adapted from Bender and Chu, 1963.)

to a metacentric chromosome by pericentric inversion. Paracentric inversion occurs when a piece of chromosome breaks out, rotates, and rejoins the original chromosome without involvement of the centromere.

Symmetrical translocation may lead to a reduction in chromosome number through fusion after the breakage has occurred. Occasionally very tiny chromosomes are produced in this way and may be lost without damage to the organism. Centric fusion occurs when translocation of the long arm of an acrocentric chromosome to a nonhomologous chromosome results in a single chromosome which contains most of the genetic material of the two originals. Tandem fusion occurs when the translocation is between a piece broken from the tip of a chromosome and one that breaks close to the centromere.

The number of chromosomes may be increased in three ways—nondisjunction, centromeric misdivision, and polyploidy. Nondisjunction of single chromosomes, known as trisomy, will add one chromosome to a karyotype.

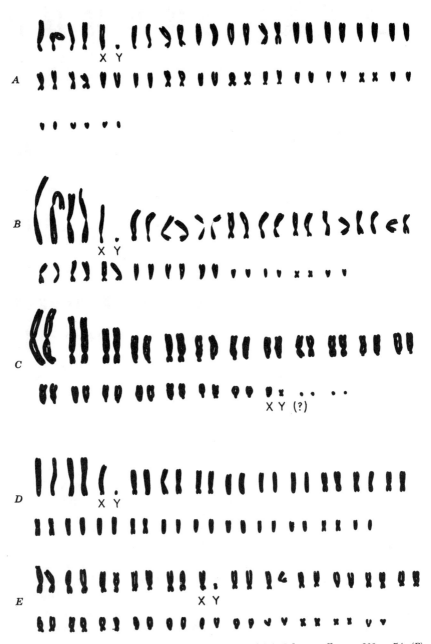

Fig. 25.12.   Chromosomes of Ceboidea. Karyotypes of (A) *Cebus apella*, ♂, 2N = 54; (B) *Saimiri sciurea*, ♂, 2N = 44; (C) *Alouatta seniculus*, ♂, 2N = 44; (D) *Callicebus cupreus*, ♂, 2N = 46; (E) *Callithrix chrysoleucos*, ♂, 2N = 46. (Adapted from Bender and Chu, 1963.)

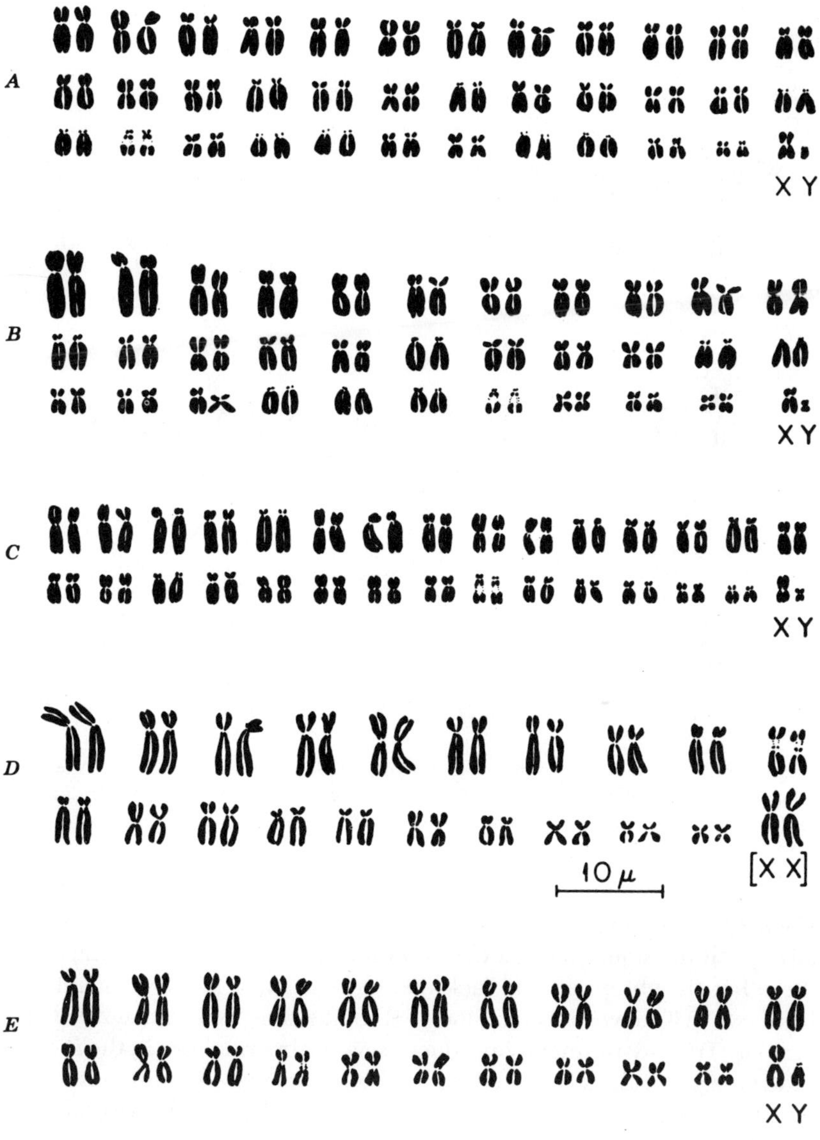

Fig. 25.13.   Chromosomes of Cercopithecoidea. Karyotypes of (A) *Cercopithecus mitis*, ♂, 2N = 72; (B) *Cercopithecus mona*, ♂, 2N = 66; (C) *Cercopithecus aethiops*, ♂, 2N = 60; (D) *Papio cynocephalus*, ♀, 2N = 42; and (E) *Colobus* [*polykomos*], ♂, 2N = 44. (Adapted from Bender and Chu, 1963.)

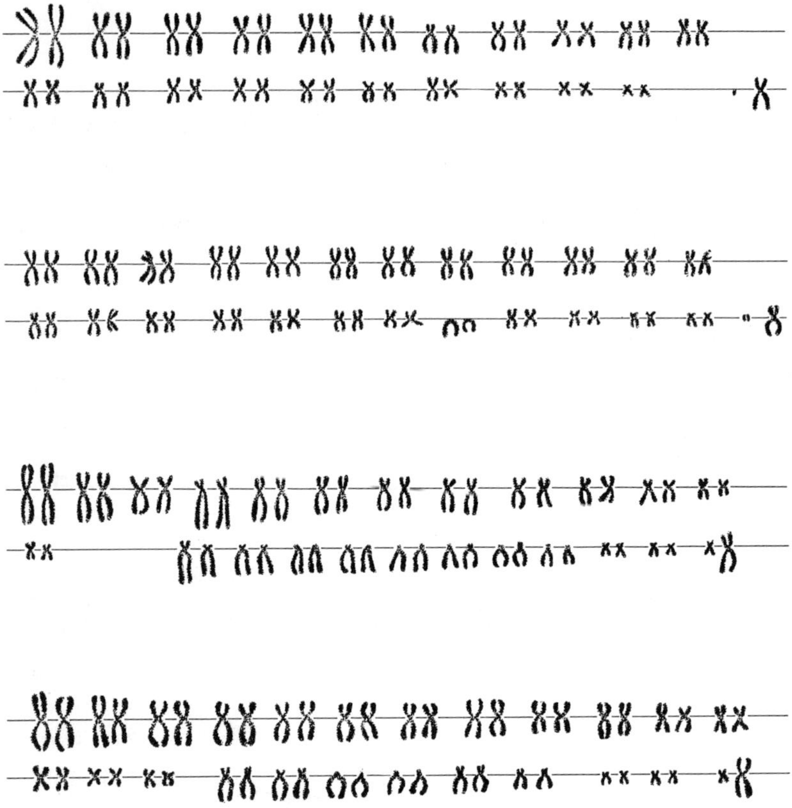

Fig. 25.14.   Chromosomes of Pongidae. Karyotypes of (A) *Hylobates lar*, ♂, 2N = 44; (B) *Hylobates syndactylus*, ♂, 2N = 50; (C) *Pongo pygmaeus*, ♂, 2N = 48; (D) *Pan gorilla*, ♂, 2N = 48. (Drawn from photographs, courtesy of H. P. Klinger.)

When two homologous (sister) chromosomes fail to move to opposite poles during cell division, an extra chromosome will appear in one daughter cell. The other daughter cell will lack one chromosome. All descendant cells of these two will have fewer or more chromosomes than the normal diploid number. Trisomy frequently produces severe abnormalities in the individual. Centromeric misdivision may also lead to an increase in chromosome number. The centromeres may split across rather than lengthwise at meiosis. The chromosome is broken in two, and each new chromosome has a functioning centromere. Polyploidy, increase by complete sets of the haploid number (N) of chromosomes (3N, 4N, etc., chromosomes), is highly unlikely to occur in the Primates. Animals with a chromosomal sex determining mechanism almost always are infertile if additional X- or Y-chromosomes are present. Human polyploid forms would be unlikely to reproduce themselves.

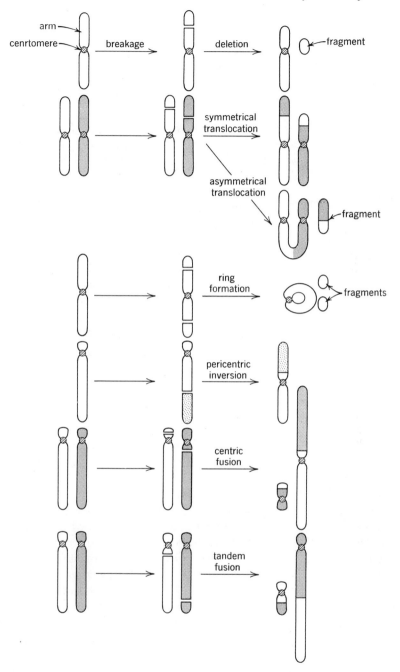

Fig. 25.15.   Diagrammatic representation of various ways in which chromosomes may rearrange themselves.

Primate Chromosomes

Much work has been done in the last few years to increase our knowledge of the chromosomes of the Primates. Some chromosomal characteristics of the major taxa are worth summarizing (Tables 25.4-25.7), for they indicate the extent to which evolution of the karyotypes has occurred during primate evolution.

A few karyotypes of Tupaiiformes have been published (Fig. 25.9). These are summarized in Table 25.4. That there are two diploid numbers ($2N = 60$ and $2N = 62$) and two karyotypes from specimens identified as *Tupaia glis* illustrates some of the problems raised by current notions of primate classification. Each of the different diploid numbers and karyotypes is correct— each was found in several animals, and the results were confirmed by repeated analyses in more than one laboratory. The cytogenetic data suggest that chromosomal polymorphism exists in the species *Tupaia glis*. Chromosomal polymorphism does occur among mammals. The most notable example is the population of common shrews near Harwell, England. It is also possible that the animals recognized as *T. glis* actually are members of two species very similar morphologically, perhaps sibling species reproductively isolated through karyotype evolution or some other mechanism.

Lemuriformes have a large number of different chromosome numbers and karyotypes (Table 25.4 and Fig. 25.10). There is considerable variation within the genus *Lemur*. The diploid numbers vary from $2N = 44$ to $2N = 60$, and chromosome numbers vary within a single presently recognized species. The morphology of the chromosomes is most variable within the subfamily Lemurinae. The lengths of the chromosomes range from 14μ to 0.3μ ($1\mu = 10^{-4}$ cm). This variability in size is greater than that displayed by other primate chromosomes. The chromosome with a length of 0.3μ is so small that it is just within the resolving power of the best light microscopes; it is the smallest primate chromosome yet observed. Chu and Bender have described the possible ways in which this variety of karyotypes may have evolved. Comparisons of karyotypes of the living primates may not reveal all the mechanisms which led to the observed similarities and differences, but they do enable us to suggest some minimum hypotheses about transformations of karyotypes, provided we know that two species whose karyotypes are being compared are related. *Lemur variegatus* has 46 chromosomes ($2N = 46$). It is the only Old World primate with the same diploid number as man but with very different karyotype. Although both *H. sapiens* and *L. variegatus* are primates, they are not closely related, and comparisons of their karyotypes are not apt to answer the questions we are asking.

Chu and Bender discussed the relation of the karyotypes of *L. fulvus fulvus* and *L. macaco*. The former has a diploid number $2N = 48$, the latter a diploid number $2N = 44$. Chromosome pairs of the two species generally correspond in gross morphology. *L. macaco* has two pairs of small meta-

TABLE 25.4
Chromosomes of Prosimii [a]

Species	Diploid number (2N)	Classification[b] Somatic chromosomes			Sex chromosomes	
		A	M	S	X	Y
Tupaiiformes						
Tupaia glis	62	48	12	–	M	A
Tupaia glis	60	46	14	0	nd[c]	nd
Urogale everetti	44					
Tarsiiformes						
Tarsius bancanus	80	66	14	–	nd	nd
Lemuriformes						
Microcebus murinus	66	64	0	2	nd	nd
Cheirogaleus major	66					
Lemur mongoz	60	54	0	4	A	A
Lemur fulvus albifrons	60	54	0	4	A	A
Lemur fulvus rufus	60	54	0	4	A	A
Lemur fulvus collaris[d]	52	38	8	4	M	A
Lemur fulvus collaris[e]	48					
Lemur fulvus fulvus	48	30	10	6	A	A
Lemur variegatus	46	26	14	4	M	A
Lemur macaco	44	22	12	8	A	A
Lemur catta	56	44	6	4	A	A
Hapalemur griseus olivaceous	58	52	2	4	nd	nd
Hapalemur griseus griseus	54	42	4	6	A	A
Propithecus verreauxi verreauxi	48					
Propithecus verreauxi coquereli	48					
Lorisiformes						
Galago crassicaudatus	62	54	0	6	S	A
Galago senegalensis	38	6	20	10	S	A
Perodicticus potto	62	36	12	12	S	A
Nycticebus coucang	50	0	48	–	M	M

[a]Based on Bender and Chu (1963), Klinger (1963), and other sources.

[b]Chromosomes are classified as either A, acrocentric chromosomes, M, metacentric, and S, subterminal, or A, acrocentric, and M, metacentric (see p. 443).

[c]nd = not determined and included in somatic count.

[d]White-collared variety.

[e]Red-collared variety.

centrics which do not correspond to any pairs of *L. fulvus fulvus*. *L. fulvus fulvus* has four pairs of acrocentrics which have arm lengths approximately the same as the total length of the arms of the small metacentric pairs of *L. macaco*. A reciprocal translocation near the centromeres (centric fusion) of two nonhomologous acrocentric chromosomes produces a large metacentric chromosome and a small fragment with a centromere. This fragment may be lost. The transformation of the karyotype of *L. fulvus fulvus* into that of *L. macaco* could have occurred in this manner. Centric fusion occurs fairly often among other mammals, and it is easier to lose a centromere than to gain one. Chu and Bender have suggested that reduction in chromosome number through centric fusion played an important part in karyotype evolution among the Prosimii and, indeed, among all the Primates. Karyotypes of other Lemurinae may be derived from each other in the same way as the karyotype of *L. macaco* may be derived from that of *L. fulvus fulvus*. Chu argued that, if centric fusion and reduction in chromosome number was the mechanism of karyotype evolution among lemurs, the observed karyotypes could most easily be derived from an ancestral type with 64 or 66 acrocentric chromosomes. Shortly after he said this he had the opportunity to examine the karyotypes of two members of the Lemuriformes, *Microcebus* and *Cheirogaleus* of the subfamily Cheirogaleinae. Each has a karyotype with 66 chromosomes, 64 of which are acrocentrics.

Chu and Bender suggested a scheme of prosimian phylogeny based upon the possible derivations of one karyotype from another, and this scheme is shown in Fig. 25.16. There is nothing in this scheme that is inconsistent with established views of primate phylogeny. These analyses of karyotypes have suggested that a revision of the taxonomy of the subfamily Lemurinae and, especially, of the genus *Lemur* is called for.

Analysis of the chromosomes of the Lorisiformes suggests that the karyotypes of two species of the genus *Galago* (*G. senegalensis* and *G. crassicaudatus*, Fig. 25.11) may be derived from the same hypothetical form from which the karyotypes of *Cheirogaleus* and *Microcebus* may come (Fig. 25.16). The available information on the chromosomes of the Lorisiformes is listed in Table 25.4.

Chromosomes of *Tarsius* have been examined by Klinger. The diploid number $2N = 80$ is the largest chromosome number recorded for any mammal. Most of the chromosomes, 66 of the 80, are acrocentric. The sex chromosomes have not yet been identified.

The chromosomes of the Ceboidea (Table 25.5 and Fig. 25.12) are less variable than those of the Lemuriformes. As far as we know, the morphological variation within a genus is negligible, and there is no variation in diploid number within a genus. Chu and Bender have pointed out that the least specialized ceboids appear to have the largest number of chromosomes and the largest number of acrocentrics, whereas the more specialized, *Ateles* for example, have the smallest number of acrocentrics and the smallest diploid numbers.

The relatively low degree of variability in the morphology of ceboid chromosomes may be a reflection of the restricted ecozones in which these animals live. If selective pressures are strict, the chance is small that new configurations of chromosomes will survive. There appears to be an association between highly specialized animals and a small diploid number among the Ceboidea and, perhaps, among other groups of primates. A highly specialized animal is assumed to live in a very restricted ecozone. Survival of the species depends upon maintenance of adaptive relationships with the

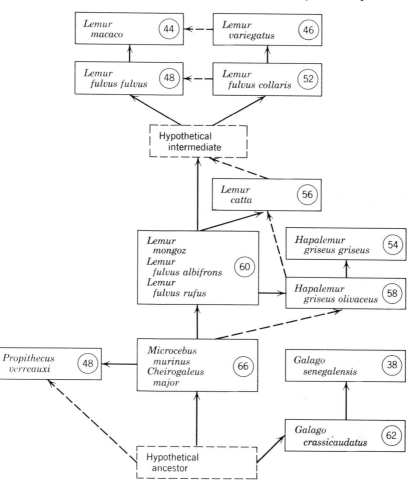

Fig. 25.16. Scheme of prosimian phylogeny based on evolution of karyotypes. The numbers in circles are the diploid numbers (2N) of the species. (Adapted from Chu and Bender, 1962.)

environment. A smaller diploid number means that more groups of genes segregate together at meiosis. This lowers the chance that the genome of offspring will deviate from that of the parental generation.

Chromosomes of the Cercopithecoidea (Table 25.6 and Fig. 25.13) are quite variable, with several groups readily distinguishable. Among Cerco-pithecinae the terrestrial forms—*Papio, Macaca, Cercocebus*—have very similar karyotypes and identical diploid numbers 2N = 42. The existence of viable, fertile, hybrid offspring of baboons and macaques (*Papio* × *Macaca*) suggests that the gene complements of the two species, the chro-mosomal material, are extremely similar. They are probably more alike than

TABLE 25.5
Chromosomes of Ceboidea
(From Bender and Chu, 1963.)

Species	Diploid number (2N)	Classification[a]				
		Somatic chromosomes			Sex chromosomes	
		A	M	S	X	Y
Cebidae						
Cebus apella	54	26	6	20	A	A
Cebus capucinus	54	27	6	20	A	nd[b]
Saimiri sciurea	44	12	16	14	S	A
Alouatta seniculus	44	30	6	6	A	M
Ateles paniscus	34	2	12	18	M	A
Ateles belzebuth	34	2	12	18	M	A
Ateles geoffroyi	34	2	12	18	M	A
Lagothrix ubericola	62					
Aotus trivirgatus	54					
Callicebus cupreus	46	24	10	10	S	A
Callimico goeldii	48	16	2	30	nd	nd
Pithecia pithecia	46					
Cacajao rubicundus	46					
Callithricidae						
Callithrix chrysoleucos	46	10	4	30	S	M
Leontideus illigeri	46	10	4	30	S	M

[a]A, acrocentric chromosomes; M, metacentric; and S, subterminal.
[b]nd = not determined and included in somatic count.

the complements of horses (*Equus caballus*) and asses (*Equus asinus*). Horses have 64 chromosomes and asses have 62. The two species are interfertile and produce viable, useful offspring—mules. Mules have 63 chromosomes; horse gametes have 32, ass gametes 31. The chromosomes of the hybrid are probably not sufficiently homologous to form pairs successfully during meiosis. The separation of the chromosomes at anaphase probably does not produce gametes with a normal complement of genes, or the normal complement is produced so rarely that the hybrid is almost always sterile. New combinations of chromosomes will serve as isolating mechanisms in the formation of a species. A zygote is not apt to be viable unless the genic complement it possesses is balanced. A hybrid produced by the gametes of two species will develop into a functioning "normal" animal if the two species are very similar in their gene complements. We have argued that baboons and macaques are, on genetic grounds, clearly members of the same genus (Chapter 4). The preceding discussion of the mechanics of chromosomal evolution and its consequences indicates why we feel that the genetic makeup of the two groups is so similar.

The chromosome numbers of the members of the genus *Cercopithecus* fall into four groups—2N = 72, 2N = 66, 2N = 60, 2N = 54. Each of

TABLE 25.6

Chromosomes of Cercopithecoidea[a]

Species	Diploid number (2N)	Classification[b]				
		Somatic chromosomes			Sex chromosomes	
		A	M	S	X	Y
Cercopithecinae						
Cercopithecus l'hoesti	72	18	28	24	M	M
Cercopithecus mitis	72	18	28	24	M	M
Cercopithecus mona	66	12	28	24	M	M
Cercopithecus nictitans	66	12	28	24	M	M
Cercopithecus aethiops	60	6	18	34	S	M
Cercopithecus diana	60					
Cercopithecus neglectus	60					
Cercopithecus nigroviridis	60					
Cercopithecus patas	54	10	18	24	S	M
Cercocebus albigena	42					
Cercocebus aterrimus	42	0	40	–	M	M
Cercocebus galeritus	42					
Cercocebus torquatus	42				S	M
Macaca cyclopis	42					
Macaca fuscata	42					
Macaca mulatta	42	0	18	22	M	M
Macaca nemestrina	42					
Papio cynocephalus	42	0	20	20	M	M
Papio gelada	42					
Papio hamadryas	42					
Papio sphinx	42					
Colobinae						
Colobus polykomos	44	0	28	14	M	A
Presbytis entellus	44	2	28	12	S	S

[a]From Bender and Chu (1963) and other sources.

[b]Chromosomes are classified as either A, acrocentric chromosomes, M, metacentric, and S, subterminal, or A, acrocentric, and M, metacentric (see p. 443).

these groups differs regularly from the next one in the series by 6 chromosomes, and each number is a multiple of 6. Another regular feature is the number of acrocentric chromosomes in three of these groups. Species of the genus *Cercopithecus* with $2N = 72$ have 9 pairs of acrocentrics; when $2N = 66$ there are 6 pairs of acrocentrics; and when $2N = 60$ there are 3 pairs of acrocentrics. Reciprocal translocation between 2 nonhomologous chromosomes can account for the transition from one of these karyotypes to the other. The species with $2N = 54$, *Cercopithecus patas*, has 5 pairs of acrocentrics.

There are three reasons why polyploidy can be ruled out as an explanation for the evolution of the regularities observed in the karyotypes of the Cercopithecinae ($2N = 42, 54, 60, 66, 72$). The amount of DNA in cells

TABLE 25.7
Chromosomes of Hominoidea
(From Bender and Chu, 1963, and Hamerton et al., 1963.)

| Species | Diploid number (2N) | Classification[a] | | | | |
| | | Somatic chromosomes | | | Sex chromosomes | |
		A	M	S	X	Y
Pongidae						
Hylobates lar	44	0	38	6	nd[b]	nd
Hylobates lar	44	0	42	–	M	A
Hylobates moloch	44				M	M
Hylobates syndactylus	50	2	46	–	M	M
Pongo pygmaeus	48	20	26	–	M	M
Pan gorilla	48	16	30	0	M	M
Pan troglodytes troglodytes	48	12	34	–	M	A
Pan troglodytes troglodytes	48	0	36	10	M	M
Hominidae						
Homo sapiens	46	10	16	18	M	A

[a]Chromosomes are classified as either A, acrocentric chromosomes, M, metacentric, and S, subterminal, or A, acrocentric, and M, metacentric (see p. 443). Chromosomes of *Hylobates lar* and *Pan troglodytes troglodytes* are classified in both systems.

[b]nd = not determined and included in somatic count.

of *Macaca mulatta* (2N = 42), *Cercopithecus patas* (2N = 54), and *C. aethiops* (2N = 60) is essentially the same. There is no evidence of repetition of chromosomes, especially the single chromosome with a secondary constriction. All of the animals are fertile; the sex-determining mechanism works.

There are few reports of karyotypes from the subfamily Colobinae (Table 25.6 and Fig. 25.13). The reported diploid numbers and karyotypes are distinct from those of the Cercopithecinae.

Pongid chromosomes are of great interest to anthropologists, as are many features of the genetics of man's closest living phylogenetic relatives. None of the apes has the same diploid number as man (Table 25.7). Although there are many similarities between the karyotypes of pongids and men (Fig. 25.14 and 25.17), it is still not clear on the basis of the chromosomes whether we should greet one ape as brother and the rest as cousins! The chromosomes of chimpanzees (*Pan troglodytes*) seem to be identical with human chromosomes except for the pair numbered 22 (Fig. 25.17) and for one extra pair; in chimpanzees 2N = 48. The chromosomes of the Asiatic apes are not as similar to man's as are those of chimpanzees.

The different diploid numbers found among the Primates probably came about through mechanical changes in a basic chromosome complement. It is believed that it is generally easier to reduce the number of chromosomes than it is to increase the number. Evidence from animals other than Primates suggests that reduction is much the more common during evolution. In the karyotypes of primates, there is an association between haploid or diploid number and number of acrocentric chromosomes—a large haploid or diploid number is associated with a large number of acrocentrics, and a small haploid or diploid number with a small number of acrocentrics. A number of investigators have pointed out that this association is consistent with the suggestion that centric fusion is an important mechanism that operated to change karyotypes during primate evolution.

The primate species with the smaller numbers of chromosomes may well have evolved farther from ancestral forms than those with larger numbers. Species with the larger numbers of acrocentric chromosomes may be younger, in an evolutionary sense, than those with few acrocentrics. The older species have had more evolutionary time in which translocation and inversion may have operated to change acrocentric chromosomes to other types. Not all of these generalizations we have made are completely consistent with each other. But the inconsistencies are more apparent than real. *Microcebus* with 66 chromosomes ($2N = 66$), 64 of which are acrocentric, cannot have an ancestral karyotype of the Lemuridae and also be a recently evolved form, as its large number of acrocentric chromosomes suggests. This should serve as a warning that hypothetical schemes are not realities. The karyotype of *Microcebus* shows us that the kind of karyotype which might produce other lemur karyotypes as descendants is part of the genetic potential of Lemuriformes. *Microcebus* is not the ancestor of the

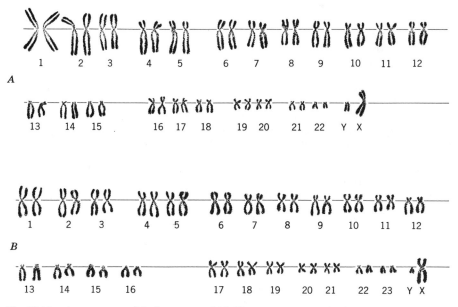

Fig. 25.17. A comparison of the karyotypes of (A) *Homo sapiens*, ♂, and (B) *Pan troglodytes*, ♂. (Drawn from photographs, courtesy of H. P. Klinger.)

Lemuriformes. None of the existing species are the ancestors of the others. But some have changed—evolved—less than others. We are using the products of a long history of evolutionary change in our attempt to reconstruct, however tentatively, the events of that history.

Suggested Readings and Sources

Anfinsen, C. B., *The Molecular Basis of Evolution*. Wiley, New York (1959).

Avery, O. T., C. M. MacLeod, and M. McCarty, Studies on the chemical nature of the substance inducing transformation of pneumococcal types. *J. Exptl. Med.* **79**, 137 (1944).

Bender, M. A, and E. H. Y. Chu, The chromosomes of Primates. *In* J. Buettner-Janusch (Ed.), *Evolutionary and Genetic Biology of Primates*, Vol. I, p. 261. Academic Press, New York (1963).

Bender, M. A, and L. E. Mettler, Chromosome studies of primates. *Science* **128**, 186 (1958).

Böök, J. A., et al., A proposed standard system of nomenclature of human mitotic chromosomes. *Am. J. Human Genet.* **12**, 384 (1960).

Chu, E. H. Y., and M. A Bender, Chromosome cytology and evolution in primates. *Science* **133**, 1399 (1961).

Chu, E. H. Y., and M. A Bender, Cytogenetics and evolution of primates. *Ann. N. Y. Acad. Sci.* **102**, 253 (1962).

Chu, E. H. Y., and B. A. Swomley, Chromosomes of lemurine lemurs. *Science* **133**, 1925 (1961).

Crick, F. H. C., The genetic code. *Sci. Am.* **207**, No. 4, 66 (1962).

Crick, F. H. C., L. Barnett, S. Brenner, and R. J. Watts-Tobin, General nature of the genetic code for proteins. *Nature* **192**, 1227 (1961).

Davidson, J. N., *The Biochemistry of the Nucleic Acids* (Fourth edition). Methuen Monograph, Wiley, New York (1960).

Dobzhansky, T., *Heredity and the Nature of Man*. Harcourt, Brace and World, New York (1964).

Ferguson-Smith, M. A., Chromosomes and human disease. *In* A. G. Steinberg (Ed.), *Progress in Medical Genetics*, Vol. I, p. 292. Grune and Stratton, New York (1961).

Ford, C. E., J. L. Hamerton, and G. B. Sharman, Chromosome polymorphism in the common shrew. *Nature* **180**, 392 (1957).

Ford, C. E., K. W. Jones, O. J. Miller, U. Mittwoch, L. S. Penrose, M. Ridler, and A. Shapiro, The chromosomes in a patient showing both mongolism and the Klinefelter syndrome. *Lancet* **1**, 709 (1959).

Fruton, J. S., and S. Simmonds, *General Biochemistry* (Second edition). Wiley, New York (1958).

Hamerton, J. L., H. P. Klinger, D. E. Mutton, and E. M. Lang, The somatic chromosomes of the Hominoidea. *Cytogenetics* **2**, 240 (1963).

Hsu, T. C., and M. L. Johnson, Karyotypes of two mammals from Malaya. *Am. Naturalist* **97**, 127 (1963).

Jones, O. W., Jr., and M. W. Nirenberg, Qualitative survey of RNA codewords. *Proc. Natl. Acad. Sci. U. S.* **48**, 2115 (1962).

Kleinfeld, R., and E. H. Y. Chu, DNA determinations of kidney cell cultures of three species of monkeys. *Cytologia (Tokyo)* **23**, 452 (1958).

Klinger, H. P., The somatic chromosomes of some primates (*Tupaia glis, Nycticebus coucang, Tarsius bancanus, Cercocebus aterrimus, Symphalangus syndactylus*). *Cytogenetics* **2**, 140 (1963).

Klinger, H. P., J. L. Hamerton, D. Mutton, and E. M. Lang, The chromosomes of the Hominoidea. *In* S. L. Washburn (Ed.), *Classification and Human Evolution*, p. 235. Aldine, Chicago (1963).

Leder, P., and M. Nirenberg, RNA codewords and protein synthesis, II. Nucleotide sequence of a valine RNA codeword. *Proc. Natl. Acad. Sci. U.S.* **52**, 420 (1964).

Leder, P., and M. W. Nirenberg, RNA codewords and protein synthesis, III. On the nucleotide sequence of a cystine and a leucine RNA codeword. *Proc. Natl. Acad. Sci. U.S.* **52**, 1521 (1964).

Lejeune, J., The 21 trisomy—current stage of chromosomal research. *In* A. G. Steinberg and A. G. Beam (Ed.), *Progress in Medical Genetics*, Vol. III, p. 144. Grune and Stratton, New York (1964).

Lejeune, J., M. Gautier, and R. Turpin, Étude des chromosomes somatique de neuf enfants mongoliens. *Compt. Rend. Acad. Sci.* **248**, 1721 (1959).

Lengyel, P., J. F. Speyer, and S. Ochoa, Synthetic polynucleotides and the amino acid code. *Proc. Natl. Acad. Sci. U.S.* **47**, 1936 (1961).

Levan, A., and W. W. Nichols, Human chromosome lengths for use in distribution studies. *Hereditas* **51**, 378 (1964).

Levinthal, C., and C. A. Thomas, The molecular basis of genetic recombination in phage. *In* W. D. McElroy and B. Glass (Ed.), *A Symposium on the Chemical Basis of Heredity*, p. 737. Johns Hopkins Press, Baltimore (1957).

Loewy, A. G., and P. Siekevitz, *Cell Structure and Function*. Holt, Rinehart and Winston, New York (1963).

McElroy, W. D., and B. Glass (Ed.), *A Symposium on the Chemical Basis of Heredity*. Johns Hopkins Press, Baltimore (1957).

Mirsky, A. E., and H. Ris, The desoxyribonucleic acid content of animal cells and its evolutionary significance. *J. Gen. Physiol.* **34**, 451 (1951).

Nirenberg, M. W., The genetic code: II. *Sci. Am.* **208**, No. 3, 80 (1963).

Nirenberg, M. W., and J. H. Matthaei, The dependence of cell-free protein synthesis in *E. coli* upon naturally occurring or synthetic polyribonucleotides. *Proc. Natl. Acad. Sci. U.S.* **47**, 1588 (1961).

Painter, T. S., Studies in mammalian spermatogenesis. II. The spermatogenesis of man. *J. Exptl. Zool.* **37**, 291 (1923).

Speyer, J. F., P. Lengyel, C. Basilio, and S. Ochoa, Synthetic polynucleotides and the amino acid code, II. *Proc. Natl. Acad. Sci. U.S.* **48**, 63 (1962).

Sutton, H. E., *Genes, Enzymes, and Inherited Diseases*. Holt, Rinehart and Winston, New York (1961).

Taylor, J. H., The organization and duplication of genetic material. *Proc. Intern. Congr. Genet. 10th Montreal*, Vol. I, p. 63. Univ. of Toronto Press (1959).

Tjio, J. H., and A. Levan, The chromosome number of man. *Hereditas* **42**, 1 (1956).

Ushijima, R. N., F. S. Shininger, and T. I. Grand, Chromosome complements of two species of primates: Cynopithecus niger and Presbytis entellus. *Science* **146**, 78 (1964).

Watson, J. D., and F. H. C. Crick, A structure for deoxyribose nucleic acid. *Nature* **171**, 737 (1953).

Watson, J. D., and F. H. C. Crick, Genetical implications of the structure of deoxyribonucleic acid. *Nature* **171**, 964 (1953).

White, M. J. D., *Animal Cytology and Evolution* (Second edition). Cambridge Univ. Press, London (1954).

Wilkins, M. H. F., Molecular configuration of nucleic acids. *Science* **140**, 941 (1963).

The Human
Blood Group Systems

26

THE POPULATION GENETICS of various systems in the living human organism plus comparative study of the same systems in man's primate relatives are essential parts of the story of human evolution. The tissue which has come to be a favorite of physical anthropologists is the blood. Blood is very easy to obtain from most individuals, and it is easy to handle in the laboratory. We are delighted to work with something so readily available. Several examples of the use of the human blood group systems in defining origins of certain populations have already been cited, and mention of aberrant human hemoglobins has been made. In this and the following chapters we shall describe in detail some of the features of blood which are useful in getting at man's evolutionary history and in characterizing differences among populations.

Blood

The blood is a very specialized tissue. It transports oxygen and food to the other tissues of the body, and it carries substances that are part of the body's system of defense against agents of disease and other invading foreign substances. The blood carries waste products to the kidneys, intestines, lungs, and skin. It helps to maintain proper fluid, salt, and acid-base balances and

proper temperature of the body. Depending on the size of the animal, the blood makes up between 5 and 10 per cent of the total weight of the body.

Blood is composed of two types of material—formed elements and liquid. The formed elements are the red blood cells—the erythrocytes—and the white blood cells—the leukocytes. The blood group substances on the surface of the red cells (or an integral part of the cell membrane) and hemoglobin, one of many proteins contained inside the red cells, are the two major subjects treated in this and succeeding chapters. We only mention here that the white blood cells constitute one of the body's main defenses against disease.

The clear yellow fluid portion of the blood is the serum or the plasma. It is called serum when the clotting factors have not been inhibited and the red and white cells have been excluded in a jelly-like almost solid clot. It is called plasma when the clotting mechanism has been inhibited. Normally when blood is removed from the body, either intentionally or through a scratch or wound, it coagulates or clots. The formation of a clot can be prevented by the use of any of a number of chemicals known as anticoagulants— heparin, sodium citrate, potassium oxalate. The blood-clotting mechanism is also disturbed, as we have mentioned, in individuals with hemophilia, a recessive sex-linked inherited defect (Chapter 23).

If a blood sample taken with anticoagulant is allowed to settle in a tube or is centrifuged, the formed elements separate from the plasma; the red cells settle to the bottom of the tube; the white cells form a layer above these; and the plasma is on the top (Fig. 26.1). Fibrinogen, a fibrous protein in solution in the serum, is converted to a gel (the clot) when blood is taken without

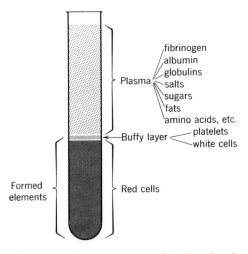

Fig. 26.1. The components of blood, as found after centrifugation of a sample taken with an anti- coagulant.

anticoagulant, and this gel contains the red and white cells. The serum is excluded from the clot and is present as a clear yellow liquid. The total amount of fibrinogen and other substances involved in the clotting mechanism is small, and there are no important differences, for our purposes, between plasma and serum. The proteins in plasma can be separated by means of various physicochemical methods into several fractions—α-globulins, β-globulins, γ-globulins, and albumin. Each of these fractions contains subfractions.

Many of the constituents of serum and red cells exist in different forms in man and show interspecific differences among the Primates. There are blood group polymorphisms; there are hemoglobin polymorphisms; and many of the serum proteins are polymorphic. The value that studies of blood have for anthropologists will become apparent as we discuss the blood group substances, the hemoglobins, and some serum proteins of man and the other primates.

Human Blood Group Substances

The human blood group substances are mucoproteins or mucopolysaccharides, large protein molecules to which various sugar molecules (saccharides) are attached. These blood group substances are also called blood group antigens. An antigen is defined as a protein (or other large molecule) that causes production of an antibody or reacts with an antibody already present in the serum. Antibodies are large proteins, part of the γ-globulin fraction of the serum. An organism often responds to the presence of a foreign substance by producing an antibody to it. The particular antibodies in which we are most interested are those which agglutinate red blood cells carrying antigenic groups on their surfaces. Almost all of the blood group antigens are present at the time of birth, and antibodies to certain of them are found circulating in the blood at birth.

Blood groups are classificatory categories. For example, erythrocytes from an individual may be classified as belonging to blood group A because they are agglutinated, clumped together, when reacted with an antibody known as anti-A. The agglutination of these erythrocytes is due to the interaction of an antigen on the red cell with an antibody which reacts specifically with that particular antigen. The study of blood groups is often called serology, for the specific reagents, the antisera, are found in serum.

Blood grouping techniques—the determination of an individual's blood groups—are simple in principle, and almost anyone can learn to perform the necessary tests, given the proper reagents. Most of our information about blood groups depends upon what is known as a simple agglutination test. A serum which contains an antibody is added to a suspension of red cells in salt solution, isotonic saline (normal physiological saline which contains 8.5 g of sodium chloride per liter). If the red cells carry the blood group antigen for which the antibody is specific, the red cells get stuck together in clumps; they are agglutinated. If the red cells do not contain the blood group antigen for which the antibody is specific, they will not be agglutinated. It is also possible to perform these agglutination tests in reverse, so to speak. It is possible to identify an antibody which is suspected to be present in a serum by adding red cells of known blood group to the serum. When the suspension of red cells of known blood group is agglutinated, the antibody is identified. The agglutination reaction may vary among individual blood samples, and the potencies of various antibody preparations may differ. Even with the most careful controls in the laboratory, such individual variation occurs. Whether or not this variation in reactivity is due to differences in the antigen is not known.

Many complex variations and extensions of these simple agglutination procedures have been devised. Not all of the blood group systems operate in a simple fashion, and various intermediate experimental steps must be taken to demonstrate some blood groups. The technical details are discussed in several of the references listed at the end of this chapter.

A blood group system includes the group specific substances usually identified on the red cells, the antibodies that make identification possible, the alleles that control the system, and anything else related to these things. The blood group antigens are often called group specific substances, for it turns out that they are found in many tissues of the body, not only in the blood. Most of the significant blood group systems of man are listed in Table 26.1.

The discovery of the ABO blood groups, the first blood group system fully described, was based upon certain well-known facts. If whole blood from two different individuals is mixed together, the red cells of one or both individuals are very often agglutinated. But sometimes they will not be. If a transfusion of whole blood is given an individual and agglutination of the cells of the donor occurs, the transfused person, the recipient, will suffer a severe reaction which usually leads to death. The agglutination of the donor's cells blocks the capillaries and other small blood vessels of the recipient and interrupts the normal circulation of the blood. If a transfusion is given, it should be from an individual whose blood cells do not become agglutinated when mixed with the blood of the recipient. Today blood transfusions are seldom made unless the bloods of donor and recipient are cross matched. This means that the blood plasma of the recipient is tested against the cells of the donor and vice versa.

Landsteiner, in 1900, carefully analyzed the pattern of agglutination reactions between the cells and plasma of various individuals. He discovered that there were some red cells which were not agglutinated by the plasma of any individuals. The plasma from these individuals agglutinated the red cells of all individuals whose red cells could be agglutinated by human plasma. Subdivisions of those whose red cells could be agglutinated were found. If plasma from one individual agglutinated the cells of another, the plasma from the latter would almost always agglutinate the cells of the former. The

TABLE 26.1
Major Human Blood Group Systems

System	Antigens	Antibodies	Phenotypes°	Genotypes
ABO	A, B, [AB]	anti–A anti–B	O, A, B, AB	*OO, AA, BB, AB,* *AO, BO*
Lewis†	Le^a, Le^b	anti–Le^a anti–Le^b	Le (a+b−), Le (a−b+), Le (a−b−)	*Le^{a}Le^a, Le^{b}Le^b, LeLe*
Rh‡				
MNSs	M, N, S, s	anti–M anti–N anti–S anti–s	M, N, MN, S, s, Ss	*MS/MS, MS/Ms, Ms/Ms,* *MS/NS, MS/Ns, Ms/NS,* *Ms/Ns, NS/NS, NS/Ns,* *Ns/Ns*
P	P$_1$, P$_2$	anti–P$_1$ anti–P + P$_1$ anti–P	P$_1$, P$_2$, p	*P$_1$P$_1$, P$_1$P$_2$, P$_2$P$_2$,* *P$_1$p, P$_2$p, pp*
Lutheran	Lu^a, Lu^b	anti–Lu^a anti–Lu^b	Lu (a+b−), Lu (a−b+)	*Lu^{a}Lu^a, Lu^{a}Lu^b,* *Lu^{b}Lu^b*
Kell	K (Kell) k (Cellano)	anti–K anti–k anti–Kp^a anti–Kp^b	K+k−, K+k+, K−k+, [K−k−]	*KK, Kk, kk*
Duffy	Fy^a, Fy^b	anti–Fy^a anti–Fy^b	Fy (a+b−), Fy (a+b+), Fy (a−b+), Fy (a−b−)	*Fy^{a}Fy^a, Fy^{a}Fy^b,* *Fy^{b}Fy^b, FyFy*
Kidd	Jk^a, Jk^b	anti–Jk^a anti–Jk^b	Jk (a+b−), Jk (a+b+), Jk (a−b+), [Jk (a−b−)]	*Jk^{a}Jk^a, Jk^{a}Jk^b,* *Jk^{b}Jk^b*
Diego	Di^a	anti–Di^a	Di (a+), Di (a−)	*Di^{a}Di^a, Di^{a}Di, DiDi*
Sutter	Js^a	anti–Js^a	Js (a+), Js (a−)	*Js^{a}Js^a, Js^{a}Js,* *JsJs*
Auberger	Au^a	anti–Au^a	Au (a+), Au (a−)	*Au^{a}Au^a, Au^{a}Au,* *AuAu*
Xg	Xg^a	anti–Xg^a	Xg (a+), Xg (a−)	*Xg^{a}Y, XgY, Xg^{a}Xg^a,* *Xg^{a}Xg, XgXg*

°Phenotypes are defined by reactions of red cells with specific antisera.
†See also Table 26.5.
‡See Tables 26.6 and 26.7.

pattern of agglutination reactions proved to be regular and constant. This pattern is now known as the ABO blood group system. This blood group system consists of group specific substances on the red cells and reciprocal group specific antibodies in the serum, that is, a person of group A has anti-B antibodies in his serum.

Other blood group systems were discovered when animals, such as rabbits and monkeys, synthesized specific antibodies if human red cells were injected into them. These are called immune antibodies. The animals were being immunized against red cell antigens just as we are immunized against smallpox, yellow fever, or poliomyelitis virus when we receive a vaccination. Antisera made from the plasma of immunized animals, when tested against red cells from various individuals, are used to discriminate between people whose red cells do and those whose red cells do not contain an antigen. The red cells with the antigen are agglutinated and those without are not agglutinated by the antiserum containing the specific antibody.

Another way in which blood group systems have been discovered is by testing the serum of individuals who have been given blood transfusions. Those who have received many blood transfusions because of illness or surgery often are immunized by antigens on the red cells of the blood from one or another of the blood donors. Tests of the donor's red cells with the serum of the recipient often reveal a new blood group antigen.

Still another situation which leads to the discovery of new blood group systems is the birth of infants with hemolytic disease. Antibodies not only agglutinate red cells, they may, in certain cases, break them up (hemolyze them). Hemolytic disease of the newborn occurs when the red cells of the growing fetus are destroyed. In many cases this hemolysis is caused by an antibody circulating in the mother's blood. This antibody may be one normally in the mother's circulation, such as anti-B. In other cases it may have been synthesized by the mother's antibody-forming mechanism under stimulation of an antigen on erythrocytes from the fetus which entered the mother's circulation.

The identification of apparently new red cell antigens and antisera is, for serologists, merely the beginning of a long process of testing and retesting. Series of tests must be made to determine if what seem to be newly discovered blood groups are indeed new or whether they are isolated occurrences of blood groups already described in the literature. Family studies must be made and pedigrees drawn up in order to determine the mode of inheritance. Many blood group systems are first characterized and described by use of a single antiserum. Later discoveries of other antisera that react with the products of alleles to the gene which produced the first antigen may expand our knowledge of any particular blood group system.

The ABO Blood Group System

The original classification of four blood groups—O, A, B, and AB—was based on the presence of blood group substance A or B on red cells. Red cells are classed in group A if they are agglutinated by anti-A antisera, in group B if agglutinated by anti-B antisera, and in group O if they lack both antigen A and antigen B, that is, are not agglutinated by anti-A or anti-B. Red cells are classified as AB if they are agglutinated by both anti-A and anti-B antibodies. Both the antibodies and the antigens of the ABO system are present from an early stage of fetal life and apparently do not change thereafter. The

ABO blood group system is unlike many others because the antibodies, anti-A and anti-B, occur in the serum of normal individuals from the time of birth. That is, a person whose erythrocytes are classified as blood group A has anti-B normally circulating in his blood. A person whose erythrocytes are classified as group B normally has anti-A in his circulation. Persons whose erythrocytes do not react with either anti-A or anti-B and whose red cells, thus, are classified as group O have both anti-A and anti-B in their circulation. Persons classified in blood group AB have neither anti-A nor anti-B. Thus, persons of group O are universal donors and persons of group AB are universal recipients. The reactions of the ABO blood group system are summarized in Table 26.2.

TABLE 26.2
Reactions in the ABO Blood Group System

Serum	Red cells			
	O	A	B	AB
O	−	+	+	+
A	−	−	+	+
B	−	+	−	+
AB	−	−	−	−

The pattern of reactions of red cells and serum leads to the hypothesis that the ABO blood groups are determined by three alleles, two of them *A* and *B* dominant to a third *O* but not dominant to each other. Various tests of this hypothesis show that this is the case. The best tests come from analysis of pedigrees. A critical mating in the test occurs when only one parent is blood group AB. This mating cannot produce children of phenotype O. In a mating where one parent is blood group O and the other is blood group AB, no children of blood group AB or O will be produced. No unreasonable exceptions have been discovered. Where apparent exceptions have been reported, further checking has shown that the actual paternity of the child is different from that originally reported. In other cases technical errors have resulted in mistaken classification of erythrocytes.

A more formal test of the hypothesis involves comparing the actual (observed) frequencies of the genes for the ABO blood groups with the frequencies expected if there are three alleles in the system. In a population in Hardy-Weinberg equilibrium, if p is the frequency of gene A, q the frequency of gene B, and r the frequency of gene O, $p + q + r = 1$. Bernstein has shown that this relationship is satisfied within the limits of error imposed by the sample. In order to show this, estimates of the gene frequencies must be made from the frequencies of the phenotypes in a population, since a direct count of the three alleles is not possible. Genotype AO is classed as phenotype A and genotype BO as B in the serology laboratory. These estimates are made using the formulae in Table 26.3. In all cases tested, the frequencies of the three alleles support the hypothesis.

TABLE 26.3

*Formulae for Estimating Gene Frequencies
in the ABO Blood Group System*

$$p = 1 - \sqrt{B + O}$$
$$q = 1 - \sqrt{A + O}$$
$$r = \sqrt{O}$$

p = frequency of gene A	A = frequency of phenotype A
q = frequency of gene B	B = frequency of phenotype B
r = frequency of gene O	O = frequency of phenotype O
$p + q + r = 1$	

Genotype	Phenotype
OO	O
AA	A
AO	A
BB	B
BO	B
AB	AB

Genetic analysis of the ABO blood group system is more difficult than that of a simple three-allele system. Alleles A and B are not dominant to each other, but they are both dominant to allele O. Further complications in understanding and analyzing this blood group system occurred when new antisera were obtained and it was discovered that blood group A could be divided into two groups. The experiments made it clear that what was originally called antigen A consists of two distinct and separate antigens A_1 and A_2. Anti-A antisera react with both antigen A_1 and antigen A_2. Anti-A_1 antisera, which are relatively rare, react only with antigen A_1. If red cells are agglutinated by anti-A and anti-A_1 antisera, they are classified as A_1. If red cells are agglutinated by anti-A and do not react with anti-A_1, they are classified as A_2. (There is no anti-A_2 antiserum.) The A_1A_2 genotype cannot be detected at present, and it is possible that A_2B individuals are often classified as B, for the A_2 antigen reacts very weakly in that combination. These two subdivisions A_1 and A_2 are inherited as discrete and separate traits. A_1 is dominant to A_2 and to O; A_2 is dominant to O; but neither A_1 nor A_2 is dominant to B. Distinction between the two antigens, which most anti-A antisera recognize only as A, had to await the discovery of anti-A_1 antiserum.

The original theory that the ABO blood group system is determined by three alleles has been expanded today to take into account the A_1 and A_2 antigens. It is now believed that four alleles determine the system—A_1, A_2, B, and O. The pattern of reactions with the four antisera is consistent with the four-allele hypothesis. The genotypes and phenotypes for the four-allele system are described in Table 26.4.

TABLE 26.4
Genotypes and Phenotypes in the ABO Blood Group System
when Antigens A_1 and A_2 Are Distinguished

	Spermatozoa				Phenotypes (determined by serological reactions)	Genotypes
	O	A_1	A_2	B	O	OO
Ova O	OO	A_1O	A_2O	BO	A_1	A_1A_1 A_1O A_1A_2
Ova A_1	A_1O	A_1A_1	A_1A_2	A_1B	A_2	A_2A_2 A_2O
Ova A_2	A_2O	A_1A_2	A_2A_2	A_2B	B	BB BO
Ova B	BO	A_1B	A_2B	BB	A_1B A_2B	A_1B A_2B

The frequencies of the ABO phenotypes and alleles have been studied in thousands of human groups. These frequencies vary considerably among human groups. Occasionally one of the alleles is absent. For example B is not present among American Indians. Few population surveys have been made with anti-A_1 antisera, since they are rare. Red cells must be tested with anti-A and anti-A_1 antisera before they may be classed as A_1 or A_2. Our information about the distribution of A_1 and A_2 is limited. Among those populations tested, there are some which appear to have only the A_1 allele. Populations of many Pacific Islands, Australian aborigines; many groups living on islands of the Indonesian archipelago; several populations in India; a number of American Indian groups; and Eskimos appear to have the A_1 and not the A_2 allele. This includes several small isolated populations of Asia and Europe. So far as is known today the populations that lack the A_2 allele are relatively small and isolated. Sampling error (genetic drift) may account for the presence of A_1 and absence of A_2. The A_2 allele is usually quite infrequent, but among the Lapps of northern Scandinavia it is reported with frequencies as high as 50 per cent.

There are some extremely important questions that must be asked about the ABO blood group system. What are the blood group specific substances? How do they interact with each other? What is their function? Why are human populations polymorphic for these systems? We will take up these questions when we discuss natural selection and the blood groups. But first we must describe some other important blood group systems. Several of these appear to interact with each other.

The mucoproteins or mucopolysaccharides, the chemical group specific substances which give the specific reactions for the A and B blood group antigens, are found in most tissues of the body with the major exception of the brain. They occur in an alcohol- and a water-soluble form. The former is found in all individuals tested so far; the latter is not. The blood group specific substances are found in most of the body fluids and secretions, particularly in saliva, gastric secretions, bile, milk, and in pleural, pericardial, peritoneal, seminal, and amniotic fluids. Very large amounts of these blood group specific substances can be extracted from certain kinds of ovarian cysts and from meconium, fecal matter found in the intestines at the time of birth. But all individuals do not secrete water-soluble A and B blood group specific substances in these various fluids. It has been discovered that individuals fall into two classes—secretors and nonsecretors.

The ability to secrete water-soluble A and B group specific substances is determined genetically. The best explanation of the mode of inheritance of this trait is to say that it is controlled by two alleles, one dominant to the other. The allele for secretion (*Se*) is dominant to the allele for nonsecretion (*se*). The alleles that determine the secretion of the ABO group specific substances are independent of those that determine the ABO blood groups on the erythrocytes. It is clear that two different loci are involved.

As we said earlier, the original classification of red blood cells into four categories (O, A, B, and AB) was based on the presence or absence of the A and B antigens. It seemed somewhat strange to many workers that the absence of a trait would be controlled by an allele. In other words was blood group O actually the absence of a chemical or the absence of appropriate antisera for detecting it? It turns out that there is a substance now called H present on cells classified as blood group O. This substance may be the product of allele *O*. The procedures for demonstrating the occurrence of a product of the *O* allele are complex. Even today we are not certain that it is the product of the *O* allele which is recognized by the reagents used.

Certain very carefully selected blood sera from normal cattle cause agglutination of red cells of blood group O from certain individuals. A large proportion of group O individuals have a substance in their saliva that neutralizes these particular anti-O sera from cows. At first it was believed that these were true anti-O sera and that the individuals had a gene for secretion of an O group specific substance in their saliva.

The pattern of reactions expected from a normal anti-O antiserum, analogous to the reactions of the other antibodies of the ABO system, did not appear. If this were a true anti-O antibody, we would expect to use it to distinguish the *AO* and *BO* heterozygotes from the *AA* and *BB* homozygotes. The *AO* and *BO* genotypes *cannot* be distinguished by use of the cattle sera. Erythrocytes of group AB should not be agglutinated by an anti-O antiserum, but they are by these cattle sera. Secretors of blood type AB should not have a substance in their saliva which reacts with anti-O sera, yet many do. Saliva from secretors of group AB contains not only substances with A and B group specific activity but also something which reacts with the anti-O serum. This is good evidence that the original sera that reacted with O cells are not related to the allele for O in the same way that the anti-A and anti-B antisera are related to the alleles for A and B.

Extracts made from seeds of certain plants, notably *Ulex europaeus* and *Lotus tetragonolobus*, will agglutinate red cells of group O and react with a group specific substance in the saliva of O individuals who are secretors. Since these extracts agglutinate red cells from A, B, and AB individuals as well and react with the A and B group specific substance in their saliva, this antigen cannot be a direct product of the allele O. Today this group specific substance is called H, and the reagents that specifically recognize it are anti-H. The H symbolizes the presumed heterogenetic origin and character of the substance.

It appears that the amount of H substance present on red cells is in some way related to the amount of A_1, A_2, or B. The degree to which anti-H reagents agglutinate red cells of different ABO blood groups follows, approximately in order: $O > A_2 > B > A_1$. Anti-H reagents cause strongest agglutination of O cells and weakest agglutination of A_1 cells. The amount of H substance present in secretors of each of the ABO phenotypes also varies in a similar fashion. One of the most puzzling things about the ABO system is the relationship between allele O and the H antigen. The following is a hypothetical account of the probable steps in the synthesis of the blood group specific antigens. A gene H directs the synthesis of a mucopolysaccharide now identified as the H antigen. This H substance is then converted by the A and B alleles into specific A and B antigens. The allele O does not convert H substance. If an allele h, recessive to gene H, exists, it will fail to direct the synthesis of H substance, and A and B alleles will not have anything from which to make the A and B antigens. This path from precursor substance to group specific A, B, and H antigens is illustrated in Fig. 26.2. There is some evidence that this is a reasonable hypothesis. The chemical similarity of the A, B, and H specific substances suggests that the production of H precedes that of A and B. If the production of the basic H substance is under control of a pair of alleles H and h, independent of the ABO system, a recessive homozygote hh should occur which has no A, B, or H antigen. Such a phenotype was discovered in India and was once called the Bombay phenotype. This phenotype has red cells which are not agglutinated by anti-A, anti-B, or anti-H antisera. The blood serum of the Bombay phenotype contains anti-A, anti-B, and anti-H antisera. One possibility is that a suppressor or inhibitor gene is preventing action of the ABO locus in synthesizing the group specific antigen. The scheme of Fig. 26.2 seemed the more reasonable when another individual of the Bombay phenotype was discovered. Thorough analysis of pedigrees indicated that this case was

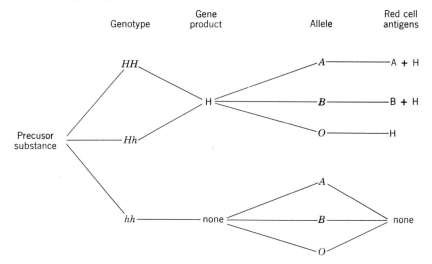

Fig. 26.2.   Hypothetical pathways from a precursor substance to A, B, and H group specific antigens. The allele *H*, in this scheme, must be present for the production of the red cell
antigens.

clearly one in which an allele for B should have been present but was not
producing B antigen.

The chemistry of the A, B, and H group specific antigens is complex. We
shall briefly describe some of the salient features, for there are many lines of
research in physical anthropology which require that we pay heed to more
than simple plus and minus blood group reactions of erythrocytes. The blood
group specific substances are mucopolysaccharides or mucoproteins of high
molecular weight, of the order of 3×10^5. The A, B, H, and Lewis (Le^a)
group specific substances are very similar; they are made up of amino acids,
sugars, and amino sugars. At least eleven of the common amino acids have
been shown to occur in these substances. There are two sugars regularly
present, D-galactose and L-fucose, and two amino sugars, D-glucosamine and
D-galactosamine. The D and L refer to optical rotatory properties of the
molecules. It appears that the basic structures of the A, B, H, and Le^a group
specific mucopolysaccharides are very similar. The group specific characteristics, which are identified by the serological reactions already mentioned,
are due to rather short sequences of sugars (oligosaccharides). These oligosaccharides apparently produce different three-dimensional configurations
on the surface of the molecule. It is presumably these spatial configurations
which confer specificity on antibody-antigen reactions. There is some evidence that the specific activity of the A substance is related to a particular
oligosaccharide unit on the surface of the molecule, N-acetyl-D-galactosamine.
The specificity of the B antigen is associated with D-galactose and of the H
antigen with L-fucose. The specificity of the Le^a antigen has not yet been

unequivocally associated with a specific oligosaccharide unit, but direct chemical evidence that it is different from H antigen has been obtained.

One special peculiarity of the A, B, and H group specific substances occurs in persons who are secretors and who are genotype *AB*. The blood group specific antigen in their secretions does not react like a mixture of A and B antigens. If purified antigens A and B are mixed together, the A substance can be removed with anti-A antiserum and the B substance with anti-B antiserum. But, if purified substance from an AB secretor is treated with anti-A antiserum, all of the substance is removed, and nothing in the solution will react with anti-B antiserum. There is also no reaction with anti-A if AB substance is treated first with anti-B antiserum. This suggests that the blood group specific substance in secretors of genotype *AB* is qualitatively different from the A substance and the B substance.

One promising line of research in physical anthropology would be to study the function of the blood group specific substances. But it is not likely that we shall understand their function unless we understand how they act, what they are chemically, how they are synthesized, and what roles they play in the tissues of the organism. We still do not know why they are on the surface of the red cell, and we have relatively few notions of why they are in other places.

The Lewis Blood Group System

The Lewis blood group system, discovered and named in 1946, is closely related in some way to the ABO system. It is described in Tables 26.1 and 26.5. Its most important feature for our purposes is the relationship it has to

TABLE 26.5

The Relationship between the Antigens of the Lewis Blood Group System and the Secretor Genotypes
(From Race and Sanger, 1962.)

Genotype	Antigens in saliva[*]				Lewis phenotype (antigen on red cells)
	ABH	Le^a	Le^{bL}	Le^{bH}	
sese LeLe *sese Lele*	−	+	−	−	Le $(a+b-)$
SeSe LeLe *SeSe Lele* *Sese LeLe* *Sese Lele*	+	+	+	+	Le $(a-b+)$
SeSe lele *Sese lele*	+	−	−	+	Le $(a-b-)$
sese lele	−	−	−	−	Le $(a-b-)$

[*]Antigens are identified by their serological reactions ($+$ or $−$) with specific antisera. Two kinds of anti-Le^b antisera are recognized.

secretors and nonsecretors of the A, B, and H water-soluble substances. At the present time it seems that adult individuals who have Le^a red cells are nonsecretors of the A, B, and H group specific substances. Most of the adults who are Le(a −), that is, those who do not have the Le^a substance on their red cells, turn out to be secretors of the A, B, or H substances. These secretors do not have the Le^a substance on their erythrocytes, but most of them nonetheless secrete the Le^a substance in their saliva. The Le^a substance on human erythrocytes behaves differently, in some respects, from the A, B, and H blood group specific substances. The A, B, and H substances appear to be very closely attached to the cell wall or cell membrane of the erythrocyte. The ABO group specific reactivity of a red cell does not change during the duration of the life of the cell; it is not changed by blood transfusion; and it is probably not possible to change it without destroying the red cell. Le^a reactivity of red cells on the other hand can be modified or removed by repeated washings of the cells with physiological saline, by incubation at temperatures slightly higher than body temperature, or under other special conditions. The results of these treatments seem to demonstrate a characteristic of the secretion of certain group specific substances. When these secretions are in sufficiently high concentration to be carried in the blood (in the serum), the Le^a substance can be absorbed on the surface of the red cell. The ABO substances on the other hand are present on the cell not only from the birth of the individual but from the time of formation of the cell and are apparently formed and attached to the cell wall during differentiation of the erythrocyte.

There is a considerable amount of evidence that suggests that the Lewis blood group substances are under fairly simple genetic control, but the exact mode of inheritance and the exact nature of the relationship between the gene and the product, the Le^a substance, is not yet clearly understood. There is an apparently high correlation between nonsecretion of A, B, and H substances and the Le(a+) red cells. This is not yet explained. The original data suggested that individuals whose red cells reacted with anti-Le^a antisera, that is, Le(a+) individuals, are homozygous for a recessive gene. A little later a very high frequency of Le(a+) red cells was found among newborn infants. This suggested to some that the trait was expressed in the heterozygous state among infants and was lost in adulthood, but this hypothesis did not hold up either.

At present the most interesting hypothesis about the nature of the Lewis blood groups is one proposed by Ceppellini, who postulated a mechanism that relates the secretor alleles to the Le^a substance (Table 26.5). This hypothesis states that the *Le^a* gene has a locus quite different from the ABO or secretor locus. The *Le^a* gene leads to the formation of Le^a substance in either heterozygous or homozygous individuals. The amount of Le^a substance produced and where it is found depend on whether an individual is a secretor or not. Homozygous nonsecretors of A, B, or H group specific substance will produce larger amounts of Le^a substance than individuals who are secretors, and the Le^a substance will be absorbed on the red cells of the nonsecretors. In formal terms the hypothesis asserts that there is an interaction between the allele for producing Le^a group specific substance and the alleles at the secretor locus.

The whole situation has been further complicated by discovery of another group specific substance called Le^b. This is not found on Le(a+) cells but is found in the majority of individuals whose red cells are Le(a−). It is absent from the saliva of nonsecretors of A, B, or H group specific substance who are Le(a+), but it is present in secretors who have Le^a substance in the saliva. Since the supply of specific anti-Le^b antisera is limited, it will probably be some time before Le^b is thoroughly studied.

Today there seem to be three separate genetic loci—ABO, secretor, and Lewis—involved in determining the nature of the ABO group specific water-soluble mucopolysaccharides in the saliva and other body fluids. The following observations relate the Lewis system to the other two. Red cells that are Le(a+b−) belong to nonsecretors of the A, B, or H substance; red cells that are Le(a−b+) belong to secretors; red cells that are Le(a−b−) almost always belong to secretors.

The Rh Blood Group System

The Rh blood groups were first reported in 1939 by Levine and Stetson who described the severe hemolytic reaction a woman had when she received a blood transfusion from her husband. This woman was suffering from the aftereffects of having given birth to a stillborn fetus. It was discovered that the serum from this woman's blood agglutinated her husband's erythrocytes and the erythrocytes from about 80 per cent of the potential blood donors against which her serum was tested. It was clear from the work of Levine and Stetson that this agglutination reaction was due to a blood group system independent of previously discovered systems. Later the name Rh was given to this system by Landsteiner and Wiener who independently discovered it. This blood group system was found when rabbits and guinea pigs were immunized with red cells of the rhesus monkey *Macaca mulatta.* Landsteiner and Wiener discovered that the antibodies produced in the rabbits and guinea pigs not only agglutinated the red cells of rhesus monkeys but also those of many men and women. It was predicted on the basis of sampling statistics that these antibodies would agglutinate the red cells of about 85 per cent of the people of European descent who lived in New York City. This prediction was borne out in subsequent years. Those erythrocytes that were agglutinated by the anti-rhesus serum prepared in the rabbit were classified as Rh positive by Landsteiner and Wiener and those that were not as Rh negative. The Rh symbol was derived, obviously, from the word rhesus. Wiener and his co-workers demonstrated that the anti-Rh antibody made in the rabbit against rhesus monkey red cells was the same as the antibodies found in the sera of many persons who had transfusion reactions after receiving compatible ABO blood.

The antibody which produced the hemolytic reaction in the mother in the famous case of Levine and Stetson proved to be the same as the rabbit anti-rhesus antibody found by Landsteiner and Wiener. The antibody in the mother was produced as a response to an Rh substance on the red cells of her developing fetus. In other words she was immunized by the blood of her own fetus. Since then it has been demonstrated that many cases of hemolytic disease and jaundice in newborns are due to the reaction of the antibodies stimulated within an Rh negative mother by the Rh positive cells of the fetus (Fig. 26.3). Some cases of hemolytic reactions of the newborn can be cured by a total exchange transfusion. The blood with destroyed red cells, the hemolyzed blood of the newborn, is replaced by whole blood from Rh positive healthy donors.

The Rh system is complex. Because the Rh blood groups are so important in discussing natural selection in genetics and evolution, we shall attempt to present them in reasonably complete and possibly comprehensible form. The reader must be warned that the Rh blood groups have been the center of a rather intriguing and vigorous controversy. The controversy has been as much over nomenclature as over the facts of the system. We shall refer to this blood group system as the Rh system, and we shall generally use the

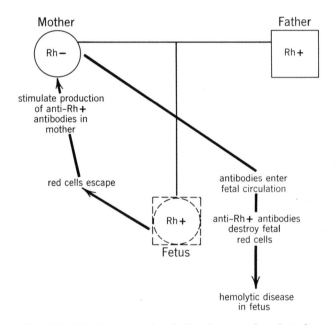

Fig. 26.3.   The formation of antibodies that cause hemolytic disease of the new born when maternal and paternal Rh phenotypes are immunologically incompatible.

notation proposed by Fisher and Race (Table 26.6). Today a reagent corresponding to the anti-d antibody is the only one that has not been demonstrated. The other antibodies have all been demonstrated, and they react as indicated in the Table. Thus an individual who is *CDE* will have red cells agglutinated by anti-C, anti-D, and anti-E antibodies. He will have *CDE/CDE* chromosomes. This person is completely Rh positive in the original sense. The completely Rh negative person has a *cde/cde* chromosomal arrangement, and his red cells are agglutinated by anti-c, anti-e, and, if it exists, anti-d. An important point to remember is that each individual has two chromosomes, so that a person who is completely *cde* must have received from each parent one chromosome with these three loci on it. It is females with *cde/cde* who often produce infants suffering from hemolytic disease.

TABLE 26.6
Reactions of the Rh Blood Group System
(Fisher-Race Notation)

Antibodies	Gene complexes[a]							
	CDE (R_z)	*CDe* (R_1)	*cDE* (R_2)	*cde* (r)	*cDe* (R_0)	*cdE* (R'')	*Cde* (R')	*CdE* (R_y)
Anti–C	+	+	−	−	−	−	+	+
Anti–D	+	+	+	−	+	−	−	−
Anti–E	+	−	+	−	−	+	−	+
Anti–c	−	−	+	+	+	+	−	−
(Anti–d)[b]	(−)	(−)	(−)	(+)	(−)	(+)	(+)	(+)
Anti–e	−	+	−	+	+	−	+	−

[a]Common gene complexes as reported for European populations.
[b]This antibody has not been demonstrated.

Shortly after the discovery of the Rh blood group system a number of workers recognized that reactions observed in the laboratory could not be due to blood group substances controlled by a simple diallelic (two-allele) system of genes. For a number of years it looked as if the products of the Rh system were really due to three very closely linked genes present on the same chromosome. Various combinations of these genes *CDE/cde* would lead to eight different chromosomes. The mode of inheritance of the Rh antigens is not completely understood. Regardless of whether there are three closely linked allelic genes or three separate genes or some other complex system, the present evidence has been interpreted to show that *C* acts as if it were an autosomal dominant to *c, D* to *d* and *E* to *e*. But genes for these antigens are passed on from generation to generation in groups of three. One does not inherit *C* or *c, D* or *d*, and *E* or *e* alleles. One inherits a complex, *CDE, Cde, cDe*, etc., and it is believed that this complex is inherited in a group of three on a single chromosome. Thus if tests with the appropriate antisera show that a person is C+, c+, D−, E+, e−, the most probable chromosomes are *CdE/cdE.* More complicated situations are often encountered.

Wiener, one of the co-discoverers of the Rh system, has developed concepts and notation that differ from those of the CDE or Fisher-Race hypothesis. He distinguishes a gene, an antigen or agglutinogen which is presumably the product of the gene, and a blood factor or factors present on the antigen (agglutinogen). The blood factor is a serological attribute and may be shared by two agglutinogens. The same agglutinogen may react with more than one antiserum. This implies that more than a single blood factor occurs on a single antigen. The relationship between some of the notation which follows from Wiener's hypotheses and the Fisher-Race notation is shown in Table 26.7. The Rh-Hr system of labeling blood factors is based upon an assumption which has not been demonstrated. The Rh blood group substances are assumed to have more than one antigenic specificity, for more than one antiserum will agglutinate the red cells with a single antigen on its surface. In technical language more than one antigenic determinant is characteristic of each Rh blood group specific substance. An antigenic determinant is that part of an antigen which elicits an immune response (production of antibodies) for the antigen.

TABLE 26.7

The Rh Blood Group System: Alternatives

CDE System			Rh-Hr System		
Gene complex[a]	Shorthand symbol	Antigens	Gene[a]	Agglutinogen	Blood factors
CDE	R_z	C,D,E	R^z	Rh_z	rh′, Rh_0, rh″
CDe	R_1	C,D,e	R^1	Rh_1	rh′, Rh_0, hr″
cDE	R_2	c,D,E	R^2	Rh_2	hr′, Rh_0, rh″
cde	r	c,d,e	*r*	rh	hr′, hr″, hr
cDe	R_0	c,D,e	R^0	Rh_0	hr′, Rh_0, hr″, hr
cdE	R″	c,d,E	*r″*	rh″	hr′, hr″
Cde	R′	C,d,e	*r′*	rh′	rh′, hr″
CdE	R_y	C,d,E	r^y	rh^y	rh′, rh″

[a]Common gene complexes or genes as reported for European populations.

The problem of defining the Rh system has hinged on whether a single set of closely linked loci, perhaps in a single cistron, or a number of independent allelic loci are responsible for the various observed serological reactions. The multiplicity of antigens and antisera that belong in this blood group system has added to problems in interpretation. Race and Sanger have recently suggested that compound antigens of the Rh system and specific antisera which react with them are performing a *cis-trans* test (Chapter 25). These reactions demonstrate that *C*(or *c*) and *E*(or *e*) are in a single cistron. Anti-ce antiserum appears to react with erythrocytes when the factors c and e are in

TABLE 26.8
A Cis-Trans Test on Loci of the Rh Blood Group System
(From Race and Sanger, 1962.)

Loci	Erythrocytes	Reaction with antisera	
		Anti–ce	Anti–Ce
CDE/cDe	CE/ce	+	–
CDe/cDE	Ce/cE	–	+

the *cis* position but not when they are in the *trans* position on the chromosomes. When the *c* and *e* loci are in the *cis* position they produce the compound antigen ce which they cannot produce when they are in the *trans* position. The occurrence of a compound anti-Ce reagent produces some confirmation of this idea. The reactions presented in Table 26.8 may actually be a *cis-trans* test.

The conventional way British serologists have explained the Rh-CDE system is illustrated in Fig. 26.4. The order of the closely linked loci is DCE, and no evidence that some other order is possible has been produced.

As more and more antisera were discovered in the Rh system, it became clear that the genetic hypotheses commonly accepted were not wholly warranted by the data. Some basic aspects of blood groups had been forgotten. Human blood group substances are chemical entities on the erythrocytes.

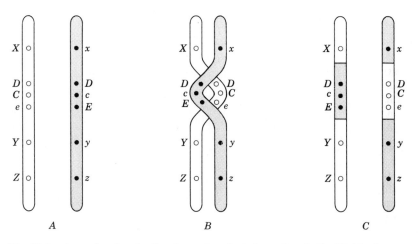

Fig. 26.4. An explanation for the observed serological reactions in the Rh blood group system. The Rh loci (*DCe/DcE*) act as a single unit. (A) Three alleles on two homologous chromosomes, (B) crossing over, (C) two chromosomes recombined.

The A, B, and H substances have been isolated and the chemical structures that are the basis for the specific antibody response of each have been investigated. This is not the case with the Rh and most other blood group substances. We do not know if one or many Rh substances produce the multiplicity of reactions discovered in the serology laboratory.

The Rh antibodies are almost always formed by an individual's immune response to antigens which are on erythrocytes introduced into his circulation. They are not naturally occurring antibodies. They are usually obtained from persons immunized by pregnancies or blood transfusions, not from laboratory animals, and no control over their production is possible. No tests of the antibodies against purified and standardized Rh blood group substances are possible at present. An unambiguous relationship between antigens and antibodies, such as is demonstrable for the ABO system, cannot yet be shown in the Rh system.

The Rh antigenic determinants are clearly transmitted by one segment of a single pair of autosomal chromosomes (Fig. 26.4). The segregation and recombination of the various Rh antigenic determinants have not been observed. Analysis of human pedigrees has not yet made possible demonstration of crossing over within this Rh chromosomal segment. Further elaborations of the genetic hypotheses of the Rh system may be intriguing, but they are probably not wholly warranted.

These considerations, plus the vast literature on Rh serology and genetics, led to the proposal by Rosenfield and his colleagues that only the reactions observed in the serology laboratories be recognized in the notation (Table 26.9). The observations upon which the Rh system is based are purely serological or immunological. There are over twenty Rh reagents, antisera, reacting in a way which suggests that a different genetically controlled antigen is recognized by each. Rosenfield has suggested that Rh reagents be numbered and the reactions recorded as plus ($+$) or minus ($-$). If an agglutination reaction ($+$) is weak, it should be recorded as *w*. Phenotypes would be reported as presented in Table 26.10. For example cde or rh becomes Rh: -1, -2, -3, 4, 5, 6, -7, -8, -9, -10, -12. If we adopt the notation proposed by Rosenfield, we merely record the serological reactions; we do not multiply the genetic assumptions involved. The newer notation will have relatively little immediate effect upon the critical clinical problems involved—hemolytic disease of the newborn and transfusion reactions. But for physical anthropologists and others who wish to characterize, in as precise and unambiguous a manner as possible, the gene pools of the populations they study, the notation described in Table 26.10 conveys all the information required and does not imply a mode of inheritance.

Although the Rh system is one of the most complex genetic systems investigated in man, it has already been of considerable value. It has been used in medicine, in problems of paternity, in describing the gene pools of various human populations, and in illuminating all aspects of research in human genetics.

TABLE 26.9
The Rh Blood Group System: a Comparison of Notations
(From Rosenfield et al., 1962.)

Antigen	Blood factor	Reagent
D	Rh_0	Rh1
C	rh'	Rh2
E	rh''	Rh3
c	hr'	Rh4
e	hr''	Rh5
f,ce	hr	Rh6
Ce	rh	Rh7
C^w	rh^{w1}	Rh8
C^x	rh^x	Rh9
V,ce^s	hr^v	Rh10
E^w	rh^{w2}	Rh11
G	rh^G	Rh12
–	Rh^A	Rh13
–	Rh^B	Rh14
–	Rh^C	Rh15
–	Rh^D	Rh16
–	Hr_0	Rh17
–	Hr	Rh18
–	hr^s	Rh19
Vs,e^s	–	Rh20
C^G	–	Rh21

TABLE 26.10
Phenotypes and Alleles as Determined by Five Reagents of the Rh Blood Group System

Phenotype[a] (reaction with numbered reagent)	Alleles		
	Proposed	CDE	Rh-Hr
Rh: 1, 2, 3, −4, −5	R 1, 2, 3,−4,−5	CDE	R^z
Rh: 1, 2, −3, −4, 5	R 1, 2,−3,−4, 5	CDe	R^1
Rh: 1, −2, 3, 4, −5	R 1,−2, 3, 4,−5	cDE	R^2
Rh: −1, −2, −3, 4, 5	R−1,−2,−3, 4, 5	cde	r
Rh: 1, −2, −3, 4, 5	R 1,−2,−3, 4, 5	cDe	R^0
Rh: −1, −2, 3, 4, −5	R−1,−2, 3, 4,−5	cdE	r''
Rh: −1, 2, −3, −4, 5	R−1, 2,−3,−4, 5	Cde	r'
Rh: −1, 2, 3, −4, −5	R−1, 2, 3,−4,−5	CdE	r^y

[a]Common phenotypes in European populations.

The MNSs Blood Group System

Landsteiner and Levine were the first to establish the existence of the MN blood group system and the nature of its inheritance. At first it was believed that it was controlled by two nondominant autosomal alleles M and N. The three genotypes—MM, MN, and NN—lead to three distinguishable phenotypes, M, MN, and N. Somewhat later it was discovered that another antibody, now called anti-S, distinguished an antigen associated with MN. Eventually an anti-s antibody was found and a relatively clear-cut system for distinguishing genotypes developed. Research showed that S could not be produced by an allele of M and N but seemed to be related to M and N as C, D, and E are related in the Rh system. The best explanation for the MNSs system is that it is produced by a series of very closely linked alleles or linked loci, similar to the Rh system. The interpretation of the MNSs system is described in Table 26.1.

On the basis of the Hardy-Weinberg equilibrium formula, matings between individuals who are MN × MN should produce offspring in the ratio of 1 MM : 2 MN : 1 NN. Since the alleles for the MN system are not dominant to each other, it is possible to identify the heterozygote MN. However it was noticed that the frequency of MN offspring from MN × MN matings was higher than expected on the basis of Hardy-Weinberg equilibrium. At first it was thought that there were technical reasons for this increased frequency of heterozygotic offspring of double heterozygotic matings, that is, incorrect techniques in the laboratory or imperfect antisera. Recently, however, this point has been examined in detail by a number of competent researchers, and it is clear that the hypothesis of technical error can be ruled out. There does seem to be selection in favor of heterozygotes. In other words the MN blood groups are a case of genetic polymorphism, possibly balanced polymorphism due to selection in favor of the heterozygotes.

The inclusion of the Ss antigens in the MNSs system means that there are ten genotypes. In practice two of these cannot be distinguished, MS/Ns and Ms/NS, with the four antisera. We cite this to show again that a considerable amount of detailed and refined genetic information can be obtained from human data and from the study of human populations if the requisite reagents are on hand.

There are several other antigens now known to be related to the MNSs system. The Hunter and Henshaw antigens, named after the West African individuals in whom they were discovered, are now included. Five other antigens, tentatively labeled Mi^a, Vw, Vr, Ma, and M^g, have been identified as being related to the MNSs system. As the complexity of the MNSs system becomes clear, it is obvious that much fundamental serological work remains to be done.

The P Blood Group System

The P blood groups were discovered by Landsteiner and Levine at the same time they discovered the MN blood groups. The sera of certain immunized rabbits agglutinated human erythrocytes which had already been classified by ABO and MN reagents, that is, classification of these blood samples, agglutinated by the special rabbit antisera, could not be done on the basis of the ABO or MN blood groups. Therefore it was decided that a new blood group system was involved, and the two categories were labeled P+ and P−.

After a while it was discovered that an antigen Tj^a, reported independently, was actually part of the P system and those people who were declared P− were actually not without an antigen. The fact that they had one required a change in the terminology. Instead of P+ and P−, the terminology for the two phenotypes is now P_1 and P_2, and the antisera are anti-P_1 and anti-P + P_1 (Table 26.1). The mode of inheritance of the P_1 antigen, demonstrated by Landsteiner and Levine, is that of an autosomal dominant. All work since their discovery of the system has supported this.

Anti-Tj^a is a naturally occurring antibody in the P system. This antibody, now known as anti-P + P_1, occurs without prior immunization. This means that the P blood group system is similar to the ABO system. Anti-P + P_1 seems to produce a more severe hemolytic reaction than anti-A or anti-B.

Another complication of interest was introduced to this system by the work of Race and Sanger who have demonstrated another antigen in the system, p. At present p seems to be extremely rare but promises to add to the usefulness and importance of the P blood group system in paternity cases and in expansion of our knowledge of the blood groups as functional, biological traits. The P blood group system also has subgroups and special features that occur so seldom that we need not discuss them here. The reader should know, however, that the blood group antigen Q, reported from Japan, is identical with the antigen P_1.

The Lutheran Blood Group System

In 1945 an antibody, part of what is now known as the Lutheran blood group system, was reported in a patient suffering from a severe dermatological and systemic disorder, *lupus erythematosus diffusus*. The antibody appeared after a blood transfusion, and the donor proved to have had his erythrocytes agglutinated by this antiserum. At present the system consists of two alleles, *Lu^a* and *Lu^b*. The phenotypes are Lu(a+) and Lu(a−). There are two antibodies, anti-Lu^a and anti-Lu^b. Available evidence indicates that the Lu^a antigen is inherited as an autosomal dominant. The blood group system appeared to be completely straightforward and simple. However in 1961 the Lutheran system, just like all the other blood group systems, was shown to be complicated. With the discovery of a phenotype that reacted like Lu(a−b−) to the known antibodies, the complications began. The most authoritative workers in the field believe that the allele responsible for the manifestation of a negative reaction (no agglutination) to both the anti-Lu^a and anti-Lu^b antibodies is a dominant. Since Lu(a−b−) has been found only in a single family, research will have to continue into the future. As yet we cannot use the Lutheran system efficiently in population studies.

The Kell Blood Group System

The Kell blood group system also appeared to be straightforward and simple at first. Race and Sanger have pointed out that the age of simplicity lasted longer for this blood group system than for any other. The original analysis was that there were reactions with anti-K and anti-k antibodies. It was believed that the Kell system consisted of nondominant autosomal alleles K and k. Recently two other antigens have been found which unquestionably belong to the Kell blood group system. They were defined by two antibodies anti-Kp^a and anti-Kp^b. The Kp^a antigen is sometimes called Penney; the Kp^b antigen is sometimes called Rautenberg, after the families in which they were found. The interpretation of the Kell system has not been completed yet because the complications are many. The present evidence is such that it is quite likely that the genetics of the Kell system will prove to be rather similar to that of the Rh system. It is possible also that the Kell system will be of interest in population studies. The antigen K has been found almost exclusively among European populations or individuals of European descent. At one time it was considered a European trait. Recently sporadic appearances of the K antigen have been reported from Africa and among some American Indians.

The Duffy Blood Group System

The Duffy system was discovered in an individual who was suffering from hemophilia and had received a number of blood transfusions over a period of 20 years. The anti-Duffy antibody was discovered in his serum. The gene for the antigen has been labeled Fy^a and its allele Fy^b. The phenotypes are Fy(a+b−), Fy(a+b+), and Fy(a−b+). The Duffy alleles Fy^a and Fy^b appear to be inherited as nondominant autosomal alleles. A third allele in the Duffy system was postulated when a survey of human populations for frequencies of the Duffy phenotypes uncovered a large number of blood samples that reacted as if the phenotype were Fy(a−b−). Most of these Duffy negative individuals were of African descent. Recently a large number of individuals from Yemen proved to be Fy(a−b−). This phenotype seems to be due to another Duffy allele Fy. If this is the case, the Fy allele is inherited as an autosomal recessive.

The Kidd Blood Group System

Discovery and naming of the Kidd blood groups was the result of hemolytic disease in a newborn, a child of a Mrs. Kidd. At present tests have been made with two antibodies, anti-Jk^a and anti-Jk^b. The fact that the two antisera will recognize four phenotypes suggests that the genes are inherited as nondominant autosomal alleles. The calculation of expected phenotype frequencies among the offspring of various Kidd mating types supports this view. It must be noted that most work is done with anti-Jk^a antisera, for the supply of reliable anti-Jk^b is small, and most population data are reported as phenotypes Jk(a+) and Jk(a−). At present there do not seem to be any special peculiarities about the Kidd blood group system, but this may be because it was discovered relatively recently and extensive work on it has not yet been done. It is believed that the Kidd system will become of interest in studies of human populations, for there seem to be striking differences in the frequencies of the phenotype Jk(a+). Among people of West Africa, this phenotype has a frequency of 95 per cent, among Chinese about 50 and among Europeans 77 per cent.

Other Blood Group Systems

The Diego blood groups were first found in Venezuela, where a new antibody anti-Di^a was found in a woman who gave birth to a child suffering from hemolytic disease. Considerable work with this antibody indicates that it is extremely rare in persons of European descent but moderately common among South American Indians and present among Chinese, Japanese, and certain North American Indians. Some authorities believe it is an antigen of so-called mongoloid peoples, that is, the inhabitants of Asia and their descendants the North and South American Indians. We have found it in a few individuals on New Guinea, and it has been reported as absent among Eskimos. Therefore we shall not call it a mongoloid or Indian gene for the time being. The Diego antigen seems to be inherited as a simple autosomal dominant. It is believed that further work with it will prove to be of great interest for physical anthropologists.

The Sutter blood groups were discovered in 1958. Their discovery appears to be the first step in defining a new blood group system. The antigen Js^a was found during cross matching of a hospital patient prior to blood transfusion. He had previously been transfused and apparently had developed anti-Js^a to the antigen on the donor's cells. The present evidence is that Js^a is inherited as an autosomal dominant. The Sutter blood groups appear to be most frequent among Africans or those of African descent. They have not yet been found in American Indians, Eskimos, or Asians and have been found in low frequencies among persons of European descent.

The Auberger blood groups were discovered in France in 1961. The antigen Au^a seems to be inherited as a simple autosomal dominant. If it proves to be independent of all other blood groups, and it seems that it will, it will be a most useful addition to the blood group systems used in studying human populations. It is already clear that two blood groups can be distinguished among European populations. About 82 per cent of those tested are classified as Au+ and 18 per cent as Au−.

One sex-linked blood group system, the Xg system, has been discovered. Since it was first reported in 1962, it has become the subject of much genetic research. One antibody anti-Xg^a is known, and it identifies two phenotypes Xg(a+) and Xg(a−). The antigen Xg^a appears to be inherited as a nondominant sex-linked trait. The Xg^aY males and Xg^aXg^a females have red cells that react strongly with anti-Xg^a. Females who are clearly heterozygous Xg^aXg have erythrocytes which react variably with the antibody. The frequency of the Xg(a+) phenotype is about the same in the populations tested so far—European, North American, Jamaican. Xg^a is a common sex-linked trait, relatively simply determined in serological laboratories, and as such it has great potential value for human geneticists.

There are some other blood group antigens which are so frequent that it is possible they are present in all humans and therefore might be called species antigens. These are often called public antigens for the same reason. It is possible, of course, that these are part of already known blood group systems. There is no point in discussing them in detail, but it is possible to list them (Table 26.11). A number of them have been demonstrated to be independent of other blood group systems, and the weight of authoritative opinion right now is that they will indeed prove to be independent.

There are other uncommon blood groups called private antigens because they may, indeed, be found only in single families. They are listed in Table 26.11. We need not discuss them here except to point out that they do exist and they can be of great value in working out particular genetic problems.

Applications

The human blood group systems are extremely interesting genetically controlled traits of man. We will explore the effect of natural selection on these systems in the next chapter. We have already seen that they are of use in studying the migration of human populations or the gene flow among human groups (Chapter 24), but more details will be given.

The evidence present in the literature suggests that rather little gene flow occurred between Americans of African descent and American Indians. Many of the comparisons made between American Indians and Americans of African descent used American Indian groups who did not have much opportunity to mix with Americans of African descent, for instance, Chip-

TABLE 26.11
Some Very Common (Public) and Very Rare (Private) Antigens in Man

Public antigens	Incidence
Vel	> 99 per cent
Yt^a	> 99
Ge	> 99
Lan	> 99
Sm	> 99

Private antigens	How discovered
Levay	Transfusion reaction
Jobbins	Hemolytic disease of newborn
Becker	Hemolytic disease of newborn
Ven	Hemolytic disease of newborn
Wr^a(Wright)	Hemolytic disease of newborn
Be^a	Transfusion reaction and hemolytic disease
Rm	Hemolytic disease of newborn
By	Hemolytic disease of newborn
Chr^a	Cross matching
Sw^a(Swann)	Cross matching
Good	—
Bi(Biles)	Cross matching
Tr^a	Cross matching

pewa Indians of Minnesota. Pollitzer and his colleagues compared blood group allele frequencies of Americans of African descent with those of the Cherokees of North Carolina, an American Indian tribe which had far more opportunity to mix with the descendants of African slaves than the Chippewas. Once again the data are consistent with the view that there has been a negligible amount of gene flow between American Indians and Americans of African descent. But this conclusion should not, in all probability, be generalized to all local populations.

The total amount of genetic change in the population of Americans of African descent since they came from Africa can be estimated by using blood group alleles that are absent or very infrequent among Americans of European descent and American Indians and that have a relatively high frequency in African populations. The frequency of the antigen, V,ce^S (Rh system, Table 26.9) is estimated as 0.225 in Africa; among Americans of African descent, the frequency is estimated as 0.144; among Americans of European descent the frequency is about 0.002; and V,ce^S is almost absent among American Indians. We find that the total amount of mixture m (Chapter 24) in the past six generations from non-African American populations into Americans of African descent is 36 per cent (0.36). This example shows that where gene flow is occurring between populations a blood group allele that is rare or absent in some of the populations can be extremely useful in estimating the extent to which the gene pool of a migrant group has been diluted.

The Pygmies of Africa and of southeast Asia prove not to be unique populations that must be explained with special histories and unusual speculations. The European Gypsies very likely originated in India, for the blood group frequencies of Gypsies are like those of the inhabitants of India and unlike those of the Hungarians among whom they lived. The Lapps of northern Scandinavia and the Basques of the Pyrenees become less difficult to understand when the blood group alleles are studied. Each has been considered a remnant of an ancient population, Asiatic and original European respectively. Careful study of the blood group allele frequencies shows that each is an isolate. Each has evolved into a relatively distinctive group, probably because of isolating mechanisms. Each is a European population. There is a gradient of allele frequencies into those of populations that live around them, an observation which suggests that the isolation has been less than absolute.

It would be most interesting to see what would happen if those aboriginal peoples reported to have rules of preferential mating involving relatives do indeed practice the consanguinity implied by such rules. Societies with a preference or a rule for cousin marriage do not produce an abnormally large number of cousin matings—there are not enough cousins to go around (Chapter 24). There is some evidence from a few studies of blood group alleles and other genetically controlled biochemical traits that there is no necessary correspondence in nonwestern, nonurban societies between genetic and terminological kinship. This is not exactly a new idea, but it is one which should be examined in greater detail in the future. The importance of paying some attention to such detail is illustrated in a study of a small Peruvian Indian group, the Shipibo. Seven exclusions were found; all but one were maternity exclusions, and one was both a maternity and a paternity exclusion. The implication is that genetic and terminological motherhood are not identical in this Indian group.

We can also illustrate some important aspects of gene action with odd discoveries which turn up in laboratories of serology. The minus-minus blood group phenotypes, as Race and Sanger call them, are individuals whose red cells do not react with any of the antisera of a particular blood group system. If the loci for any of the alleles in the system are present in such individuals they are not producing antigens. We may ask: Why are they not? Alleles which do not have a discernible gene product, such as the allele O, are often called amorphs. An amorph is an allele that does not do a thing!

The activity of a genetic locus may be affected by other loci which can suppress or modify the expression or activity of that locus. The piece of the chromosome on which the allele is located may be deleted, inverted, or translocated. If a piece of chromosome is deleted sometime during cell division, many peculiarities of phenotype are expected. Translocation and inversion simply mean that the locus has been moved through some rearrangement of chromosomal material and in its new position does not produce its normal product. Suppressor or modifier genes and deletions, translocations, and inversions of chromosomal material have long been recognized among fruit flies, bacteria, and others. They are not usually easily identifiable in human material. The discovery of minus-minus blood group phenotypes (Table 26.12) requires a genetic explanation. Minus-minus phenotypes do not generally seem to be biologically or medically abnormal. Certain blood group loci, if they are present, produce no gene product. Here we may have an example in man of a suppressor gene or a chromosomal alteration.

TABLE 26.12
Examples of Minus-Minus Phenotypes in Some Blood Group Systems

System	Phenotype	Probable genotype	Negative reaction to antisera
ABO	O	*OO*	anti-A, anti-B
Rh	- - -	- - -/- - -	all Rh antisera
Lutheran	Lu(a−b−)	*LuLu*	anti-Lu^a, anti-Lu^b

One of the most obvious applications of the blood group systems is a medicolegal one used in cases of disputed paternity. The mechanism of mammalian reproduction makes it unlikely that a child's mother can remain unrevealed. If she accuses a particular man of fathering her child, examination of the blood types of the mother, child, and putative father may aid the courts. A man cannot be identified as the father of a child on the basis of the blood groups, but he may be excluded. One obvious example is the man of blood type O who is accused of fathering a child of blood type AB. An AB child must have either one parent of group A and the other parent B or both parents AB or one parent AB and the other A or B. Table 26.13 illustrates the expectable phenotypes of offspring of various matings for several blood group systems. Anthropological applications of this feature of exclusion with blood groups have not been frequent.

A large number of human blood group systems have been described, and they are among the best known genetically controlled traits of man. The blood group systems provide us with a powerful tool for study of the genetics of human populations; they give us a means of studying the interaction between marriage rules and genetics of primitive societies; they have made possible safe and routine use of blood transfusions in medical practice, and many cases of hemolytic disease of the newborn are better understood and are susceptible to treatment because of our knowledge of these blood groups. Lawyers have an additional line of evidence in defending or prosecuting putative fathers. For anthropologists one of the most important of all the applications of the blood group systems is the determination of frequencies of various blood group alleles in many human populations. Not only can we apply this information to analyzing the history of many human

TABLE 26.13

Matings and Offspring in Selected Blood Group Systems

Mating	Offspring	
	Possible phenotypes	Impossible phenotypes
ABO System[a]		
O × O	O	A, B, AB
O × A	O, A	B, AB
O × B	O, B	A, AB
O × AB	A, B	O, AB
A × A	O, A	B, AB
A × B	O, A, B, AB	none
A × AB	A, B, AB	O
B × B	O, B	A, AB
B × AB	A, B, AB	O
AB × AB	A, B, AB	O
MNSs System[b]		
M × M	M	N, MN
M × MN	M, MN	N
M × N	MN	M, N
MN × MN	M, MN, N	none
MN × N	MN, N	M
N × N	N	M, MN
Duffy System[c]		
Fy(a+) × Fy(a+)	Fy(a+),Fy(a−)	none
Fy(a+) × Fy(a−)	Fy(a+),Fy(a−)	none
Fy(a−) × Fy(a−)	Fy(a−)	Fy(a+)

[a]Phenotypes defined by anti-A and anti-B antisera.
[b]Phenotypes defined by anti-M and anti-N antisera.
[c]Phenotypes defined by anti-Fy[a] antisera.

groups, but we can also use it to study the action of natural selection on man.

SUGGESTED READINGS AND SOURCES

Boorman, K. E., and B. E. Dodd, *Blood Group Serology*. Little, Brown, Boston (1957).

Boyd, W. C., *Genetics and the Races of Man*. Little, Brown, Boston (1950).

Boyd, W. C., *Fundamentals of Immunology* (Third edition). Interscience, Wiley, New York (1956).

Boyd, W. C., Blood groups and soluble antigens. *In* W. J. Burdette (Ed.), *Methodology in Human Genetics*, p. 335. Holden-Day, San Francisco (1962).

Boyd, W. C., Four achievements of the genetical method in physical anthropology. *Am. Anthropol.* **65**, 243 (1963).

Buettner-Janusch, J., J. R. Bove, and N. Young, Genetic traits and problems of biological parenthood in two Peruvian Indian tribes. *Am. J. Phys. Anthropol.* **22**, 149 (1964).

Ceppellini, R., On the genetics of secretor and Lewis characters: a family study. *Proc. 5th Intern. Congr. Blood Transf.*, p. 207 (1955).

Ceppellini, R., Physiological genetics of human blood factors. *In* G. E. W. Wolstenholme and C. M. O'Connor (Ed.), *Biochemistry of Human Genetics*, p. 242. Little, Brown, Boston (1959).

Chung, C. S., E. Matsunaga, and N. E. Morton, The MN polymorphism in Japan. *Japan. J. Human Genet.* **6**, 11 (1961).

Cushing, J. E., and D. H. Campbell, *Principles of Immunology*. McGraw-Hill, New York (1957).

Hiraizumi, Y., Are the MN blood groups maintained by heterosis? *Am. J. Human Genet.* **16**, 375 (1964).

Hirszfeld, L., and H. Hirszfeld, Serological differences between the blood of different races. The result of researches on the Macedonian front. *Lancet* **2**, 675 (1919).

Kabat, E. A., *Blood Group Substances, Their Chemistry and Immunochemistry*. Academic Press, New York (1956).

Landsteiner, K., Zur Kenntnis der antifermentativen, lytischen und agglutinierenden Wirkungen des Blutserums und der Lymphe. *Zentr. Bakteriol.* **27**, 357 (1900).

Landsteiner, K., and P. Levine, A new agglutinable factor differentiating individual human bloods. *Proc. Soc. Exp. Biol. N. Y.* **24**, 600 (1927).

Landsteiner, K., and A. S. Wiener, An agglutinable factor in human blood recognized by immune sera for rhesus blood. *Proc. Soc. Exp. Biol. N. Y.* **43**, 223 (1940).

Levine, P., and R. E. Stetson, An unusual case of intragroup agglutination. *J. Am. Med. Assoc.* **113**, 126 (1939).

Mann, J. D., A. Cahan, A. G. Gelb, N. Fisher, J. Hamper, P. Tippett, R. Sanger, and R. R. Race, A sex-linked blood group. *Lancet* **1**, 8 (1962).

Matsunaga, E., Some evidences of heterozygote advantage in the polymorphism of MN blood groups. *Proc. 8th Congr. Intern. Blood Transf.*, Tokyo, p. 126 (1962).

Morgan, W. T. J., and W. M. Watkins, Some aspects of the biochemistry of the human blood group substances. *Brit. Med. Bull.* **15**, 109 (1959).

Morton, N. E., and C. S. Chung, Are the MN blood groups maintained by selection? *Am. J. Human Genet.* **11**, 237 (1959).

Mourant, A. E., *The Distribution of the Human Blood Groups*. Blackwell, Oxford (1954).

Mourant, A. E., A. C. Kopec, and K. Domaniewska-Sobczak, *The ABO Blood Groups. Comprehensive Tables and Maps of World Distribution*. Blackwell, Oxford (1958).

Pollitzer, W. S., R. C. Hartmann, H. Moore, R. E. Rosenfield, H. Smith, S. Hakim, P. J. Schmidt, and W. C. Leyshon, Blood types of the Cherokee Indians. *Am. J. Phys. Anthropol.* **20**, 33 (1962).

Race, R. R., The Rh genotypes and Fisher's theory. *Blood* 3, suppl. 2, 27 (1948).

Race, R. R., and R. Sanger, *Blood Groups in Man* (Fourth edition). Blackwell, Oxford (1962).

Rege, V. P., T. J. Painter, W. M. Watkins, and W. T. J. Morgan, Isolation of a serologically active, fucose-containing, trisaccharide from human blood group Le^a substance. *Nature* **204**, 740 (1964).

Rosenfield, R. E., F. H. Allen, S. N. Swisher, and S. Kochwa, A review of Rh serology and presentation of a new terminology. *Transfusion* **2**, 287 (1962).

Schiff, F., and H. Sasaki, Der Ausscheidungstypus, ein auf serologischen wege Nachweisbares Mendelndes Merkmal. *Klin. Wochschr.* **34**, 1426 (1932).

Thieme, F. P., and C. M. Otten, The unreliability of blood typing aged bone. *Am. J. Phys. Anthropol.* **15**, 387 (1957).

Watkins, W. M., and W. T. J. Morgan, Possible genetical pathways for the biosynthesis of blood group mucopolysaccharides. *Vox Sanguinis* **4**, 97 (1959).

Wiener, A. S., Genetic theory of the Rh blood types. *Proc. Soc. Exp. Biol. N. Y.* **54**, 316 (1943).

Wiener, A. S., E. B. Gordon, and J. Moor-Jankowski, The Lewis blood groups in man. *J. Forensic Med.* **11**, 67 (1964).

Natural Selection and Blood Groups

27

WE STATED EARLIER that differences in allele frequencies in different human groups are mainly the result of the action of natural selection on the gene pools of the population (Chapter 24). One way we may describe the gene pools of human groups is by the use of relatively simple blood grouping techniques to survey various populations and determine the frequencies of blood group alleles.

We have already discussed a method of estimating the frequencies of the *A*, *B*, and *O* alleles from the observed frequencies of the phenotypes. We shall not discuss statistical methods for estimating frequencies of alleles in the other blood group systems, except to note that they are available.

Before any of the more refined and accurate statistical methods for estimating frequencies may be applied with profit to blood group data, the reliability of the data must be assured. A very important problem in studies of the general significance of blood group allele frequencies in populations is whether the population sample accurately represents the population. We cannot make a digression here to discuss sampling statistics, but the reader is warned that it is of the greatest importance that proper precautions be taken in choosing samples within populations and in attributing character-

istics to populations from samples. But there is more than a problem of statistical method involved.

There are problems associated with work in remote and tropical areas which are inhabited by many aboriginal populations studied by anthropologists. When we are fortunate enough to get into these areas and obtain blood samples, we must make all sorts of arrangements to preserve the whole blood and ship the samples to well-equipped serology laboratories. The blood group research of the physical anthropologist thereby becomes increasingly dependent upon local transportation and refrigeration facilities, airline schedules, customs and public health personnel, and many other unexpected contingencies. Major delays and improper handling in transit may lead to damage of the samples, and, consequently, misleading results will be obtained.

Another source of error lies in the laboratory itself. If crude serological methods are used or if there are clerical or technical errors in the laboratory, misleading results will be published. The reader should be aware that blood typing is fraught with difficulties, fascinating and complex. Each published frequency of an allele is the result of a great amount of work, much of it designed to circumvent the difficulties mentioned here and many others.

The greatest body of human genetic data is probably found in the published frequencies of the ABO blood group phenotypes and alleles. The development of blood transfusions as a major medical advance and the growth of blood banks required the testing of thousands of individuals. Mass casualties in the great wars of this century required ABO blood typing, and the blood types of millions of people became known. When blood transfusions are to be given, the absolute minimum information required is the ABO type of both donor and recipient.

Distribution of Blood Groups

The average worldwide gene frequencies for the three alleles of the ABO system are approximately as follows: $O = 0.623$, $A = 0.215$, $B = 0.162$. Such an average is rather misleading because the A, B, and O alleles have a striking, nonuniform distribution (Fig. 27.1-27.3). We shall cite here only some of the more unusual features of the worldwide distribution.

Blood groups A and B are believed to have been absent from the aboriginal Indian inhabitants of Central and South America. However some of the highest frequencies known for allele A have been reported for North American Indian groups. Allele B is absent from North American Indians. Many workers are of the opinion that reported cases of blood groups A and B among Indians of Central and South America are the result of interbreeding of the Indian ancestors with persons of European origin.

Allele A shows a high frequency in parts of Europe and Asia and among certain aboriginal groups in Australia. Allele B is absent from most Aus-

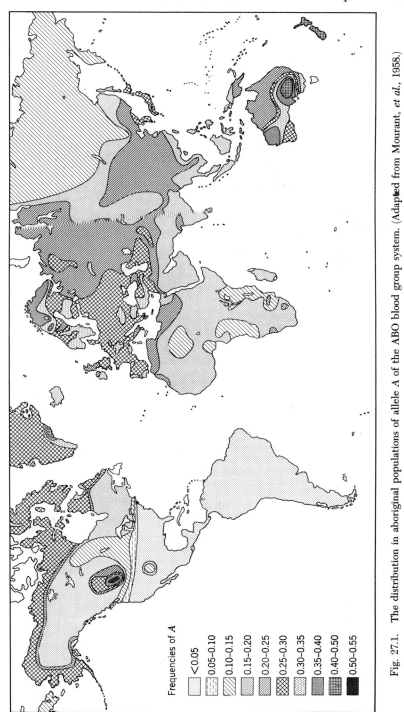

Fig. 27.1.   The distribution in aboriginal populations of allele *A* of the ABO blood group system. (Adapted from Mourant, *et al.,* 1958.)

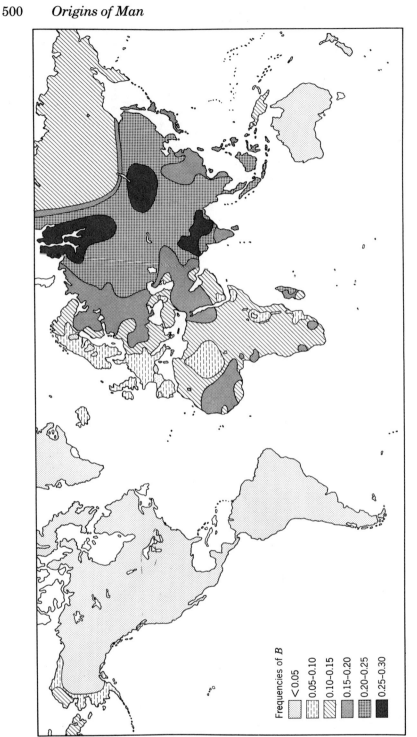

Fig. 27.2.   The distribution in aboriginal populations of allele *B* of the ABO blood group system. (Adapted from Mourant, *et al.*, 1958.)

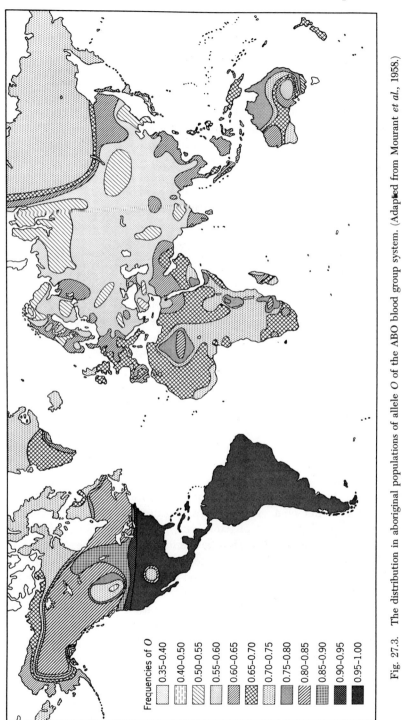

Frequencies of *O*

	0.35–0.40
	0.40–0.50
	0.50–0.55
	0.55–0.60
	0.60–0.65
	0.65–0.70
	0.70–0.75
	0.75–0.80
	0.80–0.85
	0.85–0.90
	0.90–0.95
	0.95–1.00

Fig. 27.3.   The distribution in aboriginal populations of allele *O* of the ABO blood group system. (Adapted from Mourant *et al.*, 1958.)

tralian aboriginals, although it has been reported in some tribes that live in the extreme northeastern part of the continent. It is believed its presence is due to admixture with Europeans. The maximum frequency of allele *B* is found in northern India and central Asia. It is possible to visualize a geographical gradient which shows that the frequency of allele *A* increases and the frequency of allele *B* decreases as one moves westward from the Pacific coast of Asia to the Atlantic coast of Europe. This gradient follows, with some consistency, changes in altitude and environment (Fig. 27.2). It has also been discovered that statistically significant differences in frequencies of the alleles for this system occur between populations of adjacent villages in Sicily, groups in neighboring regions of Italy, and groups in parts of Europe. The distribution of the ABO blood group alleles provides evidence for a large amount of local variation. The extreme polymorphism of the ABO blood group alleles is the principal reason we believe that selection operates on them.

The distribution of the frequencies of the M and N antigens is shown in Fig. 27.4. Most of the populations tested for the MNSs system show a frequency for allele *M* of 0.5 to 0.6. However a higher frequency of *M* has been found among American Indians, Eskimos, the inhabitants of certain parts of European Russia, and the peoples of southern Asia, and the frequency of *N* is higher than *M* throughout the Pacific area. The observation that an unexpectedly high proportion of the offspring of the double heterozygote *MN* × *MN* matings were of MN type suggested that selection operates on the heterozygote and maintains a balanced polymorphism. The *S* allele appears to be extremely rare among Australian aboriginals, whereas it has a relatively high frequency, perhaps 20 per cent, among aboriginal peoples of New Guinea.

A number of other antigens have been shown to be part of the MNSs system. These may be significant in anthropological studies when the frequencies have been determined in a sufficient number of populations. For example in the groups already surveyed, the Hunter antigen has been found in less than 1 per cent of Americans of European descent, whereas it has been found in 7 per cent of Americans of African descent and in 22 per cent of West Africans. The Henshaw antigen has also been found in West Africa, but not among Europeans tested.

Discussion of the distribution of the alleles of the Rh system is best left until such time as the genetic mechanisms controlling this system are better understood. The most significant generalization we can make is that Rh negative (*cde/cde*) individuals are more frequent among Europeans and persons of European descent than among Africans and persons of African descent. The *cDe* locus has very high frequencies (up to 89 per cent) in some African populations tested.

At present the distribution of the ABO and MN blood groups are certainly the best known. It is for this reason that the distribution maps (Fig. 27.1-27.4) are limited to alleles in these two systems. The distribution

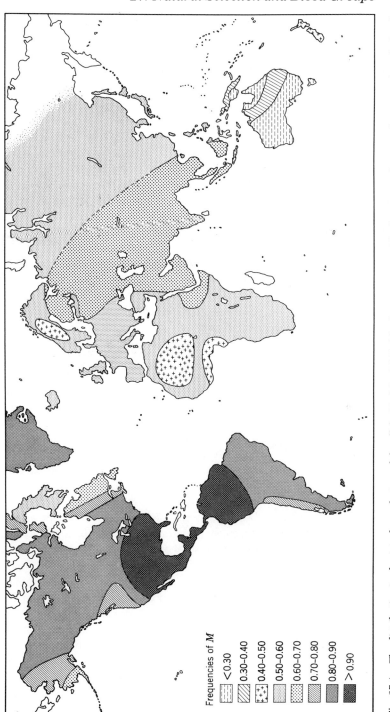

Fig 27.4. The distribution in aboriginal populations of the allele *M* of the MNSs blood group system. The frequency of *N* in each area can be calculated simply as 1.00 minus the frequency of *M*. (Adapted from Mourant, 1954.)

Frequencies of *M*

- <0.30
- 0.30–0.40
- 0.40–0.50
- 0.50–0.60
- 0.60–0.70
- 0.70–0.80
- 0.80–0.90
- >0.90

of frequencies of these alleles shows that man is indeed polymorphic at these loci. We will continue to derive more information about all blood group systems as larger quantities of reliable antisera become available and as more populations are surveyed.

Natural Selection and Nonadaptive Traits

The role played by natural selection in establishing and maintaining the observed frequencies of human blood group genes has become, in the past decade, a major research problem for physical anthropologists. The frequencies of the four phenotypes of the ABO system—A, B, AB, and O—are very different in human populations. The relative frequencies are clearly examples of polymorphisms, genetic polymorphisms. The known distribution of frequencies (Fig. 27.1–27.3) fits the hypothesis that there is some relationship between the environment and the incidence of the various ABO blood group alleles. The variance and the geographical gradient for alleles *A* and *B* are consistent with the hypothesis that both of these genes are affected by environmental selection. The fact that allele *B* decreases in frequency in human populations as one moves from east to west in Eurasia is an indication that we should examine the blood groups of the populations in this area in considerable detail.

The view that the blood groups are nonadaptive traits, that they are unaffected by environmental selection, was held for many years. This view was held particularly tenaciously during the first half of this century and is still with us. There is a historical reason for this. Nonadaptive traits were believed to be the best kinds to use in making racial classifications. It was believed that if traits were unaffected by the environment they would reflect the history of the groups, the populations, which were being classified. If they were nonadaptive, they would not change and hence would reflect the composition of the populations which were the ancestors of contemporary populations. But as Boyd and many others have pointed out, to say that the traits which distinguish human populations from each other are nonadaptive is to reject the simplest of the hypotheses—natural selection—which accounts for the origins of the physical differences among human populations.

The demonstration that several blood group systems are involved in conditions which have a differential effect upon the survival or fertility of individuals led anthropologists and others to reconsider the whole question. The discovery of the Rh system, for example, was partly the result of hemolytic disease caused by the Rh type of an infant. This suggested at once that such factors as disease and incompatibility between a mother and her fetus might be involved in maintaining the observed genetic polymorphisms of the blood groups.

How does selection act to maintain the observed frequencies of the blood group genes in human populations? Some of the most promising lines of investigation and some of the hypotheses proposed to answer this question will be discussed in the pages that follow.

The ABO Blood Groups and Disease

The relationship of ABO blood group phenotypes to disease has been discussed on and off almost since the blood groups were first discovered in 1900. Despite the fact that some excellent studies of the relationship between blood type and susceptibility or resistance to certain diseases were published between 1921 and 1953, it was not until the 1950's that major interest in this problem was revived. Aird, Bentall, and Roberts in 1953 demonstrated a statistically significant association between blood group A (phenotype A) and cancer of the stomach. Shortly thereafter, a statistically significant association was demonstrated between persons of phenotype A and pernicious anemia. Pernicious anemia may be a precursor of cancer of the stomach. It is clear from studies made so far that, at least in European, urban, industrialized societies, persons of phenotype A are more apt to develop cancer of the stomach than are O and B persons. The increased likelihood is as high as 20 per cent. A question that must eventually be answered is: What is the relationship between persons of blood type AB and cancer of the stomach? The number of AB individuals tested so far is too small for any statistical results to be significant.

Another association of statistical significance was discovered between duodenal ulcers and individuals of phenotype O. Various studies, particularly those carried out in England, demonstrate that duodenal ulcers are almost 40 per cent more common among persons of blood group O than in the other ABO phenotypes. No difference in the frequency of duodenal ulcer among phenotypes A, B, and AB has been demonstrated. Gastric ulcers, too, are more common among O phenotypes than among the other three. An interesting discovery, and one of great significance, was that duodenal ulcers were much more common among nonsecretors of the A, B, or H group specific substances than among secretors.

These attempts to find associations between blood groups and disease have produced one general finding that may prove to be of great importance. There are a number of significant associations between diseases of the gastrointestinal tract and one or another of the ABO phenotypes.

A number of other pathological conditions have been associated statistically with one or another blood group phenotype in some studies but not in others. Among the more interesting conditions examined where inconclusive results have been obtained are diabetes mellitus, tumors of the salivary gland tissue, and cancers of the female genitalia. In one study a significant excess of nonsecretors was found among those who suffered from the rheumatic consequences of streptococcal infections. Bronchopneumonia among very young children and infants appeared to be related to phenotype A in one study. A large variety of attempts to associate blood groups and disease have been made, and these are summarized in Table 27.1. It is difficult to say what associations of the sort described here mean. These studies which test the significance, in a statistical sense, of the associations are not studies which determine possible functional relationships between the ABO groups and the conditions listed.

TABLE 27.1
Statistical Associations between Blood Group Phenotypes and Disease

Disease	Associated ABO or secretor phenotype	Number of studies	Countries
Duodenal ulcer	O	8	England, Scotland, U.S.A., Denmark, Norway, Austria
Gastric ulcer	O	8	England, Scotland, U.S.A., Denmark, Norway, Austria
Cancer of stomach	A	8	England, Scotland, U.S.A., Norway, Austria, Australia, Switzerland
Pernicious anemia	A	4	England, Scotland, U.S.A., Denmark
Rheumatic fever[a]	Excess of nonsecretors	1	Great Britain
Paralytic poliomyelitis	Excess of nonsecretors, B reduced	11	England, U.S.A., Denmark, Italy, Germany, France
Diabetes mellitus	A	5	England, Scotland, Austria
Salivary gland tumors	A	2	U.S.A.
Cancer of cervix	A	3	England, Austria, Italy, Germany
Tumors of ovary	A	1	U.S.A.
Adenoma of pituitary	O	1	U.S.A.
Cancer of pancreas	A	1	England
Tender-minded[b]	A	1	U.S.A., Italy
Tough-minded[b]	AB	1	U.S.A., Italy

[a]The association for this and the following entries are inconclusive.
[b]Personality trait.

It has been suggested by many that these statistical associations imply no causal relationship whatsoever. Some have said that the associations are due to secondary stratifications of the populations studied. This means that there might be two things happening at the same time which are confused by the statistical study—a high frequency of the blood group allele *O* and an unusually high susceptibility to duodenal ulcer. Other segments of the population might at the same time show unusually high frequencies of allele *A* and unusual susceptibility to cancer or unusually high resistance to duodenal ulcers. In general, on logical grounds, such hypotheses are unlikely. First, they violate the rule that we should not multiply postulated and unknown factors unduly. Second, if such stratification were indeed the case, it is highly unlikely that it would occur in as many different populations as have already been studied. Third, there is no evidence that the large, urbanized, industrialized populations of Great Britain and North America can be segmented into smaller populations descended from homogeneous immigrant groups. Finally, certain studies have conclusively demonstrated that a significant excess of duodenal ulcers in nonsecretors of blood type O is definitely not a case of stratification.

There have been suggestions that the associations between disease and blood group genes are due to pleiotropic effects of the genes. This simply means that the blood group alleles have more than one effect. In other words the genes not only determine the blood group specific substance, but they also determine susceptibility or resistance to disease. This does not answer any questions; it simply restates the problem in different words.

As Race and Sanger have pointed out, the argument that there is an association between blood groups and disease has been vindicated, and the genetic polymorphism which Ford and others noted, as long ago as 1942, requires a functional explanation. Further demonstration of the reasonableness of the statistical associations is not required. What is required now is the study of the functional relationship between the diseases noted or any diseases and the blood group specific substances. It is no longer reasonable to dismiss the blood group phenotypes and genes as neutral.

The diseases we have mentioned here probably have little effect on the survival of various genes in the population. We must point out that at the very best the diseases produce second-order effects, that is, most of the conditions we have discussed affect persons who are into or past the age of reproduction. Therefore selection will have little effect on their contribution to the gene pool of the next generation.

The statistical studies really give us problems for further research. We do not mean to derogate from such studies, but no statistical study will tell us much about the function of the antigens and the selective mechanisms involved, no matter how elaborate it is, no matter how carefully controlled, how sophisticated, and how large and well conducted. The demonstrated statistical associations make it possible for us to speculate about some mechanisms involved in the associations between blood groups and disease.

It is interesting to note that a number of the diseases that are related to ABO blood groups are disorders of the gastrointestinal tract. Since the gastrointestinal tract in secretor individuals contains large amounts of A, B, or H substance, it may be that the chemistry of the group specific antigens is involved in determining susceptibility to carcinoma and duodenal ulcers. Associations between ABO blood group genes and types of diet have been reported, but experts in the field of nutrition are skeptical of their validity. At any rate the results suggest that some attention be paid in the laboratory to the possible meaning these associations may have for the known high levels of A, B, or H antigen in the gastric mucosa of secretors.

The observation that secretors of phenotype B are much less exposed to the risk of the rheumatic sequelae of streptococcal infections of the throat suggests that these group specific substances may make the streptococcus bacteria completely antigenic and more susceptible to the normal immunological defenses of the body. The ABO antigens have marked chemical similarity to certain pathogenic organisms such as streptococci. The hypotheses that group B individuals are less susceptible to streptococcal infections, group A individuals to plague, and group O to syphilis should be testable in laboratory and in clinical studies.

We have concentrated on the relationship between the ABO blood group alleles and disease because there is a vast amount of information available about these blood groups. If we had the same amount of information about the other blood group systems, we would probably find that other alleles are also associated with various disease conditions. Further studies of statistical associations between ABO blood groups and other traits provide us with relatively less and less information, and we believe that functional studies should now be undertaken in this area of research.

Blood Group Incompatibilities and Diseases of Newborns

The most direct effect selection has upon blood groups can be discerned by analyzing hemolytic disease of newborns. This disease is so severe in many cases that stillbirth results or the infant dies shortly after birth. The mechanism by which fetal disease is related to blood groups, specifically the ABO and the Rh blood groups, is a relatively simple one. Nonetheless it took many years of research to find out the way in which this mechanism worked. Hemolytic disease of the newborn, sometimes called *erythroblastosis fetalis*, is caused by the production of antibodies by the mother in response to the presence of antigens which the fetus has inherited from the father and which the mother lacks. The antigen usually gets into the mother's circulation on fetal red cells which penetrate the placenta. The antibody produced by the mother then filters back into the circulation of the fetus and damages or destroys the red cells of the fetus (Fig. 26.3). (For a long time it was believed that Rh incompatibility was more significant than any other in producing hemolytic reactions in the fetus.) Hemolytic disease of the newborn proves that a mechanism for selection against certain blood group phenotypes exists.

Once the mechanism was demonstrated, its effect on ABO and Rh phenotypes was studied statistically. Surveys were made to obtain the frequencies of phenotypes among offspring of mothers of various phenotypes and of matings of various phenotypes. These observed frequencies of phenotypes were compared with frequencies predicted for Hardy-Weinberg equilibrium to determine whether there were significant losses of any phenotype among offspring. Two kinds of matings are defined, homospecific and heterospecific. Homospecific matings are those in which the female either has the same red cell antigen as the male or has no specific antibody in her blood serum for the antigens of the male and is not capable of producing one. In heterospecific matings the female either has an antibody present in her blood serum for the red cell antigen of the male or is capable of producing such an antibody. Heterospecific pregnancies are those in which the female either has or can produce an antibody to the red cell antigens of the fetus she carries. Heterospecific and homospecific matings and pregnancies of the ABO blood group system are listed in Table 27.2. A heterospecific mating, for example, is one between a male of blood type A and a female of blood type O or B. The female carries in her serum anti-A antibody which will agglutinate the red cells of her husband, and, if she has a fetus of blood type A, she carries an antibody antagonistic to the red cells of her fetus. This is also called ABO incompatibility. The matings are sometimes called compatible or incompatible rather than homospecific or heterospecific.

TABLE 27.2
Homospecific and Heterospecific Matings
and Heterospecific Pregnancies of the ABO Blood Group Phenotypes

Homospecific matings	Heterospecific matings	Heterospecific pregnancies	
♂ · × ♀	♂ × ♀	Mother	Fetus
O × O	A × O	O	A
O × A	A × B	O	B
O × B	B × O	A	B
O × AB	B × A	A	AB
A × A	AB × O	B	A
A × AB	AB × A	B	AB
B × B	AB × B		
B × AB			
AB × AB			

There is an interaction between the Rh blood group system and the ABO blood group system in incompatible mothers and fetuses. Mothers are able to eliminate ABO incompatible fetal red cells from their circulation and in this way protect themselves against incompatibility due to the Rh system. The mechanism operates as follows: A woman of blood type O, Rh negative who is married to a man of blood type A, Rh positive may have a fetus of blood type A, Rh positive. If red blood cells from the fetus get into the maternal circulation, they can stimulate production of anti-Rh antibodies and lead to severe hemolytic disease of the newborn. However, the mother since birth has had anti-A antibodies in her circulation. Thus if the fetal red cells are type A as well as Rh positive, the red cell will be destroyed in all probability by the anti-A antibody of the mother. Production of anti-Rh antibodies, in this case, will not be stimulated. It is quite true, of course, that circulating natural anti-A antibodies may very well penetrate the placental barrier and eventually cause jaundice or severe hemolytic disease in the fetus of phenotype A. As early as 1905 Dienst attributed fetal loss and illness of certain pregnant women to an incompatibility between the blood of the mother and the blood of the fetus. The concept of the placental barrier was used at that time as evidence that maternal-fetal incompatibility was contrary to laws of nature. The demonstration that maternal antibodies hemolyze the red blood cells or otherwise damage the fetus clearly shows that the placenta is penetrable by some kinds of chemical substances. The placental barrier is obviously not an absolute barrier.

Mass surveys of the numbers of children of various blood types born to mothers of various ABO and Rh blood types and surveys of fetal deaths according to blood types of mothers have been made. These studies suggest that the postulated mechanism does indeed protect fetuses. There is significantly less hemolytic disease of the newborn in matings that are both ABO and Rh incompatible than there is in a variety of other kinds of matings.

Statistical surveys indicate that the postulated mechanism can have a significant effect on human populations. The problems with which all the studies are concerned may be stated as follows. Is there an increased risk of fetal death associated with certain maternal ABO and Rh genotypes? Are there significant discrepancies between the observed and expected number of children of various blood groups born to heterospecific matings? Do females with antibodies to ABO and Rh antigens (O, Rh negative mothers, for example) have the expected number, fewer, or more children of incompatible blood type?

A recent study of the risk of fetal death, carried out in New York City, increases our appreciation of the magnitude of the effect of ABO and Rh incompatibility between a mother and her fetus. The risk of fetal death is greater when there are differences in ABO blood type than when there are differences in Rh type. Thus a woman of blood type O is more likely to risk loss of a fetus than is an Rh negative woman. The woman of blood type O has anti-A and anti-B antibodies in her blood from birth, and these can attack the fetal red cells at once, if they are type A, B, or AB. In the case of Rh incompatibility, the Rh negative woman must be actively immunized against the antigen of an Rh positive fetus. The first and even second Rh positive fetus she carries may not stimulate a sufficiently strong antibody production to damage the fetus. Thus an Rh negative woman may have one or more Rh positive children before she begins to produce enough antibody to do damage. When the ABO and Rh systems were considered together, AB, Rh negative women had the largest number of stillbirths. Risk of fetal loss was highest in women who had no antibodies of the ABO system which might have protected them and the fetuses against immunization by an Rh positive antigen.

If the AB, Rh negative mothers in this same study were not included in the sample, a very simple interaction between the ABO and Rh loci apparently occurred. Mothers over 24 years of age had a uniformly greater risk of fetal loss as the number of antigenically active alleles in the mother's genotype decreased. Thus not counting AB, Rh negative mothers, the risk of fetal loss was greatest among O, Rh negative mothers. It is not wholly correct to characterize the *O* and the Rh negative (*cde*) alleles as antigenically inactive, although a woman of genotype *OO*, Rh negative will have potent antibodies for the A and B antigens and can be stimulated to produce anti-Rh antibodies. Yet it is also the case that the antibodies to A and B antigens will eliminate fetal red cells of types A, Rh positive and B, Rh positive. Further consideration shows that a woman of *OO*, Rh negative genotype will have, potentially at least, antibodies antagonistic to O, Rh positive fetuses and any A or B fetus regardless of fetal Rh type. The potential interaction of ABO and Rh systems in heterospecific pregnancies are listed in Table 27.3.

TABLE 27.3
The Interaction of ABO and Rh Phenotypes
in Maternal-Fetal Incompatibility

Maternal phenotype[a]	Incompatible fetal phenotypes
AB −[b]	A+, B+, AB+
O −	O+, A+, B+, A−, B−
B −	O+, A+, B+, AB+, A−, AB−
O +	A+, B+, A−, B−
A −	O+, A+, B+, AB+, B−, AB−
B +	A+, AB+, A−, AB−
A +	B+, AB+, B−, AB−
AB +	none

[a]Maternal phenotypes are listed in order of descending risk of fetal death.
[b]Rh phenotypes are designated negative (−) or positive (+).

One question has plagued students of ABO hemolytic disease: Why do not all incompatible pregnancies result in abortion, or at least jaundice, of the newborn? The answer is that the antisera which give the usual anti-A and anti-B reactions are composed of more than one kind of antibody. Detailed analysis reveals that only some ABO antibodies are able to pass the placental barrier and enter the fetal circulation. When these antibodies are isolated from sera of women who have had erythroblastotic infants, they show one special feature not found in ABO antibodies of normal individuals. These antibodies do not become inhibited by, that is, they do not combine with, water-soluble A, B, or H blood group substances. This suggests that there is a difference between the red cell antigen or antigenic determinants for normal circulating anti-A and anti-B antibodies and the antigenic determinants on the water-soluble A, B, or H substances of secretors. Once again we see that detailed study of the chemistry of the substances involved will expand our understanding of the situations discovered in statistical studies.

Blood Group Incompatibilities and Fertility

A very thorough study has been made of Rh negative mothers already known to have been stimulated to produce Rh antibodies. These women were married to Rh positive men. The frequency of various ABO phenotypes in the offspring of these matings deviated significantly from the expected. The differences were of the sort expected if ABO incompatibility of the fetus prevented Rh sensitization of the mother. Prezygotic selection has been suggested as one way in which predicted numbers of each genotype in the ABO system are not produced. A variety of mechanisms, such as unequal production of spermatozoa by heterozygotic males, have been proposed. A direct effect upon the spermatozoa was suggested when the presence of ABO antibodies was detected in uterine secretions. Since the sperm carry haploid complements of chromosomes, a male of genotype *AO* will, in theory, produce one-half of his sperm with the chromosome carrying the allele *A* and the other half carrying the allele *O*. It was suggested that the spermatozoa with the *A* allele might carry A antigen on their surfaces and be affected by uterine anti-A antibodies. The effect on the frequency of ABO phenotypes among offspring would in this case be the same as that due to maternal-fetal incompatibility.

A number of observations support the suggestion that an interaction between spermatozoa and uterine secretions occurs in heterospecific matings. Group O women were shown to have significantly fewer group A children than expected. One study of couples who were infertile for no apparent reason revealed a very large excess of heterospecific matings.

The demonstration of a mechanism by which spermatozoa are agglutinated or otherwise affected by uterine ABO antibodies is difficult. A variety of studies have been made to determine if, for example, anti-A antibodies react with sperm from A phenotypes but not with sperm from O phenotypes. At one time it looked as if this was indeed possible, the results of some studies suggesting that A sperm were so affected, but further work has shown that the experimental results are inconclusive. In other words an antigenic difference between haploid sperm with the A locus and those with the O locus does not seem to exist. But this does not rule out the possibility that an ABO antigen-antibody reaction occurs between spermatozoa and uterine secretions. Secretors of the A, B, or H antigen have the antigen in most tissues and body fluids. It is not unreasonable to suggest that males of the secretor phenotype produce A, B, or H antigen that becomes attached to spermatozoa in the same way as it does to other tissue cells. The spermatozoa would then be affected by some kind of reaction with ABO antibodies. Despite the fact that the effects on the frequencies of ABO phenotypes would be the same as those of maternal-fetal incompatibility and in a statistical sense would be indistinguishable, research into this problem is well worth the effort.

A Model System

In the preceding pages we have presented evidence that the ABO blood groups are subject to factors that affect the survival and fertility of the genotypes. The genes of the ABO blood group system are not neutral genes. There is ABO polymorphism in man. How are the genes in this system maintained at their observed frequencies? What maintains the observed polymorphisms in various populations? These are fundamental questions, and at present the answers are theoretical.

A balanced polymorphism in a two-allele system is maintained by positive selection on the heterozygotes, as we have seen (Chapter 24). The relative fitness of the heterozygote is greater than that of the two homozygotes. In a three-allele system balanced polymorphism, stable equilibrium, exists when the relative fitnesses of the heterozygotes are in general greater than the fitnesses of the homozygotes. A mathematical statement of this has been given by Mandel, and we shall summarize his conclusions.

In a three-allele system there are six genotypes. Let us assign to each of these a selection coefficient which we call a, b, c, d, e, and f. The selection coefficients for the three homozygotes are a, b, and c; for the three heterozygotes, d, e, and f (Table 27.4). Mathematical manipulations lead to the conclusion that the selection coefficients of the heterozygotes are greater than the selection coefficients of the homozygotes for the three-allele system in stable equilibrium. There are four essentially different sets of mathematical relationships, any one of which specifies the conditions for stable equilibrium in a three-allele system in a population (Table 27.4). First, the selection coefficients of all of the heterozygotes are greater than the selection coefficients of all of the homozygotes. Second, only one heterozygote has a smaller selection coefficient than two homozygotes, an associated homozygote and the other homozygote. In the ABO blood group system, for example, the selection coefficient of genotype *AO* must be less than the selection coefficient of *AA* or *OO* (associated homozygotes) and *BB* (nonassociated homozygote). Third, one of the heterozygotes has a smaller selection coefficient than one of the two associated homozygotes. Fourth, one of the heterozygotes has a smaller selection coefficient than the nonassociated homozygote. These relationships are abstract and theoretical. We cannot test their validity in the ABO blood group system, but we can indicate some of the reasons for the difficulties in trying to test them, and thus indicate some fruitful lines of research for the future.

We believe that the genes of the ABO system are in equilibrium in most

TABLE 27.4
*Conditions Necessary to Maintain Stable Equilibrium
in the ABO Blood Group System*

Selection coefficient	Genotype
a	OO
b	AA
c	BB
d	AO
e	BO
f	AB

Necessary conditions

(1) $d > a, b, c$ and $e > a, b, c$ and $f > a, b, c$

(2) $d < c$ and a or b and b or $a < d$
    or $e < b$ and a or c and c or $a < e$
    or $f < a$ and b or c and c or $b < f$

(3) $d < a$ or b
    or $e < a$ or c
    or $f < b$ or c

(4) $d < c$
    or $e < b$
    or $f < a$

populations, but we cannot identify all of the heterozygotes; we can identify only the *AB* genotype. It is extremely difficult to assess the mechanisms of selection maintaining the equilibrium in this system. We have discussed two of the selective agents—disease and heterospecific matings. Only further research will demonstrate the magnitude of the selective pressure of these and other agents.

If we could identify the *AO* and *BO* heterozygotes, we still would have no easy task, for the determination of the selection coefficients is difficult. A large number of individuals of each genotype should be carefully studied throughout their entire lives. This is not possible now, partly because some of the genotypes cannot be detected and partly for the obvious technical problems attached to obtaining a large sample. Another possibility is to determine the changes in gene frequencies in various age groups in various populations. Again, this is difficult if not impossible at present. Another approach is to study selected families or pedigrees intensively, pedigrees in which a very high probability of determining the ABO genotypes of individuals exists. When and if soundly based values for relative fitness or selection coefficients of the genotypes are determined, we can apply Mandel's system to the ABO blood group problem.

Several specific lines of research are indicated. Certainly any developments, such as reagents, antisera, special tests, which enable us to identify the heterozygotes *AO* and *BO* so that we may distinguish unequivocally the four genotypes—*AO, AA, BO,* and *BB*—will be a major advance. Another and perhaps the most important research is into the chemistry of the group specific substances. When we know more about what they are, how they differ from each other, how they react, and what their physiological functions are, we shall be in a better position to understand the evolutionary dynamics of the ABO system. Far too much of our information has come from statistical surveys, arguments based upon indirect evidence and abstract mathematical and statistical models. These activities are not fruitless. But as the number of statistical surveys and asssociations grows, we cannot but think that some of the arguments against such activity, as put forth by Wiener in the reference cited, have a point.

Some of the adverse critical comments are worth presenting, for they illustrate some of the long-held misconceptions about the study of natural selection in man. The most telling comment of the critics is that the associations demonstrated are trivial. The critics' barbs are directed against what appear to be total irrelevancies. The association between a propensity to fracture the femur and blood group A implies no obvious functional relationship. Neither does the suggestion that personality type is significantly associated with ABO phenotype. Furthermore these associations do not suggest, at least to some, that further research will show a functional relationship. However such trivialities do not mean that the association between gastric carcinoma and phenotype A or between duodenal ulcers and phenotype O are irrelevant. Trivial associations are to be expected if a large number of statistical surveys are made. If a large number of possible associations are tested, it is probable that a small number will, solely on statistical grounds, prove to be significant. It is for this reason we have reiterated that statistical surveys of blood groups and disease must be the first step in studies of the effects of natural selection on human populations. If there is any biological meaning to the association found between one of the blood groups and a disease, the causal relation will only become clear when we know *how* the gene or gene product affects the disease.

The nonuniform distribution of the frequencies of the ABO phenotypes is explained, by some who deny the importance of natural selection, by sampling error (genetic drift), inbreeding, isolation, and migration. We have shown that genetic drift, migration, and inbreeding have relatively small effects upon gene frequencies and their effects are usually demonstrable only in small isolated populations (Chapter 24). The differences in frequencies of the *A, B,* and *O* alleles among human populations are too great to be explained by chance or by migration. Both chance and migration may have had *some* effect, but only natural selection can account for the magnitude of the observed differences and the maintenance of the observed balanced polymorphism.

In conclusion, then, we can say that the ABO blood group system in man will be a useful one for understanding evolutionary dynamics. There are several promising fields of investigation open—search for better serological reagents to identify genotypes in populations, clinical and statistical studies to determine the magnitude of the effects of natural selection, long-term population surveys to obtain estimates of fitness, and chemical investigations to establish the structure and function of the blood group specific substances.

SUGGESTED READINGS AND SOURCES

Aird, I., H. H. Bentall, and J. A. F. Roberts, A relationship between cancer of stomach and the ABO blood groups. *Brit. Med. J.* **1**, 799 (1953).

Aird, I., J. A. H. Lee, and J. A. F. Roberts, ABO blood groups and cancer of esophagus, cancer of pancreas and pituitary adenoma. *Brit. Med. J.* **1**, 1163 (1960).

Azevêdo, E., H. Krieger, and N. E. Morton, Smallpox and the ABO blood groups in Brazil. *Am. J. Human Genet.* **16**, 451 (1964).

Behrman, S. J., J. Buettner-Janusch, R. Heglar, H. Gershowitz, and W. L. Tew, ABO(H) incompatibility as a cause of infertility: a new concept. *Am. J. Obstet. Gynecol.* **79**, 847 (1960).

Boyd, W. C., Four achievements of the genetical method in physical anthropology. *Am. Anthropol.* **65**, 243 (1963).

Brues, A. M., Selection and polymorphism in the A-B-O blood groups. *Am. J. Phys. Anthropol.* **12**, 559 (1954).

Brues, A. M., Stochastic tests of selection in the ABO blood groups. *Am. J. Phys. Anthropol.* **21**, 287 (1963).

Bryce, L. M., R. Jakobowicz, N. McArthur, and L. S. Penrose, Blood group frequencies in a mother and infant sample of the Australian population. *Ann. Eugenics* **15**, 271 (1950).

Buckwalter, J. A., and L. A. Knowler, Blood donor controls for blood group disease researchers. *Am. J. Human Genet.* **10**, 164 (1958).

Buettner-Janusch, J., Natural selection in man: the ABO(H) blood group system. *Am. Anthropol.* **61**, 437 (1959).

Buettner-Janusch, J., The study of natural selection and the ABO(H) blood group system in man. *In* G. E. Dole and R. Carneiro (Ed.), *Essays in the Science of Culture*, p. 79. Crowell, New York (1960).

Carter, C., and B. Heslop, ABO blood groups and broncho-pneumonia in children. *Brit. J. Prevent. Social Med.* **11**, 214 (1957).

Cattell, R. B., H. B. Young, and J. D. Hundleby, Blood groups and personality traits. *Am. J. Human Genet.* **16**, 397 (1964).

Chung, C. S., and N. E. Morton, Selection at the ABO locus. *Am. J. Human Genet.* **13**, 9 (1961).

Clarke, C. A., Blood groups and disease. *In* A. G. Steinberg (Ed.), *Progress in Medical Genetics*, Vol. I, p. 81. Grune and Stratton, New York (1961).

Clarke, C. A., J. W. Edwards, D. R. W. Haddock, A. W. Howel-Evans, and R. B. McConnell, The relationship of the ABO blood groups to duodenal and gastric ulceration. *Brit. Med. J.* **2**, 643 (1955).

Clarke, C. A., R. B. McConnell, and P. M. Sheppard, ABO blood groups and duodenal ulcer. *Brit. Med. J.* **1**, 758 (1957).

Creger, W. P., and A. T. Sortor, The incidence of blood group A in pernicious anemia. *Arch. Internal Med.* **98**, 136 (1956).

Dienst, A., Das Eklampsiegift. *Zentr. Gynäkol.* **29**, 353 (1905).

Dienst, A., Die Pathogenese der Eklampsie und ihre Beziehungen zur normalen Schwangerschaft, zum Hydrops und zur Schwangerschaftsniere. *Arch. Gynaekol.* **86**, 314 (1908).

Ford, E. B., Polymorphism and taxonomy. *In* J. Huxley (Ed.), *The New Systematics*, p. 493. Oxford Univ. Press, London (1940).

Glynn, A. A., L. E. Glynn, and E. J. Holbrow, The secretor status of rheumatic-fever patients. *Lancet* **2**, 759 (1956).

Grubb, R., and S. Sjöstedt, Blood groups in abortion and sterility. *Ann. Human Genet.* 19, 183 (1955).

Gullbring, B., Investigation on the occurrence of blood group antigens in spermatozoa from man, and serological demonstration of the segregation of characters. *Acta Med. Scand.* 159, 169 (1957).

Halbrecht, I., Role of hemo-agglutinins anti-A and anti-B in pathogenesis of jaundice of the newborn (icterus neonatorum praecox). *Am. J. Diseases Children* 68, 248 (1944).

Hartmann, G., *Group Antigens in Human Organs.* Munksgaard, Copenhagen (1941).

Hirszfeld, L., and H. Zborowski, Gruppenspezifische Beziehungen zwischen Mutter und Frucht und elektive Durchlässigkeit der Placenta. *Klin. Wochschr.* 1, 1152 (1925).

Kirk, R. L., J. W. Shield, N. S. Stenhouse, L. M. Bryce, and R. Jakobowicz, A further study of A-B-O blood groups and differential fertility among women in two Australian maternity hospitals. *Brit. J. Prevent. Social Med.* 9, 104 (1955).

Levene, H., and R. E. Rosenfield, ABO incompatibility. *In* A. G. Steinberg (Ed.), *Progress in Medical Genetics,* Vol. I, p. 120. Grune and Stratton, New York (1961).

Levine, P. A., Serological factors as possible causes in spontaneous abortions. *J. Heredity* 34, 71 (1943).

McArthur, N., and L. S. Penrose, World frequencies of the O, A and B blood group genes. *Ann. Eugenics* 15, 302 (1951).

Mandel, S. P. H., The stability of a multiple allelic system. *Heredity* 13, 289 (1959).

Manuila, A., Distribution of A-B-O genes in eastern Europe. *Am. J. Phys. Anthropol.* 14, 577 (1956).

Matsunaga, E., Intra-uterine selection by the ABO incompatibility of mother and foetus. *Am. J. Human Genet.* 7, 66 (1955).

Matsunaga, E., and S. Itoh, Blood groups and fertility in a Japanese population with special reference to intra-uterine selection due to maternal-foetal incompatibility. *Ann. Human Genet.* 22, 111 (1958).

Mourant, A. E., *The Distribution of the Human Blood Groups.* Blackwell, Oxford (1954).

Mourant, A. E., A. C. Kopec, and K. Domaniewska-Sobczak, *The ABO Blood Groups.* Blackwell, Oxford (1958).

Newcombe, H. B., Risk of fetal death to mothers of different ABO and Rh Blood types. *Am. J. Human Genet.* 15, 449 (1963).

Osborne, R. H., and F. V. De George, The ABO blood groups in neoplastic disease of the ovary. *Am. J. Human Genet.* 15, 380 (1963).

Pettenkofer, H. J., and R. Bickerich, Über Antigen-Gemeinschaften zwischen den menschlichen Blutgruppen ABO und den Erregern gemeingefährlicher Krankheiten. *Zentr. Bakteriol.* 179, 433 (1960).

Pettenkofer, H. J., B. Stoss, W. Helmbold, and F. Vogel, Alleged causes of the present-day world distribution of the human ABO blood groups. *Nature* 193, 445 (1962).

Reed, T. E., Tests of models representing selection in mother-child data on ABO blood groups. *Am. J. Human Genet.* 8, 257 (1956).

Roberts, J. A. F., ABO blood groups and duodenal ulcer. *Brit. Med. J.* 1, 758 (1957).

Schiff, F., and H. Sasaki, Der Ausscheidungstypus, ein auf serologischen wege Nachweisbares Mendelndes Merkmal. *Klin. Wochschr.* 34, 1426 (1932).

Springer, G. F., and A. S. Wiener, Alleged causes of the present-day world distribution of the human ABO blood groups. *Nature* 193, 444 (1962).

Struthers, D., ABO groups of infants and children dying in the west of Scotland (1949–1951). *Brit. J. Prevent. Social Med.* 5, 223 (1951).

Vogel, F., J. Dehnert, and W. Helmbold, Über Beziehungen zwischen den ABO-Blutgruppen und der Säuglingsdyspepsie. *Humangenetik* 1, 31 (1964).

Waterhouse, J. A. H., and L. Hogben, Incompatibility of mother and foetus with respect to the iso-agglutinogen A and its antibody. *Brit. J. Prevent. Social Med.* 1, 1 (1947).

Wiener, A. S., *Advances in Blood Grouping.* Grune and Stratton, New York (1961).

Blood Groups
of Nonhuman Primates

28

SOME INFORMATION about the evolutionary origin of the human blood groups may be obtained by examining the red cells of nonhuman primates for the same or related antigens. Blood group substances resembling those of man have been found in a variety of nonhuman primates by a number of investigators who used antisera that react with human red cells. Among certain primates some blood group systems are unequivocally demonstrated, just as they are in man, by the agglutination of erythrocytes by specific antisera. Among other primates the occurrence of the same blood group systems is deduced from tests made on saliva.

The reactivities of primate red cells with blood group specific human antisera are different. These individual variations in reactivity appear to be little greater than individual variation observed among samples of human red cells. The range of reactivities of primate erythrocytes overlaps the range of reactivities of human erythrocytes when tested with the same antisera.

Studies involving the use of antibodies, red cells, tissue cells, or other materials from various members of the Pongidae suggest that there is ABO serological identity between *Homo sapiens* and the pongids. Although there are no significant qualitative differences in reactions to anti-A and anti-B antisera, it is not necessary to postulate ABO genetic identity among the Hominoidea. It is significant that there is a very close relationship which suggests that the ABO blood group system is not unique to man but is shared with some of his nonhuman primate relatives. This does not mean that there is chemical identity between the A-like, B-like, and H-like group specific substances of the apes and man.

Among the Cercopithecoidea, Ceboidea, and Prosimii the ABO antigens are not as clearly related to those of *Homo sapiens* as are those of the great apes. Anti-A and anti-B antibodies agglutinate the red cells of various monkeys. A number of lemurs whose red cells were tested with anti-A and anti-B antisera appear to have a B-like antigen on their red cells. Some Old World monkeys have anti-A or anti-B antibodies in their sera, and it is possible to show that the A and B antigens are present on tissue cells and in the saliva of certain individuals. The A-like, B-like, and H-like antigens on the red cells and in the saliva of monkeys and prosimians do react with human anti-A and anti-B antisera, but the reactions are clearly different from those of human antigens A and B with these antisera. The reactions of red cell antigens of the great apes are much more like human than monkey and prosimian reactions of red cell antigens. Some investigators are convinced that the ABO specific antigens in monkeys can only be deduced from tests on saliva.

The Rh antigens among nonhuman primates have also been investigated in detail. No polymorphisms among nonhuman primates have been found with the standard blood grouping techniques, using antisera that react with human Rh antigens. Red cells of chimpanzees appear to have the D and c antigens, for their erythrocytes react in the same manner as human erythro-

519

cytes with anti-D and anti-c. Other Rh factors or antigens have not been demonstrated on red cells of chimpanzees. There is some evidence that the Rh antigens appear on red cells of gorillas. When they occur in nonhuman primates, they are found in every member of the species tested.

Studies of the MN blood groups among nonhuman primates suggest that antigens very similar to the M and N blood group substances of human red cells are present. The M antigen has been found in a large number of different species of primates. The N antigen has been found, so far, only in gorillas and chimpanzees. Wiener found that there are two different and distinctly separate MN types among chimpanzees. These two types are distinguished by an anti-N reagent prepared from plant seeds. The usual reagents suggested that chimpanzees were all of type MN, but the special anti-N reagent divided this group into MN_1 and MN_2.

A number of tests with antisera specific for other human blood group systems have been performed. Few generalizations can be made about them yet, but the demonstration that gibbons have an antigen indistinguishable from the sex-linked human antigen Xg^a bears mention here. Xg^a was demonstrated in seven out of thirteen gibbons tested with two different human anti-Xg^a antisera. The positive reactions of these gibbon erythrocytes were indistinguishable from those given by human $Xg(a+)$ cells. No evidence that Xg^a is a sex-linked trait in *Hylobates* is available. If it proves to be X-linked, as it is in *Homo*, one may speculate that potent selection among the Primates has kept this locus on the X-chromosome.

Many of the investigators who have studied the serology of nonhuman primates have used the data to support or discuss various views of phyletic relationships within the order. It is difficult at the present time to make any but extremely naive statements about such relationships when the data from blood group studies are used. One reason for this is that there are relatively few data on specific primate blood group systems. But this does not mean that primate serology will not prove to be a valuable adjunct in our phylogenetic and taxonomic studies of the primates. For example it is clear that red cells of chimpanzees give reactions, with the antibodies of the ABO system, more like the reactions given by man than the reactions of blood from orangutans. Whether this means that the same antigen is present in two species, chimpanzee and man, cannot yet be demonstrated. It is only when the antigen, that is, the blood group specific substance, is isolated and its chemical structure and spatial configuration are determined that we will know whether these antigens are the same or different. Furthermore even if the antigen is the same in two species, this does not mean that the extinct lineage from which the two living species evolved had this particular antigen. Neither does the presence of the same antigen in two species mean that the two are more closely related to each other than they are to other species. If two or three living species have been derived from a common ancestral lineage, it is not necessary to postulate that the ancestral gene pool has been retained in its original form in each of the descendant species. Natural selection will not operate in the same fashion on the various descendant species. There is no reason to suppose selection should not lead to differences among blood group antigens. It is possible that some genes may be lost or changed so completely that their products will be very different among two or more species descended from the same ancestral species. The A and B blood group specific substances, the A and B antigens, are probably a case of this kind. Each of the species of pongids studied proves to be polymorphic for the ABO antigens, although no single species has been shown to have all three alleles. The lowland gorillas appear to have the B antigen in tissue cells and in secretions, but agglutination of red cells of gorillas with anti-B antiserum does not occur. The ABO serological similarities of man and various great apes are many, but the example just cited suggests that there has been an evolutionary change in the primate ABO system. *Homo sapiens* went in one direction and the other Hominoidea went in another.

The fact that the frequencies of ABO blood groups of man differ in various parts of the world led some to postulate that man had a polyphyletic origin. The different frequencies of the ABO blood groups were presumed to have developed through crosses between races which were originally entirely blood group A, B, or O. Since we now know that the great apes, like man, are polymorphic for these antigens, it is simpler to suppose that the ABO blood groups are part of man's primate heritage. Furthermore it does not seem reasonable to argue any longer over whether blood group A or O was the original human ABO blood group and whether group B arose from a mutation after the phyletic branching of *Homo*. Since the B antigen has been found among orangutans, gibbons, siamangs, and gorillas, it too appears to be part of the gene pool which man inherited. Studies of nonhuman primate blood groups are in their infancy. A direct chemical study of the mucoprotein molecules of the various primate blood group systems should vastly expand our understanding of the evolution of the living primates. Studies of primate blood group systems provide us, at present, with some neat and knotty problems in taxonomy, but the potential value of primate serology for evolutionary studies is enormous.

SUGGESTED READINGS AND SOURCES

Buchbinder, L., The blood grouping of Macaca rhesus. *J. Immunol.* **25**, 33 (1933).

Butts, D. C. A., Hemagglutinogens of the chimpanzee. *Am. J. Phys. Anthropol.* **11**, 215 (1953).

Candela, P. B., New data on the serology of the anthropoid apes. *Am. J. Phys. Anthropol.* **27**, 209 (1940).

Dahr, P., Zur Frage der serologischen Verschiedenheit von Altweltaffen (Catarrhini) und Neuweltaffen (Platyrrhini). *Z. Immunitätsforsch.* **90**, 376 (1937).

Franks, D., The blood groups of primates. *Symp. Zool. Soc. London* **10**, 221 (1963).

Franks, D., R. R. A. Coombs, T. S. L. Beswick, and M. M. Winter, Recognition of the species of origin of cells in culture by mixed agglutination. III. Identification of the cells of different primates. *Immunology* **6**, 64 (1963).

Gavin, J., J. Noades, P. Tippett, R. Sanger, and R. R. Race, Blood group antigen Xgᵃ in gibbons. *Nature* **204,** 1322 (1964).

Landsteiner, K., and C. P. Miller, Serological studies on the blood of the primates. I. The differentiation of human and anthropoid bloods. *J. Exptl. Med.* **42,** 841 (1925).

Landsteiner, K., and C. P. Miller, Serological studies on the blood of the primates. II. The blood groups in anthropoid apes. *J. Exptl. Med.* **42,** 853 (1925).

Landsteiner, K., and A. S. Wiener, On the presence of M agglutinogens in the blood of monkeys. *J. Immunol.* **33,** 19 (1937).

Levine, P., F. Ottensooser, M. J. Celano, and W. Pollitzer, On reactions of plant anti-N with red cells of chimpanzees and other animals. *Am. J. Phys. Anthropol.* **13,** 29 (1955).

Moor-Jankowski, J., and A. S. Wiener, Primate blood groups and evolution. *Science* **148,** 255 (1965).

Moor-Jankowski, J., A. S. Wiener, and C. M. Rogers, Human blood group factors in non-human primates. *Nature* **202,** 663 (1964).

Nuttal, G. H. F., *Blood Immunity and Relationship.* Cambridge Univ. Press, Cambridge (1904).

Wiener, A. S., and E. B. Gordon, Blood groups in chimpanzees. ABO groups and M-N types. *Am. J. Phys. Anthropol.* **18,** 301 (1960).

Wiener, A. S., and E. B. Gordon, The blood groups of chimpanzees. The Rh-Hr (CDE/cde) blood types. *Am. J. Phys. Anthropol.* **19,** 35 (1961).

Wiener, A. S., E. B. Gordon, and J. Moor-Jankowski, The Lewis blood groups in man. *J. Forensic Med.* **11,** 67 (1965).

Wiener, A. S., and J. Moor-Jankowski, The V-A-B blood group system of chimpanzees: a paradox in the application of the 2×2 contingency test. *Transfusion* **5,** 64 (1965).

Wiener, A. S., J. Moor-Janowski, and E. B. Gordon, Blood groups of monkeys and apes. IV. The Rh-Hr blood types of anthropoid apes. *Am. J. Human Genet.* **16,** 246 (1964).

Wiener, A. S., J. Moor-Jankowski, and E. B. Gordon, Blood groups of apes and monkeys. V. Studies on the human blood group factors, A, B, H, and Le in Old and New World monkeys. *Am. J. Phys. Anthropol.* **22,** 175 (1964).

Zmijewski, C. M., R. S. Metzgar, and C. M. Rogers, Production of chimpanzee isohemagglutinins by immunization with human erythrocytes. *Transfusion* **5,** 71 (1965).

Human Hemoglobins

29

HEMOGLOBIN, THE RED RESPIRATORY PROTEIN found in mammalian erythrocytes, is one of the most informative molecules in primate blood. It comprises between 90 and 95 per cent of the protein in a red cell. Hemoglobin exists in alternative forms in man, and these forms are under genetic control. The frequencies of some of the alleles which control hemoglobins vary markedly among human populations. Chemical studies reveal the exact nature of the structural differences among the various human and primate hemoglobins. Hemoglobins are easy to work with, for they are readily prepared in large quantities from relatively small volumes of whole blood and are reasonably stable when handled.

Normal Hemoglobins

A hemoglobin sample prepared from whole blood of a normal adult proves, upon analysis, to be composed of different hemoglobins known to be under genetic control. The principal technique used to show that there is more than one hemoglobin in a sample or that two hemoglobins are different is electrophoresis, the movement of charged particles in an electric field. In the electrophoresis of hemoglobin, a solution containing the protein

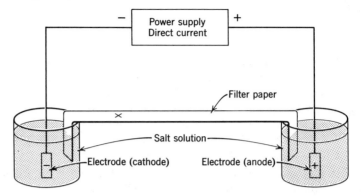

Fig. 29.1. A diagram showing essential features of apparatus used for electrophoretic separations. A sample is applied at X; the filter paper is moistened; the power is turned on; negatively charged molecules migrate toward the anode (+); positively charged molecules migrate toward the cathode (−); and neutral molecules remain near the point of application. In practice, elaborate variations of the scheme shown here have been developed. For example paper strips or large sheets of filter paper are pressed between two pieces of glass, or starch gels in plastic trays are substituted for paper. Often the starch gels are set up vertically instead of horizontally to give improved resolution. (See Fig. 29.2.)

is applied to a gel made of starch in an appropriate salt solution or to a strip of filter paper. Both ends of the starch gel or the wet paper are immersed in the salt solution, and an electric current is passed through the gel or paper (Fig. 29.1 and 29.2). The molecules of hemoglobin migrate in the electric field, and the distance that they migrate from the point of application depends on the electric charge on the hemoglobin molecules, the size of the hemoglobin molecules, the hydrogen ion concentration of the salt solution, the nature of the supporting medium (paper, starch gel), the temperature, the strength of the electric field, the length of time, and many other things. The technique of electrophoresis makes it possible for us to resolve the various component hemoglobins of an individual and to show differences between hemoglobins of two individuals. It is well for the reader to note that electrophoresis and many other analytical techniques will only demonstrate that hemoglobins are different; such techniques cannot be used to prove that two components or two hemoglobins are the same. We do not worry unduly about this when we examine the hemoglobins of members of the same species, but we must not assume that the hemoglobin from two different species, such as a South American monkey and man, are the same because they move the same distance during electrophoresis.

All primate hemoglobins are composed of globin and four heme groups. Globin is a protein made up of various integral amounts of the twenty common amino acids (Table 25.2). Heme is a large organic molecule with an atom of iron at its center. (A heme group is also called an iron proto-

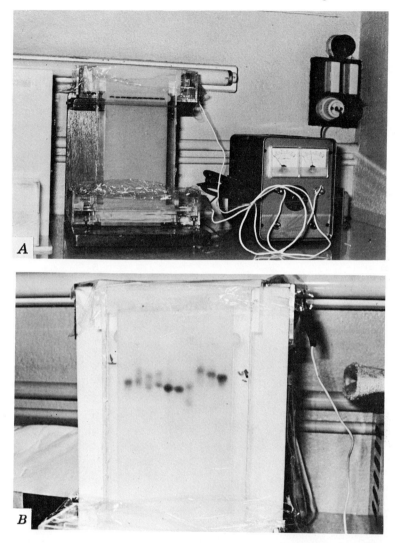

Fig. 29.2.   Photographs of apparatus used for vertical starch-gel electrophoresis in the author's laboratory in Madagascar. (*A*) A complete setup for electrophoresis of hemoglobins; the dark areas near the top of the gel are slots filled with hemoglobin solutions. The anode (+) is at the bottom. (*B*) Positions of various hemoglobins in the gel after several hours of electrophoresis. Human hemoglobin A is on the left; human hemoglobin S is on the right; the others are prosimian hemoglobins.

porphyrin.) The iron in the heme groups reacts with oxygen and carries it from the lungs to the tissues. The primary and perhaps sole function of hemoglobin is the transport of oxygen to and the removal of carbon dioxide from the tissues.

Normal adult human hemoglobin is actually composed of two different hemoglobins which can readily be separated by electrophoresis. The major portion of hemoglobin molecules is called hemoglobin A; the minor portion is called hemoglobin A_2. The hemoglobin A molecule, aside from the heme group, contains a total of 574 amino acids. All of the amino acids except isoleucine are present. These amino acids are joined in peptide bonds (Fig. 25.3) and form two pairs of two different polypeptide chains, α and β chains. The amino acid compositions of the α and β chains are presented in Table 29.1, and the sequences of amino acids are shown in Fig. 29.3. There

TABLE 29.1

*The Amino Acid Composition of the Polypeptide Chains
of Normal Human Adult and Fetal Hemoglobins*

Amino acid[a]	Number of amino acid residues per mole		
	α chain	β chain	γ chain
Aspartic[b]	12	13	13
Threonine	9	7	10
Serine	11	5	11
Glutamic[c]	5	11	12
Proline	7	7	4
Glycine	7	13	13
Alanine	21	15	11
Valine	13	18	13
Methionine	2	1	2
Isoleucine	0	0	4
Leucine	18	18	17
Tyrosine	3	3	2
Phenylalanine	7	8	8
Lysine	11	11	12
Histidine	10	9	7
Arginine	3	3	3
Tryptophan	1	2	3
Cysteine	1	2	1
Total	141	146	146

[a]The amino acids are listed in the order in which they are determined in chromatographic systems commonly used.

[b]Includes aspartic acid and asparagine.

[c]Includes glutamic acid and glutamine.

are two α chains and two β chains in each molecule, and these are coiled in helical fashion in the intact molecule. Each α chain contains 141 amino acids; each β chain contains 146. Analysis of hemoglobin A_2 shows that it is similar to hemoglobin A and contains two α chains and two other chains that are designated as δ chains which are distinct from α and β chains.

Hemoglobin of a human fetus or a newborn baby contains still a different hemoglobin named F. Hemoglobin F is characteristic of the human fetus and is replaced by adult hemoglobin A in normal individuals within the first 2 months after birth. Hemoglobin F consists of two α chains, the same as those found in hemoglobin A, and two others which are distinct, designated as γ chains (Table 29.1 and Fig. 29.3). The γ chains contain all twenty amino acids including isoleucine. Large segments of the α, β, and γ (and δ) chains contain homologous or invariant sequences, that is, the order of the amino acids is identical in many portions of the chains (Fig. 29.3).

The notation for the human hemoglobins is based upon the fact that the two polypeptide chains which make up the molecules are controlled by separate nonallelic genes. The notation for hemoglobin A is $Hb\alpha_2{}^A\beta_2{}^A$, which is a shorthand way of saying it contains two normal α chains and two normal β chains. Hemoglobin A_2 is symbolized as $Hb\alpha_2{}^A\delta_2{}^{A2}$ and hemoglobin F as $Hb\alpha_2{}^A\gamma_2{}^F$.

Fetal hemoglobin $Hb\alpha_2{}^A\gamma_2{}^F$ has two interesting properties that serve to identify it. One, the amino acid isoleucine is present. Two, exposure to strong alkali will not denature, that is not destroy, $Hb\alpha_2{}^A\gamma_2{}^F$, whereas $Hb\alpha_2{}^A\beta_2{}^A$ is completely denatured when treated with strong alkali. A quantitative estimate of the amount of adult and fetal hemoglobin present in a sample can be made by determining the total amount of hemoglobin, then treating the sample with alkali and determining the amount of hemoglobin remaining, which is the amount of fetal hemoglobin. The difference is the amount of adult hemoglobin. The amount of hemoglobin in a solution may be measured readily by comparison with a standard solution of known concentration. These comparisons can be made with the naked eye, but accurate determinations are made under carefully controlled conditions with optical instruments—colorimeters or spectrophotometers.

Abnormal or Variant Hemoglobins

One of the most interesting mutant human hemoglobins has been designated hemoglobin S, $Hb\alpha_2{}^A\beta_2{}^S$. The notation signifies that the mutation has affected the β chain. Hemoglobin S may be separated and distinguished from hemoglobin A by means of starch-gel or paper electrophoresis. In homozygous individuals, $Hb\alpha_2{}^A\beta_2{}^S$ and small amounts of $Hb\alpha_2{}^A\delta_2{}^{A2}$ are found. In heterozygous individuals three components are found—$Hb\alpha_2{}^A\beta_2{}^S$, $Hb\alpha_2{}^A\beta_2{}^A$, and, usually, $Hb\alpha_2{}^A\delta_2{}^{A2}$. Hemoglobin S may also be demonstrated by performing a sickling test on fresh blood. The red cells are placed in an atmosphere from which oxygen has been excluded, and the reduced oxygen tension leads to the crinkling of red cells into sickle shapes when they contain $Hb\alpha_2{}^A\beta_2{}^S$. Individuals whose red cells do this are called sicklers, and the disease is called sickle cell anemia or sicklemia. The sickle cell trait is inherited as if it were a nondominant autosomal allele. Both heterozygotes and homozygotes have anemia of varying degrees of severity. The red cells from both genotypes sickle in an oxygen-free atmosphere; the sickling of red cells of heterozygotes is less severe, and the sickling of red cells of homozygotes is more severe. The homozygous individuals are severely α chain

β

γ

α
^{1}
Val—Leu—Ser—Pro—Ala—Asp—LYS—Thr—Asn—Val^{10}—Lys—Ala—Ala—TRY—GLY—LYS—VAL^{20}—Gly—Ala—His—Ala—Gly—Glu—Tyr—GLY—Ala—GLU—

β
^{1}
Val—His—Leu—Thr—Pro—Glu—Glu—LYS—Ser—Ala—Val—Thr—Ala—Leu—TRY—GLY—LYS—VAL—Asn—Val^{20}—
Asp—Glu—Val—GLY—Gly—GLU—

γ
Gly—His—Phe—Thr—Glu—Glu—Asp—LYS—Ala—Thr—Ileu—Thr—Ser—Leu—TRY—GLY—LYS—VAL—Asn—Val—
Glu—Asp—Ala—GLY—Gly—GLU—

α
Ala—LEU—Glu—ARG—Met—Phe—Leu—Ser—Phe—PRO—Thr—THR—
Lys—Thr—Tyr—Phe—Pro—His—Phe—Asp—LEU—SER—His—Gly—Ser—ALA—

β
Ala—LEU—Gly—ARG—Leu—Leu—Val—Val—Tyr—PRO—Try—THR—^{30}
Gln—Arg—Phe—Phe—Glu—Ser—Phe—Gly—Asp—LEU—SER—^{50} Thr—Pro—Asp—ALA—Val—

γ
Thr—LEU—Gly—ARG—Leu—Leu—Val—Val—Tyr—PRO—Try—THR—
Gln—Arg—Phe—Phe—Asp—Ser—Phe—Gly—Asn—LEU—SER—Ser—Ala—Ser—ALA—Ileu—

α
Gln—VAL—LYS—Gly—HIS—GLY—LYS—LYS—VAL—Ala—Asp—Ala—Leu—Thr—Asn—Ala—Val—Ala—HIS—Val—ASP—Asp—Met—Pro—

β
Met—Gly—Asn—Pro—Lys—VAL—LYS—Ala—HIS—GLY—LYS—LYS—VAL—Ala—Asp—Ala—Leu—Thr—Asn—Ala—Val—Ala—HIS—Val—ASP—Asp—Met—Pro—^{60}
Met—Gly—Asn—Pro—Lys—VAL—LYS—Ala—HIS—GLY—LYS—LYS—VAL—Leu—Gly—Ala—Phe—Ser—As₂—Gly—Leu—Ala—HIS—Leu—ASP—Asn—Leu—Lys—^{80}

γ
Met—Gly—Asn—Pro—Lys—VAL—LYS—Ala—HIS—GLY—LYS—LYS—VAL—Leu—Thr—Ser—Leu—Gly—Asp—Ala—Ileu—Lys—HIS—Leu—ASP—Asp—Leu—Lys—

α
Asn—Ala—Leu—Ser—Ala—LEU—SER—Asp—LEU—HIS—Ala—His—LYS—LEU—Arg—VAL—ASP—PRO—Val—ASN—PHE—Lys—LEU—LEU—Ser—His—Cys—LEU—^{90}...^{100}

β
Gly—Thr—Phe—Ala—Thr—LEU—SER—Glu—LEU—HIS—Cys—Asp—LYS—LEU—His—VAL—ASP—PRO—Glu—ASN—PHE—Arg—LEU—LEU—Gly—Asn—Val—LEU—^{110}

γ
Gly—Thr—Phe—Ala—Gln—LEU—SER—Glu—LEU—HIS—Cys—Asp—LYS—LEU—His—VAL—ASP—PRO—Glu—ASN—PHE—Lys—LEU—LEU—Gly—Asn—Val—LEU—

α
Leu—Val—Thr—LEU—ALA—Ala—His—Leu—Pro—Ala—GLU—PHE—THR—PRO—Ala—VAL—His—ALA—Ser—Leu—Asp—LYS—Phe—Leu—Ala—Ser—VAL—Ser—^{130}

β
Val—Cys—Val—LEU—ALA—His—His—Phe—Gly—Lys—GLU—PHE—THR—PRO—Pro—VAL—Gln—ALA—Ala—Tyr—Gln—LYS—Val—Val—Ala—Gly—VAL—Ala—^{120} ^{130}

γ
Val—Thr—Val—LEU—ALA—Ileu—His—Phe—Gly—Lys—GLU—PHE—THR—PRO—Glu—VAL—Gln—ALA—Ser—Try—Gln—LYS—Met—Val—Thr—Gly—VAL—Ala—

α
Thr—Val—LEU—Thr—Ser—Lys^{140}—TYR—Arg^{141}

β
Asn—Ala—LEU—Ala—His—Lys^{140}—TYR—His^{146}

γ
Ser—Ala—LEU—Ser—Ser—Arg—TYR—His

Fig. 29.3. The sequence of amino acids in the α, β, and γ chains of normal human hemoglobin A(Hbα₂ᴬβ₂ᴬ) and hemoglobin F(Hbα₂ᴬγ₂ᶠ). The sequences are numbered beginning at the NH₂-terminus of each chain. The amino acids shown in capital letters (e.g., LYS at position 7 in α chain and position 8 in β and γ chains) are identical or homologous in all three chains. So-called gaps have been inserted to emphasize the homologies among the chains. These spacings may occur when a complete hemoglobin molecule containing two pairs of chains is folded in its three-dimensional helical configuration. If only two of the three chains are compared, the number of homologies increases.

affected and usually die before reaching maturity. The presence of the sickle cell gene, as the allele is often named, produces a large number of clinical symptoms, among which are enlarged spleen, rheumatism, impaired mental functions, and the aforementioned anemia. The consequences of this mutation are dramatic (Fig. 29.4).

The observation that hemoglobin S and hemoglobin A are electrophoretically different (the two hemoglobins do not migrate the same distance on paper or starch) immediately suggests that there is a difference between the two in amino acid composition and primary structure. We have already noted that the difference lies in the β chain, and a demonstration of the specific biochemical defect in hemoglobin S was made when Ingram performed his famous "fingerprint" experiments. With further careful deter mination of the primary structure (sequence of amino acids) of hemoglobins A and S, the only difference between the two proved to be a change in one amino acid in each β chain of Hb$\alpha_2{}^A\beta_2{}^S$. There are only two changes in the 574 amino acids that make up the intact molecule. (There is only one change in the 146 amino acids that make up a single β chain.)

The first step in analyzing the structure of any hemoglobin is to separate the two polypeptide chains (α and β, α and γ, α and δ, etc.). Then each chain is broken systematically into smaller units (peptides) by enzymatic action of trypsin, a pancreatic enzyme. The resulting pieces of α chain, β chain, γ chain, δ chain, etc., are called tryptic peptides. Trypsin is an especially useful enzyme for it always breaks the same peptide bonds— those between lysine and another amino acid and those between arginine and another amino acid (Fig. 29.5).

Tryptic peptides obtained from a protein may then be separated and a peptide pattern obtained. Ingram's "fingerprint" of hemoglobin is the pattern that the peptides make on filter paper when subjected to electrophoresis and chromatography. The mixture of peptides is applied on a piece of filter paper and then subjected to an electric current. The peptides move toward the anode ($+$) or the cathode ($-$) according to the net charge each has. This step alone does not give a sufficiently fine resolution of the peptides, so further separation is obtained by chromatography performed in a direction at a right angle to the direction of electrophoresis. Paper chromatography (Fig. 29.6) is a method by which a mixture of compounds applied to a piece of filter paper can be separated when a solvent flows over the paper. Chromatography is based on the observation that different compounds have different solubilities in the same solvent and, because of this, different mobilities in the direction of solvent flow. The distance that a component moves in a particular solvent system depends on many factors —its solubility, its adsorption to the paper or other medium, temperature, and time, among other things. After electrophoresis and chromatography the colorless peptides are located by staining the paper with appropriate reagents and obtaining colored products which appear as spots on the paper. These spots form a pattern which is characteristic for a hemoglobin (Fig.

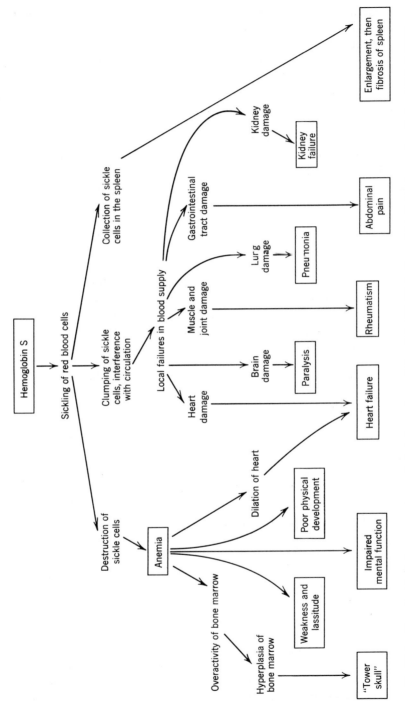

Fig. 29.4. The physiological consequences and clinical symptoms of the allele (*Hb βˢ*) for hemoglobin S. (Redrawn after Neel and Schull, 1954.)

(βT-1)
Val—His—Leu—Thr—Pro—Glu—Glu—Lys—Ser—Ala—Val—Thr—Ala—Leu—Try—Gly—Lys— →
↓ (βT-2)

(βT-3)
Val—Asn—Val—Asp—Glu—Val—Gly—Gly—Glu—Ala—Leu—Gly—Arg—Leu—Leu—Val—Val—
↓ (βT-4)

↓ (βT-5)
Tyr—Pro—Try—Thr—Gln—Arg—Phe—Phe—Glu—Ser—Phe—Gly—Asp—Leu—Ser—Thr—Pro—

↓ (βT-6)   ↓ (βT-7)   ↓ (βT-8) →
Asp—Ala—Val—Met—Gly—Asn—Pro—Lys—Val—Lys—Ala—His—Gly—Lys—Lys— . . . . .

↓ (βT-14)   ↓ (βT-15)
. . . . . . —Val—Val—Ala—Gly—Val—Ala—Asn—Ala—Leu—Ala—His—Lys—Tyr—His

Fig. 29.5. A portion of the β chain of human hemoglobin A(Hbα$_2^A$β$_2^A$). The peptide bonds broken by the action of trypsin are indicated by arrows. The peptides are numbered starting from the NH$_2$-terminus of the chain.

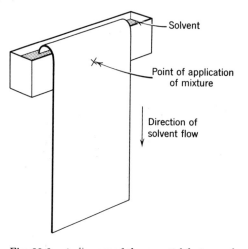

Fig. 29.6. A diagram of the essential features of descending paper chromatography. The paper acts as a wick and the solvent moves down the paper. Since various components in a sample have different solubilities in the solvent, the various components migrate down the paper at different rates. In practice the trough containing the solvent is supported on a rack in a tightly covered container. The atmosphere in the container is saturated with vapor from the solvent. Many of the solvents used in paper chromatography of amino acids and proteins are toxic and/or foul-smelling organic compounds, e.g., phenol and pyridine. It is usual to keep the chromatographic containers in a well-ventilated room.

29.7 and 29.8) and for each chain of a hemoglobin. The reagent most commonly used for staining is ninhydrin, which produces purple spots with peptides and amino acids. These unique paper peptide patterns have been called fingerprints.

The paper peptide pattern of hemoglobin S is almost the same as that of hemoglobin A (Fig. 29.8). A single peptide appears in a different position. The difference between the two hemoglobins is in one peptide of the β chain. The difference consists of the substitution of one amino acid for another. This substitution is a valine in the β chain of $Hb\alpha_2^A\beta_2^S$ for a glutamic acid at position six, counting from the NH_2-terminus of the β chain (Fig. 29.9).

Peptide patterns on paper are not conclusive evidence that two hemoglobins are the same. For example if two peptides are composed of different but similarly charged amino acids or if the same two amino acids are not in the same sequence, the electrophoretic and chromatographic properties may be the same. The resulting paper peptide patterns will not differ. The amino acid composition of each peptide must be determined, and the

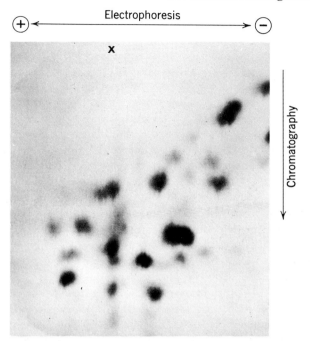

Fig. 29.7.   Paper peptide pattern or fingerprint of human hemoglobin A. This is a photograph made after the paper was stained
with ninhydrin. The mixture of peptides was applied at X.

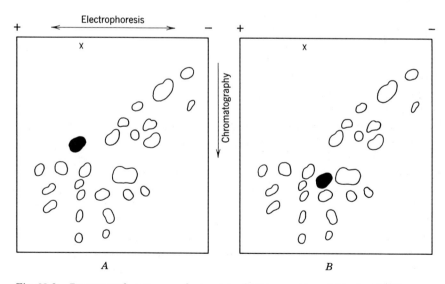

Fig. 29.8.   Paper peptide patterns or fingerprints of (A) human hemoglobin A and (B) human
hemoglobin S. These drawings are tracings of fingerprints. The mixture of peptides was applied
at X. The dark spot in each fingerprint is the peptide that is different in the two hemoglobins.

$$\beta^A \quad \overset{1}{\text{Val}}\text{—His—Leu—Thr—}\overset{5}{\text{Pro}}\text{—GLU—Glu—Lys—Ser—}\overset{10}{\text{Ala}}\text{—Val—}$$

$$\beta^S \quad \text{Val—His—Leu—Thr—Pro—VAL—Glu—Lys—Ser—Ala—Val—}$$

Fig. 29.9. A comparison of the NH_2-terminal sequences of β chain of human hemoglobin A (normal) and hemoglobin S (sickle cell). The only difference between the two hemoglobins is at position six of the β chain.

sequence of amino acids must be demonstrated before two hemoglobins can be called identical. To do this is laborious and time consuming. Although the methods are well worth the effort, we shall not describe them further here. The reader must remember that it is necessary to demonstrate the primary structures (the sequence of amino acids) of hemoglobins, or other proteins, before we can be certain about the similarities and differences between two of them. We can extract much primary genetic information from a protein for it is presumably one of the direct products of the genetic message carried by DNA. But we must be careful that we do not jump to conclusions before the structure has been worked out.

The importance of determining primary structures of proteins cannot be overstressed. The sequence of amino acids in a particular protein is a reflection of the sequence of nucleotides in DNA. The sequence of amino acids is about as close as we can get at present, in mammalian systems, to the genetic material, the DNA. Each amino acid presumably has a one-to-one correspondence with a nucleotide triplet in the DNA of the chromosomes. The sequence of amino acids is, thus, an excellent indicator of the sequence of words in the genetic message carried by DNA, and there are well-established laboratory procedures which enable us to determine these sequences. We hope eventually that methods of determining sequences of bases in DNA will be found. At present there are no such methods available, but we refer to a publication of Hoyer, McCarthy, and Bolton, who have developed a method of comparing sequences in DNA from various organisms without actually determining the order of the bases. It is very likely that study of sequences in proteins will continue to be one of the best means we have of investigating the genetic material.

The biochemical studies of hemoglobin S and hemoglobins of nonhuman primates (Chapter 30) are but two examples of current researches into the processes which go on in the nucleus of the cell and which have a direct effect on the genotype and phenotype of the organism. In hemoglobin S a valine replaces a glutamic acid of hemoglobin A. One of the triplets which codes glutamic acid is GUA, and the triplet for valine is GUU (Table 25.1). Adenine is replaced by uracil. A change of one nucleotide in RNA (or a change of a single pair of complementary bases in DNA) produces marked changes in properties of the hemoglobin and marked physiological effects in the organism.

The differences between hemoglobin A and a number of other aberrant human hemoglobins (Table 29.2) are also single amino acid substitutions

TABLE 29.2
Abnormal Human Hemoglobins

Hemoglobin	Designation	Abnormality		
H	$\beta_4{}^A$	All β chains		
Bart's	$\gamma_4{}^F$	All γ chains		

			Specific mutation (Amino acid replacement)		
		Chain	Position	Normal $\rightarrow$	Abnormal
S	$\alpha_2{}^A\beta_2{}^S$	β	6	Glu	Val
C	$\alpha_2{}^A\beta_2{}^C$	β	6	Glu	Lys
E	$\alpha_2{}^A\beta_2{}^E$	β	26	Glu	Lys
D_β	$\alpha_2{}^A\beta_2{}^D$	β	121	Glu	Gln
F_{Texas}	$\alpha_2{}^A\gamma_2{}^{F_{Texas}}$	γ	5 or 6	Glu	Lys
$G_{Honolulu}$[a]	$\alpha_2{}^G\beta_2{}^A$	α	30	Glu	Gln
$G_{Philadelphia}$[b]	$\alpha_2{}^G\beta_2{}^A$	α	68	Asn	Lys
$G_{San\ Jose}$	$\alpha_2{}^A\beta_2{}^G$	β	7	Glu	Gly
I	$\alpha_2{}^I\beta_2{}^A$	α	16	Lys	Asp
M_{Boston}	$\alpha_2{}^M\beta_2{}^A$	α	58	His	Tyr
$M_{Milwaukee-1}$	$\alpha_2{}^A\beta_2{}^M$	β	67	Val	Glu
$M_{Saskatoon}$[c]	$\alpha_2{}^A\beta_2{}^M$	β	63	His	Tyr
O_{Arabia}	$\alpha_2{}^A\beta_2{}^O$	β	121	Glu	Lys
$O_{Indonesia}$	$\alpha_2{}^O\beta_2{}^A$	α	116	Glu	Lys
Hikari	$\alpha_2{}^A\beta_2{}^{Hik}$	β	61	Lys	Asn
Norfolk	$\alpha_2{}^{Nor}\beta_2{}^A$	α	57	Gly	Asp
Shimonoseki	$\alpha_2{}^{Shimo}\beta_2{}^A$	α	54	Gln	Arg
Zürich	$\alpha_2{}^A\beta_2{}^Z$	β	63	His	Arg

α chain variants[d]	β chain variants[d]	Other variant hemoglobins
D_α	D^e	Hopkins 1
G_{Ibadan}	G^e	Stanleyville II
K	J^e	
M^e	L	
Q	M^e	
Beilinson	N	
Hopkins 2	R	
Russ	Lepore	
Stanleyville I		

[a]Same as $G_{Singapore}$ and $G_{Hongkong}$.

[b]Same as $G_{Bristol}$ and $D_{\alpha St.\ Louis}$.

[c]Same as M_{Emory}.

[d]Only the aberrant chain has been identified; the amino acid replacements have not been specified.

[e]May include several different hemoglobins.

in the β chain. Other variant human hemoglobins differ from $Hb\alpha_2{}^A\beta_2{}^A$ by substitutions of amino acids in the α chain. For example hemoglobin I has a substitution in the α chain, and it is designated as $Hb\alpha_2{}^I\beta_2{}^A$. The methods used for demonstrating the difference between hemoglobin S and hemoglobin A have been applied equally successfully to other variant human hemoglobins and to hemoglobins of nonhuman primates.

Most of the mutant hemoglobins of *Homo sapiens* have been demonstrated to be under simple genetic control of the sort described for $Hb\alpha_2{}^A\beta_2{}^S$. A number of these variant hemoglobins are associated with or produce abnormal clinical signs. Many variant human hemoglobins have no specially noticeable effects in heterozygous individuals, such as genotypes $Hb_\beta{}^A/Hb_\beta{}^C$, $Hb_\alpha{}^A/Hb_\alpha{}^I$, and $Hb_\beta{}^A/Hb_\beta{}^E$. In the homozygous state, in genotypes $Hb_\beta{}^C/Hb_\beta{}^C$ and $Hb_\beta{}^E/Hb_\beta{}^E$ for example, the consequences for the individual do not appear to be especially severe. Mild hemolytic conditions resulting in anemia, abnormal erythrocytes, and hyperplasia (enlargement) of the bone marrow are often encountered. These individuals are usually not severely incapacitated.

The relatively high frequencies and the distributions of some aberrant hemoglobins provide fascinating problems for the physical anthropologist. These variant hemoglobins, all of them deleterious to some extent, give us opportunities to view the action of natural selection and gene flow in maintaining genetic polymorphisms in human populations. $Hb\alpha_2{}^A\beta_2{}^S$ reaches relatively high frequencies in a large number of human populations (Fig. 29.10), despite the severe effect it has on individual viability and fertility. We have already demonstrated how the allele is maintained at relatively high frequencies in human populations (Chapter 24). Hemoglobin C, $Hb\alpha_2{}^A\beta_2{}^C$, has a very special distribution (Fig. 29.11). It reaches its highest frequencies on the Volta plateau in northern Ghana. The frequency drops in all directions from this plateau. It is found among descendants of African slaves in Surinam, Curaçao, Venezuela, and the United States. Hemoglobin E, $Hb\alpha_2{}^A\beta_2{}^E$, is found in southern Asia. It has frequencies as high as 27 per cent in Cambodia (Fig. 29.11), and sporadic examples of it are found in Greece and Turkey. A rather large number of hemoglobin variants have been discovered in very low frequency or in single families. Some of these are among the hemoglobins listed in Table 29.2.

Hemoglobins D, G, M, and others were originally distinguished from each other by their electrophoretic mobilities. Each consists of several hemoglobins which were only discovered when attempts were made to determine amino acid sequences. It may no longer be proper to speak of them as hemoglobin D or hemoglobin G, since there are several different hemoglobins included in each.

The hemoglobins labeled M are a slightly different case. One of them was the first abnormal hemoglobin to be described. It was found in 1948 in a patient suffering from an hereditary anemia, methemoglobinemia. The anemia in such cases is due to the presence of methemoglobin (hence M).

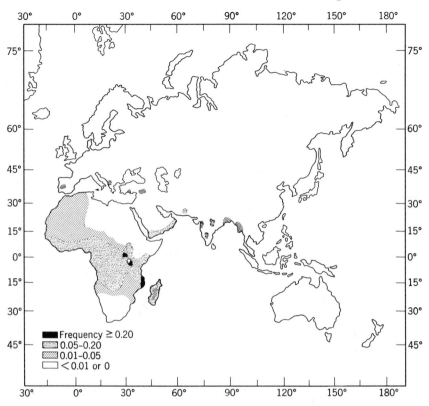

Fig. 29.10.   The distribution of the allele for hemoglobin S in the Old World.

Methemoglobin is hemoglobin in which two of the iron atoms in the molecule are in the oxidized (ferric) rather than in the usual reduced (ferrous) state. This derivative of hemoglobin is incapable of combining with oxygen, for it already has combined—it is oxidized. Thus it cannot carry out its normal function of combining with oxygen in the lungs and transporting oxygen to the tissues. Many of the individuals with one of the hemoglobin M's turn out to be heterozygotes, with both Hb$\alpha_2{}^A\beta_2{}^A$ and one particular hemoglobin M. Some of the hemoglobin M's have an amino acid substitution in the α chain, others in the β chain (Table 29.2). Since only two of four iron atoms are involved, one-half of the chains in any particular methemoglobin molecule are presumably capable of combining with oxygen.

Hemoglobin Norfolk and hemoglobin Zürich are not associated with anemia or other clinical abnormalities. The amino acid substitutions in these two hemoglobins occur at the same positions or very close to the positions at which the substitutions are found in the hemoglobin M's (Table 29.2). These replacements occur in a region of the chain that is close to the iron atom. The amino acid replacements in methemoglobins (Tyr and Glu) form such stable complexes with ferric iron that the normal reduction

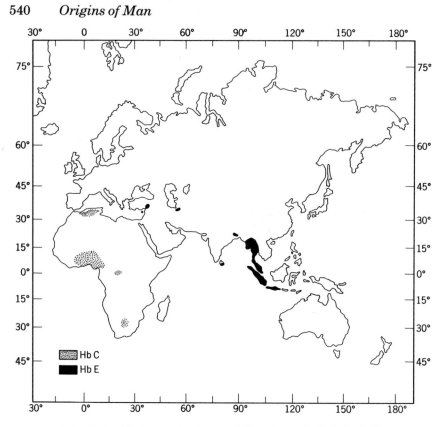

Fig. 29.11.   The distribution of hemoglobins C and E in the Old World.

of iron cannot take place. The amino acid replacements in hemoglobin Norfolk (Asp) and in hemoglobin Zürich (Arg) allow for normal oxidation-reduction of the iron atoms.

A close relationship between several of the variant hemoglobins and an inherited anemia, thalassemia, was once believed to exist. Thalassemia, caused by an inherited deficiency in the production of hemoglobin, is frequently encountered in individuals who have one or another of the variant hemoglobins.

Hemoglobin S and Malaria

The sickle cell trait occurs with highest frequency in tropical Africa (10 to 40 per cent). It is also found with high frequency (5 to 30 per cent) in groups living in India, Greece, and southern Turkey. Lower frequencies (less than 10 per cent) have been reported for populations living around the Mediterranean—Palestine, Tunisia, Algeria, and Sicily. The distribution of the sickle cell trait is continuous throughout these areas (Fig. 29.10). The trait is also found in populations which are presumed to have emigrated from Africa and the Mediterranean regions. It is not found in northern Europe, most of Asia, and Australia. In the New World the sickle cell trait is found only among descendants of persons who emigrated from the Old; it is not found among New World Indians.

The distribution of the sickle trait in Africa suggests that some unusual circumstance must maintain the high frequencies of this deleterious trait in African populations. The homozygotes ($Hb_\beta S/Hb_\beta S$) almost never live to maturity. Thus there is a constant loss of the allele $Hb_\beta S$ in every generation. Since this allele is found in frequencies as high as 0.20, something operates to maintain the allele in the population and to compensate for its loss through death of homozygotes. Positive selection on the heterozygotes is the mechanism which maintains genetic polymorphism of this sort.

The unusual circumstance responsible for this hemoglobin polymorphism is believed to be malaria. The heterozygotes are considered more resistant to malaria of the type produced by one specific parasite, *Plasmodium falciparum*, transmitted to man by the *Anopheles* mosquito. Allison has presented three types of evidence to support the hypothesis that $Hb_\beta A/Hb_\beta S$ heterozygotes are more resistant to falciparum malaria than are $Hb_\beta A/Hb_\beta A$ and $Hb_\beta S/Hb_\beta S$ homozygotes. First, in Africa the distribution of the $Hb_\beta S$ allele paralleled the distribution of hyperendemic (constant, high frequency) falciparum malaria (Fig. 29.10 and 29.12). High frequencies of $Hb_\beta S$ were correlated with high rates of infection with falciparum malaria. Second, some observations showed that the number of parasites was smaller in sicklers than in nonsicklers when both had malaria. The statistical difference between the two groups reached only low levels of significance, but the difference was in the direction predicted by the hypothesis. Third, sicklers were less susceptible to experimentally induced malaria than were non-sicklers. Allison used thirty African volunteers, fifteen with the sickle cell trait and fifteen without. He injected each subject with malaria parasites, *P. falciparum*. Fourteen of the nonsicklers developed malaria as indicated by the presence of the parasites in blood samples taken from peripheral veins. Only two of the sicklers had parasites in their peripheral blood. The numbers are small, but the results support the hypothesis, although they do not prove it. Most of this evidence is indirect and circumstantial, but it does indicate that the rate of infection with *P. falciparum* is lower among sicklers than it is among nonsicklers. The infection among sicklers who have malaria appears to be less severe.

Direct evidence in support of the malaria-sickle cell hypothesis is difficult to obtain. It would consist of carefully controlled studies of the relations between deaths due to falciparum malaria and the sickle cell trait. It would also consist of a demonstrated physiological basis for the protection of sicklers against malaria. Some direct evidence is available. Several studies have shown that there were fewer sicklers among children in Africa who died

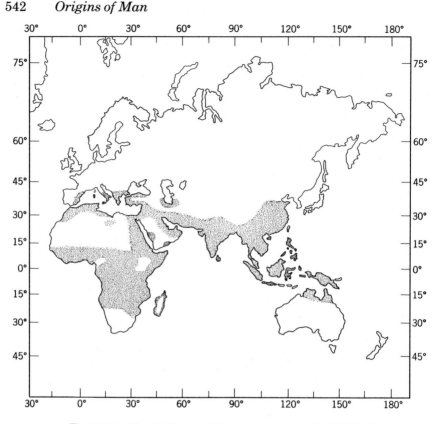

Fig. 29.12.   The distribution of falciparum malaria in the Old World.

of malaria, particularly of the virulent cerebral form, than there were among children in control groups.

A large number of reports followed Allison's first publications. They should be studied with care by physical anthropologists, for they show the difficulties faced by investigators who wish to determine whether there is an association between a genetically controlled trait and a disease. Mortality is high in the areas of Africa in which malaria and the sickle cell trait are common. Mortality is especially high among infants and young children. It is very difficult to specify the extent to which malaria is responsible for the differential mortality because infants and young children die of multiple causes. Total selection, furthermore, is almost always most intense where population density is high, as it is in many parts of tropical Africa. It is not always easy to resolve the total selection on a population into components due to various specific causes such as malaria or malnutrition.

The malaria parasite is transmitted to man by the *Anopheles* mosquito. This mosquito is an essential part of the life cycle of the parasite, for without it the parasite does not survive. The human host provides red cells in which the malaria parasite can develop. A number of investigators have suggested that hemoglobin S is not a suitable medium in which the malaria parasites can develop. At least it is a less favorable medium than normal hemoglobin. Another suggestion is that the parasitized cells are differentially affected by hemoglobin S and hemoglobin A. It has been found that erythrocytes with the malaria parasite have a tendency to stick to the walls of the blood vessels when the parasite is in various stages of its life cycle. Such cells would not be able to pick up oxygen from the lungs, and those with $Hb\alpha_2^A\beta_2^S$ would sickle when the amount of oxygen in the vicinity of the cells is reduced. Sickle cells would then be destroyed at a greater rate than normal cells, and the parasite in them would be killed. If a significant portion of parasitized red cells were destroyed, the parasites in the host would be severely handicapped.

One other kind of investigation gave support to the hypothesis. A powerful effect upon the frequencies of the alleles which lead to hemoglobin A and hemoglobin S would be exerted by an increased fertility of the heterozygotes Hb_β^A/Hb_β^S. Relatively few data about this are available. However an examination of the Black Carib population of Central America demonstrated a significant difference between the fertility of homozygous and heterozygous females. The female heterozygotes (Hb_β^A/Hb_β^S) had a relatively greater fertility than the normal homozygotes (Hb_β^A/Hb_β^A). The difference was of the right order of magnitude to account for the frequency of the sickle cell trait. One explanation is that homozygous mothers (Hb_β^A/Hb_β^A) may have a higher rate of fetal loss. The mechanism suggested is that the placenta of homozygous mothers (Hb_β^A/Hb_β^A), who are not immune to malaria, is predisposed during attacks of malaria to some kind of mechanical injury which results in fetal death. The placentas of heterozygous mothers (Hb_β^A/Hb_β^S) are believed less subject to mechanical injury as a result of malarial infection.

At the present time the hypothesis that *Plasmodium falciparum* malaria has a major role in maintaining the observed polymorphism of the sickle cell trait has widespread support. There are objections to the malaria-sickle cell hypothesis, and we are not wholly convinced of its validity. If differential susceptibility to *P. falciparum* malaria were the only factor that maintained sickle cell polymorphism, the mortality from malaria alone would be very large in the population. Experts in the study of malaria have stated that the mortality rate, which must be assumed, seems unnecessarily high. Other disorders may affect sicklers less than nonsicklers. Then the polymorphism might be the result of a combination of factors. Furthermore in Africa in areas where malaria is prevalent, there are many populations that have little or no hemoglobin S. There are also populations with relatively high frequencies of Hb_β^S that are almost free of falciparum malaria. As we shall see, it is possible to explain away these apparent contradictions. Nonetheless we shall retain our agnostic attitude toward the malaria-sickle cell hypothesis until the mechanism by which hemoglobin S prevents malaria is demonstrated.

At one time it was suggested that the sickle cell trait was maintained by recurrent mutation of $Hb_\beta{}^A$ to $Hb_\beta{}^S$. The mutation rate required to maintain a frequency of 0.20 for a deleterious trait would be abnormally high. From what we know of human mutation rates (Chapter 24), a sickle cell-mutation hypothesis is not reasonable.

The problem is to explain the various observed frequencies of the allele for hemoglobin S in African populations. The sickle cell trait, if it confers an advantage, must do so, according to the malaria hypothesis, when falciparum malaria is endemic in the human environment. Livingstone has suggested that the spread of the gene for hemoglobin S and the spread of falciparum malaria may well have been concomitant events in Africa.

West Africa is the critical area in which many studies of the sickle cell trait have been made. The observed frequencies of the sickle cell trait show that the genes are not in stable equilibrium in many populations despite the presence of endemic malaria. How did the allele $Hb_\beta{}^S$ get into these populations, *and* how is it maintained? Livingstone suggests, from his studies in West Africa, that gene flow and natural selection acted together. In groups where $Hb_\beta{}^S$ is now present at low frequencies, the allele was introduced by relatively recent migration. The allele is a newcomer in the gene pool of some West African groups. Natural selection, that is, malaria, has had a short time in which to operate and produce a stable equilibrium. It has been operating, but the effect of gene flow has not yet been obscured.

Plasmodium falciparum, the particular parasite that produces the kind of malaria to which sicklers are resistant, is believed to be the most recent species of *Plasmodium* to have developed a parasitic relationship with man. It has, in the perspective of evolutionary time, only recently become a part of the human environment. The destruction of large segments of tropical rain forests by slash-and-burn agriculture was a necessary condition for the development of an environment in which the mosquito which carries the malaria parasite could live. The *Anopheles* mosquito, the intermediate host for the malaria parasite, does not appear to be part of the normal fauna of the tropical forest. It must have still open pools of water in which to breed. Its econiche is found near forest settlements and agricultural plots. The slash-and-burn agriculture of many African peoples provides a reasonable environment for this mosquito.

The so-called agricultural revolution probably occurred less than 10,000 years ago. It spread to or occurred independently in Africa somewhat later than that. In Africa it produced changes in the environment for both man and mosquito. Men and mosquitos were partners in a complex adaptive cycle. It is not unlikely, as Livingstone put it, that one factor in the adaptive process was man's blundering into the tropical rain forest and producing an environment in which the parasite could adapt to him. The second factor in this adaptive cycle was the larger size of human populations. Settled communities with an ensured food supply provided the largest and most numerous animals (men) on which the mosquito carrying the parasite could feed.

Livingstone marshalled archeological, linguistic, epidemiological, and genetic evidence to support his reconstruction. Given that malaria is the principal factor resisted better by $Hb_\beta{}^A/Hb_\beta{}^S$ heterozygotes than by either homozygote, we have an impressive story. The spread of agriculture increases the area in which the sickle cell trait is at an advantage and, hence, leads to a spread of the gene. Livingstone's hypothetical reconstruction is extremely plausible, given the initial premise that the sickle cell trait confers relative immunity to malaria.

Thalassemia

Thalassemia has often been mentioned in association with human hemoglobin variants. A genetic relationship between certain hemoglobins and thalassemia, once postulated, is not considered to be the case today. Thalassemia, often called Cooley's anemia, appears in two clinical states—thalassemia major and thalassemia minor. The former is presumed to be the homozygous expression of the trait, the latter heterozygous. Distribution of thalassemia (Fig. 29.13) suggests that, if the trait is due to a simple nondominant allele, the heterozygote has some advantages in environments where malaria is hyperendemic.

The symptoms of the two clinically recognized forms of thalassemia are listed in Table 29.3. Thalassemia minor, the heterozygous or carrier state, exhibits mild microcytosis, that is, a relatively large number of abnormally small red cells are found in the circulation. These red cells also appear to be more resistant to changes in osmotic pressure (changes in salt concentration of the blood) than normal red cells. The anemia these changes produce is mild and is characterized by reduction in amount of hemoglobin, hence this is called hypochromic (less-colored) anemia.

Thalassemia major is a severe hemolytic anemia, and afflicted individuals cannot survive unless they receive blood transfusions regularly. The spleen becomes enlarged, and there is hyperplasia of the bone marrow. Some of the red cells contain both adult and fetal hemoglobins. Thalassemia major is a set of symptoms which are produced by a block in the synthesis of hemoglobin. The interruption of hemoglobin synthesis or the reduction in amount of hemoglobin synthesized leads to the production of microcytic erythrocytes which in turn results in hemolytic anemia, for the microcytes are rapidly destroyed. The pressure of the anemia strains the tissues that produce red cells, and there is an increase in the number of red cells produced. The stress on the red cell forming mechanism leads to hyperplasia of the bone marrow and increased production of the cells which synthesize

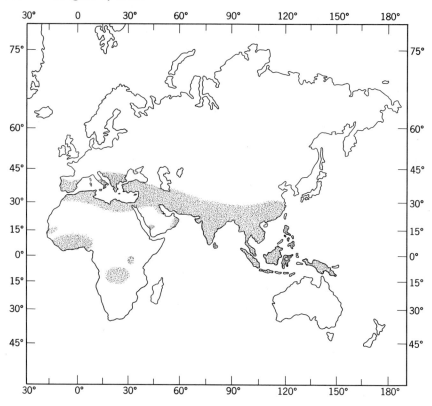

Fig. 29.13.   The distribution of thalassemia in the Old World.

fetal hemoglobin ($Hb\alpha_2{}^A\gamma_2{}^F$) postnatally. The severe hemolysis forces cells to continue to produce $Hb\alpha_2{}^A\gamma_2{}^F$ instead of switching from production of $Hb\alpha_2{}^A\gamma_2{}^F$ to normal production of $Hb\alpha_2{}^A\beta_2{}^A$.

The genetics of thalassemia is growing less obscure from month to month. There is both genetic and chemical evidence that thalassemia is due to suppression of the synthesis of either the α chains or the β chains of $Hb\alpha_2{}^A\beta_2{}^A$. This means that the clinical syndrome is produced by two different genetic loci, and the terms α-thalassemia and β-thalassemia are now used to refer to α and β chain defects, respectively. Thalassemia major is the clinical entity found in individuals homozygous for the α or β chain defect, $Hb_\alpha{}^{Th}/Hb_\alpha{}^{Th}$ or $Hb_\beta{}^{Th}/Hb_\beta{}^{Th}$, where Th denotes thalassemia. Thalassemia minor occurs in individuals heterozygous for either α- or β-thalassemia, $Hb_\alpha{}^N/Hb_\alpha{}^{Th}$ or $Hb_\beta{}^N/Hb_\beta{}^{Th}$ or, conceivably, $Hb_\alpha{}^N/Hb_\alpha{}^{Th}$ plus $Hb_\beta{}^N/Hb_\beta{}^{Th}$ where N denotes normal production of α or β chains and Th denotes thalassemia.

The basic defect is not an abnormal hemoglobin, but a block in the synthesis of normal hemoglobin. Ingram proposed a theoretical model to ex-

TABLE 29.3
Some Clinical Signs in Thalassemia

	Thalassemia major	Thalassemia minor
Anemia	Severe	None to moderate
Hemoglobin F	Present (10–90 per cent of total hemoglobin)	Absent or present ($<$ 10 per cent)
Hemoglobin A_2	Normal or increased amount	Normal or increased amount
Jaundice	Extreme	None or mild
Spleen size	Enlarged	Normal or slightly enlarged
Liver size	Enlarged	Normal
Bone changes	Extreme	None
Red blood cell count	Decreased	Normal or increased
Red blood cell size	Small	Normal or small
White blood cells	Abnormal sizes and numbers	Normal

plain the molecular basis for this disorder. Because so many aspects of modern physical anthropology will and have come down to research on molecules, we briefly summarize Ingram's model. It is based on the model for protein synthesis presented in Chapter 25.

In thalassemia the messenger RNA that conveys instructions from DNA to the ribosomes for synthesis of α and/or β chains of hemoglobin is considered defective. The defective messenger RNA blocks enough ribosomes so that hemoglobin production is reduced. Reduced production of hemoglobin A leads to the variety of clinical signs which together make up the thalassemia major syndrome (Table 29.3). Thalassemia minor occurs when the defect in the messenger RNA is related to only one locus, so that about half as many α or β chains as usual will be synthesized. When synthesis of β chains is interrupted (β-thalassemia), there is an excess of α chains, but there is normal synthesis and production of δ chains. The observed increase in $Hb\alpha_2{}^A\delta_2{}^{A2}$ in many cases of thalassemia minor is a result of the excess number of α chains available for combination with δ chains. Synthesis of α chains is interrupted in individuals with α-thalassemia. A hemoglobin composed of four β chains, $Hb\beta_4$ (hemoglobin H), is sometimes found in such patients. Fetal hemoglobin $Hb\alpha_2{}^A\gamma_2{}^F$ is also found in persons with β-thalassemia, as is the unusual $Hb\gamma_4$ (Bart's).

The distribution of thalassemia (Fig. 29.13) cannot be fully explained until the genetics and biochemistry of the trait are clearly understood. We must eventually distinguish the types of thalassemia included in the frequencies reported for various populations. Thalassemia is quite frequent in certain parts of the world. The highest frequencies are generally correlated with endemic malaria (Fig. 29.12). It has been suggested that thalassemia has survived in human populations only because selection is maintaining it in balanced polymorphism. How could so disadvantageous an allele survive? Because the heterozygous individuals have increased resistance to malaria. There is not much direct evidence that individuals with thalassemia minor are resistant to malaria, but there is little doubt that thalassemia reaches its greatest frequencies among populations which live in environments that include endemic malaria.

Hemoglobins S, C, and E and Thalassemia

Among the fascinating problems which the abnormal homoglobins present is the one of interaction of alleles for β chain variants. A number of hemoglobins besides $Hb\alpha_2{}^A\beta_2{}^S$ occur in sufficiently high frequencies to rule out mutation as the sole means by which these frequencies are maintained. We shall consider the interaction of hemoglobin $C(Hb\alpha_2{}^A\beta_2{}^C)$, hemoglobin $E(Hb\alpha_2{}^A\beta_2{}^E)$, or thalassemia with $Hb\alpha_2{}^A\beta_2{}^S$.

There is no population in which the frequency of $Hb\alpha_2{}^A\beta_2{}^C$ reaches the relative frequency achieved by $Hb\alpha_2{}^A\beta_2{}^S$. The observed frequencies suggest that a balanced polymorphism does not exist, regardless of whether the frequencies are maintained by selection. After the malaria-sickle cell hypothesis received such strong support, attempts were made to relate malaria to hemoglobin C, just as attempts were made to relate malaria to thalassemia. A study of the hemoglobin C trait among the Yoruba of Nigeria showed that matings in which either the mother or the father was a heterozygote ($Hb_\beta{}^A/Hb_\beta{}^C$) had slightly more offspring per mating than matings between normal homozygotes. On the other hand no direct or good inferential support for the hypothesis that malaria is the agent of selection on hemoglobin C has been reported. A study of the descendants of African slaves in the Netherlands Antilles and in Dutch Guiana (Curaçao and Surinam) suggested that $Hb_\beta{}^S$ confers a greater advantage than $Hb_\beta{}^C$ when there is endemic falciparum malaria (Surinam). In Curaçao there is no malaria, and the frequency of $Hb_\beta{}^C$ is about as high as that of $Hb_\beta{}^S$. Many of the population data are difficult to interpret, for the $Hb_\beta{}^S$ allele is also present in many of the human groups that have significant frequencies of $Hb_\beta{}^C$. It is possible that the $Hb_\beta{}^S$ allele is more efficient at producing increased resistance to malaria than the $Hb_\beta{}^C$ allele.

The genotypes which have mutations at the Hb_β locus have great differences in fitness, meaning that changes in frequencies of some of the genotypes will occur in much less time than they would occur at most other loci. The $Hb_\beta{}^S$ and $Hb_\beta{}^C$ alleles are found in many of the same populations in West Africa (Fig. 29.14). The frequency of the $Hb_\beta{}^S$ allele usually is higher than the frequency of the $Hb_\beta{}^C$ allele in these populations. The distribution of these two mutant alleles at the Hb_β locus can be interpreted as the early stages of a replacement of one allele by the other. The $Hb_\beta{}^S$ allele appears to be replacing the $Hb_\beta{}^C$ allele in West Africa, for the heterozygote $Hb_\beta{}^A/Hb_\beta{}^S$ is more fit than the heterozygote $Hb_\beta{}^A/Hb_\beta{}^C$. Livingstone has called the $Hb_\beta{}^S$ allele "predatory" because of its apparent tendency to re-

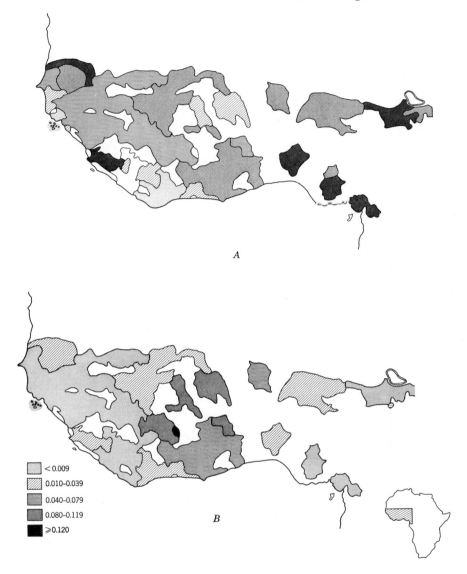

Fig. 29.14.   The distribution of the alleles for (A) hemoglobin S $Hb_\beta{}^S$ and (B) hemoglobin C $Hb_\beta{}^C$ in West Africa. (Adapted from Rucknagel and Neel, 1961.)

place or eliminate other variant hemoglobin alleles, if malaria is in the environment. Even though the fitness of $Hb_\beta{}^S/Hb_\beta{}^S$ is essentially zero and the fitness of $Hb_\beta{}^C/Hb_\beta{}^C$ may be as high as 0.6, the fitness of $Hb_\beta{}^A/Hb_\beta{}^S$ is greater than the fitness of $Hb_\beta{}^A/Hb_\beta{}^C$. The fitness of the hemoglobin S heterozygotes is greater than the fitness of the hemoglobin C heterozygotes by an amount sufficient to make $Hb_\beta{}^S$ predatory. In considering $Hb\alpha_2{}^A\beta_2{}^A$,

Hba$_2$^{A}β$_2$^{S}, and Hba$_2$^{A}β$_2$^{C}, Livingstone has shown that the Hb_β^S allele will replace the Hb_β^C allele within 50 generations, even if the initial frequency of Hb_β^C is as high as 0.14. Almost total elimination of Hb_β^C will occur in less than 150 generations. Graphic representations of Livingstone's mathematical calculations are shown in Fig. 29.15.

Thalassemia is also involved in the assumed predatory relationship with Hb_β^S. The thalassemia alleles have a complementary distribution to the Hb_β^S alleles in some areas (Fig. 29.10 and 29.13). This may be interpreted as indicating that one allele is replacing the other. Thalassemia is relatively frequent among Arab populations, and in some which have the Hb_β^S allele as well the relative frequency of thalassemia is very low. This is indirect evidence in support of the predation hypothesis.

If the allele Hb_β^S in malarial environments is predatory against the other β chain hemoglobin alleles, we must ask whether the worldwide distribution of the Hb_β^S allele is consistent with this view. To a large extent it is. The frequency of the Hb_β^S allele is in equilibrium with Hb_β^A in most East African populations studied, and the Hb_β^C and the Hb_β^{Th} alleles are absent. Presumably Hb_β^S has had a longer time in which to reach equilibrium with Hb_β^A and replace Hb_β^C and Hb_β^{Th} in East than in West Africa. The Hb_β^S allele may have originated as a mutation in East Africa and spread from there to West Africa and elsewhere. The mutation very likely occurred in other parts of the world also (India) and spread in a similar manner.

The general scheme we have outlined for the spread of the Hb_β^S allele in human populations depends upon the malaria hypothesis. The fact that the Hb_β^S allele confers a selective advantage upon the heterozygote cannot be

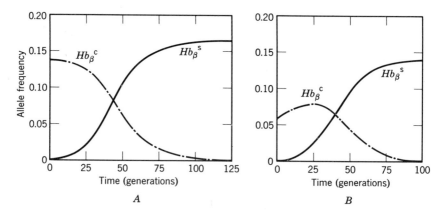

Fig. 29.15.   Theoretical estimates of the time necessary for elimination of the allele for hemoglobin C Hb_β^C by the allele for hemoglobin S Hb_β^S. (A) Initial frequency of Hb_β^C = 0.14. Relative fitnesses of genotypes are: Hb_β^A/Hb_β^S = 1.25; Hb_β^A/Hb_β^C = 1.08; Hb_β^S/Hb_β^C = 0.5; Hb_β^C/Hb_β^C = 0.6; Hb_β^S/Hb_β^S = O. (B) Initial frequency of Hb_β^C = 0.06. Relative fitnesses are: Hb_β^A/Hb_β^S = 1.20; Hb_β^A/Hb_β^C = 1.04; Hb_β^C/Hb_β^C = 0.8; Hb_β^S/Hb_β^C = 0.5; Hb_β^S/Hb_β^S = 0. (Data from Livingstone, 1964.)

doubted. This is the only reasonable explanation for its observed frequencies. But whether malaria maintains it is not proven. We already have cited the evidence which supports the malaria hypothesis. The reconstruction of the spread of the $Hb_\beta S$ allele and its predation against other β-chain alleles is circumstantial support of this hypothesis. What we now look forward to is direct demonstration of the physiological mechanism by which the heterozygotes ($Hb_\beta A/Hb_\beta S$) resist malaria.

The malaria hypothesis is also called upon to explain the distribution of the variant hemoglobin E, found in southeast Asia and throughout the Indonesian archipelago (Fig. 29.11). Livingstone suggests that the distribution of the $Hb_\beta E$ allele in the Indonesian archipelago is correlated with the spread of agriculture. *Anopheles,* malaria, and hemoglobin E moved through part of southeast Asia as the primeval forest was cleared for root crops and rice. The $Hb_\beta E$ allele apparently did not cross to the islands of Timor and Borneo. We assume that migration did not carry it to Celebes where hemoglobin O is found. Hemoglobin O ($Hb\alpha_2 O \beta_2 A$) is found on the periphery of the distribution of hemoglobin E in southeast Asia, and the suggestion has been made that the relationship of hemoglobin O to E is analogous to that of hemoglobin C to S in a malaria environment. Furthermore thalassemia has been reported on New Guinea where there is malaria. The suggestion has been made that malaria has selected the hemoglobin O heterozygotes on Celebes and thalassemia heterozygotes in New Guinea. Again we assume that gene flow from Asia has not occurred, or at least that no flow of $Hb_\beta E$ alleles into Celebes or New Guinea has provided a potential genetic immunity to malaria. This conclusion seems premature, particularly since hemoglobin O found in this part of the world is an α chain mutant, and it is the β chain mutants that appear to have the greater fitness when challenged by malaria.

Current widely accepted views about reasons for the presence of some β chain mutant hemoglobins in relatively high frequencies in some human populations may be summarized as follows. Many populations have more than one of the alleles in significant numbers, and at the same time another inherited hemoglobin disorder, thalassemia, occurs. The unifying theme is malaria. The human species, it is suggested, has been developing a genetic immunity to a relatively new parasite in its environment. A number of different mutations in the β chain allele of human hemoglobin appears to provide an unfavorable environment for the parasite. This deleterious effect upon the parasite outweighs the less than favorable effect the mutations have upon the human host.

The absence of β chain mutant hemoglobins from many parts of the world in which malaria is endemic is not explained by the hypothesis. Mutations should occur randomly; they should occur in a number of places inhabited by the human species; and there is no good reason to suppose that β chain mutations did not occur in other areas. But if they did, why did they not persist? One answer is that there are small, even minute, differences in human environments that affect the survival of mutant phenotypes. We must consider the effect of malaria in each human microenvironment. This is not a wholly satisfactory answer. To postulate gene flow or absence of gene flow whenever the distribution of the β chain mutants does not follow from the distribution of malaria is also unsatisfactory. Despite such agnostic remarks about the malaria-sickle cell hypothesis, it is one of the most challenging and fruitful theories proposed by physical anthropologists and geneticists.

Our descendants three or four generations hence may have the final proof of the hypothesis. If malaria is eradicated, and it appears that the efforts of various governments to eradicate it will be successful, the frequency of $Hb_\beta S$ will drop markedly in the next two or three generations, for the major selection pressure in its favor will have been removed.

SUGGESTED READINGS AND SOURCES

Allison, A. C., The distribution of the sickle-cell trait in East Africa and elsewhere and its apparent relationship to the incidence of subtertian malaria. *Trans. Roy. Soc. Trop. Med. Hyg.* **48**, 312 (1954).

Allison, A. C., Notes on sickle-cell polymorphism. *Ann. Human Genet.* **19**, 39 (1954).

Allison, A. C., Protection afforded by sickle-cell trait against subtertian malarial infection. *Brit. Med. J.* **1**, 290 (1954).

Allison, A. C., The sickle-cell and haemoglobin C genes in some African populations. *Ann. Human Genet.* **21**, 67 (1956).

Barkhan, P., M. E. Stevenson, G. Pinker, N. Dance, and E. M. Shooter, Haemoglobin Lepore trait. *Brit. J. Haematol.* **10**, 437 (1964).

Barnicot, N. A., A. C. Allison, B. S. Blumberg, G. Deliyanniṣ, C. Krimbas, and A. Ballas, Haemoglobin types in Greek populations. *Ann. Human Genet.* **26**, 229 (1963).

Beaven, G. H., and W. B. Gratzer, A critical review of human haemoglobin variants. Part II. Individual haemoglobins. *J. Clin. Pathol.* **12**, 101 (1959).

Block, R. J., E. L. Durrum, and G. Zweig, *A Manual of Paper Chromatography and Paper Electrophoresis* (Second edition). Academic Press, New York (1958).

Blumberg, B. S. (Ed.), *Proceedings of the Conference on Genetic Polymorphisms and Geographic Variation in Disease.* Grune and Stratton, New York (1962).

DeVries, A., H. Joshua, H. Lehmann, R. L. Hill, and R. E. Fellows, The first observation of an abnormal haemoglobin in a Jewish family: haemoglobin Beilinson. *Brit. J. Haematol.* **9**, 484 (1963).

Firschein, I. L., Population dynamics of the sickle-cell trait in the Black Caribs of British Honduras, Central America. *Am. J. Human Genet.* **13**, 233 (1961).

Foy, H., A. Kondi, G. L. Timms, W. Brass, and F. Bushra, The variability of sickle-cell rates in the tribes of Kenya and the southern Sudan. *Brit. Med. J.* **1**, 294 (1954).

Gerald, P. S., and M. L. Efron, Chemical studies of several varieties of Hb M. *Proc. Natl. Acad. Sci. U. S.* **47**, 1758 (1961).

Hammond, D., P. Sturgeon, W. Bergren, and A. Caviles, Jr., Definition of Cooley's trait or thalassemia minor: classical, clinical and routine laboratory hematology. *Ann. N. Y. Acad. Sci.* **119**, 372 (1964).

Hanada, M., and D. L. Rucknagel, The abnormality of the primary structure of hemoglobin Shimonoseki. *Biochem. Biophys. Res. Commun.* **11**, 229 (1963).

Hill, R. J., W. Konigsberg, G. Guidotti, and L. C. Craig, The structure of human hemoglobin. I. The separation of the α and β chains and their amino acid composition. *J. Biol. Chem.* **237**, 1549 (1962).

Hill, R. L., Methods for the structural analysis of human hemoglobins. *In* W. J. Burdette (Ed.), *Methodology in Human Genetics*, p. 304. Holden-Day, San Francisco (1962).

Hoyer, B. H., B. J. McCarthy, and E. T. Bolton, A molecular approach in the systematics of higher organisms. *Science* **144,** 959 (1964).

Ingram, V. M., A specific difference between the globins of normal human and sickle cell anaemia haemoglobin. *Nature* **178,** 792 (1956).

Ingram, V. M., Abnormal human haemoglobins. I. The comparison of normal human and sickle-cell haemoglobins by "fingerprinting." *Biochim. Biophys. Acta* **28,** 539 (1958).

Ingram, V. M., *Hemoglobin and Its Abnormalities*. Charles C Thomas, Springfield, Ill. (1961).

Ingram, V. M., *The Hemoglobins in Genetics and Evolution*. Columbia Univ. Press, New York (1963).

Ingram, V. M., A molecular model for thalassemia. *Ann. N. Y. Acad. Sci.* **119,** 485 (1964).

Ingram, V. M., Control mechanisms in hemoglobin synthesis. *Medicine* **43,** 759 (1964).

Jensen, W. N., D. L. Rucknagel, and W. J. Taylor, Relationship of sickling phenomenon to blood oxygen content at various sites in the body (studied by venous catheterization) as contrasted with in vitro sickling. *J. Lab. Clin. Med.* **48,** 822 (1956).

Jonxis, J. H. P., and J. F. Delafresnaye (Ed.), *Abnormal Haemoglobins*. Charles C Thomas, Springfield, Ill. (1959).

Jonxis, J. H. P., and T. H. J. Huisman, *A Laboratory Manual on Abnormal Haemoglobins*. Blackwell, Oxford (1958).

Lehmann, H., and A. B. Raper, Maintenance of high sickling rate in an African community. *Brit. Med. J.* **2,** 333 (1956).

Livingstone, F. B., Anthropological implications of sickle cell gene distribution in West Africa. *Am. Anthropol.* **60,** 533 (1958).

Livingstone, F. B., The distribution of the sickle cell gene in Liberia. *Am. J. Human Genet.* **10,** 33 (1958).

Livingstone, F. B., Balancing the human hemoglobin polymorphisms. *Human Biol.* **33,** 205 (1961).

Livingstone, F. B., Aspects of the population dynamics of the abnormal hemoglobins and glucose-6-phosphate dehydrogenase deficiency genes. *Am. J. Human Genet.* **16,** 435 (1964).

Miller, M. J., J. V. Neel, and F. B. Livingstone, Distribution of parasites in the red cells of sickle-cell trait carriers infected with *Plasmodium falciparum*. *Trans. Roy. Soc. Trop. Med. Hyg.* **50,** 294 (1956).

Neel, J. V., and W. J. Schull, *Human Heredity*. Univ. of Chicago Press, Chicago (1954).

Pauling, L., H. J. Itano, S. J. Singer, and I. C. Wells, Sickle cell anemia, a molecular disease. *Science* **110,** 543 (1949).

Roberts, D. F., and A. E. Boyo, On the stability of haemoglobin gene frequencies in West Africa. *Ann. Human Genet.* **24,** 375 (1960).

Rucknagel, D. L., and J. V. Neel, The hemoglobinopathies. *In* A. G. Steinberg (Ed.), *Progress in Medical Genetics*, Vol. I, p. 158. Grune and Stratton, New York, 1961.

Schneider, R. G., and R. T. Jones, Hemoglobin F_{Texas}: gamma-chain variant. *Science* **148,** 240 (1965).

Schroeder, W. A., The hemoglobins. *Ann. Rev. Biochem.* **32,** 301 (1963).

Schroeder, W. A., J. R. Shelton, J. B. Shelton, J. Cormick, and R. T. Jones, The amino acid sequence of the γ chain of human fetal hemoglobin. *Biochemistry* **2,** 992 (1963).

Shibata, S. T. Miyaji, I. Iuchi, S. Ueda, and I. Takeda, Hemoglobin Hikari ($\alpha_2^A \beta_2^{61\ AspNH_2}$): a fast-moving hemoglobin found in two unrelated Japanese families. *Clin. Chim. Acta* **10,** 101 (1964).

Singer, K., A. I. Chernoff, and L. Singer, Studies on abnormal hemoglobins. II. Their identification by means of the method of fractional denaturation. *Blood* **6,** 429 (1951).

Smithies, O., Zone electrophoresis in starch gels: grouped variations in the serum proteins of normal human adults. *Biochem. J.* **61,** 629 (1955).

Whalley, P. J., J. A. Pritchard, and J. R. Richards, Jr., Sickle cell trait and pregnancy. *J. Am. Med. Assoc.* **186,** 1132 (1963).

Hemoglobins of
Nonhuman Primates

30

POPULATIONS OF PRIMATES evolve; bone and muscle complexes evolve; pelves evolve; teeth evolve; hands evolve; and molecules evolve. Hemoglobin is a particularly good molecule with which to study evolution at the molecular level. Hemoglobins of *Homo sapiens* in all their normal and variant forms have been well studied and described genetically and biochemically. The story of the evolution of hemoglobin in the Primates is far from complete, but what has been written tells us much about evolutionary processes at both the level of the cell and the level of the whole organism. The amount of genetic and evolutionary information contained in a sequence of amino acids is enormous. We refer the reader to our discussion on pages 536–538, and here we reemphasize that a study of the sequence of amino acids in a protein produces some of the best genetic data available today. If we can analyze a single protein in its homologous forms in a variety of related organisms, we can determine indirectly the kinds and magnitude of genetic changes which took place during the period since the organisms concerned branched from a common stock.

At the molecular level, there are a number of logically distinct approaches to a study of evolution. We can sort most of the theoretical approaches, the discussions, and, indeed, the polemics into two groups. First, there are the theories or approaches which take for granted, or insist, that molecular

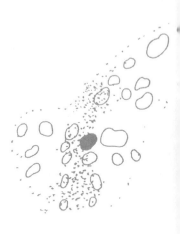

evolution can be studied independently of organisms. A basic assumption is that molecules evolve as such and that we can reconstruct this evolution. The question asked in this approach is: Can classifications of the Primates be made or unmade on the basis of the present studies of protein molecules? Enthusiastic proponents of this view are convinced that phylogenies of organisms, even classifications, may be made on the basis of molecular events. They imply that disagreements about taxonomy can be resolved with the molecular approach.

More cautious is the second view that molecules evolve as parts of organisms, indeed as parts of populations of organisms. There is no question that we can analyze the evolutionary changes in molecules themselves without referring in detail to the organisms, but we must remember that molecules evolve as *parts* of complex systems. Therefore we must ask whether molecular evolution can be observed as part of an accepted phylogeny of organisms. Can we correlate the evolutionary changes of hemoglobin with an accepted or acceptable classification of the Primates?

Analysis

We have been fortunate in being able to obtain hemoglobins from representative members of the order Primates and apply to them the methods used for describing human hemoglobins. Our ideal is unequivocally demonstrated, complete sequences of amino acids in the hemoglobin polypeptide chains from representative primates or from every species. This is an arduous task, and there are many steps, preliminary to elucidation of complete sequences, that allow us to arrive at valid conclusions without spending our whole life at routine parts of the task. Among these are electrophoresis of hemoglobins from a large number of species, fingerprints (paper peptide patterns), and amino acid compositions of separated polypeptide chains. Isolation of tryptic peptides from fingerprints and determination of their composition give us additional information. At this stage we can compare the bits and pieces of nonhuman primate hemoglobins with homologous bits and pieces of human hemoglobins. Nonhuman primate hemoglobins appear to be very like human hemoglobins. They are composed of two pairs of polypeptide chains (α-like and β-like) and four heme groups. The inheritance of hemoglobin in nonhuman primates is probably controlled by two nonallelic autosomal genes, just as it is in man. There is hemoglobin polymorphism in some species, and hemoglobin of normal individuals of some has two or more components in reasonably large quantities. The complete sequence of amino acids has been established for human α, β, and γ chains (Fig. 29.3). By comparing the compositions and the electrophoretic and chromatographic properties of peptides from nonhuman primate and human hemoglobins, we learn a great deal about the structure of the hemoglobins of nonhuman primates.

By the use of reagents specifically designed to remove one amino acid at a time from either the NH_2–terminus or the COOH–terminus, a complete sequence of a peptide can be determined. The remaining job of hooking all of the peptides together to demonstrate the sequence in an α or β chain is again time-consuming. The usual procedure is to treat the whole chain with an enzyme which breaks it at points other than lysine or arginine. The resultant peptides are examined in the same way as the tryptic peptides and have sequences which overlap the sequences of the tryptic peptides. All of the peptides can be put into their proper sequence by shuffling and shifting, and by some clever guessing. The enzymes customarily used for these studies are chymotrypsin and pepsin, obtained, respectively, from pancreatic and gastric secretions. For the nonhuman primate hemoglobins, we have not yet gone through all the steps necessary to demonstrate sequences unequivocally, and the reader must remember that the data we have provide only part of the story of the evolution of hemoglobin in the Primates.

Characteristics

Electrophoresis of hemoglobins (Fig. 29.2) gives us some essential information—the hemoglobins of the Primates are not all alike. They migrate different distances on paper or starch. Hemoglobins from at least twenty genera have been surveyed by use of electrophoretic techniques. These genera and *Homo* are representatives of seven of the eight major taxa in the order Primates (Table 5.1). No one, to our knowledge, has examined the hemoglobin of *Tarsius,* the only living genus of the Tarsiiformes.

The resolution of primate hemoglobins by electrophoresis shows that some species exhibit hemoglobin polymorphisms. Hemoglobin polymorphisms have been found in species of the Lemuriformes, Lorisiformes, Cercopithecoidea, and Pongidae. Little is known yet about the nature of these polymorphisms. We do not know what the frequencies of many of the various hemoglobins are in natural populations, although some are definitely rare. We do not know anything at all about possible selective mechanisms which maintain these polymorphisms. But we assume that all the hemoglobins found in these nonhuman primates are functional. An animal cannot survive unless its hemoglobin functions and supplies oxygen to the tissues. Anemias, even mild anemias, would seriously hamper the ability of most if not all nonhuman primates to breed and survive in their natural habitats. Methemoglobinemia, not inevitably fatal in man, is inconceivable in a primate surviving in the wild.

Electrophoresis of hemoglobins does not provide us with data of overwhelming phylogenetic import. However we are able to make some descriptive statements about the hemoglobins of the nonhuman primates, bearing in mind that migration of molecules in an electric field is a crude measure of the properties of the molecules.

The hemoglobin of *Tupaia* (Tupaiiformes) moves a shorter distance on electrophoresis than the hemoglobin of any other primates we have tested. The major feature of the hemoglobins of *Loris, Nycticebus,* and *Perodicticus* (Lorisiformes) is the occurrence of two components in nearly equal amounts. Some species of *Galago* also have two-component hemoglobin; other species of this genus have three-component hemoglobin. The hemoglobins of Lemuriformes show widely varying mobilities. Those of *Lemur catta* and *Hapalemur* have two components.

Hemoglobins of most Ceboidea and Cercopithecoidea have similar electrophoretic patterns. They migrate much like human hemoglobin. But this does not imply that they are similar in structure. The hemoglobin of baboons, for example, is very unlike human hemoglobin (page 561). We have found that hemoglobins from most individuals of a species migrate the same distance. The fact that hemoglobins from many individuals of a species have the same electrophoretic mobility does not prove they are identical. The human hemoglobins D's and M's warn us to use more elaborate analytical methods than electrophoresis.

Some polymorphisms have been discovered among the Cercopithecoidea. A large number of baboon (*Papio*) hemoglobin samples were analyzed electrophoretically. Three samples out of 584 contained more than a single component. A number of samples from rhesus monkeys (*Macaca*), out of many hundreds tested, showed more than one component after separation by electrophoresis. These apparent hemoglobin polymorphisms occurred in animals that are members of highly inbred, laboratory colonies. These rare findings among captive animals suggest that mutations at the hemoglobin loci probably seldom survive in populations of wild primates.

Electrophoretic properties of the hemoglobins of the apes (Pongidae) resemble those of *Homo*. At least some of the apes, like man, exhibit hemoglobin polymorphisms. Two electrophoretically distinct hemoglobin components have been found in gibbons and orangutans.

Hemoglobins of adult and newborn Prosimii have one property that distinguishes them from hemoglobins of Anthropoidea. All the prosimian hemoglobins which we have tested contain large proportions (as high as 75 per cent) of alkali-resistant hemoglobin. In *Homo* and in the other Anthropoidea, hemoglobin which is not denatured by treatment with alkali is present in fetuses and newborns (Chapter 29) and in some individuals with pathological conditions. There is no obvious relation between presence of or amount of alkali-resistant hemoglobin and multiple-component hemoglobin. There does not appear to be a relation between presence of isoleucine and resistance to alkali. This property serves to distinguish the hemoglobins of Prosimii from those of Anthropoidea, but it does not serve to distinguish Primates from other orders of animals. High levels of alkali-resistant hemoglobin have been found in, for example, elephant (*Loxodonta,* order Proboscidea), hyrax (*Procavia,* order Hyracoidea), tenrec and shrew

(*Hemicentetes, Rhynchocyon,* and *Petrodromus,* order Insectivora), and chicken (order Galliformes).

The amino acid compositions of the α- and β-like chains of representative species of the Lorisiformes, Lemuriformes, Cercopithecoidea, and Pongidae have been determined. These compositions are, not unexpectedly, very like the compositions of polypeptide chains of *Homo,* and they show no unusual features. The compositions of the chains of *Galago crassicaudatus, Lemur fulvus, Lemur catta, Papio cynocephalus,* and *Pan gorilla* are given in Table 30.1.

Evolutionary Comparisons

We have selected only a few members of the Primates (Table 30.2) whose hemoglobins we shall discuss in some detail. These animals represent stages from the most primitive to the most advanced, and they represent Paleocene, Eocene, Oligocene, Miocene, and Pleistocene developments in man's evolutionary history (Fig. 5.1, 7.1, and 7.2). These contemporary primates are products of the many adaptive, selective, and genetic events which have occurred since the animals in each lineage diverged from a common stock; they are modern forms, not Paleocene, Eocene, Oligocene, Miocene, or Pleistocene forms.

When we compare primate hemoglobins, the referents (standards) we use are the α, β, and γ chains of human $Hb\alpha_2{}^A\beta_2{}^A$ and $Hb\alpha_2{}^A\gamma_2{}^F$. When an α, β, or γ chain of human hemoglobin is digested with trypsin, 14 (from α chain), 15 (from β chain), or 16 (from γ chain) tryptic peptides are obtained. We number these, starting at the NH_2-terminus of the chain, αT-1, αT-2, αT-3, . . . αT-14; βT-1, βT-2, βT-3, . . . βT-15; or γT-1, γT-2, γT-3, . . . γT-16. Some tryptic peptides of human β chain are indicated in Fig. 29.5. The COOH-terminal amino acid of each peptide, except the last one in the chain, is always lysine or arginine (Chapter 29). We compare the peptides from the nonhuman primate hemoglobins with the analogous peptides from human hemoglobins; we compare amino acid compositions and sequences. We assume that the greater the number of amino acid identities between the chains of a primate hemoglobin and human hemoglobin the smaller the number of evolutionary events that separate the two hemoglobins. Each amino acid replacement or substitution is the result of an effective mutation. An effective mutation is, in this case, the successful substitution of one amino acid for another in the hemoglobin of all the individuals in a population.

The α-like chains of primate hemoglobins seem to have evolved rather little. The β-like chains of the nonhuman hemoglobins exhibit much more variability. The evidence for this is apparent if we examine the data presented in Table 30.3. We find some apparently immutable segments of the

TABLE 30.1
The Amino Acid Compositions of the Polypeptide Chains of Nonhuman Primate Hemoglobins[a]

Amino acid	α-like chains[b]					β-like chains[b]				
	Galago crassi-caudatus	Lemur fulvus	Lemur catta	Papio cyno-cephalus	Pan gorilla	Galago crassi-caudatus	Lemur fulvus	Lemur catta	Papio cyno-cephalus	Pan gorilla
Aspartic	12	13	12	12	13	17	12	13	15	13
Threonine	8	10	10	8	9	6	5	8	4	7
Serine	10	11	11	10	10	10	10	10	4	5
Glutamic	5	7	6	8	4	8	13	14	13	11
Proline	7	7	7	6	7	7	5	7	7	7
Glycine	7	8	10	8	7	13	12	13	15	13
Alanine	21	18	19	17	21	16	16	13	16	15
Valine	12	12	10	13	13	14	15	13	15	18
Methionine	2	2	2	2	2	1	1	1	1	1
Isoleucine	0	0	3	0	0	0	0	2	0	0
Leucine	19	18	16	19	18	17	19	18	19	18
Tyrosine	3	2	2	3	3	2	2	2	2	3
Phenylalanine	7	7	8	7	7	8	9	10	7	8
Lysine	11	10	10	12	11	10	11	9	10	11 (12)
Histidine	11	10	10	10	10	8	10	8	7	9
Arginine	2	3	3	3	3	2	2	2	2	2
Tryptophan	1	1	1	1	1	3	3	3	3	2
Cysteine	2	1	1	1	1	2	1	1	3	2
Total	140	140	141	140	140	144	146	147	143	145 (146)

[a]The compositions of α, β, and γ chains of human hemoglobins are given in Table 29.1.
[b]Compositions expressed as amino acid residues per mole. Data taken from Bradshaw et al. (1965) and Zuckerkandl and Schroeder (1961).

hemoglobin molecule in both the α and β chains (Tables 30.4 and 30.5). These findings are not unusual; many mammalian proteins have interspecific identities in portions of the molecule—insulin and ACTH (adrenal cortico-trophic hormone) are two examples. In the β-like chain of hemoglobin of *Lemur fulvus*, the hemoglobin examined in most detail thus far, we find that there is a minimum of 23 differences when compared with human β chain and 36 differences compared with human γ chain. The positions in the sequence at which the hemoglobin of *L. fulvus* differs from both human β and γ chains are more often the positions at which β and γ chains differ from each other (Fig. 30.1).

The similarities and differences we have found among primate hemo-globins correspond in large part to the accepted phylogeny of the Primates (Table 5.1). The hemoglobins of members of the two suborders, the Anthro-poidea and the Prosimii, differ from each other more than the hemoglobins of primates within each of these groups. The evidence suggests that the hemoglobins of all the Anthropoidea are quite similar, including those of *Homo sapiens*. Hemoglobins of the Prosimii appear to vary among them-selves far more. The baboon, *Papio*, is an exception. Clearly this anthropoid hemoglobin differs from human hemoglobin a great deal more than the hemoglobin of any other of the Anthropoidea examined to date. Peptide fingerprint patterns alone, of *Papio*, show as many differences from human hemoglobin as do those of some prosimians.

The hemoglobin of *Tupaia glis* differs considerably from human hemo-globin, on the basis of fingerprints. It differs more than the hemoglobin of most of the other primates studied, but probably not much more than that of some of the Lemuriformes. *Tupaia* hemoglobin fingerprints also differ greatly from the hemoglobin peptide patterns of certain Insectivora, *Rhyn-chocyon* and *Petrodromus*, elephant shrews of East Africa. Hemoglobin of various Lemuriformes are more like each other than they are like human hemoglobin or hemoglobin of primates in other taxa. There are many more similarities between the hemoglobins of Lorisiformes and Lemuriformes than there are between the hemoglobins of either and those of the other primates.

Hemoglobins of Ceboidea and Cercopithecoidea appear to resemble human hemoglobin rather closely. In this respect the Ceboidea are most interesting for they are not closely related to man. They appear as a com-pletely distinct lineage in Miocene deposits of South America. Their den-tition separates them from all other living primates. They may be related to an Eocene fossil group, the Omomyidae, which is found in many parts of the world but not in South America (Chapter 7). The apparent similarity of ceboid hemoglobins and human hemoglobin A should not be given any particular phylogenetic significance. Even if the primary structures of ceboid and human hemoglobins were extremely similar, we still would insist that this is not evidence of close relationship between the two groups. We dis-

TABLE 30.2
Primates Chosen for Studies of Evolution of Hemoglobin

Taxon	Epoch of divergence	Selected genera
Tupaiiformes	Paleocene	*Tupaia*
Tarsiiformes	Eocene	—
Lorisiformes	Eocene	*Galago, Perodicticus*
Lemuriformes	Eocene	*Lemur, Propithecus*
Ceboidea	Miocene	*Cacajao, Saimiri*
Cercopithecoidea	Oligocene	*Cercopithecus, Papio*
Pongidae	Oligocene	*Pongo, Hylobates*
Hominidae	Pleistocene	*Homo*

TABLE 30.3
Amino Acid Replacements in the Hemoglobins of Some Nonhuman Primates
(From Buettner-Janusch and Hill, 1965.)

Species	α-like chains			β- or γ-like chains			
	Number of peptides examined	Number of amino acids in peptides	Minimum number of amino acid replacements[a]	Number of peptides examined	Number of amino acids in peptides	Minimum number of amino acid replacements	
						compared with β[b]	compared with γ[c]
Lemur fulvus	10	101	6	12	134	23	36
Lemur catta	2	33	0	5	29	3	2
Lemur variegatus	4	24	3	11	96	23	25
Propithecus verreauxi	4	21	4	3	30	4	10
Galago crassicaudatus	2	33	1	10	87	9	21
Perodicticus potto	3	37	0	5	49	8	9
Papio cynocephalus	—	—	—	6	64	3	18
Hylobates lar	5	53	0	7	65	0	18

[a]The α chain of human hemoglobin A is the standard.
[b]The β chain of human hemoglobin A is the standard.
[c]The γ chain of human hemoglobin F is the standard.

TABLE 30.4
Invariant Segments in α Chains of Primate Hemoglobins
(From Buettner-Janusch and Hill, 1965.)

Tryptic peptides[a]

αT-1	αT-2	αT-5	αT-6	αT-7	αT-9
Homo	*Homo*	*Homo*	*Homo*	*Homo*	*Homo*
Hylobates	*Hylobates*	*Lemur fulvus*	*Hylobates*	*Hylobates*	*Hylobates*
Lemur fulvus	*Lemur fulvus*	*Propithecus*	*Lemur fulvus*		*Lemur catta*
Lemur variegatus	*Lemur catta*				*Perodicticus*
Perodicticus	*Perodicticus*				*Galago*

[a]The primates listed in each column are those whose hemoglobins yield certain tryptic peptides which are probably identical in composition and sequence.

TABLE 30.5

Invariant Segments in β Chains of Primate Hemoglobin
(From Buettner-Janusch and Hill, 1965.)

Tryptic peptides[a]

βT-1, 2, and 5	βT-3	βT-4	βT-6	βT-7	βT-14	βT-15
Homo	Homo	Homo	Homo	Homo	Homo	Homo
Hylobates	Hylobates	Homo γT-4	Homo γT-6	Homo γT-7	Lemur fulvus	Homo γT-15
	Perodicticus	Hylobates	Hylobates	Hylobates	Galago	Lemur fulvus
	Papio	Perodicticus	Papio	Papio		Lemur variegatus
		Papio	Perodicticus	Lemur fulvus		Galago
		Lemur fulvus	Lemur fulvus	Lemur variegatus		
		Lemur variegatus	Lemur variegatus	Lemur catta		
		Lemur catta	Lemur catta	Propithecus		
		Propithecus	Propithecus	Galago		
		Galago	Galago			

[a]The primates listed in each column are those whose hemoglobins yield certain tryptic peptides which are probably identical in composition and sequence.

βLemur ¹Thr—Leu—Leu—Ser—Ala—GLU—Glu—Asp—Ala—His—Val—THR—Ser—LEU—TRY—GLY—LYS—VAL—ASN—²⁰VAL—GLY—
βA   Val—His—Leu—Thr—Pro—GLU—Glu—Lys—Ser—Ala—Val—THR—Ala—LEU—TRY—GLY—LYS—VAL—ASN—VAL—Asp—GLY—
γF   Gly—His—Phe—Thr—Glu—GLU—Asp—Lys—Ala—Thr—Ileu—THR—Ser—LEU—TRY—GLY—LYS—VAL—ASN—VAL—Glu—Asp—Ala—GLY—

βLemur GLY—GLU—Ala—LEU—GLY—Arg—LEU—LEU—VAL—VAL—(TYR, PRO, THR, THR, Glu, ³⁰ARG, PHE, PHE, Glu, SER, PHE, GLY, Asp)—(LEU,
βA   GLY—GLU—Ala—LEU—GLY—ARG—LEU—LEU—VAL—VAL—TYR—PRO—TRY—THR—Gln—ARG—PHE—PHE—Glu—SER—PHE—GLY—Asp—LEU—
γF   GLY—GLU—Thr—LEU—GLY—ARG—LEU—LEU—VAL—VAL—TYR—PRO—TRY—THR—PRO—THR—PHE—Asp—SER—PHE—GLY—Asn—LEU—

βLemur ⁵⁰SER, Ser, Pro, Ser, ALA, Val, MET, GLY, Asp, PRO, ⁶⁰LYS)—(VAL, LYS)—(ALA, HIS, GLY, LYS)—(VAL, LEU, Ser, Ala, Phe, Ser,
βA   SER—Thr—Pro—Asp—ALA—Val—MET—GLY—Asn—PRO—LYS—VAL—LYS — ALA—HIS—GLY—LYS—LYS — VAL—LEU—Gly—Ala—Phe—Ser—
γF   SER—Ser—Ala—Ser—ALA—Ileu—MET—GLY—Asn—PRO—LYS—VAL—LYS — ALA—HIS—GLY—LYS—LYS — VAL—LEU—Thr—Ser—Leu—Gly—

βLemur Glu, Gly)—(Leu, His, HIS, LEU, ASP, Asp, LEU, LYS, GLY, THR, PHE, ALA, Ala, LEU, SER, ⁹⁰GLU, LEU, HIS, CYS, Val, Ala, LEU,
βA   Asp—Gly — Leu—Ala—HIS—LEU—ASP—Asn—LEU—LYS—GLY—THR—PHE—ALA—Thr—LEU—SER—GLU—LEU—HIS—CYS—Asp—Lys—LEU—
γF   Asp—Ala — Ileu—Lys—HIS—LEU—ASP—Asp—LEU—LYS—GLY—THR—PHE—ALA—Gln—LEU—SER—GLU—LEU—HIS—CYS—Asp—Lys—LEU—

βLemur HIS, VAL, ASP, ¹⁰⁰PRO, GLU, Asp, PHE, Lys, LEU, LEU, GLY, Asp, Ser, ¹¹⁰LEU, ALA, Asp, HIS, PHE, GLY, ¹²⁰LYS)
βA   HIS—VAL—ASP—PRO—GLU—Asn—PHE—Arg—LEU—LEU—GLY—Asn—Val—LEU—Val—Cys—VAL—LEU—ALA—His—His—PHE—GLY—LYS—
γF   HIS—VAL—ASP—PRO—GLU—Asn—PHE—Lys—LEU—LEU—GLY—Asn—Val—LEU—VAL—Thr—VAL—LEU—ALA—Ileu—His—PHE—GLY—LYS—

βLemur ¹³³Val....—VAL—Ala—GLY—VAL—(ALA, ¹⁴⁰LEU, Ala, His, Lys)—(TYR, ¹⁴⁶HIS)
βA   Val....—VAL—Ala—GLY—VAL—ALA—LEU—Ala—His—Lys — TYR—HIS
γF   Met....—VAL—Thr—GLY—VAL—ALA—Ser— ALA—LEU—Ser—Ser—Arg — TYR—HIS

Fig. 30.1. Portions of the β-like chain of hemoglobin of *Lemur fulvus* and β and γ chains of human hemoglobins A and F. The homologous portions of the three chains are indicated by capital letters (e.g., GLU at position 6). The homologous portions of β and γ chains are in boldface type (e.g., **Thr** at position 4). The sequences in parentheses are tentative; they have not been unequivocally demonstrated. In *Lemur* both aspartic acid (Asp) and asparagine (Asn) have been determined as Asp; glutamic acid (Glu) and glutamine (Gln) have been determined as Glu. (Data from Buettner-Janusch and Hill, 1965.)

cussed the hemoglobin of one of the Cercopithecoidea above (*Papio*). Most other cercopithecoid hemoglobins are quite similar to those of man.

Pongid hemoglobins are very much like those of man. The information available suggests that many pongid hemoglobins, if the sources were unknown, might be lost among the many variant human hemoglobins. The hemoglobin of chimpanzee appears to be identical with human hemoglobin; the hemoglobin of gorilla shows only four differences in amino acid composition.

The molecular approach to systematics has been enthusiastically embraced by many investigators, but the hemoglobin data, at this stage, are not crucial for reorganizing phylogeny. These data and data on other proteins bring us closer to the genetic basis for similarities and differences in two species than does the analysis of most morphological characters. When we wish to explain the presence or absence of a genetic trait, even a single amino acid substitution in a polypeptide chain of hemoglobin, we must consider the population of organisms. How does the changed molecule react in the environment to improve the survival of the organisms? We cannot yet answer these questions for the hemoglobins. Just what are the advantages of threonine (in *Lemur*) or valine (in *Homo*) as the NH_2-terminal amino acid of the β chain (Fig. 30.1)? It makes sense chemically to say that one acidic or basic or neutral amino acid may be substituted for another acidic or basic or neutral amino acid in a protein without altering the properties of the molecule. But the chemistry of the amino acid is not the point at issue here. The question is: Are threonine and valine biologically equivalent? Is this substitution neutral to natural selection? If the substitution of threonine for valine is neutral, how has it become fixed in populations of lemurs? We cannot avoid suggesting that the observed amino acid replacements have some effects—we need only cite the profound effect which hemoglobin S has on the organism (Chapter 29). A selectively neutral trait might become fixed in a population through accidents of genetic drift (sampling error). Since there are many amino acid replacements in the β chain of hemoglobin of *Lemur fulvus*, we would have to postulate a very large number of fortuitous sampling accidents to accept the hypothesis that fixing of mutations at the hemoglobin loci is due to genetic drift. Thus we return to our original suggestion that the action of natural selection has fixed the random mutations which we observe as amino acid replacements. And we trust that biochemists can explain the functional differences between threonine and valine and the functional differences between the other observed amino acid replacements which allow one species to have a reproductive advantage over another.

When we view the hemoglobin data within a valid phylogeny of the Primates, we can draw some conclusions about the molecule. The relative conservatism or lack of change displayed by the α chains (Table 30.3) suggests that they have restrictions placed on them. We know that there are a large number of different chains which combine with the α or α-like chain, and these combinations form functional hemoglobins—$Hb\alpha_2{}^A\beta_2{}^A$, $Hb\alpha_2{}^A\gamma_2{}^F$, $Hb\alpha_2{}^A\delta_2{}^{A2}$, and the nonhuman primate hemoglobins.

The observation that both α and β chains contain invariant sequences (Tables 30.4 and 30.5) suggests that any replacements in these segments disrupt the synthesis or function of hemoglobin. We observe no mutations in certain parts of the molecule. Any mutations which might have occurred were not preserved. Since millions of mutations undoubtedly have occurred in the millions of animals that have lived as part of any one of the evolutionary lineages of primates, it is remarkable how few mutations have been passed by the censor of natural selection.

An often-asked question is: How fast did something evolve? When we consider the hemoglobin of *Lemur fulvus* and human hemoglobins, we can calculate some rates from the number of changes in amino acids. The maximum time which separates lemurs from men is 55 million years. The lineage which led to the present-day lemurs probably originated in the Eocene epoch. The maximum estimate of the time that has elapsed since the Lemuriformes branched phylogenetically from the ancestral stock common to all the Primates is thus 55 million years. The fossil evidence (Chapter 7) is not so extensive that we can be positive the phylogenetic divergence of the Lemuriformes occurred then. It may have been a later event; if so, our estimates will be reduced.

There are 6 replacements in α chain of *Lemur* when compared with α chain of man. To fix an effective mutation in the α chain of a population of lemurs requires an average of 9.1 million years (55 million years for the 6 replacements, 55 million years/6 = 9.1 million years). There are 23 replacements in β-like chain of *Lemur* when compared with β chain of man. This gives a minimum of 2.4 million years to fix a mutation. If we compare the β-like chain of *Lemur* with human γ chain, the average number of years to fix a mutation is 1.5 million. These calculations suggest that the rate at which one genetic locus evolves is 4 to 6 times the rate at which the other evolves. It is reasonable to say that the two rates are not independent, for the synthesis of a functional hemoglobin molecule requires that the products of the genes (α chain and β chain) at two nonallelic loci join together successfully. These calculations of mutation rates suggest that it is not very meaningful to estimate rates of evolution from amino acid replacements alone. The rate at which amino acids are replaced is dependent upon the rate at which random changes (mutations) occur in the genetic material and the pressure of natural selection.

We have used two very simple assumptions in this discussion of measuring evolutionary rates by comparing primary structures of proteins. The mutations occurred regularly or at least randomly since the lemurs diverged from the common ancestor, and the common ancestor of man and lemur had a hemoglobin identical to that of modern man. These are not particularly good assumptions, but they enable us to calculate the rates simply. There are other assumptions that we could use, most of which would complicate the calculation of evolutionary rates even more. Some of the more realistic assumptions include the following. Some of the mutations occurred in each lineage (lemur and man). The mutations were not regularly distributed in time. If we postulate that half of the mutations occurred in each lineage, the rate at which an effective mutation is fixed is one-half of the rate calculated originally. If we postulate that the mutations were not regularly distributed through the 55 million year period, we must decide what fraction of that time to use, and we have no criteria on which to base a decision.

There is another possibility which we should consider. The identities at many positions in both α and β chains of the two hemoglobins may not be the result of the common ancestry of lemurs and man. They may be due to parallel selection in the two groups of animals. If parallel selection is responsible for amino acid identities in the chains of human and lemur hemoglobins, these acquired identities represent two replacements. Introducing the postulate of parallel selection into the calculations would increase the estimated evolutionary rates. But we have no way of distinguishing those amino acids in the two chains that are identical because of descent and those that are identical because of parallel selection. We do not intend to resolve these problems, for we do not yet have the necessary data. The same data may be used to calculate many different rates of evolution for homologous molecules in a group of organisms. The rate is clearly not linear, and it is a function of many variables, most of which we cannot yet control or specify.

A somewhat unfortunate emphasis in the past 4 or 5 years on the possible phylogenetic and taxonomic implications of molecular data has obscured the value, to anthropologists, of thinking about and investigating the functional aspects of the problem. When a larger body of relevant molecular data, structural and functional, is available, we may begin to tamper with concepts of primate phylogeny and taxonomy. Our concepts and definitions of a species, a genus, or a family are based on many characters, of which, at present, the hemoglobin molecule is but one. The similarity of the hemoglobin of gorillas to that of man, for example, should not by itself be given too much significance. The two species cannot be compared phylogenetically on the basis of their hemoglobin molecules alone. Gorilla hemoglobin is not just an abnormal human hemoglobin. The hemoglobin molecules can be viewed phylogenetically as part of two related species. For the time being, we shall learn more about hemoglobin evolution through such comparisons than we shall learn about primate evolution.

The molecular approach to the study of evolution has been used successfully with hemoglobin and serum proteins (Chapter 31) of the Primates. There is a vast amount of information contained in the structure of molecules such as hemoglobin. Since we accept evolution and natural selection, we must not stop our analyses when we have determined the evolutionary changes in the chemical structure of hemoglobin. The differences in amino acid sequences and compositions among various primate hemoglobins are not likely to have persisted and become the most common or the only hemoglobin of a species unless the changes were selectively advantageous. The individuals in whom the mutations appeared must have had a reproductive advantage over their fellows. The hemoglobin molecule is only one example of the significant use of the molecular approach to the study of evolution. All morphological differences among organisms are, basically, due to molecular differences. We widen our understanding of all facets of primate evolution when we study the relation between the structure and the function of molecules and when we are able to determine why certain molecular structures are more advantageous than others.

SUGGESTED READINGS AND SOURCES

Beaven, G. H., and W. B. Gratzer, Foetal haemoglobin in the monkey. *Nature* **184**, 1730 (1959).

Bradshaw, R. A., L. A. Rogers, R. L. Hill, and J. Buettner-Janusch. The amino acid composition and the amino terminal end groups of the α- and β-chains of four lemur hemoglobins. *Arch. Biochem. Biophys.* **109**, 571 (1965).

Buettner-Janusch, J., Hemoglobins and transferrins of baboons. *Folia Primat.* **1**, 73 (1963).

Buettner-Janusch, J., and V. Buettner-Janusch, Haemoglobins of *Galago crassicaudatus*. *Nature* **197**, 1018 (1963).

Buettner-Janusch, J., and V. Buettner-Janusch, Hemoglobins of primates. *In* J. Buettner-Janusch (Ed.), *Evolutionary and Genetic Biology of Primates*, Vol. II, p. 75. Academic Press, New York (1964).

Buettner-Janusch, J., V. Buettner-Janusch, and J. B. Sale, Plasma proteins of the African elephant and the hyrax. *Nature* **201**, 510 (1964).

Buettner-Janusch, J., and R. L. Hill, Molecules and monkeys. *Science* **147**, 836 (1965).

Buettner-Janusch, J., and J. B. Twichell, Alkali-resistant haemoglobins in prosimian Primates. *Nature* **192**, 669 (1961).

Buettner-Janusch, J., J. B. Twichell, B. Y.-S. Wong, and G. van Wagenen, Multiple haemoglobins and transferrins in a macaque sibship. *Nature* **192**, 948 (1961).

Goodman, M., Man's place in the phylogeny of the Primates as reflected in serum proteins. *In* S. L. Washburn (Ed.), *Classification and Human Evolution*, p. 204. Aldine, Chicago (1963).

Ingram, V. M., *The Hemoglobins in Genetics and Evolution*. Columbia Univ. Press, New York (1963).

Jacob, G. F., and N. C. Tappen, Abnormal haemoglobins in monkeys. *Nature* **180**, 241 (1957).

Jacob, G. F., and N. C. Tappen, Haemoglobin in monkeys. *Nature* **181**, 197 (1957).

Lie-Injo Luan Eng, M. Mansjoer, and H. W. A. Donhuysen, Two types of hemoglobin in Macaca irus mordax (Thos. & Wrought) monkeys. *Commun. Vet. Bogor, Indonesia* **4**, 59 (1960).

Riggs, A., Molecular adaptation in hemoglobins: nature of the Bohr effect. *Nature* **183**, 1037 (1959).

Riggs, A., and M. Wells, Oxygen affinity of hemoglobin S and A. *Biochim. Biophys. Acta* **50**, 243 (1960).

Schleyer, F., Über Versuche zur Blutartunterscheidung aus dem Verlauf der Alkalidenaturierung. *Deut. Z. Ges. Gerichtl. Med.* **51**, 173 (1961).

Schmidt-Nielsen, K., and B. Gjonnes, Oxygen dissociation of mammalian blood in relation to body size. *Federation Proc.* **11**, 140 (1952).

Sen, N. N., K. C. Das, and B. K. Aikat, Foetal haemoglobin in the monkey. *Nature* **186**, 977 (1960).

Simpson, G. G., Organisms and molecules in evolution. *Science* **146**, 1535 (1964).

Zuckerkandl, E., Perspectives in molecular anthropology. *In* S. L. Washburn (Ed.), *Classification and Human Evolution*, p. 243. Aldine, Chicago (1963).

Zuckerkandl, E., R. T. Jones, and L. Pauling, A comparison of animal hemoglobins by tryptic peptide analysis. *Proc. Natl. Acad. Sci. U. S.* **46,** 1349 (1960).

Zuckerkandl, E., and W. A. Schroeder, Amino-acid composition of the polypeptide chains of gorilla haemoglobin. *Nature* **192,** 984 (1961).

Biochemical Variations
in Man and
Other Primates

31

SERUM OR PLASMA, the clear yellow fluid portion of the blood, contains in solution many different proteins (Fig. 26.1). Many of them are under relatively simple genetic control, and many appear in different molecular forms in man. They provide genetic polymorphisms for us to study, and fortunately most of the alternative forms of the serum proteins are fairly easy to detect in the laboratory. There is little evidence yet which bears upon the nature of these polymorphisms. Some are probably balanced; others are fortuitous; and others are probably due to occasional mutations. Many, probably all, of these proteins are involved in vital functions of the organism. But it is not clear at the present time whether there are significant functional differences among the various molecular forms of each protein.

There are a number of techniques for demonstrating the protein constituents present in a blood sample. One of the most efficient and widely used is that of electrophoresis, especially vertical starch-gel electrophoresis (Fig. 29.2 and 31.1). After a sample of serum, red cell hemolysate, or tissue extract has been subjected to electrophoresis in a starch gel, the gel is stained with a dye which combines with proteins to yield easily detectable colored products. Some dyes will stain almost all proteins and large peptides. A number of special stains or reactions are used to identify specific proteins. Occasionally several proteins migrate as a single band and differential staining may be used to identify the components.

Under the usual conditions for electrophoresis of serum proteins, when the salt solutions (buffers) used for making the gels are slightly alkaline, the albumins migrate farthest toward the anode (+) from the point of application of the sample, and in order of decreasing mobility are the β-globulins and the α-globulins. γ-Globulins migrate toward the cathode (−) in the commonly used systems. The advantage of vertical starch-gel electrophoresis is that the classes of proteins formerly called α- and β-globulins are resolved into a large number of components. The various protein bands expected in a starch gel after one or more samples of serum have been subjected to electrophoresis are shown in Fig. 31.1. The resolution and reproducibility of the results of experiments depend upon a large number of factors which must be carefully controlled—the hydrogen ion concentration of the buffer, the hardness and the thickness of the gel, the temperature, the strength of the electric field, the concentration of proteins in the sample, etc. The diagram in Fig. 31.1 shows the relative positions of many different serum

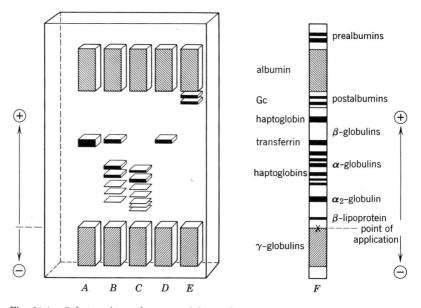

Fig. 31.1.   Relative electrophoretic mobilities of protein constituents of human serum. These diagrams show the relative positions of various components when electrophoresis is carried out in alkaline starch gels. (A), (B), (C) Haptoglobin-hemoglobin complexes for phenotypes Hp 1-1, Hp 2-1, and Hp 2-2, respectively; (D) transferrin Tf C; (E) group specific component Gc 2-1; and (F) composite, showing relative mobilities of important constituents of serum. Variations in conditions of electrophoresis lead to changes in the resolution of some of the components. The range of conditions commonly used is: voltage —7 to 20 volts/cm; temperature—5° to 25° C; time—10 to 24 hours; hydrogen ion concentration—pH 8 to pH 9; salt concentration—< 0.05M. In order to obtain reproducible results, the same set of conditions must be rigorously adhered to for each sample of serum and for each gel.

proteins. In practice one protein at a time (in all its various forms) is identified under a single set of conditions. Each group of proteins that we discuss—haptoglobins, transferrins, Gc groups, and the others—requires slightly different conditions for the best possible resolution. We have attempted to illustrate an ideal composite gel so that the reader will have some notion of what we see in a starch gel after electrophoresis of one or more samples of primate serum.

Haptoglobins

Haptoglobins are α_2-globulins (Fig. 31.1). In the organism haptoglobins combine with free hemoglobin in the serum. This property of haptoglobins probably prevents free hemoglobin from passing into the tubules of the kidney through the glomeruli (networks of capillaries). The kidney tubules are apt to be damaged by the passage of large proteins such as hemoglobin. In the laboratory the procedure for determining haptoglobin phenotypes includes adding a small amount of hemoglobin to serum prior to electrophoresis.

Three essentially different haptoglobins were demonstrated by Smithies in 1955. He named the three patterns seen on starch gels (Fig. 31.1) as haptoglobin type 1-1 (Hp 1-1), 2-1 (Hp 2-1), and 2-2 (Hp 2-2). The patterns correspond to one heterozygote (Hp^2/Hp^1) and two homozygotes (Hp^1/Hp^1 and Hp^2/Hp^2). Hp 1-1 is seen on a gel, after staining, as a single, dark band. Hp 2-2 has a number of bands that do not move as far as the Hp 1-1 band; the single, darkly stained band of Hp 1-1 is absent. The bands in Hp 2-2 vary in width and in the extent to which they stain. Hp 2-1 has one band with the same mobility as that of the Hp 1-1 band, but which stains less strongly, and Hp 2-1 contains several bands that move more slowly than the Hp 1-1 component. These are not the same as the bands found in Hp 2-2. They do not move the same distance and the degree to which they take up stain is also different. An analysis of haptoglobin types in families suggested to Smithies and Walker that the patterns were produced by two nondominant autosomal alleles, Hp^1 and Hp^2. The two alleles segregated in Mendelian fashion in families. There are individuals in whose serum no haptoglobin is found (ahaptoglobinemic individuals). These cases are most often found in African populations. In European populations ahaptoglobinemia almost never appears. The absence or deficiency of haptoglobin may be due to a total failure of haptoglobin synthesis or to a high rate of protein destruction. It is possible that there are many different reasons for the absence of haptoglobins. Some evidence is available that absence of haptoglobin is genetically controlled.

A large number of populations have been surveyed for haptoglobins, and the frequencies of the various types prove that this is another human genetic polymorphism (Table 31.1). The haptoglobin frequencies are reported as

TABLE 31.1
Frequencies of the Haptoglobin Allele, Hp[1]

Population	Number tested	Frequency[a] of Hp[1]
North America		
European descent		
Ann Arbor, Mich.	68	0.43
Seattle, Wash.	350	0.38
Canada	49	0.46
African descent		
Ann Arbor, Mich.	43	0.59
Seattle, Wash.	760	0.54
California	51	0.61
Chinese (New York)	118	0.34
Japanese	23	0.30
Orientals (Seattle, mostly Japanese)	242	0.25
Indians		
Apache	98	0.59
Navajo	263	0.45
North Athabascan	202	0.42
Tlingit	82	0.44
Eskimo	418	0.29
Eskimo	167	0.33
Anaktuvuk Eskimo	57	0.52
Central America		
Maya	414	0.59
Lacandon	31	0.93
Non-Maya	170	0.57
South America		
Venezuela		
Mestizo	208	0.55
Irapa	129	0.20
Paraujano	116	0.53
Macoita	74	0.43
Maquiritare	59	0.61
Peru		
Aymara	56	0.71
Quechua	117	0.74
Shipibo	70	0.65
Isconahua	16	0.47
Brazil Caingang	326	0.73
Chile Araucanian	34	0.75

TABLE 31.1 (*Continued*)
Frequencies of the Haptoglobin Allele, Hp[1]

Population	Number tested	Frequency[a] of Hp[1]
Europe		
Great Britain	180	0.43
Great Britain	218	0.39
Sweden	160	0.41
Sweden	220	0.44
Swedish Lapps	329	0.32
Norway	1000	0.36
Denmark	2046	0.40
Finland	891	0.36
Bavaria	273	0.46
Switzerland	920	0.40
France	406	0.40
Italy		
Berra	120	0.41
Cologna	208	0.35
Naples	93	0.34
Sardinia	147	0.37
Sicily	107	0.40
Spanish Basques	107	0.37
Asia		
Iran	34	0.25
India	74	0.18
India	33	0.17
India	219	0.09
India		
Tamil	291	0.09
Toda	89	0.35
Kurumba	49	0.18
Irula	74	0.07
Oraon	125	0.15
Ceylon		
Sinhalese	159	0.17
Tamil	140	0.13
Wanni	99	0.14
Veddah	64	0.13
West Pakistan		
Punjab	207	0.20
Pathan	185	0.24
Japan	822	0.27
Japan	349	0.24
Japan	488	0.27
Malay		
Malays	236	0.24
Chinese	167	0.28
Proto-Malays	66	0.47

TABLE 31.1 (*Continued*)
Frequencies of the Haptoglobin Allele, Hp^1

Population	Number tested	Frequency[a] of Hp^1
Thailand		
Bangkok	274	0.23
Northern Thai	139	0.26
Maeo	34	0.21
Yaeo	25	0.19
Africa		
Nigeria		
Yoruba	99	0.87
Yoruba	30	0.72
Habe	120	0.60
Fulani	111	0.76
Ibo	70	0.62
Liberia and Ivory Coast	614	0.72
Gambia	157	0.70
Congo		
Metropolitan	186	0.60
Nonmetropolitan	468	0.57
Pygmy	125	0.40
Ethiopia		
Amahara	107	0.45
Tigré	104	0.39
Billen	101	0.37
South Africa		
Cape colored	88	0.47
Hottentot	59	0.51
Zulu	116	0.53
Xhosa and Msutu	315	0.55
Bushman	113	0.29
Islands and subcontinents		
Australia		
European descent	323	0.38
Aborigines—North Queensland	463	0.17
Aborigines—Western desert	133	0.17
Oceania		
Borneo	22	0.50
Marshall Islands	52	0.58
Tonga	200	0.60
Madagascar		
Plateau	132	0.54
Lowland and coastal	150	0.66
Greenland	74	0.30
Iceland	188	0.39

[a]Based on Sutton et al. (1960) and other sources.

the incidence of the gene Hp^1. The frequency of the allele Hp^2 is simply 1.00 minus the frequency of Hp^1. The lowest frequency reported for the Hp^1 allele is from a group of Irula tribesmen of India, where $Hp^1 = 0.07$. The highest frequencies are found among populations of Central and West Africa and among Indians of Central and South America. The frequency of the Hp^1 allele appears to rise more or less regularly, if one examines its geographical distribution, from India to the northeast through Siberia and into North America. It also rises from India to the northwest across northern and western Europe. Another striking and apparently regular increase occurs from North America south to the equatorial regions of South America. The same pattern appears to occur among African populations; the highest frequencies of Hp^1 are found among groups living near the equator, and the incidence of this allele is lower in North Africa and South Africa.

The remarkable changes in frequency of the Hp^1 allele from India to the northeast and northwest and from the northern hemisphere to the equator may be attributed to environmental selection. It is known that the capacity of haptoglobin to bind hemoglobin varies considerably among types. The Hp 1-1 type has a considerably greater capacity than the others. It is possible that the relatively high incidence of hemolytic conditions, especially anemias, known to occur among tropical populations, is related to a possible advantage in the hemoglobin binding capacity which one haptoglobin type may have over others.

As work on the haptoglobins progresses, it becomes evident that, as with many blood group systems, things are not as simple as they seemed at first. The problem now consists of several parts—the variant patterns found in sera of homozygotes and heterozygotes, the absence of haptoglobins, and the nature of the product of the Hp^2 allele.

The possibility that there are several haptoglobin alleles instead of the two originally postulated led Smithies and his co-workers to investigate the product of the Hp^1 allele—purified haptoglobin. Two products of the Hp^1 allele are distinguished by their different electrophoretic mobilities. One is designated Hp 1F (fast), and the other Hp 1S (slow). These two haptoglobins appear to be the products of two alleles, Hp^{1F} and Hp^{1S}, instead of the single allele Hp^1 as originally assumed.

The question of genetic control in ahaptoglobinemics, phenotype Hp 0, also becomes more difficult to assess. Most of the evidence from studies of families suggests that the allele Hp^0 does not occur at the haptoglobin structural locus. If it does, its occurrence is extremely rare and does not account for most of the ahaptoglobinemic individuals discovered. Some young people with apparent ahaptoglobinemia (Hp 0) develop haptoglobins when they become adults. These haptoglobins can be classified as one of the known phenotypes. Thus the question of whether Hp 0 types are the result of an amorph Hp^0 or some other allele (modifier, controller, operator, suppressor) has not been answered.

Variant haptoglobin types from the Hp 2-1 heterozygotes are recognized.

They were originally considered as modified heterozygotes and received the designation Hp 2-1M. The alleles were believed to be Hp^1, Hp^2, and Hp^{2m}. The number of consistently observed variations grew, and now five patterns of the heterozygous condition are recognized, Hp 2-1a, Hp 2-1b, Hp 2-1c, Hp 2-1d, and Hp 2-1e (Fig. 31.2). Hp 2-1a is the most common of these and corresponds to phenotype Hp 2-1. The other variant patterns probably correspond to Hp 2-1M, and these are usually seen only in sera from Africans and Americans of African descent.

The problem of genetic control of the various haptoglobin patterns found among heterozygotes has been discussed by Sutton. The haptoglobins are polymers, that is, large molecules constructed from a number of similar or identical subunits. For example DNA is a polymer made up of four kinds of subunits, the purine and pyrimidine bases (Chapter 25). Hemoglobin is a relatively small polymer, for the intact, functioning molecule has only two of each of the subunits (the α chains and the β chains, Chapter 29). Haptoglobins have not yet been analyzed completely, but there is sufficient evidence to support the view that they are polymers built from varying amounts of three kinds of polypeptide chains—α^1, α^2, and β. Sutton's analysis of family data and pedigrees suggests that there are at least three alleles which correspond to the original Hp^2 gene. Combinations of these alleles with Hp^1 (1S or 1F) in heterozygotes produce qualitatively the same haptoglobins, but different relative quantities of the components are found in the original Hp 2-1 heterozygotes. The differences in the patterns of the

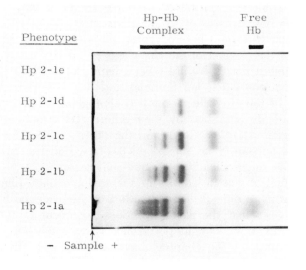

Fig. 31.2.   Photograph of a starch gel showing variations in the heterozygous haptoglobin phenotype Hp 2-1. (Photograph courtesy of H. E. Sutton, from Sutton and Karp, 1964.)

five heterozygous types are believed to be due to unequal amounts of the products of the various Hp^2 alleles. These suggestions about the alleles which control synthesis of the haptoglobins are subject to change and revision as more evidence is collected. Regardless of revisions we may make in the future, the discussion illustrates the importance of attempting to understand fully the nature of the genetic systems which we use in population studies, even if these involve studies and techniques not traditional in anthropology.

Serum Transferrins

The transferrins are part of the β-globulin fraction of the serum (Fig. 31.1). Ferric ions (oxidized iron) are carried to and from the bone marrow and other tissues by these proteins. This is an important function, for iron is essential in many vital processes. Iron is an important component of such proteins as hemoglobin, myoglobin, cytochrome, and some enzymes.

Transferrins are identified unequivocally by mixing a small amount of radioactive iron (Fe^{59}) with the serum to be tested. The transferrin combines with the iron. This radioactive label on the transferrin distinguishes it from other proteins in the serum. After electrophoresis the starch gel is placed on a piece of X-ray film. A 24-hour exposure will produce a dark band on the film at the spot to which the transferrin-radioactive iron complex has moved.

At least twelve molecular varieties of human transferrin have been identified (Fig. 31.3). Each is apparently controlled by a nondominant autosomal allele at the same locus. The C phenotype (Tf C), genotype Tf^C/Tf^C, is the most common in all populations studied so far. The D transferrins (Tf D's) move more slowly, the B transferrins (Tf B's) more rapidly than Tf C. The distribution of transferrin phenotypes has not been studied as extensively as that of the haptoglobins. Nonetheless some populations from every part of the world have been tested (Table 31.2). Some of the populations which appear to have no transferrin polymorphisms prove to be represented by rather small samples. The alleles which produce the fast-moving transferrins (Tf B's) are not widely distributed and are almost never found in Africa or the Pacific islands. The transferrins which move more slowly than C (Tf D's) have a much wider distribution. They are most common in New Guinea and Australia and are found among various African tribes and in a number of American Indian groups. European populations and their descendant populations have a very low incidence of both Tf B's and Tf D's.

The reasons for transferrin polymorphisms are not known. Suggestions have been made that the transferrin-iron complex prevents the multipli-

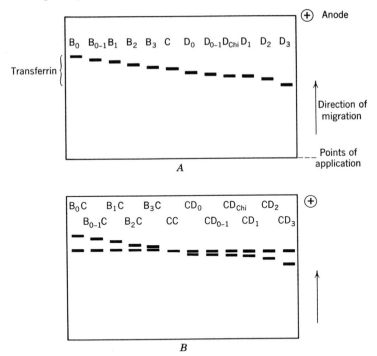

Fig. 31.3.   Relative electrophoretic mobilities, in starch gels, of human transferrins. (A) Twelve molecular forms of transferrin and (B) some heterozygous
phenotypes.

cation of viruses in the body. There is no evidence that resistance to viruses
maintains the observed transferrin polymorphisms, and neither diet, disease,
nor climate has been correlated with the distribution of the transferrin
alleles. Transferrin polymorphisms in man and his primate relatives are just
beginning to be studied.

Transferrins are large molecules; they have a molecular weight of about
90,000; and they are proteins with sialic acid groups attached. Sialic or
neuraminic acid is an organic molecule related to sugar (saccharide) and
containing nitrogen. Digestion of transferrin with trypsin produces a large
number of peptides, so large that the paper peptide patterns of different
transferring appear similar and are very difficult to analyze.

The sialic acid groups are attached to the protein portion of the transferrin molecule. An enzyme, neuraminidase, can be used to remove one
sialic acid group at a time from the transferrin molecule. The electrophoretic
mobility of Tf C is a function of the number of sialic acid groups present
on the molecule. There are five electrophoretic mobilities demonstrable.
The intact transferrin molecule has the highest mobility (moves farthest).
If one sialic acid residue is removed, the molecule has a reduced mobility.

TABLE 31.2
Frequencies of Transferrin Phenotypes[a]

Population	Number tested	Tf C	Tf B's (fast)		Tf D's (slow)	
North America						
European descent						
U.S.A.	471	0.987	$0.011 B_2C$	$0.002 B_0C$		
Canada	425	0.988	$0.012 B_2C$			
African descent						
Seattle, Wash.,					$0.081 CD_1$	$0.002 CD_3$
and						
Cleveland, O.	1362	0.912	$0.003 B_2C$		$0.001 D_1$	$0.001 CD_0$
California	48	0.875			$0.104 CD_1$	$0.021 CD_3$
New York	99	0.899			$0.091 CD_1$	$0.010 D_1$
Chinese (New York)	116	0.940			$0.060 D_{Chi}$	
Orientals (Seattle,						
mostly Japanese)	242	0.988	$0.004 B_1C$	$0.004 B_1$	$0.004 CD_1$	
Indians						
Navajo	230	0.926	$0.004 B_2C$	$0.070 B_{0-1}C$		
Alaska	49	1.000				
Eskimo	167	1.000				
Central America						
Maya	449	0.989	0.004 BC		0.007 CD	
Non-Maya	189	0.978	0.011 BD		0.011 CD	
South America						
Venezuela						
Mestizo	102	0.970	$0.010 B_2C$		$0.020 CD_1$	
Irapa	89	0.944			$0.056 CD_1$	
Paraujano	75	0.947			$0.053 CD_1$	
Macoita	42	0.905			$0.095 CD_1$	
Peru						
Aymara and						
Quechua	173	0.994	$0.006 B_2C$			
Shipibo	70	1.000				
Isconahua	16	1.000				
Europe						
England	139	0.986	$0.007 B_2C$	$0.007 B_1C$	$(0.002 CD_{0-1})$	
Sweden	1173	0.986	$0.010 B_2C$	$0.002 B_1C$	$0.002 CD_1$	
Swedish Lapps	329	0.982			$0.018 CD_1$	

TABLE 31.2 (*Continued*)
Frequencies of Transferrin Phenotypes[a]

Population	Number tested	Tf C	Tf B's (fast)	Tf D's (slow)	
Asia					
India					
Tamil	291	1.000			
Toda	89	1.000			
Kurumba	49	1.000			
Irula	74	1.000			
Oraon	125	0.936		0.064 CD$_1$	
Ceylon					
Sinhalese	159	0.988	0.006 B$_2$C	0.006 CD$_1$	
Tamil	140	1.000			
Wanni	99	1.000			
Veddah	64	0.890		0.094 CD$_1$	0.016 D$_1$
West Pakistan					
Punjabi	207	1.000			
Pathan	185	0.989	0.011 BC		
Japan	822	0.984	0.001 BC	0.015 CD	
Chinese (Formosa)	40	0.950	0.025 B$_2$C	0.025 CD$_1$	
Formosa	300	0.957		0.043 CD$_1$	
Malay					
Malays	236	0.953		0.047 CD$_1$	
Chinese	103	0.922		0.068 CD$_1$	0.010 D$_1$
Proto-Malays	66	0.970		0.030 CD$_1$	
Thailand					
Bangkok	274	0.941		0.055 CD$_1$	0.004 D$_1$
Northern Thai	139	0.892		0.101 CD$_1$	0.007 CD$_0$
Maeo	34	0.971		0.029 CD$_1$	
Yaeo	25	0.920		0.080 CD$_1$	
Africa					
Nigeria					
Habe	120	0.850		0.150 CD$_1$	
Fulani	111	0.937		0.063 CD$_1$	
Fulani	68	0.838		0.147 CD$_1$	0.015 D$_1$
Ibo	70	0.871		0.129 CD$_1$	
Liberia					
Northwest	179	0.961		0.039 CD$_1$	
Central	52	0.866	0.019 BC	0.115 CD$_1$	
Southeast	75	0.880		0.120 CD$_1$	
Mandingo	27	0.963		0.037 CD$_1$	
Gambia	157	0.975		0.019 CD$_1$	0.006 CD$_2$
Congo					
Metropolitan	186	0.957		0.043 CD$_1$	
Nonmetropolitan	446	0.933		0.067 CD$_1$	
Pygmy	121	0.934		0.066 CD$_1$	

TABLE 31.2 (*Continued*)
Frequencies of Transferrin Phenotypes[a]

Population	Number tested	Tf C	Tf B's (fast)	Tf D's (slow)	
South Africa					
Cape colored	88	0.977		$0.023\,CD_1$	
Hottentot	59	0.932		$0.068\,CD_1$	
Zulu	116	0.974		$0.026\,CD_1$	
Xhosa	265	0.977		$0.023\,CD_1$	
Msutu	218	0.908		$0.092\,CD_1$	
Bushman	113	0.876		$0.115\,CD_1$	$0.009\,D_1$
Uganda Baganda	165	0.970		$0.030\,CD_1$	
Tanganyika Bondei	60	0.917		$0.000\,CD_1$	
Kenya Masai	50	1.000			
Islands and					
subcontinents					
Australia					
Aborigines-North					
Queensland	463	0.842	$0.022\,CB_1$	$0.134\,CD_1$	$0.002\,D_1$
Aborigines-Western					
desert	130	0.554		$0.400\,CD_1$	$0.046\,D_1$
New Guinea					
Highlands	550	0.820		$0.171\,CD_1$	$0.009\,D_1$
Papua	48	0.917		$0.083\,CD_1$	
Oceania					
Marshall Islands	106	1.000			
Gilbert Islands	236	1.000			
Tonga	196	1.000			
Fiji	93	0.978		$0.022\,CD_1$	
Philippines	403	0.990		$0.010\,CD_1$	
Madagascar					
Plateau	131	0.901		$0.099\,CD_1$	
Lowland and					
coastal	149	0.879		$0.114\,CD_1$	$0.007\,D_1$

[a]Based on Giblett (1962) and other sources.

When all four sialic acid residues are removed the mobility is very much reduced (Fig. 31.4). The difference between the mobility of Tf C and a Tf B or a Tf D cannot be accounted for by presence or absence of sialic acid residues. When purified Tf B_0, B_{0-1}, B_1, B_2, D_{Chi}, D_1, and D_3 were treated with neuraminidase, the same step-like decrease in mobility of the transferrin molecule was observed in each case as one sialic acid group after the other was removed by the action of the enzyme. It is probable that the different transferrins have different amino acid compositions, primary structures, and/or spatial configurations, but these differences have not yet been demonstrated. Studies of iron-binding capacity of transferrins show small, probably not significant differences among the various types.

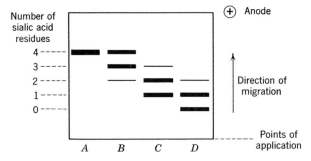

Fig. 31.4.   Relative electrophoretic mobility, in starch gels, of transferrin Tf C before and after sequential removal of the four residues of sialic acid with the enzyme neuraminidase. The relative concentrations of neuraminidase are (A) none, (B) 1, (C) 5, and (D) 20. The width of each band is proportional to the amount of protein. (Adapted from Parker and Bearn, 1961.)

Group Specific Component (Gc)

The Gc or group specific component human polymorphic system was first discovered and described by Hirschfeld in 1959. The Gc proteins migrate, in starch-gel electrophoresis, slightly more slowly than albumin (Fig. 31.1). They are part of a group of proteins once called postalbumins, for they appear behind the albumins after electrophoresis of serum in starch gels. The Gc 1 band moves more rapidly than the Gc 2 band.

Soon after the discovery of the Gc system, the frequencies of the three phenotypes were determined in a number of human groups. The phenotypes are labeled Gc 1–1, Gc 2–1, and Gc 2–2. Family studies suggest that these phenotypes are controlled by two nondominant autosomal alleles. The three genotypes are probably Gc^1/Gc^1, Gc^2/Gc^1, and Gc^2/Gc^2. Other Gc types were discovered among Chippewa Indians and Australian aborigines. These variant alleles are labeled, at present, $Gc^{Chippewa}$ and $Gc^{Aborigine}$. Since Gc is a recently discovered human polymorphism, generalizations about it must await considerable additional work. The presently known frequencies in the Gc system are listed in Table 31.3.

The Gm and Inv Serum Groups

The Gm serum factors were discovered in 1956 in a most unusual manner. Their discoverer, Grubb, was determining the amount of γ-globulin in the sera of six patients with hypogammaglobulinemia (low level of γ-globulin),

TABLE 31.3
Frequencies of the Group Specific Component Allele, Gc²

Population	Number tested	Frequency[a] of Gc^2
North America		
European descent (New York)	148	0.274
African descent (New York)	144	0.108
Chinese (New York)	117	0.231
Indians		
Navajo	245	0.023
Chippewa	159	0.223
Alabama-Coushatta	65	0.150
Eskimo	67	0.298
Europe		
England	49	0.286
Sweden	1744	0.257
Swedish Lapps	79	0.127
Denmark	126	0.230
Norway	383	0.278
Germany	208	0.264
Asia		
Israel		
Ashkenazi Jews	99	0.338
North African Jews	64	0.297
Persian Jews	49	0.245
Iraqi Jews	85	0.241
Yemenite Jews	49	0.194
Kurdish Jews	42	0.190
Arab	48	0.260
India (Bombay)	90	0.306
Japan	108	0.227
Africa		
Nigeria		
Habe	103	0.073
Fulani	100	0.050
Congo Bantu	100	0.095
Uganda Baganda	81	0.093
Tanganyika Bondei	57	0.096
Kenya Masai	50	0.070
Australian aborigines		
Western desert	126	0.040
Central	299	0.078
Western central	144	0.062
Kimberleys	134	0.123
North	49	0.112
Cape York	404	0.194

[a]From Cleve and Bearn (1962) and other sources.

when he found that one of the sera agglutinated red blood cells which were coated with an incomplete Rh antibody (a reagent which he used in measuring the amount of γ-globulin). An incomplete antibody is one which combines with red cells, coats them, but does not agglutinate them. The technique for determining the Gm group of an individual developed from this observation of Grubb.

The determination of an individual's Gm phenotype involves a series of reactions with special reagents. An incomplete Rh antibody, carefully selected, is used to coat human erythrocytes which are group O, Rh positive. Such specially coated cells will be agglutinated if they are mixed with serum from some individuals who are suffering from chronic rheumatoid arthritis. This agglutination may be inhibited if the coated cells are first mixed with the serum from certain normal people. There is present in the serum of these normal individuals a special γ-globulin component which Grubb called the Gm factor. A serum which has this Gm inhibition factor is called Gm(a+). This factor is apparently controlled by a gene, Gm^a, which is inherited as an autosomal dominant or codominant allele. Sera from other rheumatoid arthritics had their agglutinating capacity inhibited by Gm(a+) sera and many Gm(a−) sera as well. Thus another Gm factor, Gm(b+), was discovered. It is controlled by a gene, Gm^b, which is allelic to Gm^a. Three phenotypes, Gm(a+b−), Gm(a+b+), and Gm(a−b+), were recognized with the aid of these two different kinds of sera from patients with rheumatoid arthritis.

Several other Gm factors have been discovered, each of which appears to be inherited as a nondominant autosomal allele. A Gm-like factor has been discovered in some African groups and in individuals of African descent. A Gm(x+) reaction, unlike the reaction of either Gm(a+) or Gm(b+), has been described among Europeans and some Asiatic and Australian groups. The available population data and family studies suggest that five alleles occur at the Gm locus—Gm^a, Gm^b, Gm^{ab}, Gm^{ax}, and one that might be called Gm-like. A list of the Gm factors and the genotypes and phenotypes found in European populations is presented in Table 31.4. The alleles at the Gm locus and the populations in which they have been found are listed in Table 31.5. There is no question that the Gm system is a peculiar one, and the polymorphism it exhibits in human populations requires further study before we can make reasonable generalizations about it. Obvious factors of natural selection—disease, diet, or climate—have not proved to be associated with any of the various Gm phenotypes.

Different combinations of coated erythrocytes and sera from persons with rheumatoid arthritis have been used to detect other genetically controlled inhibiting factors (the Inv factors) in the γ-globulin fraction of serum. These are controlled by alleles at a locus independent of *Gm*. The alleles at this second locus are known as Inv^a and Inv^b. Some of what is known about the Inv system is summarized in Table 31.4. New Gm and Inv factors are discovered from time to time, and the alleles reported here are probably not all the alleles at the Gm and Inv loci.

TABLE 31.4

The Phenotypes and Genotypes of the Gm and Inv Serum Groups in European Populations

(After Steinberg, 1962.)

Phenotype	Genotype
Gm system	
Gm(a+b−)	Gm^a/Gm^a
Gm(a+b+)	Gm^a/Gm^b
Gm(a−b+)	Gm^b/Gm^b
Gm(ax+b−)	Gm^{ax}/Gm^{ax}, Gm^{ax}/Gm^a
Gm(ax+b+)	Gm^{ax}/Gm^b
Inv system	
Inv(a+b−)	Inv^a/Inv^a
Inv(a+b+)	Inv^a/Inv^b
Inv(a−b+)	Inv^b/Inv^b

TABLE 31.5

Alleles Postulated at the Gm Locus in Various Populations

(After Steinberg, 1962.)

Population	Alleles
Europeans	Gm^a, Gm^b, Gm^{ax}
American Indians	Gm^a, Gm^{ax}, Gm^{ab}
Eskimos	Gm^a, Gm^{ax}, Gm^{ab}
Asians	Gm^a, Gm^{ax}, Gm^{ab}
Africans	Gm^{ab}, possibly *Gm-like*
Australian aborigines	
Western desert	Gm^a, Gm^{ax}
Coastal	Gm^a, Gm^{ax}, Gm^{ab}

Glucose-6-phosphate Dehydrogenase

Hemoglobin is not the only protein in the red cell. About 5 per cent of the total protein which is obtained after red cells have been hemolyzed (the cell walls broken) to provide hemoglobin samples is a heterogeneous mixture of several enzymes with specific functions. One of the better known, anthropologically, is glucose-6-phosphate dehydrogenase (G6PD). This enzyme (G6PD) is necessary as a catalyst in a biological oxidation-reduction reaction of glucose-6-phosphate—one of the stages in the metabolism of carbohydrates.

The G6PD deficiency disease was discovered when a number of Americans of African and Asian descent were treated with certain antimalarial drugs, particularly primaquine. Primaquine produced a mild hemolysis in these persons. Investigation showed that the individuals were deficient in the red cell enzyme G6PD. The evidence now suggests that G6PD is present in these individuals, but in very low amounts. A number of other drugs, chemically related to primaquine, produce hemolysis when given to individuals deficient in G6PD. Favism, a hemolytic condition produced by eating fava beans (*Vicia fava*), is found among populations living in the Mediterranean area and is believed to be connected with G6PD deficiency.

The G6PD deficiency in red cells is inherited as an X-linked trait which appears to be incompletely dominant. Females that are clearly homozygous for the trait have as marked a deficiency as males with the trait. Heterozygous females are usually intermediate between normal individuals and those clearly G6PD deficient, and some heterozygous females are indistinguishable from those who do not have the gene.

The distribution of G6PD deficiency parallels that of falciparum malaria (Fig. 29.12 and 31.5). Resistance to malaria has been suggested as the advantage which G6PD deficiency confers and as the factor which maintains it in various populations. The resistance to malaria notion has been

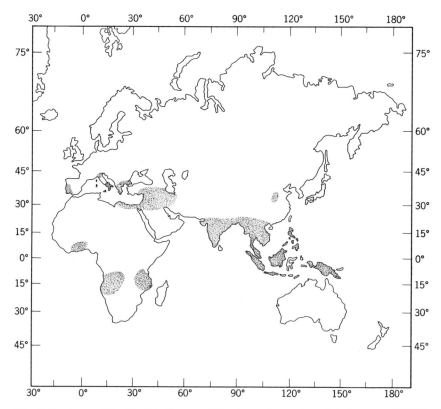

Fig. 31.5.   Areas of the Old World from which glucose-6-phosphate dehydrogenase (G6PD) deficiency disease has been reported.

challenged on the basis of the distribution of the trait and the distribution of the parasite in New Guinea. A number of populations in New Guinea are equally exposed to hyperendemic malaria but have radically different G6PD deficiency frequencies. It has been suggested that the founder effect or gene flow accounts for the origin of these differences in frequency among New Guinea groups. Analysis of the effect of gene flow or the founder principle is complicated for a sex-linked trait. There is a much greater chance that the genes for the sex-linked trait will come from females than from males; two-thirds of the X-chromosomes in the population are contributed by the females. Nonetheless the maintenance of relatively high frequencies of a deleterious gene will only occur if selection is acting in its favor. Falciparum malaria is a good candidate for the agent in all the protein abnormalities which affect red cells, for *Plasmodium falciparum* lives on the proteins contained in the cell. If the red cell of the host is unable to support the parasite or if the hemoglobin of the host cannot be metabolized by the parasite, the host certainly has an immunity or resistance to malaria. Arguments about distributions, parallel or not, are no longer apt to lead to further understanding of the evolutionary dynamics that produce relatively high frequencies of various red cell abnormalities in some human groups. As we said when we discussed the sickle cell-malaria hypothesis, the exact physiological interaction of the parasite and the abnormal condition must be studied.

Other Protein Polymorphisms

A number of genetically controlled red cell enzyme polymorphisms besides glucose-6-phosphate dehydrogenase have been described in humans. They have not yet been completely characterized, and, for most of them, only European populations have been studied. But the reader should learn of their existence. They will be the focus of considerable work in the future and will expand our list of human genetic polymorphisms. Many of these enzymes are found in tissues other than the red cells, but it is very convenient to extract them from erythrocytes, where many of them were discovered originally. Furthermore it is simpler to obtain samples of red cells from a population than it is to acquire samples of muscle, brain, or liver tissue. The enzyme polymorphisms which have been best characterized and for which good genetic data exist are: red cell acid phosphatase, phosphoglucomutase, and red cell carbonic anhydrase.

Red cell acid phosphatase occurs in three different molecular forms, each apparently controlled by a nondominant autosomal allele (Table 31.6). Only five of the six predicted phenotypes have been described. This is not surprising, for the population sample studied is small. The variant forms of the enzyme are identified in starch gels, after electrophoresis of hemolyzed red cells, with a reaction specific for acid phosphatase. Acid phosphatase is an enzyme that catalyzes the removal of phosphate groups from organic compounds such as phosphoserine. The acid phosphatases are so named because their optimum activity occurs under slightly acid conditions. (Another group of phosphatases, the alkaline phosphatases, acts best under slightly alkaline conditions.)

TABLE 31.6
The Red Cell Acid Phosphatase and Phosphoglucomutase
Phenotypes and Genotypes
(After Spencer et al., 1964.)

	Phenotype	Genotype
Acid phosphatase		
	A	P^a/P^a
	BA	P^a/P^b
	B	P^b/P^b
	CA	P^a/P^c
	CB	P^b/P^c
	(C)[a]	
Phosphoglucomutase		
	PGM 1	PGM^1/PGM^1
	PGM 2–1	PGM^2/PGM^1
	PGM 2	PGM^2/PGM^2

[a]This phenotype has not been described.

Phosphoglucomutase (PGM) is found in many mammalian tissues, and it is easily identified by subjecting red cell hemolysates to starch-gel electrophoresis. PGM is a phosphotransferase which catalyzes the transfer of a phosphate group from the carbon at position 1 to the carbon at position 6 of glucose. Three phenotypes have been described, corresponding to three patterns of PGM bands seen in starch gels, 1, 2–1, and 2. The 2–1 pattern consists of seven bands of PGM activity; the two homozygous patterns, 1 and 2, consist of five bands of PGM activity. Family studies and population data suggest that two alleles, PGM^1 and PGM^2, control the production of the proteins which produce the three patterns of PGM activity seen in the gels (Table 31.6).

Carbonic anhydrase is a red cell enzyme that is important in respiration. It acts as a catalyst in the conversion of carbon dioxide to bicarbonate. Alternative forms of the enzyme have been detected in human populations. The various patterns seen in starch gels have been named CA I, CA Ia, CA Ib, CA Ic, and CA II. The available data from family studies suggest that each type is produced by a single, nondominant autosomal allele at the same locus. The variant type CA Ic was found among native inhabitants of Guam and Saipan.

A number of other serum protein polymorphisms have been described, although little is yet known of their distribution. At least two human serum phosphatases have been identified. All individuals tested have group I phosphatase which migrates somewhat more slowly than transferrin Tf C in starch gels. Some individuals also-have a second, more slowly moving band, and these individuals are classed as group II. Studies of families and twins suggest that these two types are inherited, but the exact mode of inheritance is not known.

Serum cholinesterase occurs in two alternative forms in human populations, although one of the two types, the atypical, is rather infrequent. Cholinesterase (or acetylcholinesterase) acts as a catalyst on acetylcholine to produce acetate and choline. Acetylcholine is a chemical compound of importance in the transmission of stimuli in the nervous system. A pair of nondominant autosomal alleles appears to control the production of the normal and variant enzymes.

α_1-Acid glycoprotein, a serum protein with a large carbohydrate component, is detected in several forms in starch gels. The various patterns are probably under genetic control.

There are many other protein polymorphisms being investigated today. It takes time and effort to define these polymorphisms, to determine whether or not they are inherited, and to describe, if they are inherited, the mode of inheritance. The stability of many of these proteins makes possible the analysis of blood samples obtained from populations which can be visited only once. Many data from populations can be obtained from samples which, when properly prepared initially, have been stored for long periods of time. Many of the world's primitive peoples live in areas which are not easily accessible, and it is expensive to visit and, especially, to revisit these areas. When new polymorphic systems are reported, it is sometimes possible to test for them in populations which were studied several years prior to the discovery of a particular polymorphic system.

Other Genetic Polymorphisms in Man

Many abnormalities may be discovered when compounds excreted in urine are examined. The excretion of amino acids (aminoaciduria) is often a sign of an inherited metabolic abnormality. Occasionally individuals are found who excrete relatively large amounts of β-amino-isobutyric acid (BAIB) in their urine. This compound is a metabolic end product of pyrimidines and amino acids. Careful investigation of excretion of BAIB revealed that excretion of high levels of BAIB seems to be inherited as a simple, autosomal recessive allele.

Studies of the amount of BAIB excreted by individuals showed that traces of this compound are found in the urine of almost everyone. Thus excretion of BAIB is not a simple all-or-none trait. Rather, high levels of BAIB in the urine appear to be under genetic control. The frequency of high BAIB excretors varies among human populations (Table 31.7). The populations of

TABLE 31.7
Frequencies of High Excretors of β-amino-isobutyric Acid (BAIB) [a]

Population	Number tested	Frequency of high excretors
North America		
European descent		
Michigan	71	0.03
Texas	255	0.10
New York	218	0.10
New York	148	0.11
African descent		
Michigan	25	0.20
New York	38	0.15
Indians		
Apache	110	0.59
Apache	113	0.42
Athabascan	25	0.56
Eskimo	120	0.23
Chinese	33	0.45
Japanese	41	0.41
Central America		
African descent (Black Caribs)	285	0.32
Asia		
India	16	0
Thailand	13	0.46
Marshall Islands		
Rongelap	188	0.86
Utirik	18	0.83

[a]Based on Sutton (1960) and Blumberg and Gartler (1959).

eastern Asia have the highest frequency of high excretors. The amount of BAIB excreted varies among those who excrete small amounts as well as among those who excrete large amounts. It has been observed that increased amounts of BAIB are excreted by females during pregnancy. Its excretion does not appear to be due to a deleterious allele. Although all the reactions which produce BAIB as an end product have not been described, high excretors do not seem to have any metabolic defects.

Certain other polymorphisms found by examination of compounds excreted in the urine by the kidneys are due to rare deleterious alleles and are probably found in populations because recurrent mutations occur and because the heterozygotes are not affected. For example in phenylketonuric idiocy, high levels of phenylpyruvic acid are excreted by the kidneys. The presence of phenylpyruvic acid is due to an allele which blocks one of the steps in the normal metabolism of phenylalanine. Homozygotic individuals have severely impaired mental abilities and seldom reproduce. The heterozygotes are not affected.

The ability to taste the compound phenylthiocarbamide (PTC) is inherited as an autosomal recessive trait. Individuals who can taste PTC are of genotypes *TT* or *Tt*, and nontasters are *tt*. A large number of populations have been tested for this trait, and it is clear that this is another genetically controlled polymorphism (Table 31.8). The ability to taste PTC is not likely to be useful to most people, for the compound is not commonly found in food. But a number of chemical substances with structural similarities to PTC are also differentially discriminated by tasters and nontasters. Food aversions may be the result of an inherited ability to distinguish minute amounts of materials with unpleasant tastes. No obvious associations between the distribution of this trait and altitude, humidity, extremes of climate, or other climatic factors have been found. Some evidence is available that nodular goiter (enlargement of the thyroid gland) is significantly more frequent among nontasters than among PTC tasters. The frequencies of the tasting trait should be carefully investigated in populations that live in regions low in natural sources of iodine, for goiter is sometimes due to absence of iodine from the diet. The presence of compounds in certain foods (goitrogenic foods) may cause goiter. Populations whose diets contain goitrogenic substances should also be carefully surveyed for ability to taste PTC.

Polymorphisms in the Nonhuman Primates

Many of the biochemical polymorphisms described for man are found in his primate relatives. The genetically controlled polymorphisms are not an evolutionary development unique to *Homo*. The relative scarcity of actively breeding laboratory colonies of nonhuman primates has prevented a rapid accumulation of the kinds of genetic data so important in determining modes of inheritance—family studies, collection of pedigrees, fertility studies, and population studies. However studies of genetic polymorphisms in nonhuman primates have described many of the proteins homologous to those so valuable in various aspects of human genetics.

There are some notable differences between homologous groups of proteins of *Homo* and those of other primates. The haptoglobins, for example, do not appear in more than a single type in any one species. Only a band analogous to human Hp 1-1 (Fig. 31.1) has been observed. There is some suggestion that two forms of the single band haptoglobin, similar to human types Hp 1F and Hp 1S, may occur among some primates.

Transferrins, on the other hand, are highly variable among the nonhuman primates. Almost every species examined has produced more than one phenotype. We assume that the various transferrin bands seen after starchgel electrophoresis of primate sera are under the same kind of genetic con-

TABLE 31.8

Frequencies of Nontasters of Phenylthiocarbamide (PTC) and Related Compounds[a]

Population	Number tested	Frequency of nontasters
North America		
European descent	3643	0.30
Indians		
Navajo	264	0.02
Cree and Beaver	489	0.02
Alaska Eskimo	68	0.26
Labrador Eskimo	130	0.41
South America		
Brazil		
Ashkenazi Jews	244	0.28
Indians	163	0.01
Japanese	295	0.07
Venezuela Guajiro	100	0.40
Europe		
England	440	0.31
Wales	203	0.17
Norway	266	0.30
Norwegian and Swedish Lapps	140	0.07
Denmark	314	0.32
Finland	202	0.29
Spain	306	0.25
Spanish Basques	98	0.25
Portugal	454	0.24
Chinese (London)	66	0.11
Asia		
Israel		
Ashkenazi (European) Jews	440	0.21
Non-Ashkenazi (Balkan) Jews	101	0.22
North African Jews	340	0.15
Iraqi and Persian Jews	336	0.16
Yemenite Jews	261	0.18
Kurdish Jews	121	0.14
Gerba Jews	41	0.41
Cochin Jews	41	0.32
India		
Bombay	200	0.42
Tamil	50	0.27
Paniyan	247	0.12
Riang	401	0.16
Gujarat	1095	0.48

TABLE 31.8 *(Continued)*

Frequencies of Nontasters of Phenylthiocarbamide (PTC) and Related Compounds[a]

Population	Number tested	Frequency of nontasters
Malay		
Malays	50	0.16
Chinese	50	0.02
Negrito	50	0.18
Senoi	50	0.04
Africa		
West Africa	74	0.03
Kenya		
Bantu	208	0.04
Arab	63	0.25

[a]Based on Saldanha and Nacrur (1963) and other sources.

trol as are those of man. Tentatively, certain generalizations may be made about transferrins when electrophoretic mobility is used as a criterion. The transferrins of the African apes, chimpanzees and gorillas (*Pan*), move more slowly on starch-gel electrophoresis than human Tf C. Those of the Asiatic apes, orangutans (*Pongo*) and gibbons (*Hylobates*), are somewhat faster moving than human Tf C. The transferrins of the Cercopithecoidea, Ceboidea, and Prosimii migrate as far as or farther than human Tf B_0. Until detailed genetic and biochemical studies necessary for characterization of the transferrins of each species are completed, the phylogenetic significance of these variations in primate transferrins cannot be assessed.

Chimpanzees (*Pan*) appear to have the most unusual transferrin phenotypes, at least seven of which have been described. Sera from a few gorillas (*Pan*) have been tested; four transferrin phenotypes have been observed. Two phenotypes have been found in sera from gibbons (*Hylobates*). The transferrins of chimpanzees have been studied in sufficient detail so that we have good evidence for the mode of inheritance of this group of proteins in at least one species of pongids. The patterns which these iron-binding proteins from chimpanzees make on starch gels are more complex than those made by other primate transferrins (including human). The seven phenotypes observed among chimpanzees (Fig. 31.6) appear to be produced by various combinations of four nondominant autosomal alleles, Tf^A, Tf^B, Tf^C, and Tf^D. However, the transferrins of some chimpanzees contain more than one or two components, that is, several iron-binding bands are seen on starch gels. These additional bands are believed to be minor components of the major bands which define the phenotypes (Fig. 31.6), for they are narrower than the major bands and the radioactive label on them is not as intense. Data on related animals were obtained from records of a breeding colony of chimpanzees maintained in the United States for many years. The

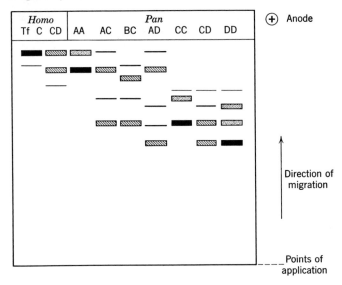

Fig. 31.6.   Transferrin phenotypes in *Pan* (chimpanzees) as determined by relative mobilities in starch-gel electrophoresis. The relative intensities of the various bands are, in decreasing order: black bar, hatched bar, stippled bar, thin bar, line. (Adapted from Boyer and Young, 1960.)

breeding data are consistent with a four-allele system for the transferrins. The data give direct support to the suggestions that the type CC is homozygous and the types AC, BC, and DC are heterozygous. No data have yet been collected which allow us to be certain of the genetic basis for phenotypes AA, AD, and DD. Phenotypes BB, AB, and BD have not been observed.

In the genus *Macaca*, 11 molecular forms of transferrin have been described by Goodman. It is assumed that each of the 11 molecular forms is produced by a separate, nondominant autosomal allele. These 11 alleles can produce 66 genotypes; the number of genotypes is $N = n(n + 1)/2$, $n = 11$ (Chapter 23). At least 34 of the 66 possible phenotypes (genotypes = phenotypes) have been described (Tables 31.9 and 31.10). The 11 molecular forms of transferrin in *Macaca* have been defined on the basis of their electrophoretic mobilities. There may be many more than 11 alleles and many more than 66 phenotypes in this genus. If two proteins have the same electrophoretic mobility, they are not necessarily identical. Goodman has shown that transferrins with almost identical mobilities are actually different. He has done this by examining the offspring of those parents who appear to have identical transferrin phenotypes. The transferrin of the offspring proved to be composed of two bands with only slight differences in mobility. Nonetheless two components were resolved.

These data on transferrins in *Macaca* illustrate the many problems which

TABLE 31.9

Transferrin Phenotypes and Alleles in Fourteen Populations of Macaca

	Observed phenotypes in 372 macaques[a]			Alleles
Tf AC	Tf CF'	Tf DH'		Tf^A
AD	CF	DH		
BC	CG	E		Tf^B
BD	CH'	F'		Tf^C
B(D)G	CH	F'G		
BE	D'F	F'H'		$Tf^{D'}$
BG	D'G	F		Tf^D
BH	D	FG		
C	DF'	G		Tf^E
CD'	DF	H'		$Tf^{F'}$
CD	DG	H		
CE				Tf^F
				Tf^G
				$Tf^{H'}$
				Tf^H

[a]The relative mobilities, in starch gels, of the various transferrins are: A(fastest), B, C, D', D, E, F', F, G, H', H(slowest). (Data taken from Goodman et al., 1965.)

must be considered when so variable a trait is investigated among nonhuman primates. There are at least eleven different transferrins which should be characterized chemically; there are many "family" studies which should be made in order to determine unequivocally the genetic basis of the traits; the frequencies of the genotypes should be determined for large numbers of individuals of the many populations in the genus *Macaca*.

We complain that genetic studies of man are difficult because matings cannot be controlled for purposes of research on specific problems, generations are long, and population surveys are difficult. Yet the difficulties are even greater with nonhuman primate studies. Very few family studies can be made, and it is very difficult to sample wild populations. When we find a genetic trait in so many forms and one as easy to identify as the transferrins, our problems are multiplied. We can account for the distribution of the transferrins in populations of *Macaca* by mutation, by gene flow, and by selection, but we cannot yet specify how each of these processes operated. For example the $Tf^{F'}$ allele (Table 31.10) may have been introduced into *Macaca mulatta* ("*M. irus*") in Thailand from contiguous groups of *M. speciosa*. The Thailand *M. mulatta* ("*M. irus*") is a large population, adapted to and living in a wide range of habitats. Gene flow between this species and other closely related species is a quite reasonable explanation for the occurrence of the variety of transferrin alleles in this population. Other evolutionary processes may also be involved.

As we said earlier, the number of species of *Macaca* seems excessive, par-

TABLE 31.10
Frequencies of Transferrin Alleles in Populations of Macaca
(After Goodman et al., 1965.)

Allele[a]	Macaca mulatta Nepal border	M. nemestrina	M. fuscata fuscata Japan	M. fuscata yakui Japan	M. irus Philippines	M. irus Thailand	M. speciosa India	M. speciosa Thailand
Tf^A	0.010	0	0	0	0	0	0	0
Tf^B	0.036	0	0	0	0	0.075	0	0
Tf^C	0.454	0.056	0	0	0.005	0.275	0	0
$Tf^{D'}$	0.010	0	0	0	0	0	0	0
Tf^D	0.217	0.444	0	0	0.995	0.400	0	0
Tf^E	0.072	0	0	0	0	0.025	0	0
$Tf^{F'}$	0.005	0	0	0	0	0.075	0	0.469
Tf^F	0.046	0.167	1.000	1.000	0	0.025	0	0
Tf^G	0.119	0.333	0	0	0	0.050	0	0
$Tf^{H'}$	0	0	0	0	0	0.075	1.000	0.531
Tf^H	0.031	0	0	0	0	0	0	0
Number examined	97	18	5	5	96	20	21	16

[a]Calculations of frequencies are based on the assumption that the genes for transferrins are nondominant autosomal alleles. The phenotype B(D)G was arbitrarily considered as Tf BG.

ticularly in light of the ease with which fertile offspring are produced when animals belonging to differently named species are bred with each other. If we adopt the splitter's view of the taxonomy of the genus *Macaca,* we are in great difficulty, for species are closed genetic systems and gene flow does not occur between them. If gene flow has not contributed to the spread of the many transferrin alleles among populations of *Macaca,* how can we explain their present distribution? Unless parallel mutations and almost identical environmental selection are postulated, we cannot explain it. And to postulate such identity between two or more groups of *Macaca* is tantamount to considering them parts of the same species. This is a good example of the importance for modern genetics and physical anthropology of understanding the problems of speciation and systematics (Chapter 4).

Many of the other polymorphic systems found in man have been examined in other primates, but most of the work is in its preliminary stages. There is evidence that G6PD deficiency occurs in some genera, *Macaca* and *Pan.* One study of the genus *Macaca* demonstrated a variant G6PD rather than an enzyme deficiency in members of one species. BAIB excretion is variable among some nonhuman primates, as is the reaction to tasting PTC. Nonhuman primates cannot tell us whether they taste PTC. When PTC is mixed with their food, some of the animals eat in their normal manner. Others react violently—they shriek; they reject the food; they retch—and we assume that these are the tasters. Red cell enzymes—acid and alkaline phosphatases and carbonic anhydrase—appear in several molecular forms among man's relatives. Several serum proteins—Gc trait, Gm groups, serum pseudocholinesterase—produce starch-gel patterns or inhibition reactions indicative of polymorphisms.

The growing body of information about genetically controlled protein polymorphisms of the Primates suggests that we turn more of our attention in the future to comparative studies of primate proteins. Phylogenetic research on living primates must include protein molecules as well as bones, muscles, and behavior.

SUGGESTED READINGS AND SOURCES

Allison, A. C., Malaria and glucose-6-phosphate dehydrogenase deficiency. *Nature* **197,** 609 (1963).

Allison, A. C., and B. S. Blumberg, Ability to taste phenylthiocarbamide among Alaskan Eskimos and other populations. *Human Biol.* **31,** 352 (1959).

Allison, A. C., B. S. Blumberg, and W. ap Rees, Haptoglobin types in British, Spanish Basque, and Nigerian populations. *Nature* **181,** 824 (1958).

Allison, A. C., B. S. Blumberg, and S. M. Gartler, Urinary excretion of β-amino-isobutyric acid in Eskimo and Indian populations of Alaska. *Nature* **183,** 118 (1959).

Armstrong, M. D., K. Yates, Y. Kakimoto, K. Taniguchi, and T. Kappe, Excretion of β-aminoisobutyric acid by man. *J. Biol. Chem.* **238,** 1447 (1963).

Azevêdo, E., H. Krieger, M. P. Mi, and N. E. Morton, PTC taste sensitivity and endemic goiter in Brazil. *Am. J. Human Genet.* **17,** 87 (1965).

Barnicot, N. A., J. P. Garlick, and D. F. Roberts, Haptoglobin and transferrin inheritance in northern Nigerians. *Ann. Human Genet.* **24**, 171 (1960).

Blumberg, B. S., A. C. Allison, and B. Garry, The haptoglobins and haemoglobins of Alaskan Eskimos and Indians. *Ann. Human Genet.* **23**, 349 (1959).

Blumberg, B. S., and S. M. Gartler, High prevalence of high-level β-amino-*iso*butyric acid excretors in Micronesians. *Nature* **184**, 1990 (1959).

Boyer, S. H., and W. J. Young, β-globulin polymorphism in chimpanzees. *Nature* **187**, 1035 (1960).

Buettner-Janusch, J., J. R. Bove, and N. Young, Genetic traits and problems of biological parenthood in two Peruvian Indian tribes. *Am. J. Phys. Anthropol.* **22**, 149 (1964).

Buettner-Janusch, J., and V. Buettner-Janusch, Hemoglobins, haptoglobins, and transferrins in the peoples of Madagascar. *Am. J. Phys. Anthropol.* **22**, 163 (1964).

Clements, F. W., Naturally occurring goitrogens. *Brit. Med. Bull.* **16**, 133 (1960).

Cleve, H., and A. G. Bearn, The group specific component of serum; genetic and chemical considerations. *In* A. G. Steinberg and A. G. Bearn (Ed.), *Progress in Medical Genetics,* Vol. II, p. 64. Grune and Stratton, New York (1962).

Connell, G. E., and O. Smithies, Human haptoglobins: estimation and purification. *Biochem. J.* **72**, 115 (1959).

Fletcher, K. A., and B. G. Maegraith, Glucose-6-phosphate and 6-phosphogluconate dehydrogenase activities in erythrocytes of monkeys infected with *Plasmodium knowlesi. Nature* **196**, 1316 (1962).

Fraser, G. R., Cretinism and taste sensitivity to phenylthiocarbamide. *Lancet* **1**, 964 (1961).

Galatius-Jensen, F., Rare phenotypes in the Hp system. *Acta Genet. Statist. Med.* **8**, 248 (1958).

Giblett, E. R., Haptoglobin types in American Negroes. *Nature* **183**, 192 (1959).

Giblett, E. R., The plasma transferrins. *In* A. G. Steinberg and A. G. Bearn (Ed.), *Progress in Medical Genetics,* Vol. II, p. 34. Grune and Stratton, New York (1962).

Giblett, E. R., C. G. Hickman, and O. Smithies, Serum transferrins. *Nature* **183**, 1589 (1959).

Giblett, E. R., and A. G. Steinberg, The inheritance of serum haptoglobin types in American Negroes: evidence for a third allele Hp^{2m}. *Am. J. Human Genet.* **12**, 160 (1960).

Goedde, H. W., and V. Riedel, Activities of pseudocholinesterases (acylcholine-acylhydrolase, EC 2.1.1.8) in *Macacca mulatta (rhesus)* and *Cercopithecus aethiops. Nature* **203**, 1405 (1964).

Goodman, M., A. Kulkarni, E. Poulik, and E. Reklys, Species and geographic differences in the transferrin polymorphism of macaques. *Science* **147**, 884 (1965).

Goodman, M., R. McBride, E. Poulik, and E. Reklys, Serum transferrins in the Orange Park chimpanzee colony classified by the Boyer and Young scheme. *Nature* **197**, 259 (1963).

Goodman, M., and E. Poulik, Serum transferrins in the genus *Macaca:* species distribution of nineteen phenotypes. *Nature* **191**, 1407 (1961).

Goodman, M., and A. Riopelle, Inheritance of serum transferrins in chimpanzees. *Nature* **197**, 261 (1963).

Grubb, R., Agglutination of erythrocytes coated with "incomplete" anti-Rh by certain rheumatoid arthritic sera and some other sera. *Acta. Pathol. Microbiol. Scand.* **39**, 195 (1956).

Harris, H., H. Kalmus, and W. R. Trotter, Taste sensitivity to phenylthiourea in goitre and diabetes. *Lancet* **2**, 1038 (1949).

Harris, H., E. B. Robson, and M. Siniscalco, β-globulin variants in man. *Nature* **182**, 452 (1958).

Hirschfeld, J., Immune-electrophoretic demonstration of qualitative differences in human sera and their relation to the haptoglobins. *Acta. Pathol. Microbiol. Scand.* **47**, 160 (1959).

Hopkinson, D. A., N. Spencer, and H. Harris, Genetical studies on human red cell acid phosphatase. *Am. J. Human Genet.* **16**, 141 (1964).

Kidson, C., and J. G. Groman, A challenge to the concept of selection by malaria in glucose-6-phosphate dehydrogenase deficiency. *Nature* **196**, 49 (1962).

Kirk, R. L., H. Cleve, and A. G. Bearn, The distribution of the Gc-types in sera from Australian aborigines. *Am. J. Phys. Anthropol.* **21**, 215 (1963).

Kirk, R. L., and L. Y. C. Lai, The distribution of haptoglobin and transferrin groups in south and south east Asia. *Acta Genet. Statist. Med.* **11**, 97 (1961).

Kirk, R. L., L. Y. C. Lai, and W. R. Horsfall, The haptoglobin and transferrin groups among Australian aborigines from North Queensland. *Australian J. Sci.* **24**, 486 (1962).

Kirk, R. L., L. Y. C. Lai, G. H. Vos, R. L. Wickremasinghe, and D. J. B. Perera, The blood and serum groups of selected populations in south India and Ceylon. *Am. J. Phys. Anthropol.* **20**, 485 (1962).

Kitchin, F. D., and A. G. Bearn, The serum group specific component in nonhuman primates. *Am. J. Human Genet.* **17**, 42 (1965).

Lange, V., and J. Schmitt, Das Serumeiweissbild der Primaten unter besonderer Berücksichtigung der Haptoglobine und Transferrine. *Folia Primat.* **1**, 208 (1963).

Lehmann, H., and J. Liddell, Genetical variants of human serum pseudocholinesterase. *In* A. G. Steinberg and A. G. Bearn (Ed.), *Progress in Medical Genetics*, Vol. III, p. 75. Grune and Stratton, New York (1964).

Lie-Injo Luan Eng, Glucose-6-phosphate dehydrogenase activity in the red blood cells of monkeys. *Nature* **195**, 1110 (1962).

Owen, J. A., H. J. Silberman, and C. Got, Detection of haemoglobin, haemoglobin-haptoglobin complexes and other substances with peroxidase activity after zone electrophoresis. *Nature* **182**, 1373 (1958).

Parker, W. C., and A. G. Bearn, Alterations in sialic acid content of human transferrin. *Science* **133**, 1014 (1961).

Parker, W. C., and A. G. Bearn, Haptoglobin and transferrin variation in humans and primates: two new transferrins in Chinese and Japanese populations. *Ann. Human Genet.* **25**, 227 (1961).

Parker, W. C., and A. G. Bearn, Haptoglobin and transferrin gene frequencies in a Navajo population: a new transferrin variant. *Science* **134**, 106 (1961).

Parker, W. C., H. Cleve, and A. G. Bearn, Determination of phenotypes in the human group-specific component (Gc) system by starch gel electrophoresis. *Am. J. Human Genet.* **15**, 353 (1963).

Saldanha, P. H., and J. Nacrur, Taste thresholds for phenylthiourea among Chileans. *Am. J. Phys. Anthropol.* **21**, 113 (1963).

Salzano, F. M., and H. E. Sutton, Haptoglobin and transferrin types of Indians from Santa Catarina, Brazil. *Am. J. Human Genet.* **17**, 281 (1965).

Shaw, C. R., F. N. Syner, and R. E. Tashian, New genetically determined molecular form of erythrocyte esterase in man. *Science* **138**, 31 (1962).

Shim, B.-S., and A. G. Bearn, The distribution of haptoglobin subtypes in various populations, including subtype patterns in some nonhuman primates. *Am. J. Human Genet.* **16**, 477 (1964).

Spencer, N., D. A. Hopkinson, and H. Harris, Quantitative differences and gene dosage in the human red cell acid phosphatase polymorphism. *Nature* **201**, 299 (1964).

Spencer, N., D. A. Hopkinson, and H. Harris, Phosphoglucomutase polymorphism in man. *Nature* **204**, 742 (1964).

Smithies, O., Grouped variations in the occurrence of new protein components in normal human serum. *Nature* **175**, 307 (1955).

Smithies, O., Zone electrophoresis in starch gels and its application to studies of serum proteins. *In* C. B. Anfinsen et al. (Ed.), *Advances in Protein Chemistry*, Vol. XIV, p. 65. Academic Press, New York (1959).

Smithies, O., An improved procedure for starch-gel electrophoresis: further variations in the serum proteins of normal individuals. *Biochem. J.* **71**, 585 (1959).

Smithies, O., and N. F. Walker, Genetic control of some serum proteins in normal humans. *Nature* **176**, 1265 (1955).

Steinberg, A. G., Progress in the study of genetically determined human gamma globulin types (the Gm and Inv groups). *In* A. G. Steinberg and A. G. Bearn (Ed.), *Progress in Medical Genetics*, Vol. II, p. 1. Grune and Stratton, New York (1962).

Steinberg, A. G., and H. Matsumoto, Studies on the Gm, Inv, Hp and Tf serum factors of Japanese populations and families. *Human Biol.* **36**, 77 (1964).

Sutton, H. E., Beta-aminoisobutyricaciduria. *In* J. B. Stanbury, J. B. Wyngaarden, and D. S. Fredrickson (Ed.), *The Metabolic Basis of Inherited Disease*, p. 792. Blakiston, McGraw-Hill, New York (1960).

Sutton, H. E., and P. J. Clark, A biochemical study of Chinese and Caucasoids. *Am. J. Phys. Anthropol.* **13**, 53 (1955).

Sutton, H. E., and G. W. Karp, Jr., Variations in heterozygous expression at the haptoglobin locus. *Am. J. Human Genet.* **16**, 419 (1964).

Sutton, H. E., G. A. Matson, A. R. Robinson, and R. W. Koucky, Distribution of haptoglobin, transferrin and hemoglobin types among Indians of southern Mexico and Guatemala. *Am. J. Human Genet.* **12**, 338 (1960).

Sutton, H. E., J. V. Neel, F. B. Livingstone, G. Binson, P. Kunstadter, and L. E. Trombley, The frequencies of haptoglobin types in five populations. *Ann. Human Genet.* **23**, 175 (1959).

Sutton, H. E., and F. M. Salzano, Haptoglobin and transferrin types in southern Brazilian Indians. *Acta. Genet. Statist. Med.* **13**, 1 (1963).

Tashian, R. E., Genetic variation and evolution of the carboxylic esterases and carbonic anhydrases of primate erythrocytes. *Am. J. Human Genet.* **17**, 257 (1965).

Vyas, G. N., H. M. Bhatia, P. K. Sukumaran, V. Balkrishnan, and L. D. Sanghvi, Study of blood groups, abnormal hemoglobins and other genetical characters in some tribes of Gujarat. *Am. J. Phys. Anthropol.* **20**, 255 (1962).

Human Evolution Today

32

THE DEVELOPMENT OF MODERN industrial technology and urban civilization has made it possible for our species to manipulate the planet to the extent that many environmental pressures are no longer as potent as they once were. Infectious disease, for example, no longer accounts for the removal of a large proportion of the individuals born each year. Nor does it eliminate a major part of the population under the age of 30. Birth and death rates are no longer a direct reflection of the potency of disease in the environment with which the human organism interacts. The increase in the average life-span since the middle Pleistocene is an example of the consequences of change in the specific factors by which natural selection molds our species (Table 32.1). The relaxation of the pressure of environmental selection is not the case for most of the people of the world. For at least two-thirds of the total human population the conditions of life have not been very much improved. The spectacular development of agriculture, industry, and medicine in North America and western Europe have, as yet, had only a small effect upon the rest of the world. For example it is unlikely that the average life-span is much longer than 35 years in many human populations. Yet the meager amount of medical, industrial, and agricultural technology, which has spread into the "underdeveloped" areas of the planet, has effected a sufficient decrease in infant mortality to produce what has often been called the population explosion. The removal or relaxation of the influence of environmental pressure on man and the enormous increase in the world's population have led many a scholar to wonder if our species is losing its biological fitness. We must attempt to understand what a fit genotype is and to determine to what extent selection operates on contemporary human groups.

Our Genetic Load and Fitness

We know that natural selection is operating on *Homo sapiens.* The evidence is the demonstration of the relative advantage of certain genotypes. It is operating on the alleles for some of the blood group systems (Chapter 27) and for some of the hemoglobins (Chapter 29). An allele that is deleterious in one environment may be advantageous in another. The abnormal human hemoglobin S is one obvious example. Positive selection on certain genotypes is not the only measure of biological fitness in a population. Biological fitness is not easy to measure, but the concepts relevant to such measurements are included in the term genetic load. The genetic load of a species is a measure of the number of deleterious traits maintained in a population or of the damage to the population by the factors under study. It may be measured as decreased average fitness, or somewhat more specifi-

TABLE 32.1
Average Life-span of Human Populations
(From Deevey, 1960.)

Population	Average life-span (years)
Neandertal	29.4
Upper Paleolithic	32.4
Mesolithic	31.5
Neolithic Anatolia	38.2
Austrian Bronze Age	38
Classic Greece	35
Classic Rome	32
England 1276	48
England 1376–1400	38
United States 1900–1902	61.5
United States 1950	70

cally, as mortality, sterility, or morbidity due to specified causes, usually deleterious alleles. We might say that estimates of genetic load are attempts to measure what would happen if the phenomenon being studied (mutation, for example) were suspended and everything else in the population remained constant. The genetic load of a species may be partially hidden and partially manifest. The genetic load depends on several variables—the occurrence of mutations, the number of detrimental mutations, the number of mutant recessive alleles, and the number of partially lethal mutant dominant alleles. Most mutations are deleterious or even lethal. The average mutation rate is roughly 10^{-5} mutations per gene per generation (Table 24.3). This means that 1 gamete per 100,000 carries a new mutant in a given gene every generation. A human gamete contains, probably, a minimum of 10,000 genes; a fertilized ovum, an embryo, possesses $2 \times 10,000 = 20,000$ genes (a minimum value). If all genes undergo mutations at the same rate (10^{-5}), then 20 per cent of the individuals born in one generation will carry a new mutant (20,000 genes $\times$ 10^{-5} mutations per gene per generation $\times$ 100 per cent = $2 \times 10^4 \times 10^{-5} \times 100 = 20$). This is a minimum estimate, for there may be many more than 10,000 genes per gamete. This estimate is crude, for we do not have enough evidence to know whether every human gene is as likely as every other gene to mutate.

Genetic load is also a function of the load of segregation and recombination. Although heterozygotes may be the best genotypes, less fit (or inferior) genotypes will recur because of segregation. Although genes in certain combinations may be more fit, recombination of these genes may lead to less fit genotypes. Another component of the genetic load is the incompatibility component, blood group incompatibility for example. There is substitution load—the result of changing fitness of genotypes in changing environments.

And there is also dysmetric load. This refers to the differential fitness of some members of a population in certain niches. Under the best conditions, there will be the right number of genotypes for each niche.

We know that the genetic load exists, and the best estimate of its size can be made by studying the mutational component. This is best done by examining consanguineous matings, as Morton, Crow, and Muller have done. Because of the restrictions on the mating or the marriage of close relatives in almost all societies of *Homo sapiens,* the genetic load is best estimated from studies of offspring of consanguineous matings, those matings in which the number and proportion of homozygous offspring are likely to be relatively large. There is a slight increase in the proportion of inherited diseases among the offspring of inbred individuals. Morton and his colleagues have compared mortality of the progeny of consanguineous and nonconsanguineous parents, and they have estimated that the average genetic load of *Homo sapiens* is between three and five lethal equivalents per person. A lethal equivalent is a single allele that in homozygous combination would cause the death of the individual, that is, reduce his Darwinian fitness to zero. Or it may be two genes, each causing death in homozygous combination one-half of the time. A lethal equivalent may be the result of a number of deleterious mutant recessive alleles, any one of which would not produce a major effect by itself. The Darwinian fitness of an individual is the degree to which he will pass his genes to the succeeding generation.

Man has a heavy genetic load. But this should not give us an inordinately gloomy perspective of man's future. If we compare estimates of man's genetic load with estimates from other animals', fruit flies for example, the prospect before us is not so dismal. The mutational component of the genetic load of man appears no greater than that of other animals. Furthermore since mutation is the source of new genes, we can look upon the heavy genetic load as the cost of continuing to evolve.

The evolution of culture and the rise of modern technological civilization have reduced the impact of the effects of the environment upon individual human organisms. A large number of recessive traits which now manifest themselves in the living homozygous offspring of heterozygous parents would have led, under conditions of life in aboriginal societies or in prehistoric times, to the early death of the child. Furthermore we now know that many, if not all, recessive alleles actually have some effect in the heterozygous state. As the conditions of human existence changed with the development of culture, it is highly likely that the deleterious effects, however slight, of many recessive alleles were lessened.

There is no question that even the rather minimal amount of preventive medicine and public health service available in many primitive and underdeveloped areas of the world has had a profound effect. The enormous rise in the live birth rate and the population explosion attest to this. The success of antimalarial programs and the wide use of vaccines against yellow fever, poliomyelitis, and smallpox are more than minimal public health ventures which have had and are continuing to have a great effect upon the populations of the tropical areas of the world. Excellent prenatal and postnatal care are given in industrialized urban societies. There has been a tremendous reduction in number of deaths from infectious diseases since antibiotics were developed. Many individuals who would normally be less able to cope with bacterial infection or disease now survive in what might be called a genetically deleterious state. The success of surgery and medicine in treating many inherited disorders has its effect too. There is little doubt that individuals who manifested retinoblastoma, inherited malignant tumors of the eyeballs, would not often have survived to the age at which this dominant allele could be passed to the next generation. But with the development of antibiotics, massive public health measures, and surgery, such individuals not only survive infancy and early childhood but manage to live until they can reproduce. Many deleterious traits which do not manifest themselves fully until early adulthood, or even later, may have an effect (presumably bad) on the individual during infancy and childhood. Under conditions of life as rigorous as they are in many aboriginal societies and as it is assumed they were in prehistoric times, it is highly likely that individuals with this kind of inherited trait would be more likely to die of infectious disease or malnutrition than their relatives without the trait.

When we speak of the fitness of a population or of the progeny of a generation, we refer only to Darwinian fitness, adaptive value, or selective value, to reproductive efficiency and fertility. We do not refer, consciously or unconsciously, to health, vigor, blond hair, rosy cheeks, or running the mile in 4 minutes or less. As we have said before, selection is operating when distinct genotypes propagate their genes to the next generation at different rates. The genotype which transmits a larger proportion of its genes to the succeeding generation is said to be subject to positive selection; it is more fit. We refer to genotypically distinct classes within a population, to genotypes at a particular genetic locus, and not to a sum of all the genetic loci. We have adequate information about a very few of the 10,000 or more loci of man. The extent to which one genotype at the expense of other genotypes (composed of alleles at the same loci) propagates its genes is a measure of the Darwinian fitness of the genotype. We are interested in assessing, if we can, the differential fertility of various classes of genotypes. We cannot measure the relative fitness of large numbers of genotypes because of the inadequacy of our data. We have much information about the action of selection on specific genetic loci in man. But we do not have sufficient detailed demographic (population) data to go with the genetic data, and we cannot measure the fitness of more than a few classes of genotypes within human populations. We must approach this problem indirectly. It is possible to make some estimate of the opportunity which exists in various human populations for selection to occur. Because we know that the differential fecundity of various genotypes is a measure of relative Darwinian fitness, we assume that the possibility for selection to act in man is related to the number of children produced by each generation. We have already pointed out, on a number of occasions, that the greater the mortality and the greater the variations in fertility, the greater is the opportunity for selection to operate. Sometimes it is argued that by reducing infant mortality we reduce the opportunity for selection to operate. This is not necessarily so. A number of studies have suggested that the opportunity for selection to operate increases as the number of children born per family decreases, if the mortality rate is assumed to be the same in the populations being compared. The variation in fertility (measured as the variance in average family size) is directly related to the total selection which acts upon a population. The total selection on a population may be represented by a quantity which Crow calls the index of total selection (I). This index consists of two parts, an index of selection due to differential fertility (I_f) and an index due to mortality (I_m). I_f and I_m can be calculated from census data, and I_m is measured as mortality before the individuals reach the age of reproduction. The contribution of differential fertility to the index of total selection, calculated for several populations (Table 32.2), appears to increase as family size decreases. The fertility component of the index of total selection is not the whole story. Variations in infant mortality and variations in fertility both play a role, and there is some difficulty in assessing their respective contributions. When infant mortality and family size decrease, and in modern urban, industralized societies they usually decrease simultaneously, the variance in family size increases. If fewer children per family are born and most of them survive, the reduction in the mortality component (I_m) of the index of total selection (I) may be compensated for, in part, by a greater fertility component (I_f). As postnatal mortality decreases, I_m contributes a smaller part to the total (Table 32.2). The belief that selection ceases to act upon populations when infant mortality is reduced is not wholly warranted.

As urbanization and industrialization are progressing, many of the small, isolated populations which have persisted in Europe and America are breaking up. They are exchanging genes more and more frequently with neighboring populations. The number of deleterious recessive alleles in the gene pool is larger than it was two or three generations ago. Some external sources of mutation, radiation for example, have increased in this generation. These consequences of cultural change and technological advance suggest that the number of deleterious phenotypes and genotypes is increasing. Whether this means that the genetic load is increasing we do not know. The number of deleterious alleles may be increasing, but these alleles may at the same time be eliminated by selection.

It is difficult, if not impossible, to apply eugenic measures to human populations in such a manner that we eliminate or significantly reduce the incidence of the most serious, genetically determined, recessive deleterious traits. The number of individuals with some of the mental deficiencies due to recessive autosomal alleles is presently increasing. It is unlikely that the relative number of these alleles is increasing, but the absolute number of viable

TABLE 32.2

Indices of Selection for Some Hypothetical Mortality Rates
(From Crow, 1958.)

Population	Average number of children	I_f	Indices[a] I		
			$(I_m = 1.00)$	$(I_m = 0.33)$	$(I_m = 0.11)$
Rural Quebec	9.9	0.20	1.40	0.60	0.33
Hutterite	9.0	0.17	1.34	0.56	0.30
Gold Coast (1945)	6.5	0.23	1.46	0.64	0.37
New South Wales (1898–1902)	6.2	0.42	1.84	0.89	0.58
United States (born 1839)	5.5	0.23	1.46	0.64	0.37
United States (born 1866)	3.0	0.64	2.28	1.19	0.82
United States (1910)	3.9	0.78	2.58	1.37	0.98
United States (1950)	2.3	1.14	3.28	1.85	1.38
Ramah Navajo	2.1	1.57	4.14	2.43	1.86

[a] I = index of total selection; I_f = index of selection due to fertility; I_m = hypothetical index of selection due to mortality. $I_m = p_d/(1-p_d)$, where p_d = hypothetical mortality rate, the proportion of individuals who die before reaching reproductive age. When $p_d = 0.5$, $I_m = 1.00$; when $p_d = 0.25$, $I_m = 0.33$; when $p_d = 0.10$, $I_m = 0.11$. $I_f = V/x_s^2$, where x_s = average number of surviving children and V = variance in number of children. $I = I_m + I_f(1-p_d)$.

individuals who are homozygous at these loci is. Present estimates suggest that within this generation the total number of genetically determined mental deficiencies may reach the astonishing figure of 6,000,000 in the United States. Many believe this means we should prevent the breeding of individuals with these traits. Once more we must emphasize that homozygotes who manifest the mental deficiencies do not contribute significantly to the gene pool of the next generation. They have a much reduced fertility compared to normal individuals in the population. Preventing the breeding of such individuals has an insignificant effect upon the incidence of these alleles in the population. A much greater number of recessive genes are carried by heterozygous individuals who are not usually identifiable. Recessive autosomal alleles are kept in a population by phenotypically normal carriers. One of the tasks of modern genetics is the development of techniques for identifying individuals who are heterozygous for deleterious recessive alleles.

Research on mental deficiencies has already developed a method for treating and curing one of the genetically determined mental deficiencies—phenylketonuric idiocy. Phenylketonuric idiocy, controlled by a recessive autosomal allele, is due to the inability of the recessive homozygous individual to metabolize, in the normal manner, the amino acid phenylalanine. The accumulation of phenylalanine or its toxic by-products leads to a severe mental deficiency. The biochemical basis of this disease is the absence of an enzyme in the liver, phenylalanine hydroxylase, which catalyzes the conversion of phenylalanine to tyrosine in normal individuals. Simple laboratory tests on the urine or blood detect the defect. If the homozygous individuals are identified at a sufficiently early age, the development of the mental deficiency can be prevented by feeding them phenylalanine-free diets. As these individuals grow older, it is probably possible to return them to a normal diet. Research of this kind into the metabolic bases of inherited diseases will produce infinitely greater benefits to mankind than programs of sterilization of the homozygous mentally deficient.

It appears to be a commonly held opinion that a condition which is determined, influenced, or controlled by genes cannot possibly be cured, alleviated, or ameliorated to enable individuals to function in reasonably normal ways. The work on phenylketonuric idiocy demonstrates that we can learn to manage our genetic heritage and should not be frightened by the words hereditary defect. Cultural achievements have, in a sense, made once deleterious traits fit in a new environment. The phenylketonuric may soon be no more at a disadvantage than the diabetic individual.

We still have not given a direct answer to the question of whether the overall Darwinian fitness of our species is being reduced by a probable increase in the genetic load of lethal or deleterious mutations. First, we are not certain that there has been a significant rise in the mutational load, though the evidence seems to indicate that a rise has occurred. Second, it is probable that the greatest threat to the future evolution of *Homo sapiens* lies in the culture which man has developed. The mutational load five generations hence is of small concern if the technological potential for self-destruction is realized. Genetic load is also of small concern when we consider the pressing problem of the control of the absolute numbers of the species. The population explosion is more likely to produce more serious problems for the survival of the species than is the genetic load. We do not suggest solutions to these problems; we merely point out that they are more serious and immediate than the problem that advocates of stringent selective breeding wish to correct.

Third, Darwinian fitness of a population or species must be measured relative to that of some other population or species. *Homo sapiens* is not at present in ecological competition with any other species of the order Primates. It is not highly probable that a serious competitor for the same econiche will appear from one of the other planets in the solar system or from outside the solar system. It is possible to declare that some other order or some other phylum, such as Arthropoda, has a higher relative fitness than man. But insects are not in the same ecozone. It is reasonable to argue that certain genotypic classes of *Homo sapiens* are more fit than others, but the eventual result of differential fitness will be a shift in gene frequencies, not an extinction of the species. The measure of Darwinian fitness is in differential fertility of classes of genotypes, not in assessment of what one or another of us may think are desirable characteristics in succeeding generations of men.

Race

Any consideration of evolution in contemporary *Homo sapiens* must include a discussion of race. Scientists can subdivide almost any other mammalian species (or any plant or animal species for that matter) into varieties, subspecies, strains, stocks, or races without arousing the passions of more than a small number of dedicated and perhaps pedantic specialists. But when races of *Homo sapiens* are discussed, a number of irrelevancies, emotional and factual, are introduced by partisans of various ideologies. Instead of dealing with the many controversies about race, we shall present here some of the questions which can be investigated today by using the techniques and methods of science.

Should we use the word race to denote the subdivisions into which the species *Homo sapiens* may be grouped, or is one of the following expressions preferable—ethnic group, geographical stock, or variety? Discussion of this question is not apt to produce fruitful research or profound statements. We believe that races are best considered Mendelian populations. This restricts our use of the word race to a wholly biological context. We do not consider that sociopsychological connotations of the word race are useful in physical anthropology or genetics. We have carefully avoided words such as negroid, mongoloid, caucasoid, and the related expressions: colored, black, yellow, white. We do not find that these are biologically useful terms. They are not usually defined with any degree of precision or consistency by those who use them. Those who use them seldom make an attempt to determine whether Mendelian populations are being referred to. These expressions are probably best kept out of the language of science. Race is a perfectly useful and valid term, and we shall and do use it. A race of *Homo sapiens* is a Mendelian population, a reproductive community of individuals which share in a common gene pool. The level at which we define the reproductive community depends upon the problem we are investigating.

Of course races exist! They exist today, and they probably existed in the Pleistocene as well. We could call upon some rather fancy semantics to explain the word exist in this context. We shall not indulge ourselves but simply note that subgroups of our species occur and are definable according to consistent, established genetic criteria. A race and a Mendelian population, from the point of view of geneticists, are the same thing. Everything we said about the separation and lack of separation between and among human populations holds for races. Races are human populations which are more or less isolated from each other genetically. Some populations are sufficiently distinct or isolated from each other so that we may view them as separate Mendelian populations on the basis of a set of alleles at a single genetic locus. The implication is that there are an infinite number of possible races within the species *Homo sapiens*. Species are closed systems; Mendelian populations or races are open systems. A genetic system which is not closed may be defined at any particular level. All members of our species belong to one Mendelian population, and its name is *Homo sapiens*. This large species-wide Mendelian population is divisible into smaller Mendelian populations, an almost infinitely large number of them.

It seems best, today, to use Garn's definitions of three kinds of Mendelian populations or races—geographical, local, and microgeographical or micro-races. A geographical race is a collection of Mendelian populations separated from other similar collections by major geographical barriers. Islands, whole continents, and such major barriers as mountain ranges and deserts serve to delimit this kind of race (Table 32.3).

TABLE 32.3
Geographical Races
(From Garn, 1961.)

Race	Geographical range
Amerindian	From Alaska, Northern Canada, and Labrador through all of the Americas to the tip of South America
Polynesian	Pacific islands, from New Zealand to Hawaii and Easter Island
Micronesian	Pacific islands, limited to area from Ulithi, Palau, and Tobi to Marshall and Gilbert Islands
Melanesian-Papuan	New Guinea and neighboring islands
Australian	Australia
Asiatic	Eastern continental Asia, Japan, Philippine Islands, Sumatra, Borneo, Celebes, Formosa
Indian	India, from the Himalayas to the tip of the Indian peninsula
European	Europe and western Asia, the Middle East and Africa north of the Sahara
African	Africa south of the Sahara

Local races are subpopulations within major geographical races, and they correspond, to a large extent, to breeding populations or breeding isolates. Local races are Mendelian populations adapted to local environmental pressures. They are maintained by social as well as physical barriers to the gene

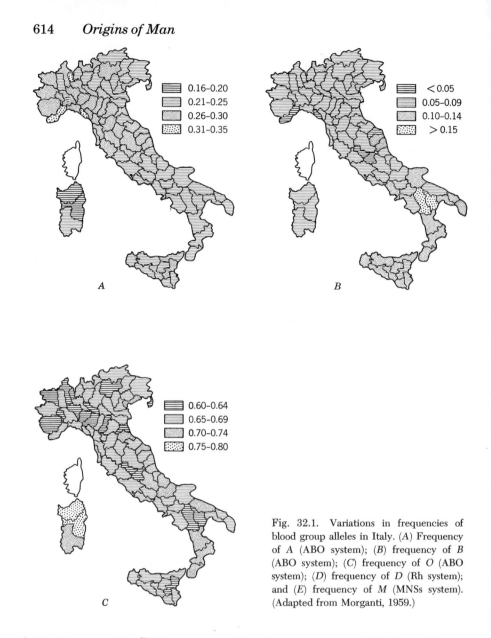

Fig. 32.1. Variations in frequencies of blood group alleles in Italy. (A) Frequency of A (ABO system); (B) frequency of B (ABO system); (C) frequency of O (ABO system); (D) frequency of D (Rh system); and (E) frequency of M (MNSs system). (Adapted from Morganti, 1959.)

flow between them. The concept of local races is very useful. It enables us to distinguish Mendelian populations within a larger geographical population. The Hopi, the Navajo, and the Chippewa are local races within the larger geographical race—Amerindian. It is easier to identify or define a local race if the population is small or isolated. It may be isolated geographically—the Lapps of northern Scandinavia, the Basques of the Pyrenees, the Eskimos of North America, or the Ainu of Japan. Or it may be isolated culturally—the Gypsies or the Yemenite Jews.

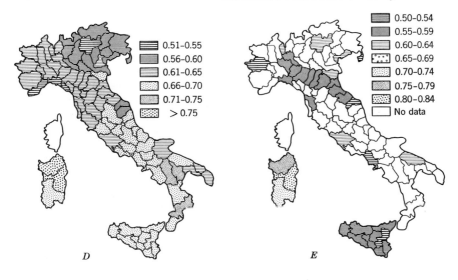

D

E

0.51–0.55
0.56–0.60
0.61–0.65
0.66–0.70
0.71–0.75
> 0.75

0.50–0.54
0.55–0.59
0.60–0.64
0.65–0.69
0.70–0.74
0.75–0.79
0.80–0.84
No data

Microraces are distinct populations which are not clearly bounded breeding populations or isolates. They are a case of statistical isolation. Their existence is demonstrated by the variations in the frequencies of various alleles in a densely populated country such as Italy (Fig. 32.1). A microrace is not as clearly a reproductive community as a geographical or local race. Geographical isolation is the major factor producing geographical races. Cultural isolation and, to a degree, geographical isolation produce local races. Microraces are products of the mating range of the human male (or female). Studies of assortative mating in man show that marriage and/or mating is most often a function of the distance which separates the birthplaces of the two individuals concerned. The majority of matings and marriages are between individuals who live close to each other. Microraces are also the result of local environmental differences. The microraces into which we may divide *Homo sapiens* are many, approaching infinity, for they are defined statistically. They are distinguished by differences in allele frequencies at one, two, three, or more genetic loci. Many physical characteristics of man, such as hair color, head shape, and pigmentation, also vary from locality to locality. Since we know there is a genetic component in these traits but cannot specify it, we can say that these traits, too, are characteristic of microraces.

Microraces must be Mendelian populations, that is, reproductive communities with a common gene pool. Geographical and local races are relatively easy to identify, for there are external criteria which set various reproductive communities apart from each other. Australian aborigines and American Indians are different reproductive communities. Navajo Indians and Chippewa Indians are different reproductive communities. But it is not so easy to determine the boundaries around the various microreproductive communities of Italy (Fig. 32.1).

Several authorities have presented racial classifications of the living populations of *Homo sapiens* (Table 32.4). The bases for these classifications are, primarily, the frequencies of blood group alleles, certain physical characteristics, and geography. As one would suspect, the classifications differ considerably. This does not mean that these classifications are bad or that they are mutually contradictory. It is very difficult to devise a classification for subgroups of a species. A species is a closed genetic system and is defined by reproductive isolation. A subspecies, a race, a Mendelian population within a species, is defined by a gene pool shared by a group of individuals. A Mendelian population is a reproductive community. We have no difficulty in putting Australian aboriginals, Eskimos, Pitcairn Islanders, and Englishmen into separate reproductive communities. But it is not so easy to put all human populations into exclusive reproductive communities. Further problems are made when historical events are considered. Populations have a history, and what do we do with the vast immigrant populations of the United States? Many would prefer to classify the residents of the United States into separate races on the basis of their ancestry—Asian, African, European, etc. As we have seen, there has been considerable gene flow between Americans of African and Americans of European descent and also between Americans of European and Americans of Asian descent. These people now make up many different reproductive communities, and we define them differently for different problems (Chapter 24). The reproductive community in Bullock County, Georgia, is defined one way if we wish to study gene flow between European and African groups, and it is defined another way if we wish to study the difference in the frequency of the $Hb_\beta S$ allele in older and younger age groups in Bullock County. Classifications are not immutable, and the ones that we use must be appropriate for the occasion.

A species as well studied as man, a species about which so much excellent genetic information is available, is not an easy subject for racial classification. Indeed, some ask, why bother making racial classifications when there are so many problems of fundamental biological importance and interest which we can study? Now that so many genetic data are available about the species *Homo sapiens*, it is arguable that it is more useful to determine why blood group allele frequencies in Spezia are different from those in Rome than it is to classify Spezians and Romans into separate races. How much more exciting it is to study the ways in which natural selection manipulates allele frequencies. How are high frequencies of aberrant hemoglobins maintained? This kind of problem overrides the earlier concern with racial classifications of the inhabitants of our planet.

How do racial differences come about? How did races originate? We have answered the first question already—by natural selection, by gene flow between populations, by genetic drift, by means of the founder principle, by the development of sociocultural and geographical isolation.

Note: B-J considers Genus Homo as having only 2 species
1) africanus: australopithecines
2) sapiens: pithecanthropines, neanderthals, mod. man

32/Human Evolution Today   617

The origin, in an historical sense, of the contemporary geographical races of *Homo sapiens* has worried many a scholar and author. A number of authorities believe that races of *Homo sapiens* evolved as separate lineages from separate ancestral fossils. There are two extreme points of view, both of which have been expressed. One is that the Asiatic peoples evolved from an ancestor much like the orangutan, African peoples from a gorilla-like ancestor, and Europeans from something else. There is no need to discuss this for no evidence exists which makes it meaningful. Another extreme point of view is that several ancestral races, represented by hominid fossils found in various parts of the Old World in separate places and in different geological time zones, crossed some kind of evolutionary threshhold and became *Homo sapiens* not once, but as many as five times. This hypothesis implies that a number of separate *species* of the fossil Hominidae were transformed in the course of evolution (during the Pleistocene) into *Homo sapiens* in separate parts of the world at different times. It is an enormously complicated proposition. A rather large number of parallel coincidences would have to have occurred in different ecozones if separate lineages of fossil Hominidae led to *Homo sapiens*.

There is nothing in the hominid fossil material to suggest that races did not exist in the past as they do today. But there is so little fossil material from any single time period that it is difficult to infer specific Mendelian populations except in a rather hypothetical and metaphorical manner. The fossil evidence is wholly consistent with the view that *Homo sapiens* developed in the Old World from the forms we have called *Homo africanus*. There is no special evidence that suggests contemporary populations of *Homo sapiens* are the product of the convergence of *separate* phyletic lines of evolution. That is, there is no unassailable evidence that separate fossil species of the genus *Homo* coexisted during the Pleistocene.

The species of the genus *Homo* which we infer from the fossils were unquestionably divided into Mendelian populations, races. The Javanese fossils of the pithecanthropine group and the Chinese fossils of the sinanthropine group are two races of middle Pleistocene *Homo sapiens*, possibly assignable to the subspecies *erectus*. The suggestion that distinct species of fossil hominids evolved separately into a single species, *Homo sapiens*, might be more acceptable if we could show that two hominid species coexisted at any time in the past.

The australopithecine fossils have been divided by some anthropologists into two groups, the robustus forms and the more gracile africanus forms. We concluded that only a single taxon could be inferred from these forms (Chapters 9 and 10), even though some anthropologists would infer two or more taxa. We noted that the morphology of the fossils does not strongly support the inference that more than a single species existed, and the argument about the ecological separation of the two forms is also weak. On the basis of the evidence available, it is difficult to reconstruct two totally different ecozones for the two presumed species. The wear patterns of the cheek

IMPORTANT - get a general impression of lists correlated w/ authors

Garn 1961

Boyd 1964

TABLE 32.4
Various Racial Classifications of Homo sapiens

Geographical[a]	Genetical[b]
Amerindian	American group
Polynesian	American Indian race
Micronesian	Pacific group
Melanesian-Papuan	Indonesian race
Australian	Melanesian race
Asiatic	Polynesian race
Indian	Australian (aboriginal race)
European	Asian group
African	Asian race
	Indo-Dravidian race
	European group
	Early Europeans
	Lapps
	Northwest Europeans
	Eastern and central Europeans
	Mediterraneans
	African group
	African race

Local[c] *(Garn 1961)*		Local[d] *(Dobzhansky 1962)*	
Northwest European	North American	Northwest European	East African
Northeast European	Central American	Northeast European	Sudanese
Alpine	South American	Alpine	Forest Negro
Mediterranean	Fuegian	Mediterranean	Bantu
Iranian	Lapp	Hindu	Bushman and Hottentot
East African	Pacific "Negrito"	Turkic	African Pygmy
Sudanese	African Pygmy	Tibetan	Dravidian
Forest Negro	Eskimo	North Chinese	Negrito
Bantu	Ainu	Classic Mongoloid	Melanesian-Papuan
Turkic	Murrayian Australian	Eskimo	Murrayian
Tibetan	Carpenterian Australian	Southeast Asiatic	Carpenterian
North Chinese	Bushman and Hottentot	Ainu	Micronesian
Extreme Mongoloid	North American colored	Lapp	Polynesian
Southeast Asiatic	South African colored	North American Indian	Neo-Hawaiian
Hindu	Ladino	Central American Indian	Ladino
Dravidian	Neo-Hawaiian	South American Indian	North American colored
		Fuegian	South African colored

[a] From Garn (1961).
[b] From Boyd (1964).
[c] From Garn (1961).
[d] From Dobzhansky (1962).

teeth (molars and premolars) of the robustus forms are believed, by some, to be due to a vegetable diet, that is, sandy grass stems and roots, whereas the wear patterns of the cheek teeth of the not-so-robustus group are believed to be the result of a carnivorous diet. It has also been suggested that the two forms lived in separate climatic zones, the robustus in damp zones, the africanus in dry. Even if these inferences are valid, the fossils could represent segments of an allopatric, variable species rather than two distinct taxa. Since we know that the living hominids, *Homo sapiens,* are highly variable, it is not unreasonable to infer that early Pleistocene hominids were variable also. The occurrence of the robustus and the not-so-robustus forms in unambiguous sympatric association would support the splitter's position, but the two forms do not occur together. In the one site in which both are found, Olduvai Gorge, they are found at different levels. Hence they were not contemporaneous there. The view that the many races into which contemporary *Homo sapiens* may be divided evolved from separate species or races of Pleistocene hominids is not supported by the fossil record.

Race and Environment

So-called racial mixture, which is nothing more than gene flow, is considered harmful by the biologically unsophisticated and uninformed. There is no evidence that breeding between populations produces harmful biological results. When we discussed the extent to which gene flow occurs (Chapter 24), we said nothing about the results of such gene flow being harmful. Gene flow between populations does not produce any new genes, but it alters the frequencies of various alleles. It alters the composition of the gene pool upon which natural selection may operate.

Certain biological parameters of a population will help us assess various notions about population mixture. One of the more useful kinds of studies is the comparison of the progeny of an endogamous population with those of an exogamous population. The expressions exogamous or endogamous are used to refer to types of mating patterns. Exogamous matings occur when there is an increase in the frequency of marriages between individuals who do not belong to the same breeding isolates. Endogamous matings occur between individuals of the same breeding isolate. The terms exogamous and endogamous are genetically somewhat less precise than the term consanguinity. Endogamous populations are those in which matings within the group are the norm. We would expect that such populations would be more inbred and, possibly, more consanguineous than exogamous populations. Since we are often not able to specify the degree to which consanguinity is practiced or the extent to which a population is inbred, the terms and concepts of exogamy and endogamy must suffice. Hulse studied this problem in Switzerland. He examined the offspring of an exogamous as opposed to the offspring of an endogamous Swiss population. He found that there was a difference. The offspring of exogamous Swiss parents were slightly taller and heavier and had slightly longer heads than the offspring of endogamous Swiss parents.

There is another factor involved which is relevant to the question of gene flow and racial purity. Animal and plant breeders have discovered that hybrid offspring produced by crossing distinct varieties, strains, races, or populations of animals and plants are more vigorous, are biologically superior, to either parental strain. They appear to be superior in disease resistance, growth, and fertility. Hybrid corn and hybrid tomatoes, for example, are often superior to parental lines. Hybrid vigor also appears to occur in some strains of domestic animals. This phenomenon of hybrid vigor or heterosis is not yet understood fully.

Is it possible to demonstrate such hybrid vigor in the offspring of exogamous human matings or in the offspring of matings between individuals or populations from separate geographical areas? A number of studies of this sort have been undertaken by anthropologists in the past. It is exceedingly difficult to manage the experimental conditions in this research. Comparable environmental conditions can be quite rigorously controlled when plants or animals are hybridized, but we must take advantage of accidental experiments when we study so-called hybridization among strains of the species *Homo sapiens.* It is unlikely that ideal conditions can be found for studies of heterosis among human populations, but it is possible to get something more than an impressionistic answer. Some human populations appear to demonstrate hybrid vigor. The inhabitants of Pitcairn Island in the Pacific, the descendants of the mutineers of the Bounty, have often been considered an example of hybrid vigor in man. They are the products of a cross between English mutineer sailors and women of Tahitian origin. They are large, rugged, healthy, vigorous individuals, although they have a high frequency of dental caries. These hybrids were not produced under controlled conditions such as those used to produce the superior strains of hybrid corn, tomatoes, and certain domestic animals. Nonetheless the Pitcairn Islanders show that matings between individuals of diverse geographical origin may produce vigorous offspring. Studies are being carried out in Hawaii on populations of mixed geographical origins. Preliminary evidence suggests that there is no factor attributable to the racial (geographical) origin of the various groups which can account for similarities and differences in such parameters as stillbirths and birth weights as measures of biological vigor. A number of other studies of this problem are listed in the references at the end of this chapter.

The term heterosis or hybrid vigor should probably be restricted to the carefully controlled studies possible with domestic plants. But one example in man springs to mind. If we consider one genetic locus, the hemoglobin locus, the many individuals who are heterozygous are an example of heterosis. The hybrids who are $Hb_\beta{}^A/Hb_\beta{}^S$ have a higher Darwinian fitness than the two homozygous classes, $Hb_\beta{}^A/Hb_\beta{}^A$ and $Hb_\beta{}^S/Hb_\beta{}^S$, but in a particular environment.

When the question of the biologically or sociologically inferior or superior hybrid offspring of human racial crossings comes up, it is almost impossible not to discuss the old problem of heredity versus environment. The question is complex and clouded by emotional and passionate expression. Most of the traits of character and personality that appear to differ among human populations are clearly culturally determined and formed. As Garn has pointed out, the apparent relationships between occupations and national origin melt away like wax when the social, educational, or economic status changes. Our great-grandparents, grandparents, parents, and contemporaries argued about the predisposition to criminal behavior (law violations) on the part of various immigrant groups in the United States. The pattern of predisposition to criminal behavior, if there is such a predisposition, has never been demonstrated to be consistent from one generation to the next. Within a generation or two, the pattern of criminal activity readjusts itself and approximates the norm for the native population.

The phenotypes of individuals are the products of a complex interaction between genome and environment from the moment of fertilization until the time of death. The genomes of individuals are variable and complex. Estimates of the number of possible genotypes in a human population, if only a small number of independent loci are considered, show us that the number of different interactions between organisms and environment are enormous. Since we use so many simplifying assumptions and sophisticated mathematical models, it is not especially fruitful to examine an either-or problem, *either* heredity *or* environment.

Physical anthropologists, geneticists, and other biologists have been discussing the facts of the biology of human populations for a great many years. Despite the enormous increase in our knowledge about the nature of the biological mechanisms that operate when populations breed with each other, there has been no noticeable reduction in race prejudice and race bigotry. It is a sad but almost inevitable conclusion that all the information in the world is going to have little if any effect upon emotional attitudes of men toward visible biological differences among and between individual members of *Homo sapiens*. Race prejudice and bigotry are subjects best handled by those competent to deal with social and individual psychology. Racial attitudes, racial prejudices, and bigotry are consequences of social, cultural, psychological, and economic parameters, not of the inherent biological features of various groups of the species.

The notion of racial purity has little significance in physical anthropology. The biological concept closest to racial purity is that of homozygosity which can be achieved only through inbreeding. Thus those who argue that racial purity is a desirable end must face the fact that there is only one way to achieve homozygosity—by the practice of inbreeding. But inbreeding reduces the genotypic variability of a population. The reduction of variability, in turn, lessens the opportunity for evolution. The future evolution of man depends, in part, on outbreeding.

Some believe that the genotypes control or produce culture—that social, psychological, and cultural differences among men are the product or the consequence of differences in genotypes. Others believe that the genome is irrelevant—that individuals are genotypically constant—when cultural differences are being investigated. Is it race or is it culture which makes the difference? The answer is that it is both. Here again it is not an either-or situation. We have shown that the number of possible genotypes considering only 23 independent loci is enormous—more than 94 billion (Chapter 23). The implication is that each individual is unique, genotypically. Then how do the great similarities among men come about? The answer is the action of the environment, cultural and physical. With our unique genomes we are born into a cultural system and a physical environment that set limits upon the expression of our genotypes. The only meaningful answer to questions of this sort is to continue investigation of the human genome and the limits set upon its expression by the environment—internal and external, individual, cultural, and physical.

SUGGESTED READINGS AND SOURCES

Anderson, O. W., Age-specific mortality differentials historically and currently; observations and implications. *Bull. Hist. Med.* **27**, 521 (1953).

Armstrong, M. D., and F. H. Tyler, Studies on phenylketonuria. I. Restricted phenylalanine intake in phenylketonuria. *J. Clin. Invest.* **34**, 565 (1955).

Blumberg, B. S. (Ed.), *Proceedings of the Conference on Genetic Polymorphisms and Geographic Variations in Disease.* Grune and Stratton, New York (1961).

Boyd, W. C., Modern ideas on race, in the light of our knowledge of blood groups and other characters with known mode of inheritance. *In* C. A. Leone (Ed.), *Taxonomic Biochemistry and Serology*, p. 119. Ronald Press, New York (1964).

Coon, C. S., *The Origin of Races.* Knopf, New York (1962).

Crow, J. F., Some possibilities for measuring selection intensities in man. *Human Biol.* **30**, 1 (1958).

Crow, J., Population genetics: selection. *In* W. J. Burdette (Ed.), *Methodology in Human Genetics*, p. 53. Holden-Day, San Francisco (1962).

Deevey, E. S., Jr., The human population. *Sci. Am.* **203**, No. 3, 194 (1960).

Dobzhansky, T., *Mankind Evolving.* Yale Univ. Press, New Haven (1962).

Garn, S. M., *Human Races.* (*Second edition.*) Charles C Thomas, Springfield, Ill. (1965).

Hulse, F. S., Exogamie et hétérosis. *Arch. Suisse d'Anthropol. Gén.* **22**, 103 (1957).

Knox, W. E., Phenylketonuria. *In* J. B. Stanbury, J. B. Wyngaarden, and D. S. Fredrickson (Ed.), *The Metabolic Basis of Inherited Disease*, p. 321. Blakiston, McGraw-Hill, New York (1960).

Laughlin, W. S., Races of mankind: continental and local. *Anthropol. Papers, Univ. Alaska* **8**, 89 (1960).

Morganti, G., Distribution of blood groups in Italy. *In* G. E. W. Wolstenholme and C. M. O'Connor (Ed.), *Medical Biology and Etruscan Origins.* Little, Brown, Boston (1959).

Morton, N. E., Genetics of interracial crosses in Hawaii. *Eugen. Quart.* **9**, 23 (1962).

Morton, N. E., J. F. Crow, and H. J. Muller, An estimate of the mutational damage in man from data on consanguineous marriages. *Proc. Natl. Acad. Sci. U. S.* **42**, 855 (1956).

Roberts, D. F., and G. A. Harrison (Ed.), *Natural Selection in Human Populations.* Pergamon, New York (1959).

Schull, W. J., Empirical risks in consanguineous marriages: sex ratio, malformation, and viability. *Am. J. Human Genet.* **10**, 294 (1958).

Shapiro, H. L., Descendants of the mutineers of the Bounty. *Mem. Bernice P. Bishop Museum* **9**, (1929).

Shapiro, H. L., *The Heritage of the Bounty.* Natural History Library, Garden City, New York (1962).

Sutherland, E. H., and D. R. Cressy, *Principles of Criminology* (Fifth edition). Lippincott, Philadelphia (1955).

Sutton, H. E., *An Introduction to Human Genetics.* Holt, Rinehart and Winston, New York (1965).

Glossary

ABDUCTION movement of a limb away from the midline of the body, movement of two limbs away from each other

ACETABULUM—cup-shaped socket in the hip bone into which the head of the thigh bone, the femur, fits (Fig. 6.10)

ACROCENTRIC—chromosome having a constriction close to one end (Fig. 25.8)

ADAPTATION—adjustment and modification for a specific environment, as applied to specific structures or functions or to whole organisms

ADAPTIVE RADIATION—rapid increase in numbers and kinds of any evolving group of organisms into several distinctive econiches to which each group is particularly adapted

ADAPTIVE VALUE—*see* fitness

ADDUCTION—movement of a limb toward the midline of the body

AGGLUTINATION—clumping; specifically, clumping of red cells

AGGLUTINOGEN—antigen (*q.v.*) agglutinated by the action of antibodies (*q.v.*)

AHAPTOGLOBINEMIA—absence of detectable amount of haptoglobin (*q.v.*)

ALBUMIN—a protein found in the serum portion of blood

ALLELE—alternative form of a gene at the same locus in one of a pair of homologous chromosomes

ALLOCHRONIC SPECIES (paleospecies)—populations of an evolving lineage living in nonoverlapping time periods; species that do not occur at the same time level

ALLOPATRIC SPECIES—a species, populations of which live in nonoverlapping, mutually exclusive, often adjacent geographical ranges

AMINO ACID—organic compound found in all living organisms; component part of a protein; composed of carbon, hydrogen, oxygen and nitrogen; contains both an acidic group (COOH) and a basic group (NH_2) attached to a central carbon atom; has both acidic and basic properties

AMNION—membrane surrounding a developing embryo or fetus and fluid

AMORPH—gene that has no discernible expression, no discernible effect in the organism

ANALOGOUS STRUCTURES—parts of the anatomy that have a similar form and a similar function in two groups of organisms and that are not related by evolutionary descent from a common ancestral form (cf., homologous structures)

ANEMIA—condition defined by a less than normal number of red blood cells or less than normal amount of hemoglobin

ANODE—positive pole in an electric field; the pole toward which anions, negatively charged ions, will move (cf., cathode)

ANTIBODY—a protein produced in the body in response to the presence of an antigen (*q.v.*); found in the serum fraction of blood

ANTICOAGULANT—anything that prevents blood from clotting

ANTIGEN—a substance, foreign or already present in the body, capable of stimulating the production of an antibody (*q.v.*) complementary to or specific for it

ANTIGENIC DETERMINANT—that part of an antigen that acts to stimulate production of an antibody

ANTISERUM—serum portion of the blood containing an antibody (*q.v.*) produced in response to introduction of a specific antigen (*q.v.*)

ARCADE, DENTAL—arch formed by teeth in the jaw (Fig. 8.1)

ARCHETYPE—original, model, ideal form of which observed organisms are imperfect copies; obsolete term

ARTICULATION—joining together of bones at the joints

ARTIFACT (artefact)—in laboratory experiments, a component whose presence is due to extraneous factors; in archeology, an object showing evidence of having been produced by human workmanship

ASSORTATIVE MATING—in human groups, the prevalence of matings between phenotypically identical or phenotypically different individuals; nonrandom matings; a statement of how often "like marries like"

AUDITORY OSSICLES—small bones of the middle ear

AUTOSOMAL TRAIT—trait carried on any chromosome except a sex chromosome

AUTOSOME—any chromosome except a sex chromosome

AXIAL SKELETON—skull, vertebral column, ribs, and breast bone

BACKCROSS—mating of a hybrid offspring with one of its parents; the mating of a hybrid offspring with a member of the group of which either parent is a member

BACTERIOPHAGE—a virus or similar substance capable of multiplying in bacteria

BACTERIUM—simplest living cell; microorganism

BINOMIAL NOMENCLATURE—scientific name of an organism designated by both a generic and a specific trivial name; first standardized by Linnaeus

BIOSPECIES—*see* genetical species

BIPEDAL—walking on two legs

BLOOD CLOT—semisolid gelatinous mass of red and white blood cells

BLOOD FACTOR—part of an antigen (*q.v.*) in the Rh blood group system

BLOOD GROUP—classificatory category defined by reactions of red blood cells with antibodies (*q.v.*)

BRACHIATION—movement by swinging arm-over-arm along a horizontal support

BREEDING ISOLATE (genetic isolate)—a population whose members seldom, if ever, mate with individuals from other populations

BUCCAL—within the mouth

CAECUM (cecum)—blind pouch of large intestine; vestigial in man

CALCARINE FISSURE—well-marked groove on the medial surface of posterior part of each hemisphere of the brain

CANINE—conical or pointed tooth in the front of the mouth (Fig. 6.2)

CARBON-14 (C^{14}) DATING—determination of the age of a specimen by measurement of the amount of a natural radioactive isotope of carbon (C^{14}) present in the specimen

CARRIER—individual heterozygous for a recessive or nondominant allele; usually, individual with a deleterious recessive allele

CATARRHINI—Old World monkeys, apes, and man

CATASTROPHISM—doctrine that explains extinction by assuming a general catastrophe

CATHODE—negative pole in an electric field; the pole toward which cations, positively charged ions, will move (cf., anode)

CAUDAL—in·the direction of the tail

C^{14} DATING—*see* carbon-14 dating

CENTROMERE—primary constriction in a chromosome (Fig. 25.8)

CEPHALIC—*see* cranial

CHROMATID—one-half of a chromosome resulting from longitudinal duplication of the chromosome (Fig. 25.8)

CHROMATOGRAPHY—specifically, paper chromatography, a laboratory technique used to separate mixtures of similar components; based on the observation that different compounds, in the same solvent, have different solubilities and move different distances on a supporting medium (Fig. 29.6)

CHROMOSOME—sausage- or rod-shaped structure in the nucleus of a cell that contains the material which serves to convey inherited traits from one generation to another; composed of nucleic acid and protein

CHRONOMETRIC DATING—determination of the absolute age in years of a specimen or a geological formation

CINGULUM—ridge around the base of the crown of a tooth

CIRCUMDUCTION—complex movement best characterized by the cone-shaped figure generated by moving one outstretched arm in a circle

CISTRON—smallest functioning genetic unit on a chromosome

CLASSIFICATION—formal system used to describe relationships among organisms

CLAVICULATE—having well-developed clavicles, collar bones

CLAWS—sharp, curved, compressed appendages on extremities, molded to the ends of extremities; composed of a deep layer and a superficial layer (cf., nails)

CODON—specific sequence of three nucleotides which determine or code one amino acid

COLON LABYRINTH—the particularly long and convoluted large intestine of Indriidae

CONDYLE—rounded process on a bone at a joint

CONES—structures in the eye, on the retina, that allow discrimination of color, texture, spatial relationships

CONGERIES—a nonrandom collection of individuals

CONSANGUINITY—mating of individuals who have one or more common ancestors

CONVERGENCE—similarity of traits and adaptive relationships of two groups of organisms which are not closely related phylogenetically; wings of butterflies and wings of bats are convergent structures (cf., parallelism)

CORTEX, CEREBRAL—the cerebrum; large rounded structure that fills cranium

CRANIAL (cephalic)—toward the head

CRANIUM—all parts of a skull except the face

CREPUSCULAR—active at twilight or dawn or in dim light, as applied to behavior of animals

CROSSING OVER—exchange of genetic material between homologous chromosomes at meiosis (Fig. 25.5)

CROSS MATCHING—process of determining blood groups, prior to blood transfusion, by testing red blood cells of donor versus serum of recipient and serum of donor versus red blood cells of recipient

CULTURE—that complex whole which includes knowledge, belief, art, morals, law, custom, and any other capabilities and habits acquired by man as a member of society (Tylor)

CUSP—projection on the occlusal surface (*q.v.*) of a tooth (Fig. 6.2)

CYTOCHROME—respiratory protein found in muscle, brain, and other tissues of the body

CYTOGENETICS—study of genetics of cells

CYTOLOGY—study of biology of cells

DARWINIAN FITNESS—*see* fitness

DECIDUOUS DENTITION—teeth lost before maturity; milk teeth

DEMOGRAPHY—study of human populations, their size, density, growth, vital statistics

DENTAL FORMULA—number of teeth of each kind in one-half of the upper jaw and the number of each kind in one-half of the lower jaw

DENTITION—the teeth, their numbers and kinds

DEOXYRIBONUCLEASE—enzyme that specifically breaks down deoxyribonucleic acid

DEOXYRIBONUCLEIC ACID (DNA)—large organic molecule composed of two intertwined strands of similar units, nucleotides; each nucleotide contains a nitrogenous base, deoxyribose, and phosphate (Fig. 25.1)

DEOXYRIBOSE—5-carbon sugar molecule

DIASTEMA—space between teeth in the jaw; usually, a space between upper incisor and canine into which lower canine fits when jaws are closed

DIASTROPHISM—movements and rearrangements in the earth's crust

DIGIT—finger or toe; the bones of a finger or a toe

DIMORPHISM—occurrence in two forms

DIPLOID NUMBER (DIPLOID COMPLEMENT, 2N)—number of chromosomes in any cell except a germinal cell; twice the number of chromosomes in a germinal cell

DISPLAY—pattern of motor activity serving to convey information, this motor activity being mediated by the cerebral cortex

DISTAL—referring to the direction away from the point of attachment of a limb (cf., proximal)

DIURNAL—active during the day, as applied to behavior of animals

DNA—*see* deoxyribonucleic acid

DOMINANCE HIERARCHY—set of ranked relationships among members of a group of animals, usually applied to groups of baboons and other monkeys; the most dominant animal(s) is at the top of the hierarchy

DOMINANT (allele or gene)—allele which always expresses itself in the phenotype in both heterozygous and homozygous condition; its allele is not distinguishable in the heterozygous condition (cf., recessive)

DORSAL—in a direction toward the backbone

DORSIFLEXION—bending in a backward direction

Drosophila—fruit fly, one of the animals commonly used in genetic studies

ECOLOGY—study of relationships between organisms and their environments

ECONICHE (ecozone)—place in the environment occupied by an organism

EFFECTIVE MUTATION—change in the genetic material that becomes fixed in a population

EFFECTIVE POPULATION SIZE—size of a breeding population, determined by the number of parents in a population, the average number of offspring per family, and the variance (*q.v.*) in number of offspring

ELECTROPHORESIS—laboratory technique used to separate mixtures of similar compounds; based upon differential migration rates of charged particles in solution in an electric field (Fig. 29.1)

EMBRYO—human organism during first eight weeks of development in the uterus (cf., fetus)

ENDEMIC—referring to a disease that is constantly present in a particular environment

ENDOCRANIAL CAST—copy or cast of the inner surface of a skull

ENDOGAMY—matings within a delineated social group (cf., exogamy)

ENZYME—protein that functions as a catalyst in the synthesis or degradation of many constituents of living organisms

EOCENE—epoch that began about 60 and ended about 40 million years ago

EPIPHYSIS—enlarged end of a bone formed from separate centers of growth or ossification

ERYTHROBLASTOSIS FETALIS—hemolytic disease of newborns

ERYTHROCYTE—red blood cell

ESTROUS CYCLE—ovulatory cycle in the female; cycle of anatomical, psychological, behavioral, endocrinological changes in the female (also *see* menstrual cycle)

ESTRUS—period of heat; the time of most intense sex drive in females; period of sexual receptivity in most mammalian females

ETHNIC GROUP—alternative term for race proposed by Ashley Montagu; a population with common cultural ties

ETHOLOGY—study of animal behavior

EUGENICS—social control of human matings in order to eliminate certain inherited traits or to improve the human species

EUTHERIAN MAMMAL—animal whose embryo and fetus is nourished by a placenta

EVOLUTION—descent with modification or change

EVOLUTIONARY SPECIES—ancestral-descendant sequence of populations evolving in a separate line and with its own evolutionary role and tendencies

EXOGAMY—matings outside a delineated social group (cf., endogamy)

EXTINCTION—disappearance of a group of organisms from the evolutionary record

FALCIPARUM MALARIA—malaria caused by a specific parasite, *Plasmodium falciparum*

FASCIA—layer of connective tissue

FAVISM—severe anemia occurring in certain individuals after ingestion of fava beans; resembles anemia occurring in some individuals after treatment with antimalarial drugs; probably under genetic control

FEMUR—long bone in the thigh (Fig. 6.18)

FETUS—human organism from eighth week of development after fertilization until birth (cf., embryo)

FINGERPRINT—array of peptides derived from a protein after mixture of peptides have been separated by electrophoresis (*q.v.*) and/or chromatography (*q.v.*) on filter paper; also called a paper peptide pattern (Fig. 29.7)

FITNESS (adaptive value, Darwinian fitness, selective value)—reproductive capacity of a population

FLEXION—bending forward

FLUORINE DATING—determination of relative age of a specimen by measurement of the amount of fluorine in it; older specimens usually contain more fluorine than younger ones

FORAMEN MAGNUM—hole on the base of the skull through which spinal cord passes (Fig. 6.6)

FOSSIL—parts of an ancient organism that have become mineralized or left impressions or casts in surrounding materials

FOSSILIZATION—process by which parts of ancient organisms become mineralized

FOUNDER PRINCIPLE—establishment of a new population by a few original migrants or founders whose gene pool may be an aberrant sample of the gene pool of the larger population from which it migrated

FOVEA—shallow pit in the retina of the eye; place of greatest visual acuity

FREQUENCY (gene frequency)—relative incidence of a given allele in a population

GAMETE—mature sperm cell or ovum

GENE—unit of hereditary material

GENEALOGY—enumeration of ancestors and their descendants (Fig. 23.5)

GENE FLOW—*see* migration

GENE POOL—the total of the genes in a breeding population

GENETICAL SPECIES (biospecies)—population or group of populations of actually or potentially interbreeding organisms that are reproductively isolated from other such groups

GENETIC CODE—sequence of nucleotides in deoxyribonucleic acid that ultimately determines the sequence of amino acids in a protein end product of cell metabolism

GENETIC DRIFT—*see* sampling error

GENETIC ISOLATE—*see* breeding isolate

GENETIC LOAD—a measure of the number of inherited deleterious traits in a population

GENETICS—study of inherited traits

GENOME—chromosomes of an organism and their genes

GENOTYPE—actual genetic composition of an organism (cf., phenotype)

GENUS—category in classification of organisms; consists of one species or of related species with a presumed common phylogenetic origin

GEOCHRONOLOGY—timing of events in the past by study of geology

GERMINAL CELL—sex cell, sperm or ovum

GESTATION—carrying of embryo and fetus in the uterus of the mother; length of time from conception until birth

GLOBIN—protein portion of hemoglobin

GLOBULIN, α-, β-, γ- —three fractions of the serum portion of blood; three groups of proteins distinguished by mobility after electrophoresis (*q.v.*) of serum

GLUCOSE-6-PHOSPHATE DEHYDROGENASE (G6PD)—red cell enzyme functioning in oxidation-reduction reactions of carbohydrate metabolism

GLUTEUS MUSCLES—three muscles of the pelvic girdle and thigh, gluteus maximus, gluteus medius, and gluteus minimus; man's gluteus maximus is essential for bipedal locomotion; man's gluteus medius and gluteus minimus are necessary for movements of the thigh (Fig. 6.14)

GRADE—level of organization, such as monkey grade or human grade

GRADIENT—regular change in rate or frequency

GROOMING—picking through fur with teeth or hands

GROOMING CLAW—modified nail on the second digit of each lower extremity of a prosimian

GROUP SPECIFIC SUBSTANCE—blood group antigen on a red cell or in some body fluids such as saliva

HALF-LIFE—as applied to radioactive isotopes, the length of time in which the measurable radioactivity decreases by one-half

HALLUX—big toe, the first digit of the lower extremity or hindlimb

HAND, CONVERGENT—typical appendage of mammalian forelimb; digits form a fan when fingers are extended; when fingers are flexed they converge

HAND, PREHENSILE—one hand with which an animal picks up and holds food, may have opposable thumb or pseudo-opposable thumb; animal habitually using only one hand for picking up and holding food has a prehensile hand

HAPLOID NUMBER (haploid complement, N)—number of chromosomes in a mature germinal cell

HAPTOGLOBIN—protein in the serum portion of blood

HARDY-WEINBERG LAW—an expression of the proportions of various genotypes in a stable population

HAREM—group of animals consisting of one male with a number of adult and subadult females and infants in more or less permanent association

HELIX—mathematically defined spiral structure having turns of constant slope from the base and constant distance from the axis

HEME—iron-containing portion of hemoglobin; in general, the prosthetic group of a conjugated protein composed of linked nitrogenous organic ring compounds and iron

HEMOGLOBIN—red respiratory protein comprising more than 90 per cent of the protein content of a red cell and functioning to transport oxygen to and from the tissues of the body

HEMOLYZE—to break cell walls and expel contents of cell into surrounding medium; hemolysis is the process of hemolyzing; hemolysate is the result of hemolyzing

HEMOPHILIA—inherited disease defined by failure of blood to clot after injury of tissues; present in several distinct clinical and genetic forms, such as hemophilia A, Christmas disease, hemophilia C, parahemophilia

HEREDITARY DEFECT—inherited abnormality

HERNIATED INTERVERTEBRAL DISC—protrusion of disc between two vertebrae

HETEROGENETIC—having more than one genetic origin

HETEROSIS—*see* hybrid vigor

HETEROSPECIFIC MATING—mating in which the female has present or can produce in her serum an antibody to a red cell antigen of the male; incompatible mating

HETEROSPECIFIC PREGNANCY—pregnancy in which the female has present or can produce an antibody to a red cell antigen of her fetus; incompatible pregnancy

HETEROZYGOTE, HETEROZYGOUS—having two different alleles at corresponding loci on a pair of homologous chromosomes (cf., homozygote)

HIERARCHY, LINNAEAN—sequential stratification of taxonomic levels in present system of classification of organisms

HOLANDRIC—trait carried on the Y-chromosome

HOMEOSTASIS—maintenance of a relatively stable internal physiological environment

HOMOIOTHERMAL (homeothermal)—maintaining a relatively constant internal body temperature independent of the environment; warm blooded (cf., poikilothermal)

HOMOLOGOUS STRUCTURES—parts of the anatomy that have a similar form and a similar function in two groups of organisms that are related by evolutionary descent from a common ancestral form (cf., analogous structures)

HOMOLOGY—similarity or identity in structure or in function due to common origins

HOMOSPECIFIC MATING—mating in which the female has the same red cell antigens as the male, has no antibody to a red cell antigen of the male, or cannot produce an antibody to a red cell antigen of the male; compatible mating

HOMOSPECIFIC PREGNANCY—pregnancy in which the female has the same red cell antigens as her fetus, does not have or cannot produce an antibody to a red cell antigen of her fetus; compatible pregnancy

HOMOZYGOTE, HOMOZYGOUS—having the identical allele at corresponding loci on a pair of homologous chromosomes (cf., heterozygote)

HORIZON—assemblage, as applied to geological epochs or cultural periods

HORMONE—physiological substance, sometimes called a chemical messenger, secreted by one organ and exerting an effect on another organ

HUMERUS—long bone of the upper arm (Fig. 6.17)

HYBRID—offspring of parents of different genetic composition

HYBRID VIGOR (heterosis)—increased reproductive advantage displayed by offspring of matings of individuals from two different groups

HYPERPLASIA—abnormal enlargement

ILIAC SPINE—landmark on the ilium, the position of which may be indicative of the arrangement of important muscles of locomotion (Fig. 6.10)

ILIOPSOAS MUSCLE—complex of two muscles of the pelvic girdle, psoas major and iliacus, which help prevent man from falling over backward when he stands erect (Fig. 6.12 and 6.13)

ILIUM—one of three bones of the pelvis (Fig. 6.10)

IMMUNITY—ability or capacity, natural or induced, to resist a disease

IMMUNOLOGY—study of the processes of antibody formation and production

IMPRINTING—a learning pattern of young in many species; the young is attracted to a moving object or to a parent

INBREEDING—mating of individuals in a group with other individuals of the same group

INCEST TABOO—preferential marriage rules; rules which forbid matings or marriages between certain defined classes of relatives

INCISOR—flat tooth, in the front of the mouth, used for biting or scraping (Fig. 6.2)

INCUS—small bone of the middle ear

INDEPENDENT ASSORTMENT—two traits, simultaneously considered, will sort and recombine independently of each other; Mendel's second law

INFRARED RADIATION—radiation outside the range of visible light; radiation with wavelengths longer than the wavelength of red light

INION—external protuberance on the occiput of the skull (Fig. 6.7)

INSERTION (of muscle)—one end of a muscle, arbitrarily designated as the distal end (cf., origin)

INTEGUMENT—covering, specifically skin and its appendages

INTERMEMBRAL INDEX—relationship between length of upper extremity, arm, and length of lower extremity, leg; expressed as 100 × the ratio of the length of humerus plus radius to the length of the femur plus tibia

IRREVERSIBILITY—as applied to evolution, a stage-by-stage return to an ancestral form does not occur; structures and functions once lost in a lineage are never regained

ISCHIUM—one of three bones of the pelvis (Fig. 6.10)

ISOLATE—*see* breeding isolate

ISOTOPE—one or more forms of the same chemical element that have the same atomic number and different atomic weights; have the same chemical properties, but usually different physical properties

JAUNDICE—yellowish color of the skin due to abnormal release of bile and excessive breakdown or hemolysis of red blood cells

K-A DATING—*see* potassium-argon dating

KARYOTYPE—description of the numbers and kinds of chromosomes in a cell (Fig. 25.7)

KEEL OF NAIL—ridge, raised portion

KINSHIP—socially determined, recognized, and named relationships among individuals in a particular human group

LANGUAGE—capacity of members of *Homo sapiens* to utilize a rich system of learned vocal symbols, largely arbitrary in their form, in communication and social interaction

LATERAL—in the direction of a side of the body

LETHAL—deleterious, causing death

LINEAGE—line, usually an ancestral-descendant line

LINKAGE—transmission of two or more alleles as a unit

LINKED GENES—alleles at two or more loci transmitted as a unit

LOCUS, GENETIC (pl. loci)—place on a chromosome occupied by a single gene

LUMBAR CURVE—curve of the vertebral column of man in the lumbar region, the middle of the back (Fig. 20.2)

LUMPER—in taxonomy, one who tends to unite related units into a single taxon; one whose criteria are such that the existing taxonomic level assigned to a given taxonomic category is lowered, as families to subfamilies and species to subspecies (cf., splitter)

MACULA LUTEA—area of acute vision on retina of eye around fovea (*q.v.*) containing yellow pigment

MALAR BONE—cheek bone or zygomatic bone

MALARIA—disease caused by a parasite, *Plasmodium,* transmitted to man and other primates by mosquitos; characterized by fever, jaundice, enlarged liver and/or spleen, etc.

MAMMAE—in mammals, the glandular organs that secrete milk

MANDIBLE—bone of lower jaw, contains lower teeth (Fig. 6.8)

MARKING—behavior pattern in which an animal rubs a surface with glandular secretions or urine

MARSUPIAL—nonplacental mammal that carries its young in a pouch

MAXILLA—bone of the face; contains teeth of upper jaw (Fig. 6.4)

MECONIUM—contents of intestine of mammalian fetus at birth

MEIOSIS—process of cell division and chromosome replication in germinal cells; reduction division (Fig. 23.4)

MELANIN—dark pigment of skin and hair of animals and of structures of plants; usually composed of repeating units of derivatives of the amino acid tyrosine

MELANOMA—tumor of the skin, usually dark in color, containing melanin

MENARCHE—onset of regular menstruation

MENDELIAN POPULATION—reproductive community of individuals which share a common gene pool

MENSTRUAL CYCLE—roughly, estrous cycle (*q.v.*) of human females and of some other higher primate females; said not to have a characteristic period of heat

MENSTRUATION (menses)—periodic bleeding from the vagina consequent to the breakdown of the lining of the uterus during the estrous cycle of human females and some other higher primate females

METACENTRIC—chromosome having a constriction at, or very close to, its center (Fig. 25.8)

METHEMOGLOBIN—reddish brown oxidized form of hemoglobin in which iron is in the ferric form; will not combine with oxygen

MICROCYTOSIS—condition in which erythrocytes are abnormally small

MICRON (μ)—unit of measure equal to 0.0001 cm or approximately 0.00004 in.

MIGRATION (gene flow)—introduction of alleles from one population into another of the same species with, probably, consequent shifts in allele frequencies; exchange of genetic material between populations due to dispersion of gametes

MILLIMICRON (mμ)—unit of measure equal to 0.0000001 cm or approximately 0.00000004 in.

MIOCENE—epoch that began about 28 and ended about 12 million years ago

MITOSIS—process of cell division and replication; nucleus divides into two daughter nuclei each with a chromosome complement the same as that of original nucleus (Fig. 23.3)

MODIFIER GENE—gene that acts to change the expression of an allele at a different locus

MOLAR—square, broad cheek tooth, complicated in form, used for grinding and chewing (Fig. 6.2)

MOLARIZATION—tendency of a tooth to become molar-like during evolution

MONOPHYLETIC—single line, referring to the ancestry of a contemporary group as being found in one, not many, ancestral species; as applied to a taxonomic category, a taxon whose contained units are part of a single immediate line of descent

MORPHOLOGY—study of form and structure of organisms

MORPHOSPECIES—species defined by morphological traits

MOSAIC EVOLUTION—differential evolution of component parts of an organism or a structure

MUCOPROTEIN—protein conjugated with carbohydrates

MUCOSA, GASTRIC—lining of the stomach or intestine

MUTAGEN—any substance, process, or agent that produces a mutation

MUTANT—organism that contains an expressed mutation

MUTATION—change in the genetic material, discernible as a change in genotype

MUTON—unit of mutation in the genetic material

MYOGLOBIN—respiratory protein of muscle

NAILS—flat, variously curved structures covering the terminal phalanx of a digit (*q.v.*); may be considered degenerate claws (*q.v.*); only the superficial layer of the claw is present, the deep layer has disappeared

NATURAL SELECTION—process, agent, or situation leading to the continuation of one group of organisms or traits and the elimination of another; process by which environment eliminates less well-adapted members of a population or process by which differential reproduction results among different genotypic classes

NEONTOLOGY—study of living or recent organisms (cf., paleontology)

NEURON—nerve cell with its processes

NEUTRAL TRAIT (nonadaptive trait)—supposedly, a trait which is not acted upon by processes of natural selection

NICHE—part of environment in which an organism or a population of organisms lives

NOCTURNAL—active at night, as applied to behavior of animals

NOMENCLATURE—assignment of names to groups recognized in a system of clasification

NONADAPTIVE TRAIT—*see* neutral trait

NONDISJUNCTION—failure of the members of a pair of chromosomes to separate and move to separate poles during cell division; produces trisomy (*q.v.*)

NONDOMINANT (allele or gene)—allele expressed in either homozygous or heterozygous combination with its corresponding allele; not masked by its corresponding allele in a heterozygote

NUCHAL CREST—well-defined bony protuberance across the back of the skull between the parietal region and the occipital region (Fig. 9.7)

NUCLEIC ACID—large organic molecule composed of repeating similar units, certain nitrogenous organic bases, sugars, and phosphates

NUCLEOTIDE—single unit of nucleic acid, composed of nitrogenous organic base, sugar, and phosphate

NUCLEUS (of cell)—large almost spherical structure, bounded by a membrane, containing chromosomes and other components necessary for cell division and transmission of inherited characters

OCCIPITAL TORUS—thickened protuberance on the occiput

OCCIPUT—bone at base and back of skull (Fig. 6.6)

OCCLUSAL SURFACE—biting or grinding surface of a tooth

OLIGOCENE—epoch that began about 40 and ended about 28 million years ago

OLIGOSACCHARIDE—large molecule containing repeating units of carbohydrates, sugars, linked together in regular fashion

OPTICAL ROTATION—angle through which a beam of polarized light is rotated after traversing an optically active chemical compound

ORBICULARIS ORIS—muscles surrounding the lips (Fig. 6.9)

ORBIT—eye socket (Fig. 6.4)

ORBITAL PLATE—triangular bone forming vault or roof of eye socket

ORIGIN (of muscle)—one end of a muscle, arbitrarily designated as the proximal end (cf., insertion)

ORTHOGENESIS—a teleological view that evolution proceeds in a straight, inevitable, preordained line culminating in a particular species

ORTHOGRADE—upright

ORTHOSELECTION—evolution in the same direction; long-term selection in the same direction

OS INNOMINATUM (os coxae)—one-half of the pelvis; consisting of ilium, ischium, and pubis (Fig. 6.10)

OSSICLE—any small bone

OSTEOLOGY—study of bones

OVALOCYTOSIS—condition characterized by large number of elliptical red cells

OVULATION—release of an egg from an ovary

PALATINE BONE—bone at back of nasal cavity just behind the maxilla

PALEOCENE—epoch that began about 75 and ended about 60 million years ago

PALEONTOLOGY—study of life of past geological periods (cf., neontology)

PALEOSPECIES—*see* allochronic species

PALMAR—referring to the inner surface, the palm, of the hand

PANMIXIS (panmixia)—random breeding population; equilibrium

PARAHEMOPHILIA—that form of hemophilia (*q.v.*) due to deficiency of the specific fraction of serum called Gc globulin

PARALLELISM—development of similar traits and adaptive relationships from the same ancestral trait in two related groups of animals; development of traits in two groups related by evolutionary descent and divergence (cf., convergence)

PARALLEL SELECTION—occurrence of the same trait in two related species because of the same or similar selective pressures

PARAMETER—independent quantity, character, or measurement that may be used to specify or determine another quantity, character, or measurement

PATELLA—kneecap

PECTORAL—referring to chest

PEDIGREE—schematic means of showing genetic relationships among individuals (Fig. 23.5)

PELAGE—hair, coat of an animal

PELVIS—the bony structure in the lower part of the trunk (Fig. 6.10)

PENTADACTYLISM—five digits on each extremity, on each hand and foot

PEPTIDE—several amino acids joined together in regular fashion, naturally occurring or an intact portion of a protein (Fig. 25.3)

PEPTIDE BOND—name given to the link between two amino acids in a protein; the acidic group (COOH) of one amino acid is attached to the basic group (NH_2) of the next, and a molecule of water (H_2O) is lost to give –CO—NH– (Fig. 25.3)

PERIANAL REGION—region surrounding the anus

PERICARDIUM—membrane enclosing the heart

PERITONEUM—membrane enclosing viscera of lower trunk; membrane enclosing the peritoneal cavity

PERNICIOUS ANEMIA—severe anemia due to conditioned nutritional deficiency of vitamin B_{12}; possibly under genetic control

PHALANX (pl. phalanges)—one bone of a digit, of a finger or toe

PHARYNX—tube, membranous and muscular, in upper part of alimentary tract, extends from under surface of skull to about sixth cervical vertebra

PHENOTYPE—observed expression of the genetic composition of an organism (cf., genotype)

PHENYLKETONURIC IDIOCY—inherited inability to metabolize the amino acid phenylalanine; leads to impaired mental function

PHOTORECEPTORS—structures in the eye, in the retina, receiving light stimuli

PHOTOSYNTHESIS—process by which plants convert water and carbon dioxide to carbohydrates in the presence of light

PHYLETIC BRANCHING—separation or division of an evolutionary lineage into two or more separately evolving lineages

PHYLOGENY—evolutionary relationships among organisms; the origin and evolution of higher categories

PLACENTA—membrane containing an embryo or fetus formed in part by embryonic tissue and in part by maternal tissue

PLACENTAL BARRIER—name given the concept that the placenta prevents the passage or exchange of deleterious substances from a mother to her fetus; not wholly realistic

PLANTAR—referring to the sole of the foot

PLANTIGRADE—referring to an animal that places the plantar surface, the sole of the foot, firmly on the ground when it walks

PLASMA—the clear fluid portion of blood from which clotting factors have not been removed

Plasmodium falciparum—a parasite that causes malaria

PLATYRRHINI—New World monkeys

PLATYSMA—muscle of expression, in the face and neck, necessary for drawing back the corners of the mouth (Fig. 6.9)

PLAY—performance of behavior patterns for their own sake, not for rewards, food, etc.; incompletely performed behavior patterns, such as play-fighting or play-biting among animals

PLEIOTROPY—multiple phenotypic expressions of an allele

PLEISTOCENE—epoch that began about 2 million and ended 10,000 to 20,000 years ago

PLIOCENE—epoch that began about 12 and ended about 2 million years ago

POIKILOTHERMAL—having a variable body temperature which fluctuates with the temperature of the environment; cold blooded (cf., homoiothermal)

POLLEX—thumb, first digit of upper extremity

POLYGYNY—marriage pattern in which a man has more than one wife

POLYMORPHISM, BALANCED—occurrence and maintenance in significant frequency and in stable genetic equilibrium of more than one allele at a specific genetic locus

POLYMORHPISM, GENETIC—occurrence together in the same locality of two or more discontinuous forms of a species or of a trait in such proportions that the rarest of them cannot be maintained merely by recurrent mutations

POLYMORPHISM, TRANSIENT—continued occurrence of alternative alleles at a locus until the more advantageous allele becomes so frequent that the alternative is preserved only by recurrent mutation

POLYPEPTIDE—large peptide (*q.v.*)

POLYPHYLETIC—having an origin in more than one lineage

POLYPLOIDY—increase in diploid number (*q.v.*) of chromosomes in a cell by a complete set, by the haploid number (*q.v.*)

POPULATION—individuals that form a potentially interbreeding community at a given locality

POSTCRANIAL—referring to any part of a skeleton except the skull

POSTORBITAL BAR—bony bar on the side of and behind the orbit in skulls of all living and most fossil primates (Fig. 7.3)

POTASSIUM-ARGON (K-A) DATING—determination of the age of a specimen by measurement of the amounts of the radioactive isotopes potassium-40 (K^{40}) and argon-40 (A^{40}) in the specimen

POWER GRIP—position of the hand when it is exerting maximum force (Fig. 20.8)

PRECISION GRIP—position of the hand when it is holding an object with maximum accuracy of control (Fig. 20.8)

PREHENSILE—grasping; an animal with a prehensile hand can grasp objects and manipulate them with one hand

PREHENSIVE GRIP—position the hand assumes when it is holding an object

PREHENSIVE PATTERN—position of the hand as it reaches for an object

PREMOLAR—cheek tooth similar in form to a molar, appearing farther toward the front of the mouth than a molar (Fig. 6.2)

PRIMAQUINE—drug used in treatment of malaria

PRIMATES—order of mammals to which man belongs

PROCUMBENT—slanting forward, as applied to front teeth

PROGNATHISM—forward protrusion of jaws; alveolar prognathism occurs when only the teeth protrude; facial prognathism occurs when all of the face below the eyes protrudes

PRONOGRADE—standing or moving on all four limbs

PROPOSITUS—the individual who is the first in a pedigree to come to the attention of those investigating an inherited disorder in the pedigree

PROTEIN—large organic molecule composed of various numbers of various amino acids linked together by peptide bonds (*q.v.*)

PROXIMAL—referring to the direction toward the point of attachment of a limb (cf., distal)

PUBIS—one of three bones of the pelvis (Fig. 6.10)

PURINE—type of nitrogenous organic ring compound, one of the components of a nucleotide

PYRAMIDAL TRACT—bundle of nerve fibers that serve to control voluntary movements of muscles

PYRIMIDINE—type of nitrogenous organic ring compound, one of the components of a nucleotide

QUADRUPEDAL—walking on all four limbs

RACE—Mendelian population separated from another by major geographical barriers; breeding isolate; a population distinguished from another by demonstration of differences in allele frequencies

RADIOACTIVE DECAY—loss of radioactivity; decreased specific activity (*q.v.*)

RADIOACTIVITY—property of emitting radiant energy

RAMUS (pl. rami)—any branch-like structure; specifically, the vertical body of the mandible (Fig. 6.5)

RECESSIVE (allele or gene)—allele which is expressed only in the homozygous condition (cf., dominant)

RECOMBINANT—product, usually an organism, obtained after crossing over and recombination of genetic material

RECOMBINATION (of genes)—process of exchange of genetic material in a pair of homologous chromosomes

RECON—smallest unit of genetic recombination

REFLECTOMETER—instrument that measures intensity of reflected light at a specified wavelength

REPLICATION—duplication; specifically, duplication of genetic material

RETE MIRABILE—terminal arterial network composed of a mass of minute vessels

RETINA—light-sensitive layer that lines posterior chamber of the eye

RHINENCEPHALON—that part of the forebrain associated with sense of smell

RIBONUCLEIC ACID (RNA)—large organic molecule composed of a large number of similar units, nucleotides; each nucleotide contains a nitrogenous base, ribose, and a phosphate group

RIBOSE—5-carbon sugar molecule

RIBOSOMES—small granules in cytoplasm of cells, rich in ribonucleic acid and protein, active in protein synthesis, found in all cells that synthesize protein

RNA—*see* ribonucleic acid

RODS—structures in the eye, on the retina, which are sensitive in very dim light

ROSTRUM—muzzle

SACRUM—group of fused vertebrae that form the upper and back part of the pelvic cavity

SAGITTAL CREST—bony ridge protruding along midline on the top of the skull; occurs in some nonhuman primates

SALINE, ISOTONIC—physiological salt solution, a solution that has the same concentration of salts as most physiological fluids; contains 0.85 gm of sodium chloride per 100 ml

SAMPLING ERROR (genetic drift)—change in allele frequencies due to random fixation; Sewall Wright effect

SATELLITE—part of a chromosome separated from the main portion of the chromosome by a secondary constriction (Fig. 25.8)

SCIATIC NOTCH—notch on the dorsal surface of the ischium (Fig. 6.10)

SCROTUM—pouch of skin containing testes

SEGREGATION OF GENES—traits are transmitted as discrete units which do not blend with or contaminate each other; Mendel's first law

SELECTION COEFFICIENT—measure of the contribution made by one genotype to the gene pool of the next generation

SELECTIVE VALUE—*see* fitness

SEMEN—product of male reproductive organs, consists of sperm and secretions of accessory glands

SEROLOGY—study of the nature of antigens and antibodies and their reactions

SERUM—clear fluid portion of blood which does not contain fibrinogen

SEX-LINKED—referring to a trait the allele for which is located on one of the sex-determining chromosomes, usually the X-chromosome

SEXUAL DIMORPHISM—appearance of a trait in one form in males of a species and another form in females of the same species

SEXUAL SKIN—portion of the skin that regularly changes in color or texture during estrous cycle of female, usually on the chest or perianal region

SHOULDER GIRDLE—bones by which the upper limbs are attached to the trunk; scapula and clavicle

SIBLING SPECIES—two or more closely related species, reproductively isolated, but morphologically indistinguishable or almost indistinguishable

SIBSHIP—brothers and sisters

SICKLE CELL ANEMIA (sicklemia)—illness characterized by red blood cells that take on a sickle shape in the absence of oxygen, by a decreased amount of hemoglobin or a decreased number of red blood cells, and by the presence of hemoglobin S, an abnormal hemoglobin, in the red cells

SICKLE CELL TRAIT—red cells take on a sickle shape in the absence of oxygen; individuals who are heterozygous for hemoglobin S have the trait

SIMIAN—pertaining to the level of organization displayed by monkeys and apes, in contrast to other primates; not a technical term

SIMIAN SHELF—bony structure behind the front teeth in the front part of the lower jaw of apes

SOMATIC CELL—any cell except a sperm or ovum

SPECIATION—processes that lead to reproductive isolation between two populations of organisms; formation of species

SPECIES—population or group of populations of actually or potentially interbreeding organisms that are reproductively isolated from other such organisms

SPECIFIC ACTIVITY—measure of the proportion of radioactive isotope in a mixture of the radioactive and nonradioactive isotopes of a chemical element

SPECTROGRAM, SOUND—chart that displays pitch of a sound in one direction and time in the other, commonly used for analysis of speech and other vocalizations

SPECTROPHOTOMETRY—technique of measuring the amount of light or other radiation absorbed or transmitted at a specific wavelength by a specimen

SPHEROCYTOSIS—presence of spherical red cells that are smaller than normal; usually found in anemic individuals

SPLITTER—in taxonomy, one who tends to divide his material more finely than the average; one who attempts to express even small differences nomenclaturally; one whose criteria are such that the existing taxonomic level assigned to a given taxonomic category is pushed upward, as subfamilies to families, subspecies to species (cf., lumper)

SPONDYLOLISTHESIS—forward displacement of a lumbar vertebra producing painful compression of nerves

STARCH GEL—jelly like material composed of starch heated in a dilute salt solution

STRATIGRAPHY—study and description of sequences of rock formations and relationships between particular rock deposits

STRATUM (pl. strata)—well-defined geological formation consisting of more or less homogeneous material

SUBFOSSIL—skeletal remains less than about 10,000 years old; usually unfossilized skeletal remains of now extinct animals of post-Pleistocene times

SUBSPECIES—subdivision of a species; a breeding isolate or population within a species sufficiently distinct from other such to warrant a separate label

SUBTERMINAL—chromosome having a constriction neither in the center nor at one end but somewhere in between (Fig. 25.8)

SUPERPOSITION—order in which rocks are placed one above another; older formations are at the bottom, younger are at the top

SUPPRESSOR GENE—allele whose effect is the suppression of the expression of another allele

SUPRAORBITAL TORUS—bony ridge over the eye sockets in a skull (Fig. 6.4)

SUTURE—line formed where two bones of the skull grow together (Fig. 6.7)

SYMPATRIC SPECIES—two or more reproductively isolated populations occupying the same or overlapping areas

SYMPHYSIS—permanent cartilaginous joint

SYSTEMATICS—scientific study of the kinds and diversity of organisms and any and all relationships among them

TAPETUM—layer in the eye, in the retina, that reflects light

TARSUS—ankle bone

TAXON (pl. taxa)—category in a classificatory scheme; a group of organisms recognized as a unit

TAXONOMY—theoretical study of classification of organisms

TELOCENTRIC—chromosome having a terminal constriction (Fig. 25.8)

TENDON—fibrous connective tissue

TERRITORIALITY—animals' habitual use of a circumscribed area; behavior characterized by recognition of, use of, and some kind of defensive reactions toward the area

TESTIS—organ that produces sperm

THALASSEMIA—clinically and genetically heterogeneous group of anemias, presently believed to be result of disruption in normal synthesis of hemoglobin

THORAX—chest, region of body containing heart and lungs

THUMB, OPPOSABLE—thumb that rotates at carpometacarpal joint so that it is opposed to digits 2 through 5

THUMB, PSEUDO-OPPOSABLE—thumb that moves only in one plane at carpometacarpal joint, does not rotate at the joint

TOOTHCOMB (tooth scraper)—comb-like structure formed by the lower front teeth of most prosimians (Fig. 2.1 and 14.4)

TOTAL MORPHOLOGICAL PATTERN—integrated combination of unitary characters that together make up the complete functional design of a given anatomical structure

TRANSFERRIN—iron-binding protein found in the serum portion of blood

TRICHOSIDERIN—iron-containing pigment found in hair

TRIPLET (nucleotide)—combination of three nucleotides which forms the code for one amino acid

TRISOMY—appearance of three chromosomes of one type instead of an homologous pair, results from nondisjunction (*q.v.*)

TRITUBERCULAR THEORY—primitive mammalian molar teeth had three cusps and evolution of the dentition included modification by adding various numbers of new cusps

TROOP—relatively stable structured group of primates who are associated during a daily cycle of activity for several seasons

TRYPSIN—enzyme from pancreas especially useful in studies of protein structure; breaks peptide bond at COOH end of lysine or arginine

TUBERCLE—small, rounded protuberance

TURBINAL BONES—bones of the nose

TYPOLOGY—study of types as representatives or models of large groups of specimens

ULTRAVIOLET RADIATION—radiation outside the range of visible light; radiation with wavelengths shorter than the wavelength of violet or blue light

UNGUICULATE—having nails instead of claws (*q.v.*)

UTERUS—in females, the organ in which an embryo and fetus develops; situated in the pelvic cavity

VARIANCE—measure of the dispersion or variation of each value in a sample from the average value; the average squared deviation from the mean

VASCULAR—referring to blood vessels

VASOCONSTRICTION—constriction of blood vessels

VASODILATION—expansion of blood vessels

VASOMOTOR—referring to constriction or expansion of blood vessels

VAULT—arched or dome-shaped structure, specifically refers to the skull cap of man and other hominids

VENTRAL—in a direction toward the belly; pertaining to the front surface of an animal that holds its trunk erect; pertaining to the under surface of an animal that moves on all fours

VERTEBRA (pl. vertebrae)—individual bone of the vertebral column or backbone (Figs. 6.19 and 20.2)

VIGILANCE—form of animal behavior that directs or stimulates group activity in situations of danger

VILLAFRANCHIAN—assemblage of fauna marking the end of the Pliocene epoch and beginning of the Pleistocene

VIRUS—agent composed of nucleic acid and protein, incapable of multiplying outside of a host organism

VISION, BINOCULAR—capacity to see the same image with both eyes because of overlap of fields of vision

VISION, COLOR—capacity to discriminate colors

VISION, PHOTOPIC—capacity to discriminate spatial relationships, colors, and textures in high intensity light

VISION, SCOTOPIC—twilight vision; capacity to see in dim light without high degree of discrimination

VISION, STEREOSCOPIC—vision in which images in the visual fields of the two eyes are transmitted to the same region of the brain

VOCALIZATION—process of uttering sounds

X-CHROMOSOME—female sex-determining chromosome; in all cells of a normal female there are two X-chromosomes and 22 pairs of autosomes

Y-CHROMOSOME—male sex-determining chromosome; in all cells of a normal male, there is one Y-chromosome, one X-chromosome, and 22 pairs of autosomes

ZYGOMATICUS—muscles of cheek and mouth (Fig. 6.9)

ZYGOTE—fertilized ovum before it undergoes differentiation; also individual that results after differentiation

Author Index

Subject Index

649